Lecture Notes in Computer Science 9125

Commenced Publication in 1973
Founding and Former Series Editors:
Gerhard Goos, Juris Hartmanis, and Jan van Leeuwen

More information about this series at http://www.springer.com/series/7407

Evripidis Bampis (Ed.)

Experimental Algorithms

14th International Symposium, SEA 2015
Paris, France, June 29 – July 1, 2015
Proceedings

 Springer

Editor
Evripidis Bampis
Université Pierre et Marie Curie LIP6
Paris
France

ISSN 0302-9743   ISSN 1611-3349 (electronic)
Lecture Notes in Computer Science
ISBN 978-3-319-20085-9   ISBN 978-3-319-20086-6 (eBook)
DOI 10.1007/978-3-319-20086-6

Library of Congress Control Number: 2015940731

LNCS Sublibrary: SL1 – Theoretical Computer Science and General Issues

Printed on acid-free paper

Springer International Publishing AG Switzerland is part of Springer Science+Business Media
(www.springer.com)

Preface

This volume contains the papers presented at the 14th International Symposium on Experimental Algorithms (SEA 2015) held during June 29–July 1, 2015, in Paris. SEA is an international forum for researchers in the area of design, analysis, and experimental evaluation and engineering of algorithms, as well as in different aspects of computational optimization and its applications. The preceding symposia were held in Riga, Monte Verita, Rio de Janeiro, Santorini, Menorca, Rome, Cape Cod, Dortmund, Ischia, Crete, Bordeaux, Rome, and Copenhagen.

In response to the call for papers, we received 76 submissions. Each submission was reviewed by at least three reviewers. The submissions were mainly judged on originality, technical quality, and relevance to the topics of the conference. On the basis of the reviews and on extensive electronic discussions, the Program Committee (PC) decided to accept 30 papers. In addition to the accepted contributions, the program also included three distinguished lectures by Alessandra Carbone (UPMC), Erik Demaine (MIT), and Kurt Mehlhorn (MPI).

We would like to thank all the authors who responded to the call for papers, the invited speakers, the members of the PC, the external reviewers and the members of the Organizing Committee. We also would like to thank the developers and maintainers of the EasyChair conference system, which was used to manage the electronic submissions, the review process, and the electronic PC meeting. We also acknowledge Springer for publishing the proceedings of SEA 2015 in their LNCS series.

April 2015   Evripidis Bampis

Organization

Program Committee

Eric Angel	Université d'Evry Val d'Essonne, France
Evripidis Bampis (Chair)	Sorbonne Universités, UPMC, France
Vincenzo Bonifaci	IASI CNR, Università di Roma "La Sapienza", Italy
David Coudert	INRIA Sophia Antipolis, France
Christoph Dürr	CNRS, Sorbonne Universités, UPMC, France
Thomas Erlebach	University of Leicester, UK
Loukas Georgiadis	University of Ioannina, Greece
Leo Liberti	Ecole Polytechnique, France
Giorgio Lucarelli	University of Grenoble-Alpes, France
Alberto Marchetti-Spaccamela	Università di Roma "La Sapienza", Italy
Sotiris Nikoletseas	University of Patras, Greece
Aris Pagourtzis	National Technical University of Athens, Greece
Marina Papatriantafilou	Chalmers University of Technology and GöteborgUniversity, Sweden
Vijaya Ramachandran	The University of Texas at Austin, USA
Nicolas Schabanel	CNRS, Université Paris Diderot, France
Maria Serna	Universitat Politecnica de Catalunya, Spain
Frits Spieksma	KU Leuven, Belgium
Clifford Stein	Columbia University, USA
Dorothea Wagner	Karlsruhe Institute of Technology (KIT), Germany
Renato Werneck	Amazon, USA
Christos Zaroliagis	University of Patras, Greece
Norbert Zeh	Dalhousie University in Halifax, Canada

Organizing Committee

Evripidis Bampis
Christoph Dürr
Giorgio Lucarelli

Subreviewers

Aris Anagnostopoulos
Alexandros Angelopoulos
Joaquim Assunção
Moritz Baum
Raphaël Bleuse

Bapi Chaterjee
Riccardo De Masellis
François Delbot
Josep Diaz
Julian Dibbelt

Pavlos Eirinakis	Thomas Pajor
Donatella Firmani	Panagiota Panagopoulou
Goran Frehse	Grammati Pantziou
Aristotelis Giannakos	Nikos Parotsidis
Georgios Goumas	Matthias Petri
Laurent Gourvès	Solon Pissis
Michael Hamann	Mauricio Resende
Spyros Kontogiannis	Ignaz Rutter
Charis Kouzinopoulos	Iosif Salem
Christian Laforest	Stefan Schirra
Luigi Laura	Christian Schulz
Arnaud Legrand	Raimund Seidel
Nikos Leonardos	Abhinav Srivastav
Dimitrios Letsios	Yannis Stamatiou
Farnaz Moradi	Ben Strasser
Matthias Müller-Hannemann	Valentin Tudor
Alfredo Navarra	Gabriel Valiente
Yiannis Nikolakopoulos	Tony Wirth
Christos Nomikos	Georgios Zois
Martin Nöllenburg	Tobias Zündorf
Eoin O'Mahony	

Contents

Transportation Networks

Other Applications II

Graph Problems II

Data Structures

Parallel Construction of Succinct Trees

Leo Ferres[1], José Fuentes-Sepúlveda[1]([✉]), Meng He[2], and Norbert Zeh[2]

[1] Department of Computer Science, Universidad de Concepción, Concepción, Chile
{lferres,jfuentess}@udec.cl
[2] Faculty of Computer Science, Dalhousie University, Halifax, Canada
{mhe,nzeh}@cs.dal.ca

Abstract. Succinct representations of trees are an elegant solution to make large trees fit in main memory while still supporting navigational operations in constant time. However, their construction time remains a bottleneck. We introduce a practical parallel algorithm that improves the state of the art in succinct tree construction. Given a tree on n nodes stored as a sequence of balanced parentheses, our algorithm builds a succinct tree representation in $O(n/p + \lg p)$ time, where p is the number of available cores. The constructed representation uses $2n + o(n)$ bits of space and supports a rich set of operations in $O(\lg n)$ time. In experiments using up to 64 cores and on inputs of different sizes, our algorithm achieved good parallel speed-up. We also present an algorithm that takes $O(n/p + \lg p)$ time to construct the balanced parenthesis representation of the input tree required by our succinct tree construction algorithm.

1 Introduction

Trees are ubiquitous in Computer Science. They have applications in every aspect of computing from XML/HTML processing to abstract syntax trees (AST) in compilers, phylogenetic trees in computational genomics or shortest path trees in path planning. The ever increasing amounts of structured, hierarchical data processed in many applications have turned the processing of the corresponding large tree structures into a bottleneck, particularly when they do not fit in memory. Succinct tree representations store trees using as few bits as possible and thereby significantly increase the size of trees that fit in memory while still supporting important primitive operations in constant time. There exist such representations that use only $2n + o(n)$ bits to store the topology of a tree with n nodes, which is close to the information-theoretic lower bound and much less than the space used by traditional pointer-based representations.

Alas, the construction of succinct trees is quite slow compared to the construction of pointer-based representations. Multicore parallelism offers one possible tool to speed up the construction of succinct trees, but little work has been done in this direction so far. The only results we are aware of focus on the construction of wavelet trees, which are used in representations of text indexes. In [10],

This work was supported by the Emerging Leaders in the Americas scholarship programme, NSERC, and the Canada Research Chairs programme.

© Springer International Publishing Switzerland 2015
E. Bampis (Ed.): SEA 2015, LNCS 9125, pp. 3–14, 2015.
DOI: 10.1007/978-3-319-20086-6_1

two practical multicore algorithms for wavelet tree construction were introduced. Both algorithms perform $O(n \lg \sigma)$[1] work and have span $O(n)$, where n is the input size and σ is the alphabet size. In [21], Shun introduced three new algorithms to construct wavelet trees in parallel. Among these three algorithms, the best algorithm in practice performs $O(n \lg \sigma)$ work and has span $O(\lg n \lg \sigma)$. Shun also explained how to parallelize the construction of rank/select structures so that it requires $O(n)$ work and $O(1)$ span for rank structures, and $O(n)$ work and $O(\lg n)$ span for select structures.

In this paper, we provide a parallel algorithm that constructs the RMMT tree representation of [19] in $O(n/p + \lg p)$ time using p cores. This structure uses $2n + o(n)$ bits to store an ordinal tree on n nodes and supports a rich set of basic operations on these trees in $O(\lg n)$ time. While this query time is theoretically suboptimal, the RMMT structure is simple enough to be practical and has been verified experimentally to be very small and support fast queries in practice [1]. Combined with the fast parallel construction algorithm presented in this paper, it provides an excellent tool for manipulating very large trees in many applications.

We implemented and tested our algorithm on a number of real-world input trees having billions of nodes. Our experiments show that our algorithm run on a single core is competitive with state-of-the-art sequential constructions of the RMMT structure and achieves good speed-up on up to 64 cores and likely beyond.

2   Preliminaries

Succinct Trees. Jacobson [15] was the first to propose the design of succinct data structures. He showed how to represent an ordinal tree on n nodes using $2n + o(n)$ bits so that computing the first child, next sibling or parent of any node takes $O(\lg n)$ time in the bit probe model. Clark and Munro [5] showed how to support the same operations in constant time in the word RAM model with word size $\Theta(\lg n)$. Since then, much work has been done on succinct tree representations, to support more operations, to achieve compression, to provide support for updates, and so on [2,9,11,13,16–19]. See [20] for a thorough survey.

Navarro and Sadakane [19] recently proposed a succinct tree representation, referred to as NS-representation throughout this paper, which was the first to achieve a redundancy of $O(n/\lg^c n)$ bits for any positive constant c. The *redundancy* of a data structure is the additional space it uses above the information-theoretic lower bound. While all previous tree representations achieved a redundancy of $o(n)$ bits, their redundancy was $\Omega(n \lg \lg n/\lg n)$ bits, that is, just slightly sub-linear. The NS-representation also supports a large number of navigational operations in constant time (see Table 1 in [19]); only the work in [9,13] supports two additional operations. An experimental study of succinct trees [1] showed that a simplified version of the NS-representation uses less space than other existing representations in most cases and performs most operations faster. In this paper, we provide a parallel algorithm for constructing this representation.

---

[1] We use $\lg x$ to mean $\log_2 x$ throughout this paper.

The NS-representation is based on the balanced parenthesis sequence P of the input tree T, which is obtained by performing a preorder traversal of T and writing down an opening parenthesis when visiting a node for the first time and a closing parenthesis after visiting all its descendants. Thus, the length of P is $2n$.

The NS-representation is not the first structure to use balanced parentheses to represent trees. Munro and Raman [18] used succinct representations of balanced parentheses to represent ordinal trees and reduced a set of navigational operations on trees to operations on their balanced parenthesis sequences. Their solution supports only a subset of the operations supported by the NS-representation. Additional operations can be supported using auxiliary data structures [17], but supporting all operations in Table 1 of [19] requires many auxiliary structures, which increases the size of the final data structure and makes it complex in both theory and practice. The main novelty of the NS-representation lies in its reduction of a large set of operations on trees and balanced parenthesis sequences to a small set of *primitive operations*. Representing P as a bit vector storing a 1 for each opening parenthesis and a 0 for each closing parenthesis, these primitive operations are the following, where g is an arbitrary function on $\{0,1\}$:

$$\mathtt{sum}(P,g,i,j) = \textstyle\sum_{k=i}^{j} g(P[k])$$
$$\mathtt{fwd\_search}(P,g,i,d) = \min\{j \mid j \geq i, \mathtt{sum}(P,g,i,j) = d\}$$
$$\mathtt{bwd\_search}(P,g,i,d) = \max\{j \mid j \leq i, \mathtt{sum}(P,g,j,i) = d\}$$
$$\mathtt{rmq}(P,g,i,j) = \min\{\mathtt{sum}(P,g,i,k) \mid i \leq k \leq j\}$$
$$\mathtt{RMQ}(P,g,i,j) = \max\{\mathtt{sum}(P,g,i,k) \mid i \leq k \leq j\}$$
$$\mathtt{rmqi}(P,g,i,j) = \operatorname*{argmin}_{k\in[i,j]}\{\mathtt{sum}(P,g,i,k)\}$$
$$\mathtt{RMQi}(P,g,i,j) = \operatorname*{argmax}_{k\in[i,j]}\{\mathtt{sum}(P,g,i,k)\}$$

Most operations supported by the NS-representation reduce to these primitives by choosing g to be one of the following three functions:

$$\pi : 1 \mapsto 1 \qquad\qquad \phi : 1 \mapsto 1 \qquad\qquad \psi : 1 \mapsto 0$$
$$0 \mapsto -1 \qquad\qquad 0 \mapsto 0 \qquad\qquad 0 \mapsto 1$$

For example, assuming the ith parenthesis in P is an opening parenthesis, the matching closing parenthesis can be found using $\mathtt{fwd\_search}(P, \pi, i, 0)$. Thus, it (almost)[2] suffices to support the primitive operations above for $g \in \{\pi, \phi, \psi\}$. To do so, Navarro and Sadakane designed a simple data structure called *Range Min-Max Tree* (RMMT), which supports the primitive operations above in logarithmic time when used to represent the entire sequence P. To achieve constant-time operations, P is partitioned into chunks. Each chunk is represented using an

---

[2] A few navigational operations cannot be expressed using these primitives. The NS-representation includes additional structures to support these operations.

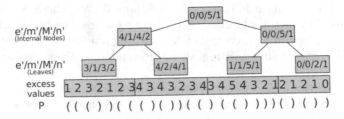

Fig. 1. Range min-max tree

RMMT, which supports primitive operations inside the chunk in constant time if the chunk is small enough. Additional data structures are used to support operations on the entire sequence P in constant time.

Next we briefly review the RMMT structure and how it supports the primitive operations for $g = \pi$. Navarro and Sadakane [19] discussed how to make it support these operations also for ϕ and ψ while increasing its size by only $O(n/\lg^c n)$. To define the version of the RMMT we implemented, we partition P into chunks of size $s = w \lg n$, where w is the machine word size. For simplicity, we assume that the length of P is a multiple of s. The RMMT is a complete binary tree over the sequence of chunks (see Figure 1). (If the number of chunks is not a power of 2, we pad the sequence with chunks of zeroes to reach the closest power of 2. These chunks are not stored explicitly.) Each node u of the RMMT represents a subsequence P_u of P that is the concatenation of the chunks corresponding to the descendant leaves of u. Since the RMMT is a complete tree, we need not store its structure explicitly. Instead, we index its nodes as in a binary heap and refer to each node by its index. The representation of the RMMT consists of four arrays e', m', M', and n', each of length equal to the number of nodes in the RMMT. The uth entry of each of these arrays stores some crucial information about P_u: Let the *excess* at position i of P be defined as $\text{sum}(P, \pi, 0, i) = \sum_{k=0}^{i} \pi(P[k])$. $e'[u]$ stores the excess at the last position in P_u. $m'[u]$ and $M'[u]$ store the minimum and maximum excess, respectively, at any position in P_u. $n'[u]$ stores the number of positions in P_u that have the minimum excess value $m'[u]$.

Combined with a standard technique called *table lookup*, an RMMT supports the primitive operations for π in $O(\lg n)$ time. Consider $\text{fwd\_search}(P, \pi, i, d)$ for example. We first check the chunk containing $P[i]$ to see if the answer is inside this chunk. This takes $O(\lg n)$ time by dividing the chunk into portions of length $w/2$ and testing for each portion in turn whether it contains the answer. Using a lookup table whose content does not depend on P, the test for each portion of length $w/2$ takes constant time: For each possible bit vector of length $w/2$ and each of the $w/2$ positions in it, the table stores the answer of $\text{fwd\_search}(P, \pi, i, d)$ if it can be found inside this bit vector, or -1 otherwise. As there are $2^{w/2}$ bit vectors of length $w/2$, this table uses $2^{w/2}\text{poly}(w)$ bits. If we find the answer inside the chunk containing $P[i]$, we report it. Otherwise, let u be the leaf corresponding to this chunk. If u has a right sibling, we inspect the sibling's m' and M' values to determine whether it contains the answer. If so, we let v be this right sibling. Otherwise, we move up the tree from u until we find a right sibling v of an ancestor of u whose corresponding subsequence P_v contains the query answer. Then we use a similar procedure to descend down the tree starting from v to look for the leaf descendant of v containing the answer and spend another $O(\lg n)$ time to determine the position of the answer inside its chunk. Since we spend $O(\lg n)$ time for each of the two leaves we inspect and the tests for any other node in the tree take constant time, the cost is $O(\lg n)$.

Supporting operations on the leaves, such as finding the ith leaf from the left, reduces to `rank` and `select` operations over a bit vector $P_1[1..2n]$ where $P_1[i] = 1$ iff $P[i] = 1$ and $P[i+1] = 0$. `rank` and `select` operations over P_1 in turn reduce to `sum` and `fwd_search` operations over P_1 and can thus be supported by an RMMT for P_1. P_1 does not need to be stored explicitly because any consecutive $O(w)$ bits of P_1 can be computed from the corresponding bits of P using table lookup.

To analyze the space usage, observe that storing P requires $2n$ bits, while the space needed to store the vectors e', m', M', and n' is $2(n/s)\lg n = 2n/w$. The space needed to store the same vectors for the RMMT of P_1 is the same. Since we can assume that $w = \Omega(\lg n)$, the total size of the RMMT is thus $2n + O(n/\lg n)$ bits.

Dynamic Multithreading (DyM) Model. In the DyM model [7, Chapter 27], a *multithreaded computation* is modelled as a directed acyclic graph $G = (V, E)$ whose vertices are instructions and where $(u, v) \in E$ if u must be executed before v. The time T_p needed to execute the computation on p cores depends on two parameters of the computation: its *work* T_1 and its *span* T_∞. The work is the running time on a single core, that is, the number of nodes (i.e., instructions) in G, assuming each instruction takes constant time. Since p cores can execute only p instructions at a time, we have $T_p = \Omega(T_1/p)$. The span is the length of the longest path in G. Since the instructions on this path need to be executed in order, we also have $T_p = \Omega(T_\infty)$. Together, these two lower bounds give $T_p = \Omega(T_\infty + T_1/p)$. Work-stealing schedulers match the optimal bound to within a factor of 2 [4]. The degree to which an algorithm can take advantage of the presence of $p > 1$ cores is captured by its *speed-up* T_1/T_p and its *parallelism* T_1/T_∞. In the absence of cache effects, the best possible speed-up is p, known as *linear speed-up*. Parallelism provides an upper bound on the achievable speed-up.

To describe parallel algorithms in the DyM model, we augment sequential pseudocode with three keywords. The **spawn** keyword, followed by a procedure call, indicates that the procedure should run in its own thread and *may* thus be executed in parallel to the thread that spawned it. The **sync** keyword indicates that the current thread must wait for the termination of all threads it has spawned. It thus provides a simple barrier-style synchronization mechanism. Finally, **parfor** is "syntactic sugar" for **spawn**ing one thread per iteration in a for loop, thereby allowing these iterations to run in parallel, followed by a **sync** operation that waits for all iterations to complete. In practice, the overhead is logarithmic in the number of iterations. When a procedure exits, it implicitly performs a **sync** to ensure all threads it spawned finish first.

3   A Parallel Algorithm for Succinct Tree Construction

In this section, we describe our new parallel algorithm for constructing the RMMT of a given tree, called the *Parallel Succinct Tree Algorithm* (PSTA). Its input is the balanced parenthesis sequence P of an n-node tree T. This is a tree representation commonly used in practice, particularly in secondary storage, and known as the "folklore encoding". For trees whose folklore encoding is not directly available, we designed a parallel algorithm that can compute such an encoding in $O(n/p + \lg p)$ time, and the details are omitted due to space constraint. Our algorithms assume that manipulating w bits takes constant time. Additionally, we assume the (time and space) overhead of scheduling threads on cores is negligible. This is guaranteed by the results of [4], and the number of available processing units in current systems is generally much smaller than the input size n, so this cost is indeed negligible in practice.

Before describing the PSTA algorithm, we observe that the entries in e' corresponding to internal nodes of the RMMT need not be stored explicitly. This is because the entry of e' corresponding to an internal node is equal to the entry that corresponds to the last leaf descendant of this node; since the RMMT is complete, we can easily locate this leaf in constant time. Thus, our algorithm treats e' as an array of length $\lceil 2n/s \rceil$ with one entry per leaf. Our algorithm consists of three phases. In the first phase (Algorithm 1), it computes the leaves of the RMMT, i.e., the array e', as well as the entries of m', M' and n' that correspond to leaves. In the second phase (Algorithm 2), the algorithm computes the entries of m', M' and n' corresponding to internal nodes of the RMMT. In the third phase (Algorithm 3), it computes the universal lookup tables used to answer queries. The input to our algorithm consists of the balanced parenthesis sequence, P, the size of each chunk, s, and the number of available threads, *threads*.

To compute the entries of arrays e', m', M', and n' corresponding to the leaves of the RMMT (Algorithm 1), we first assign the same number of consecutive chunks, ct, to each thread (line 4). We call such a concatenation of chunks assigned to a single thread a *superchunk*. For simplicity, we assume that the total number of chunks, $\lceil 2n/s \rceil$, is divisible by *threads*. Each thread then computes the *local* excess value of the last position in each of its assigned chunks, as well as the minimum and maximum local excess in each chunk, and the number of times the minimum local excess occurs in each chunk (lines 8–17). These values are stored in the entries of e', m', M', and n' corresponding to this chunk (lines 18–21). The local excess value of a position i in P is defined to be $\mathrm{sum}(P, \pi, j, i)$, where j is the index of the first position of the superchunk containing position i. Note that the locations with minimum local excess in each chunk are the same as the positions with minimum global excess because the difference between local and global excess is exactly $\mathrm{sum}(P, \pi, 0, j - 1)$. Thus, the entries in n' corresponding to leaves store their final values at the end of the loop in lines 5–21, while the corresponding entries of e', m', and M' store *local* excess values.

To convert the entries in e' into global excess values, observe that the global excess at the end of each superchunk equals the sum of the local excess values at the ends of all superchunks up to and including this superchunk. Thus, we

Input : P, s, *threads*
Output: RMMT represented as arrays
        e', m', M', n' and universal
        lookup tables

1 $o := \lceil 2n/s \rceil - 1$ // # *internal nodes*
2 $e' :=$ array of size $\lceil 2n/s \rceil$
3 m', M', $n' :=$ arrays of size
    $\lceil 2n/s \rceil + o$
4 $ct := \lceil 2n/s \rceil / threads$
5 **parfor** $t := 0$ **to** *threads* $- 1$ **do**
6      $e'_t, m'_t, M'_t, n'_t := 0$
7   **for** $chk := 0$ **to** $ct - 1$ **do**
8          $low := (t * ct + chk) * s$
9          $up := low + s$
10   **for** $par := low$ **to** $up - 1$ **do**
11              $e'_t += 2 * P[par] - 1$
12   **if** $e'_t < m'_t$ **then**
13                  $m'_t := e'_t; n'_t := 1$
14   **else if** $e'_t = m'_t$ **then**
15                  $n'_t += 1$
16   **else if** $e'_t > M'_t$ **then**
17                  $M'_t := e'_t$
18          $e'[t * ct + chk] := e'_t$
19          $m'[t * ct + chk + o] := m'_t$
20          $M'[t * ct + chk + o] := M'_t$
21          $n'[t * ct + chk + o] := n'_t$
22 $parallel\_prefix\_sum(e', ct)$
23 **parfor** $t := 1$ **to** *threads* $- 1$ **do**
24   **for** $chk := 0$ **to** $ct - 1$ **do**
25   **if** $chk < ct - 1$ **then**
26              $e'[t * ct + chk] +=$
        $e'[t * ct - 1]$
27          $m'[t * ct + chk + o] +=$
        $e'[t * ct - 1]$
28          $M'[t * ct + chk + o] +=$
        $e'[t * ct - 1]$

Algorithm 1. PSTA (part I)

1 $lvl := \lceil \lg threads \rceil$
2 **parfor** $st := 0$ **to** $2^{lvl} - 1$ **do**
3   **for** $l := \lceil \lg(2n/s) \rceil - 1$ **downto**
    lvl **do**
4   **for** $d := 0$ **to** $2^{l-lvl} - 1$ **do**
5              $i := d + 2^l - 1 + st * 2^{l-lvl}$
6              $concat(i, m', M', n')$
7 **for** $l := lvl - 1$ **to** 0 **do**
8   **parfor** $d := 0$ **to** $2^l - 1$ **do**
9          $i := d + 2^l - 1$
10          $concat(i, m', M', n')$

Algorithm 2. PSTA (part II)

1 **parfor** $x := -w$ **to** $w - 1$ **do**
2   **parfor** $y := 0$ **to** $\sqrt{2^w} - 1$ **do**
3          $i := ((x + w) << w)$ OR w
4          $near\_fwd\_pos[i] := w$
5          $p, excess := 0$
6   **repeat**
7              $excess += 1 - 2 *$
        $((y$ AND $(1 << p)) = 0)$
8   **if** $excess = x$ **then**
9                  $near\_fwd\_pos[i] := p$
10   **break**
11              $p += 1$
12   **until** $p \geq w$

Algorithm 3. PSTA (part III)

Input : i, m', M', n'

1 $m'[i] := \min(m'[2i + 1], m'[2i + 2])$
2 $M'[i] := \max(M'[2i + 1], M'[2i + 2])$
3 **if** $m'[2i + 1] < m'[2i + 2]$ **then**
4      $n'[i] := n'[2i + 1]$
5 **else if** $m'[2i + 1] > m'[2i + 2]$ **then**
6      $n'[i] := n'[2i + 2]$
7 **else if** $m'[2i + 1] = m'[2i + 2]$ **then**
8      $n'[i] := n'[2i + 1] + n'[2i + 2]$

Function concat

use a parallel prefix sum algorithm [14] in line 22 to compute the global excess values at the ends of all superchunks and store these values in the corresponding entries of e'. The remaining local excess values in e', m', and M' can now be converted into global excess values by increasing each by the global excess at the end of the preceding superchunk. Lines 23–28 do exactly this.

The computation of entries of m', M', and n' (Algorithm 2) first chooses the level closest to the root that contains at least *threads* nodes and creates one thread for each such node v. The thread associated with node v calculates the m', M', and n' values of all nodes in the subtree with root v, by applying the function *concat* to the nodes in the subtree bottom up (lines 2–6). The invocation of this function for a node computes its m', M', and n' values from the corresponding values of its children. With a scheduler that balances the work, such as a work-stealing scheduler, cores have a similar workload. Lines 7–10 apply a similar bottom-up strategy for computing the m', M', and n' values of the nodes in the top lvl levels, but they do this by processing these levels sequentially and, for each level, processing the nodes on this level in parallel.

Algorithm 3 illustrates the construction of universal lookup tables using the construction of the table *near_fwd_pos* as an example. This table is used to support the `fwd_search` operation (see Section 2). Other lookup tables can be constructed analogously. As each entry in such a universal table can be computed independently, we can easily compute them in parallel.

Theoretical Analysis. Lines 1–21 of Algorithm 1 require $O(n)$ work and have span $O(n/p)$. Line 22 requires $O(p)$ work and has span $O(\lg p)$ because we compute a prefix sum over only p values. Lines 23–28 require $O(n/s)$ work and have span $O(n/sp)$. Lines 1–6 of Algorithm 2 require $O(n/s)$ work and have span $O(n/sp)$. Lines 7–10 require $O(p)$ work and have span $O(\lg p)$. Algorithm 3 requires $O(\sqrt{2^w}\mathtt{poly}(w))$ work and has span $O(\sqrt{2^w}\mathtt{poly}(w)/p)$. As was defined in Section 2, w is the machine word size. Thus, the total work of PSTA is $T_1 = O(n + \lg p + \sqrt{2^w}\mathtt{poly}(w))$ and its span is $O(n/p + \lg p + \sqrt{2^w}\mathtt{poly}(w)/p)$. For $p \to \infty$, we get a span of $T_\infty = O(\lg n)$. This gives a running time of $T_p = O(T_1/p + T_\infty) = O(n/p + \lg p + \sqrt{2^w}\mathtt{poly}(w)/p)$ on p cores. The speedup is $T_1/T_p = O\left(\frac{p(n+\sqrt{2^w}\mathtt{poly}(w))}{n+\sqrt{2^w}\mathtt{poly}(w)+p\lg p}\right)$. Under the assumption that $p \ll n$, the speedup approaches $O(p)$. Moreover, the parallelism T_1/T_∞ (the maximum theoretical speedup) of PSTA is $\frac{n+\sqrt{2^w}\mathtt{poly}(w)}{\lg n}$.

The PSTA algorithm does not need any extra memory related to the use of threads. Indeed, it just needs space proportional to the input size and the space needed to schedule the threads. A work-stealing scheduler, like the one used by the DyM model, exhibits at most a linear expansion space, that is, $O(S_1 p)$, where S_1 is the minimum amount of space used by the scheduler for any execution of a multithreaded computation using one core. This upper bound is optimal within a constant factor [4]. In summary, the working space needed by our algorithm is $O(n \lg n + S_1 p)$ bits. Since in our algorithm the scheduler does not need to consider the input size to schedule threads, $S_1 = O(1)$. Thus, since in modern machines it is usual that $p \ll n$, the scheduling space is negligible and the working space is dominated by $O(n \lg n)$.

Note that in succinct data structure design, it is common to adopt the assumption that $w = \Theta(\lg n)$, and when constructing lookup tables, consider all possible bit vectors of length $(\lg n)/2$ (instead of $w/2$). This guarantees that the universal lookup tables occupy only $o(n)$ bits. Adopting the same strategy, we can simplify our analysis and obtain $T_p = O(n/p + \lg p)$. Thus, we have the following theorem:

Theorem 1. *A $(2n + o(n))$-bit representation of an ordinal tree on n nodes and its balanced parenthesis sequence can be computed in $O(n/p + \lg p)$ time using $O(n \lg n)$ bits of working space, where p is the number of cores. This representation can support the operations in Table 1 of [19] in $O(\lg n)$ time.*

4   Experimental Results

To evaluate the performance of our PSTA algorithm, we compare it against libcds [6] and sdsl [12], which are state-of-the-art implementations of the RMMT. Both assume that the input tree is given as a parenthesis sequence, as we do here. Our implementation of the PSTA algorithm deviates from the description in Section 3 in that the prefix sum computation in line 22 of the algorithm is done sequentially in our implementation. This changes the running time to $O(n/p + p)$ but simplifies the implementation. Since $p \ll n/p$ for the input sizes we are interested in and the numbers of cores available on current multicore systems, the impact on the running time is insignificant. We implemented the PSTA algorithm in C and compiled it using GCC 4.9 with optimization level -O2 and using the -ffast-math flag.[3] All parallel code was compiled using the GCC Cilk branch. The same flags were used to compile libcds and sdsl, which were written in C++.

We tested our algorithm on five inputs. The first two were suffix trees of the DNA (dna, 1,154,482,174 parentheses), and protein (prot, 670,721,006 parentheses) data from the Pizza & Chili corpus[4]. These suffix trees were constructed using code from http://www.daimi.au.dk/~mailund/suffix_tree.html. The next two inputs were XML trees of the Wikipedia dump[5] (wiki, 498,753,914 parentheses) and OpenStreetMap dump[6] (osm, 4,675,776,358 parentheses). The final input was a complete binary tree of depth 30 (ctree, 2,147,483,644 parentheses).

The experiments were carried out on a machine with four 16-core AMD Opteron[TM] 6278 processors clocked at 2.4GHz, with 64KB of L1 cache per core, 2MB of L2 cache shared between two cores, and 6MB of L3 cache shared between 8 cores. The machine had 189GB of DDR3 RAM, clocked at 1333MHz.

Running times were measured using the high-resolution (nanosecond) C functions in <time.h>. Memory usage was measured using the tools provided by malloc_count [3]. In our experiments, the chunk size s was fixed at 256.

---

[3] The code and data needed to replicate our results are available at http://www.inf. udec.cl/~josefuentes/sea2015.

[4] http://pizzachili.dcc.uchile.cl

[5] http://dumps.wikimedia.org/enwiki/20150112/enwiki-20150112-pages-articles.xml. bz2 (January 12, 2015)

[6] http://wiki.openstreetmap.org/wiki/Planet.osm (January 10, 2015)

Table 1. Running times of PSTA,
libcds, and sdsl in seconds

p	wiki	prot	dna	ctree	osm
libcds	33.16	44.24	75.87	140.41	339.21
sdsl	1.93	2.66	4.57	8.35	18.10
1	2.89	4.22	7.21	12.16	30.60
2	1.44	2.13	3.64	6.15	15.43
4	.73	1.10	1.87	3.18	7.98
8	.37	.57	.98	1.59	4.14
16	.25	.35	.58	.86	2.21
32	.18	.25	.39	.63	1.33
64	.27	.29	.39	.48	1.01

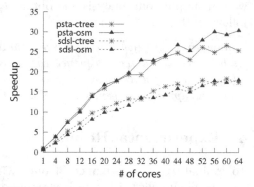

Fig. 2. Speed-up on ctree and osm data sets

Running Time and Speed-Up. Table 1 shows the wall clock times achieved by psta, libcds, and sdsl on different inputs. Each time is the minimum achieved over three non-consecutive runs, reflecting our assumption that slightly increased running times are the result of "noise" from external processes such as operating system and networking tasks. Figure 2 shows the speed-up for the ctree and osm inputs compared to the running times of psta on a single core and of sdsl.

The psta algorithm on a single core and sdsl outperformed libcds by an order of magnitude. One of the reasons for this is that libcds implements a different version of RMMT including *rank* and *select* structures, while psta and sdsl do not. Constructing these structures is costly. On a single core, sdsl was about 1.5 times faster than psta, but neither sdsl nor libcds were able to take advantage of multiple cores, so psta outperformed both of them starting at $p = 2$. The advantage of sdsl over psta on a single core, in spite of implementing essentially the same algorithm, can be attributed to (1) lack of tuning of psta and (2) some overhead with running parallel code on a single core.

Up to 16 cores, the speed-up of psta is almost linear whenever p is a power of 2 and the efficiency (speed-up/p) is 70% or higher, except for ctree on 32 cores. This is very good for a multicore architecture. When p is not a power of 2, speed-up is slightly worse. The reason is that, when p is a power of 2, psta can assign exactly one subtree to each thread (see Algorithm 2), distributing the work homogeneously across cores without any work stealing. When the number of threads is not a power of two, some threads have to process more than one subtree and other threads process only one, which degrades performance due to the overhead of work stealing.

There were three other factors that limited the performance of psta in our experiments: network topology, input size, and resource contention with the OS.

Topology. The four processors on our machine were connected in a grid topology [8]. Up to 32 threads, all threads can be run on a single processor or on two adjacent processors in the grid, which keeps the cost of communication between threads low. Beyond 32 threads, at least three processor are needed and at least two of them are not adjacent in the grid. This increases the cost of communication between threads on these processors noticeably.

Input size. For the two largest inputs we tested, osm and ctree, speed-up kept increasing as we added more cores. For wiki, however, the best speed-up was achieved with 36 cores. Beyond this, the amount of work to be done per thread was small enough that the scheduling overhead caused by additional threads started to outweigh the benefit of reducing the processing time per thread further.

Resource contention. For $p < 64$, at least one core on our machine was available to OS processes, which allowed the remaining cores to be used exclusively by psta. For $p = 64$, psta competed with the OS for available cores. This had a detrimental effect on the efficiency of psta for $p = 64$.

Memory Usage. We measured the amount of working memory (i.e., memory not occupied by the raw parenthesis sequence) used by psta, libcds, and sdsl. We did this by monitoring how much memory was allocated/released with malloc/free and recording the peak usage. For psta, we only measured the memory usage for $p = 1$. The extra memory needed for thread scheduling when $p > 1$ was negligible. Due to lack of space, we report the results only for the two largest inputs, ctree and osm. For the ctree input, psta, libcds, and sdsl used 112MB, 38MB, and 76MB of memory, respectively. For osm, they used 331MB, 85MB, and 194MB, respectively. Even though psta uses more memory than both libcds and sdsl, the difference between psta and sdsl is a factor of less than two. The difference between psta and libcds is no more than a factor of four and is outweighed by the substantially worse performance of libcds.

Part of the higher memory usage of psta stems from the allocation of e', m', M' and n' arrays to store the partial excess values in the algorithm. Storing these values, however, is a key factor that helps psta to achieve very good performance.

5 Conclusions and Future Work

In this paper, we demonstrated that it is possible to improve the construction time of succinct trees using multicore parallelism. We introduced a practical algorithm that takes $O(n/p + \lg p)$ time to construct a succinct representation of a tree with n nodes using p threads. This representation supports a rich set of operations in $O(\lg n)$ time. Our algorithm substantially outperformed state-of-the-art sequential constructions of this data structure, achieved very good speed-up up to 64 cores, and is to the best of our knowledge the first parallel construction algorithm of a succinct representation of ordinal trees.

While we focused on representing static trees succinctly in this paper, the approach we have taken may also extend to the construction of *dynamic* succinct trees (e.g., [19]), of succinct representations of *labelled* trees, and of other succinct data structures that use succinct trees as building blocks (e.g., the succinct representation of planar graphs).

Acknowledgments. We would like to thank Diego Arroyuelo, Roberto Asín, and Rodrigo Cánovas for their time and making resources available to us.

References

1. Arroyuelo, D., Cánovas, R., Navarro, G., Sadakane, K.: Succinct trees in practice. In: ALENEX, pp. 84–97. SIAM Press, Austin (2010)
2. Benoit, D., Demaine, E.D., Munro, J.I., Raman, V.: Representing trees of higher degree. In: Dehne, F., Gupta, A., Sack, J.-R., Tamassia, R. (eds.) WADS 1999. LNCS, vol. 1663, pp. 169–180. Springer, Heidelberg (1999)
3. Bingmann, T.: malloc_count - tools for runtime memory usage analysis and profiling, January 17, 2015
4. Blumofe, R.D., Leiserson, C.E.: Scheduling multithreaded computations by work stealing. J. ACM **46**(5), 720–748 (1999)
5. Clark, D.R., Munro, J.I.: Efficient suffix trees on secondary storage. In: SODA, pp. 383–391 (1996)
6. Claude, F.: A compressed data structure library, January 17, 2015
7. Cormen, T.H., Leiserson, C.E., Rivest, R.L., Stein, C.: Introduction to Algorithms, chapter. Multithreaded Algorithms, third edn., pp. 772–812. The MIT Press (2009)
8. Drepper, U.: What every programmer should know about memory (2007)
9. Farzan, A., Munro, J.I.: A uniform paradigm to succinctly encode various families of trees. Algorithmica **68**(1), 16–40 (2014)
10. Fuentes-Sepúlveda, J., Elejalde, E., Ferres, L., Seco, D.: Efficient Wavelet tree construction and querying for multicore architectures. In: Gudmundsson, J., Katajainen, J. (eds.) SEA 2014. LNCS, vol. 8504, pp. 150–161. Springer, Heidelberg (2014)
11. Geary, R.F., Raman, R., Raman, V.: Succinct ordinal trees with level-ancestor queries. In: SODA, pp. 1–10 (2004)
12. Gog, S.: Succinct data structure library 2.0, January 17, 2015
13. He, M., Munro, J.I., Satti, S.R.: Succinct ordinal trees based on tree covering. ACM Trans. Algorithms **8**(4), 42 (2012)
14. Helman, D.R., JáJá, J.: Prefix computations on symmetric multiprocessors. J. Par. Dist. Comput. **61**(2), 265–278 (2001)
15. Jacobson, G.: Space-efficient static trees and graphs. In: FOCS, pp. 549–554 (1989)
16. Jansson, J., Sadakane, K., Sung, W.K.: Ultra-succinct representation of ordered trees. In: SODA (2007)
17. Lu, H.I., Yeh, C.C.: Balanced parentheses strike back. ACM Trans. Algorithms **4**, 28:1–28:13 (2008)
18. Munro, J.I., Raman, V.: Succinct representation of balanced parentheses, static trees and planar graphs. In: FOCS, pp. 118–126 (1997)
19. Navarro, G., Sadakane, K.: Fully functional static and dynamic succinct trees. ACM Trans. Algorithms **10**(3), 16:1–16:39 (2014)
20. Raman, R., Rao, S.S.: Succinct representations of ordinal trees. In: Brodnik, A., López-Ortiz, A., Raman, V., Viola, A. (eds.) Ianfest-66. LNCS, vol. 8066, pp. 319–332. Springer, Heidelberg (2013)
21. Shun, J.: Parallel wavelet tree construction. CoRR abs/1407.8142 (2014)

Tree Compression with Top Trees Revisited

Lorenz Hübschle-Schneider[1]($\boxtimes$) and Rajeev Raman[2]

[1] Institute of Theoretical Informatics,
Karlsruhe Institute of Technology, Karlsruhe, Germany
huebschle@kit.edu

[2] Department of Computer Science, University of Leicester, Leicester, UK
r.raman@leicester.ac.uk

Abstract. We revisit tree compression with top trees (Bille et al. [2]), and present several improvements to the compressor and its analysis. By significantly reducing the amount of information stored and guiding the compression step using a RePair-inspired heuristic, we obtain a fast compressor achieving good compression ratios, addressing an open problem posed by [2]. We show how, with relatively small overhead, the compressed file can be converted into an in-memory representation that supports basic navigation operations in worst-case logarithmic time without decompression. We also show a much improved worst-case bound on the size of the output of top-tree compression (answering an open question posed in a talk on this algorithm by Weimann in 2012).

Keywords: Tree compression · Grammar compression · Top trees · XML compression

1   Introduction

Labelled trees are one of the most frequently used nonlinear data structures in computer science, appearing in the form of suffix trees, XML files, tries, and dictionaries, to name but a few prominent examples. These trees are frequently very large, prompting a need for compression for on-disk storage. Ideally, one would like specialized tree compressors to certainly get much better compression ratios than general-purpose compressors such as bzip2 or gzip, but also for the compression to be fast; as Ferragina et al. note [10, p4:25]. [1]

In fact, it is also frequently necessary to hold such trees in main memory and perform complex navigations to query or mine them. However, common in-memory representations use pointer data structures that have significant overhead—e.g. for XML files, standard DOM[2] representations are typically 8-16 times larger than the (already large) XML file [22,24]. To process such large trees, it is essential to have compressed in-memory representations that *directly* support rapid navigation and queries, without partial or full decompression.

---

[1] Their remark is about XML tree compressors but applies to general ones as well.
[2] *Document Object Model*, a common interface for interacting with XML documents.

© Springer International Publishing Switzerland 2015
E. Bampis (Ed.): SEA 2015, LNCS 9125, pp. 15–27, 2015.
DOI: 10.1007/978-3-319-20086-6_2

Before we describe previous work, and compare it with ours, we give some definitions. A *labelled tree* is an ordered, rooted tree whose nodes have labels from an alphabet Σ of size $|\Sigma| = \sigma$. We consider the following kinds of redundancy in the tree structure. *Subtree repeats* are repeated occurrences of *rooted subtrees*, i.e. a node and all of its descendants, identical in structure and labels. *Tree pattern repeats* or *internal repeats* are repeated occurrences of *tree patterns*, i.e. connected subgraphs of the tree, identical in structure as well as labels.

1.1 Previous Work

Nearly all existing compression methods for labelled trees follow one of three major approaches: *transform-based compressors* that transform the tree's structure, e.g. into its minimal DAG, *grammar-based compressors* that compute a tree grammar, and–although not compression–*succinct representations* of the tree.

Transform-Based Compressors. We can replace subtree repeats by edges to a single shared instance of the subtree and obtain a smaller Directed Acyclic Graph (DAG) representing the tree. The smallest of these, called the *minimal DAG*, is unique and can be computed in linear time [9]. Navigation and path queries can be supported in logarithmic time [3,4]. While its size can be exponentially smaller than the tree, no compression is achieved in the worst case (a chain of nodes with the same label is its own minimal DAG, even though it is highly repetitive). Since DAG minimization only compresses repeated subtrees, it misses many internal repeats, and is thus insufficient in many cases.

Bille et al. introduced tree compression with top trees [2], which this paper builds upon. Their method exploits both repeated subtrees and tree structure repeats, and can compress exponentially better than DAG minimization. They give a $\log_\sigma^{0.19} n$ worst-case compression ratio for a tree of size n labelled from an alphabet of size σ for their algorithm. They show that navigation and a number of other operations are supported in $O(\log n)$ time directly on the compressed representation. However, they do not give any practical evaluation, and indeed state as an open question whether top-tree compression has practical value.

Tree Grammars. A popular approach to exploit the redundancy of tree patterns is to represent the tree using a formal grammar that generates the input tree, generalizing grammar compression from strings to trees [5,6,14,16–18]. These can be exponentially smaller than the minimal DAG [16]. Since it is NP-Hard to compute the smallest grammar [7], efficient heuristics are required.

One very simple yet efficient heuristic method is RePair [15]. A string compressor, it can be applied to a parentheses bitstring representation of the tree. The output grammars produced by RePair can support a variety of navigational operations and random access, in time logarithmic in the input tree size, after additional processing [3]. These methods, however, appear to require significant engineering effort before their practicality can be assessed.

TreeRePair [17] is a generalization of RePair from strings to trees. It achieves the best grammar compression ratios currently known. However, navigating TreeRePair's grammars in sublinear time with respect to their depth, which

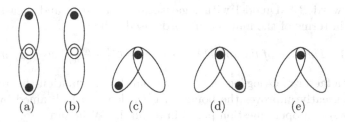

Fig. 1. Five kinds of cluster merges in top trees. Solid nodes are boundary nodes, hollow ones are boundary nodes that become internal. Source of this graphic: [2, § 2.1].

can be linear in their size [2], is an open problem. For relatively small documents (where the output of TreeRePair fits in cache), the navigation speed for simple tree traversals is about 5 times slower than succinct representations [17].

Several other popular grammar compressors exist for trees. Among them, BPLEX [5,6] is probably best-known, but is much slower than TreeRePair. The TtoG algorithm is the first to achieve a good theoretical approximation ratio [14], but has not been evaluated in practice.

Succinct Representations. Another approach is to represent the tree using near-optimal space without applying compression methods to its structure, a technique called *succinct data structures*. Unlabelled trees can be represented using $2n + o(n)$ bits [13] and support queries in constant time [21]. There are a few $n \log \sigma + \mathcal{O}(n)$ bit-representations for labelled trees, most notably that by Ferragina et al. [10], which also yields a compressor, XBZip. While XBZip has good performance on XML files *in their entirety*, including text, attributes etc., evidence suggests that it does not beat TreeRePair on pure labelled trees. As the authors admit, it is also slow.

1.2   Our Results

Our primary aim in this paper is to address the question of Bille et al. [2] regarding the practicality of the top tree approach, but we make some theoretical contributions as well. We first give some terminology and notation.

A *top tree* [1] is a hierarchical decomposition of a tree into *clusters*, which represent subgraphs of the original tree. Leaf clusters correspond to single edges, and inner clusters represent the union of the subgraphs represented by their two children. Clusters are formed in one of five ways, called *merge types*, shown in Figure 1. A cluster can have one or two *boundary nodes*, a top- and optionally a bottom boundary node, where other clusters can be attached by merging. A top tree's minimal DAG is referred to as a *top DAG*. For further details on the fundamentals of tree compression with top trees, refer to [2]. Throughout this paper, let T be any ordered, labelled tree with n_T nodes, and let Σ denote the label alphabet with $\sigma := |\Sigma|$. Let $\mathcal{T}$ be the top tree and $\mathcal{TD}$ the top DAG corresponding to T, and $n_{\mathcal{TD}}$ the total size (nodes plus edges) of $\mathcal{TD}$. We assume a standard word RAM model with logarithmic word size, and measure space complexity in terms of the number of words used. Then:

Theorem 1. *The size of the top DAG is* $n_{\mathcal{TD}} = \mathcal{O}\left(\frac{n_T}{\log_\sigma n_T} \cdot \log\log_\sigma n_T\right)$.

This is only a factor of $\mathcal{O}(\log\log_\sigma n_T)$ away from the information-theoretic lower bound, and greatly improves the bound of $\mathcal{O}(n/\log_\sigma^{0.19} n)$ obtained by Bille et al. and answers an open question posed in a talk by Weimann.

Next, we show that if only basic navigation is to be performed, the amount of information that needs to be stored can be greatly reduced, relative to the original representation [2], without affecting the asymptotic running time.

Theorem 2. *We can support navigation with the operations* Is Leaf, Is Last Child, First Child, Next Sibling, *and* Parent *in* $\mathcal{O}(\log n_T)$ *time, full decompression in time* $\mathcal{O}(n_T)$ *on a representation of size* $\mathcal{O}(n_{\mathcal{TD}})$ *storing only the top DAG's structure, the merge types of inner nodes (an integer from* [1..5]*), and leaves' labels.*

We believe this approach will have low overhead and fast running times in practice for in-memory navigation without decompression, and sketch how one would approach an implementation.

Furthermore, we introduce the notion of *combiners* that determine the order in which clusters are merged during top tree construction. Combiners aim to improve the compressibility of the top tree, resulting in a smaller top DAG. We present one such combiner that applies the basic idea of RePair [15] to top tree compression, prioritizing merges that produce subtree repeats in the top tree, in Section 3. We give a relatively naive encoding of the top tree, primarily using Huffman codes, and evaluate its compression performance. Although the output of the modified top tree compressor is up to 50 % larger than the state-of-the-art TreeRePair, it is about six times faster. We believe that the compression gap can be narrowed while maintaining the speed gap.

2   Top Trees Revisited

2.1   DAG Design Decisions

The original top tree compression paper [2] did not try to minimize the amount of information that actually needs to be stored. Instead, the focus was on implementing a wide variety of navigation operations in logarithmic time while maintaining $\mathcal{O}(n_{\mathcal{TD}})$ space *asymptotically*. Here, we reduce the amount of additional information stored about the clusters to obtain good compression ratios.

Instead of storing the labels of both endpoints of a leaf cluster's corresponding edge, we store only the child's label, not the parent's. In addition to reducing storage requirements, this reduces the top tree's alphabet size from σ^2+5 to $\sigma+5$, as each cluster has either one label or a merge type. This increases the likelihood of identical subtrees in the top tree, improving compression. Note that this change implies that there is exactly one leaf cluster in the top DAG for each distinct label in the input. To code the root, we perform a merge of type (a) (see Section 1.2 and Figure 1) between a dummy edge leading to the root and the last remaining edge after all other merges have completed.

With these modifications, we reduce the amount of information stored with the clusters to the bare minimum required for decompression, i.e. leaf clusters' labels and inner clusters' merge types.

Lastly, we speed up compression by directly constructing the top DAG during the merge process. We initialize it with all distinct leaves, and maintain a mapping from cluster IDs to node IDs in $\mathcal{TD}$, as well as a hash map mapping DAG nodes to their node IDs. When two edges are merged into a new cluster, we look up its children in the DAG and only need to add a new node to $\mathcal{TD}$ if this is its first occurrence. Otherwise, we simply update the cluster-to-node mapping.

2.2 Navigation

We now explain how to navigate the top DAG with our reduced information set. We support full decompression in time $\mathcal{O}(n_T)$, as well as operations to move around the tree in time proportional to the height of the top DAG, i.e. $\mathcal{O}(\log n_T)$. These are: determining whether the current node is a leaf or its parent's last child, and moving to its first child, next sibling, and parent. Accessing a node's label is possible in constant time given its node number in the top DAG.

Proof (Theorem 2). As a node in a DAG can be the child of any number of other nodes, it does not have a unique parent. Thus, to allow us to move back to a node's parent in the DAG, we need to maintain a stack of parent cluster IDs along with a bit to indicate whether we descended into the left or right child. We refer to this as the *DAG stack*, and update it whenever we move around in $\mathcal{TD}$ with the operations below. Similarly, we also maintain a *tree stack* containing the DAG stack of each ancestor of the current node in the (original) tree.

Decompression: We traverse the top DAG in pre-order, undoing the merge operations to reconstruct the tree. We begin with n_T isolated nodes, and then add back the edges and labels as we traverse the top DAG. As this requires constant time per cluster and edge, we can decompress the top DAG in $\mathcal{O}(n_T)$ time.

Label Access: Since only leaf clusters store labels, and these are coded as the very first clusters in the top DAG (cf. Section 2.4), their node indices come before all other nodes'. Therefore, a leaf's label index i is its node number in the top DAG. We access the label array in the position following the $(i-1)$th null byte, which we can find with a $\mathsf{Select}_0(i-1)$ operation, and decode the label string until we reach another null byte or the end.

Is Leaf: A node is a leaf iff it is no cluster's top boundary node. Moving up through the DAG stack, if we reach a cluster of type (a) or (b) from the *left* child, the node is not a leaf (the left child of such a cluster is the *upper* one in Figure 1). If, at any point before encountering such a cluster, we exhaust the

DAG stack or reach a cluster of type (b) or (c) from the right, type (d) from the left, or type (e) from either side, the node is a leaf. This can again be seen in Figure 1.

Is Last Child: We move up the DAG stack until we reach a cluster of type (c), (d), or (e) from its left child. Upon encountering a cluster of type (a) or (b) from the right, or emptying the DAG stack completely, we abort as the upward search lead us to the node's parent or exhausted the tree, respectively.

First Child and Next Sibling: First, we check whether the node is a leaf (First Child) or its parent's last child (Next Sibling), and abort if it is. First Child then pushes a copy of the DAG stack onto the tree stack. Next, we re-use the upward search performed by the previous check, removing the elements visited by the search from the DAG stack, up until the cluster with which the search ended. We descend into its right child and keep following the left child until we reach a leaf.

Parent: Since First Child pushes the DAG stack onto the tree stack, we simply reset the DAG stack to the tree stack's top element, which is removed.   □

We note here that the tree stack could, in theory, grow to a size of $\mathcal{O}(n_T \log n_T)$, as the tree can have linear height and the logarithmically sized DAG stack is pushed onto it in each First Child operation. However, we argue that due to the low depth of common labelled trees, especially XML files, this stack will remain small in practice. Even when pessimistically assuming a *very* large tree with a height of 80 nodes, with a top tree of height 50, the tree stack will comfortably fit into 32 kB when using 64-bit node IDs. Our preliminary experiments confirm this.

To improve the worst-case tree stack size in theory, we can instead keep a log of movements in the top DAG, which is limited in size to the distance travelled therein. We expect this to be significantly less than $\mathcal{O}(n_T \log n_T)$ in expectation.

2.3   Worst-Case Top DAG Size

Bille et al. show that a tree's top tree has at most $\mathcal{O}\big(n_T / \log_\sigma^{0.19} n_T\big)$ distinct clusters [2]. This bound, however, is an artifact of the proof. By modifying the definition of a *small cluster* in the compression analysis and carefully exploiting the properties of top trees, we are able to show a new, tighter, bound, which directly translates to an improvement on the worst-case compression ratio. We omit the full proof of Theorem 1 for brevity, and instead give the following essential lemmata, which we use to prove the theorem in the full version of this paper. Let $s(v)$ be the size of v's subtree, and $p(v)$ denote its parent.

Lemma 1. *Let T be any ordered labelled tree of size n_T, and let $\mathcal{T}$ be its top tree. For any node v of $\mathcal{T}$, the height of its subtree is at most $\lfloor \log_{8/7} s(v) \rfloor$.*

Lemma 2. *Let T be any ordered labelled tree of size n_T, let $\mathcal{T}$ be its top tree, and t be an integer. Then $\mathcal{T}$ contains at most $\mathcal{O}((n_T/t) \cdot \log t)$ nodes v so that $s(v) \leq t$ and $s(p(v)) > t$.*

2.4   Encoding

In the top DAG, we need to be able to access a cluster's left and right child, as well as its merge type for inner clusters or the child node's label for the edge that it refers to for leaf clusters. To realize this interface, we decompose the top DAG into a binary *core* tree and a pointer array. The core tree is defined by removing all incoming edges from each node, except for the one coming from the node with lowest pre-order number. All other occurrences are replaced by a dummy leaf node storing the pre-order number of the referenced node. Leaves in the top DAG are assigned new numbers as label pointers, which are smaller than the IDs of all inner nodes. All references to leaves, including the dummy nodes, are coded in an array of *pointers*, ordered by the pre-order number of the originating node. Similarly, the inner nodes' merge types are stored in an array in pre-order. Lastly, the core tree itself can be encoded using two bits per inner node, indicating whether the left and right children are inner nodes in the core tree.

Using this representation, all that is required for efficient navigation is an entropy coder providing constant-time random access to node pointers and merge types, and a data structure providing rank and select for the core tree and label strings. All of these building blocks can be treated as black boxes, and are well-studied and readily available, e.g. [23] and the excellent SDSL [11] library.

Simple Encoding.   To obtain file size results with reasonable effort, we now describe a very simple encoding that does not lend itself to navigation as easily. We compress the core tree bitstring and merge types using blocked Huffman coding. The pointer array and null byte-separated concatenated label string are encoded using a Huffman code. The Huffman trees are coded like the core tree above. The symbols are encoded using a fixed length and concatenated. Lastly, we store the sizes of the four Huffman code segments as a file header.

3   Heuristic Combiners

As described in the original paper [2], the construction of the top tree *exposes* internal repetitions. However, it does not attempt to maximize the size or number of identical subtrees in the top tree, i.e. its compressibility. Instead, the merge process sweeps through the tree linearly from left to right and bottom to top. This is a straight-forward cluster combining strategy that fulfills all the requirements for constructing a top tree, but does not attempt to maximize compression performance. We therefore replace the standard combining strategy with heuristic methods that try to increase compressibility of the top tree. Here, we present one such combiner that applies the basic idea of RePair to the horizontal merge step of top tree compression. (In preliminary experiments, it proved detrimental to apply the heuristic to vertical merges, and we limit ourselves to the horizontal merge step, but note that this is not a general restriction on combiners.)

We hash all clusters in the top tree as they are created. The hash value combines the cluster's label, merge type, and the hashes of its left and right children

Table 1. XML corpus used for our experiments. File sizes are given for stripped documents, i.e. after removing whitespace and tags' attributes and contents.

File name	size (MB)	# nodes	height	File name	size (MB)	# nodes	height
1998statistics	0.60	28 306	6	JST-snp.chr1	27.31	803 596	8
dblp	338.87	20 925 865	6	nasa	8.43	476 646	8
enwiki-latest-p	229.78	14 018 880	5	NCBI-gene.chr1	35.30	1 065 787	7
factor12	359.36	20 047 329	12	proteins	365.12	21 305 818	7
factor4	119.88	6 688 651	12	SwissProt	45.25	2 977 031	5
factor4.8	143.80	8 023 477	12	treebank-e	25.92	2 437 666	36
factor7	209.68	11 697 881	12	uwm	1.30	66 729	5
JST-gene.chr1	5.79	173 529	7	wiki	42.29	2 679 553	5

if these exist. As the edges in the auxiliary tree correspond to clusters in the top tree during its construction, we assign the cluster's hashes to the corresponding edges. Defining a digram as two edges whose clusters can be merged with one of the five merge types from Figure 1, we can apply the idea of RePair, identifying the edges by their hash values. In descending order of digram frequency, we merge all non-overlapping occurrences, updating the remaining edges' hash values to those of the newly created clusters.

Since this procedure does not necessarily merge a constant fraction of the edges in each iteration, we may need to additionally apply the normal horizontal merge algorithm if too few edges were merged by the heuristic. The constant upon which this decision is based thus becomes a tuning parameter. Note that we need to ensure that every edge is merged at most once per iteration.

4   Evaluation

We now present an experimental evaluation of top tree compression. In this section, we demonstrate its qualities as a fast and efficient compressor, compare it against other compressors, and show the effectiveness of our RePair-inspired combiner.

Experimental Setup. All algorithms were implemented in C++11 and compiled with the GNU C++ compiler g++ in version 4.9.2 using optimization level fast and profile-guided optimizations. The experiments were conducted on a commodity PC with an Intel Core i7-4790T CPU and 16 GB of DDR3 RAM, running Debian Linux from the sid channel. We used gzip 1.6-4 and bzip2 1.0.6-7 from the standard package repositories. Default compression settings were used for all compressors, except the -9 flag for gzip. All input and output files were located in main memory using a tmpfs RAM disk to eliminate I/O delays.

XML Corpus. We evaluated the compressor and our heuristic improvement on a corpus of common XML files [8, 12, 20, 25], listed in Table 1. In our experiments, we give file sizes for our simple encoding, which represent pessimistic results that can serve as an upper bound of what to expect from a more optimized encoding. We give these file sizes to demonstrate that even a simple encoding yields good results with regard to file size, speed, and ease of navigation (see Section 2.2).

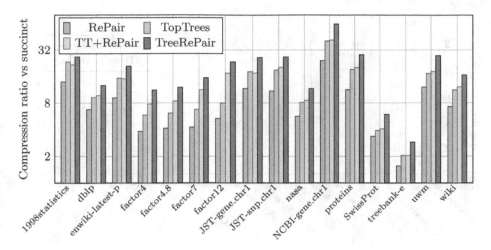

Fig. 2. Comparison of compression ratios, measured by comparing file sizes against a succinct encoding of the input file (higher is better)

Results. We use a minimum merge ratio of $c = 1.26$ for the horizontal merge step using our RePair-inspired heuristic combiner in all our experiments. This is the result of an extensive evaluation which showed that values $c \in [1.2, 1.27]$ work very well on a broad range of XML documents. We observed that values close to 1 can improve compression by up to 10 % on some files, while causing a deterioration by a similar proportion on others. Thus, while better choices of c exist for individual files, we chose a fixed value for all files to provide a fair comparison, similar to the choice of 4 as the maximum rank of the grammar in TreeRePair [17].

We use a parenthesis bitstring encoding of the input tree as a baseline to measure compression ratios. The unique label strings are concatenated, separated by null bytes. Indices into this array are stored as fixed-length numbers of $\lceil \log_2 \#labels \rceil$ bits. TreeRePair[3], which has been carefully optimized to produce very small output files, serves us as a benchmark. We are, however, reluctant to compare tree compression with top trees to TreeRePair directly, as our methods have not been optimized to the same degree.

In Figure 2 we give a compression ratios relative to the succinct encoding. We evaluated our implementation of top tree compression using the combining strategy from [2] as well as our RePair-inspired combiner. We also give the file sizes achieved by TreeRePair and those of RePair on a parentheses bitstring representation of the input tree and the concatenated nullbyte-separated label string (note that no deduplication is performed here, as this is up to the compressor). We represent RePair's grammar production rules as a sequence of integers with an implicit left-hand side and encode this representation using a Huffman code. Figure 2 shows that top tree compression consistently outperforms RePair

---

[3] https://code.google.com/p/treerepair

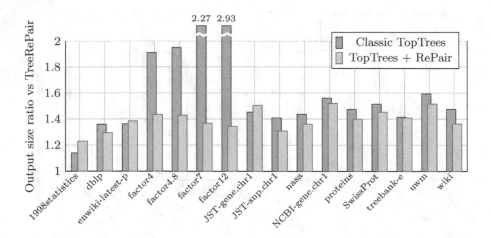

Fig. 3. Comparison of output file sizes produced by top tree compression with and without the RePair combiner, measured against TreeRePair file sizes (lower is better)

already, but does not achieve the same level of compression as TreeRePair at this stage. We can also clearly see the impact of our RePair-inspired heuristic combiner, which improves compression on nearly all files in our corpus and is studied in more detail in the next paragraph. Table 3 gives the exact numbers for the output file sizes, supplementing them with results for general-purpose compressors.

RePair Combiner. Figure 3 compares the two versions of top tree compression, using TreeRePair as a benchmark. The RePair combiner's effect is clearly visible, reducing the maximum disparity in compression relative to TreeRePair from a file 2.93 times the size (`factor12`) to one that is 52 % larger (`NCBI-gene.chr1`). This constitutes nearly a four-fold decrease in overhead (from 1.93 to 0.52). On average, files are 1.39 times the size of TreeRePair's, down from a factor of 1.64 before. On our corpus, using the heuristic combiner reduced file sizes by 10.9 % on average, with the median being a 5.0 % improvement compared to classical top tree compression. Reduced compression performance was observed on few files only, particularly smaller ones, while larger files tended to fare better.

Speed. Using our RePair-inspired combiner increases the running time of the top tree creation stage, doubling it on average. Our implementation of classical top tree compression was 10.4 times faster than TreeRePair on average over the corpus from Table 1, and still 6.2 times faster when using our RePair combiner. Detailed running time measurements are given in Table 2. In particular, classical top tree compression takes only twice as long as `gzip -9` on average, and 3.3 times when using our RePair combiner (TreeRePair: 22.4). In contrast, `bzip2` is 15.1 times *slower* than top tree compression on average, and 9.5 times when using our RePair combiner. This strikingly demonstrates the method's qualities as a fast compressor.

Table 2. Running times in seconds, median over ten iterations

File name	TopTrees	TT+RePair	TreeRePair	RePair	gzip -9	bzip2
1998statistics	0.00	0.01	0.05	0.04	0.00	0.13
dblp	6.07	11.22	45.72	39.57	2.46	74.77
enwiki-latest-p	3.97	7.15	32.98	28.33	1.29	49.12
factor12	7.21	11.60	109.47	54.19	4.48	81.86
factor4.8	2.85	4.72	46.22	21.61	1.79	33.09
factor4	2.43	3.96	39.47	17.75	1.49	28.21
factor7	4.25	6.83	67.83	31.55	2.61	48.60
JST-gene.chr1	0.04	0.06	0.38	0.54	0.03	1.27
JST-snp.chr1	0.25	0.38	2.12	3.40	0.17	6.33
nasa	0.15	0.23	0.94	0.85	0.06	1.86
NCBI-gene.chr1	0.32	0.51	2.25	3.33	0.20	7.97
proteins	7.00	11.85	50.17	53.27	2.41	81.92
SwissProt	1.14	2.12	12.35	5.74	0.50	11.15
treebank-e	1.36	1.98	12.70	4.00	2.80	3.62
uwm	0.01	0.02	0.11	0.09	0.00	0.28
wiki	0.79	1.18	5.59	4.28	0.22	9.21

Table 3. Compressed file sizes in Bytes

File name	Succinct	TopTrees	TT+RePair	TreeRePair	RePair	gzip -9	bzip2
1998statistics	18 426	788	851	692	1 327	4 080	1 301
dblp	13 740 160	1 486 208	1 416 538	1 093 533	2 037 878	2 476 347	1 116 311
enwiki-latest-p	7 901 904	516 638	525 532	379 410	866 161	1 490 278	544 606
factor12	16 402 888	2 069 437	948 167	705 740	3 092 194	6 342 947	2 913 894
factor4.8	6 565 499	1 070 045	784 519	548 853	1 587 043	2 542 773	1 168 654
factor4	5 473 158	937 660	704 105	490 945	1 429 872	2 119 269	973 463
factor7	9 571 503	1 421 376	855 063	625 094	2 248 370	3 702 132	1 700 043
JST-gene.chr1	96 159	5 332	5 523	3 672	8 273	33 316	7 027
JST-snp.chr1	547 594	29 084	27 039	20 654	50 347	194 862	49 857
nasa	341 161	42 077	39 883	29 310	60 394	83 231	34 404
NCBI-gene.chr1	721 803	17 880	17 418	11 459	29 912	199 308	47 901
proteins	17 315 832	905 613	860 366	614 892	1 537 249	3 214 663	1 141 697
SwissProt	2 343 730	598 960	574 400	305 417	699 757	829 119	398 197
treebank-e	2 396 061	1 173 463	1 170 304	830 324	1 537 334	1 858 722	1 032 303
uwm	37 491	2 177	2 070	1 366	3 101	7 539	2 102
wiki	1 242 418	110 686	102 371	75 090	171 075	247 898	93 858

5   Conclusions

We have demonstrated that tree compression with top trees is viable, and suggested several enhancements to improve the degree of compression achieved. Using the notion of combiners, we demonstrated that significant improvements can be obtained by carefully choosing the order in which clusters are merged during top tree creation. We showed that the worst-case compression ratio is within a $\log\log_\sigma n_T$ factor of the information-theoretical bound. Further, we gave efficient methods to navigate the compressed representation, and described how the top DAG can be encoded to support efficient navigation without prior decompression.

We thus conclude that tree compression with top trees is a very promising compressor for labelled trees, and has several key advantages over other compressors that make it worth pursuing. It is our belief that its great flexibility, efficient navigation, high speed, simplicity, and provable bounds should not be discarded easily. While further careful optimizations are required to close the compression ratio gap, tree compression with top trees is already a good and fast compressor with many advantages.

Future Work. We expect that significant potential for improvement lies in more sophisticated combiners. The requirements for combiners give us a lot of space to devise better merging algorithms. Combiners might also be used to improve locality in the top tree in addition to compression performance, leading to better navigation performance. Moreover, additional compression improvements should be achievable with carefully engineered output representations. Since the vast majority of total running time is currently spent on the construction of the top DAG, using more advanced encodings may improve compression without losing speed. One starting point to replace our relatively naïve representation could be a decomposition of the top DAG into two spanning trees [19].

References

1. Alstrup, S., Holm, J., Lichtenberg, K.D., Thorup, M.: Maintaining information in fully dynamic trees with top trees. ACM TALG 1(2), 243–264 (2005)
2. Bille, P., Gørtz, I.L., Landau, G.M., Weimann, O.: Tree compression with top trees. Information and Computation (2015). http://doi.org/10.1016/j.ic.2014.12.012
3. Bille, P., Landau, G.M., Raman, R., Sadakane, K., Satti, S.R., Weimann, O.: Random access to grammar-compressed strings. In: Proc. SODA, pp. 373–389. SIAM (2011)
4. Buneman, P., Grohe, M., Koch, C.: Path queries on compressed XML. In: Proc. 29th VLDB, pp. 141–152. VLDB Endowment (2003)
5. Busatto, G., Lohrey, M., Maneth, S.: Grammar-based tree compression. Tech. Rep. EPFL-REPORT-52615, École Polytechnique Fédérale de Lausanne (2004)
6. Busatto, G., Lohrey, M., Maneth, S.: Efficient memory representation of XML documents. In: Bierman, G., Koch, C. (eds.) DBPL 2005. LNCS, vol. 3774, pp. 199–216. Springer, Heidelberg (2005)
7. Charikar, M., Lehman, E., Liu, D., Panigrahy, R., Prabhakaran, M., Sahai, A., Shelat, A.: The smallest grammar problem. IEEE Trans Inf Theory 51(7), 2554–2576 (2005)
8. Delpratt, O.D.: Space efficient in-memory representation of XML documents. Ph.D. thesis, University of Leicester, supervisor: Rajeev Raman (2009)
9. Downey, P.J., Sethi, R., Tarjan, R.E.: Variations on the common subexpression problem. Journal of the ACM (JACM) 27(4), 758–771 (1980)
10. Ferragina, P., Luccio, F., Manzini, G., Muthukrishnan, S.: Compressing and indexing labeled trees, with applications. Journal of the ACM (JACM) 57(1), 4 (2009)
11. Gog, S., Beller, T., Moffat, A., Petri, M.: From theory to practice: plug and play with succinct data structures. In: Gudmundsson, J., Katajainen, J. (eds.) SEA 2014. LNCS, vol. 8504, pp. 326–337. Springer, Heidelberg (2014)
12. Hirakawa, M., Tanaka, T., Hashimoto, Y., Kuroda, M., Takagi, T., Nakamura, Y.: JSNP: a database of common gene variations in the Japanese population. Nucleic Acids Research 30(1), 158–162 (2002). http://snp.ims.u-tokyo.ac.jp/XML/Mapped/old/20060612/
13. Jacobson, G.: Space-efficient static trees and graphs. In: Proc. 30th FOCS, pp. 549–554. IEEE (1989)
14. Jez, A., Lohrey, M.: Approximation of smallest linear tree grammar. CoRR abs/1309.4958 (2013). http://arxiv.org/abs/1309.4958
15. Larsson, N.J., Moffat, A.: Off-line dictionary-based compression. Proceedings of the IEEE 88(11), 1722–1732 (2000)

16. Lohrey, M., Maneth, S.: The complexity of tree automata and XPath on grammar-compressed trees. Theoretical Computer Science **363**(2), 196–210 (2006)
17. Lohrey, M., Maneth, S., Mennicke, R.: XML tree structure compression using RePair. Information Systems **38**(8), 1150–1167 (2013)
18. Maneth, S., Busatto, G.: Tree transducers and tree compressions. In: Walukiewicz, I. (ed.) FOSSACS 2004. LNCS, vol. 2987, pp. 363–377. Springer, Heidelberg (2004)
19. Maruyama, S., Nakahara, M., Kishiue, N., Sakamoto, H.: ESP-index: A compressed index based on edit-sensitive parsing. Journal of Discrete Algorithms **18**, 100–112 (2013)
20. Miklau, G.: University of Washington XML Repository. http://www.cs.washington.edu/research/xmldatasets
21. Munro, J.I., Raman, V.: Succinct representation of balanced parentheses and static trees. SIAM Journal on Computing **31**(3), 762–776 (2001)
22. Poyias, A.: XXML: Handling extra-large XML documents (2013). http://hdl.handle.net/2381/27744
23. Pătraşcu, M.: Succincter. In: Proc. 49th FOCS, pp. 305–313. IEEE (2008)
24. Wang, F., Li, J., Homayounfar, H.: A space efficient XML DOM parser. Data & Knowledge Engineering **60**(1), 185–207 (2007)
25. Wikimedia: enwiki dump. http://dumps.wikimedia.org/enwiki/

A Bulk-Parallel Priority Queue in External Memory with STXXL

Timo Bingmann[(⊠)], Thomas Keh, and Peter Sanders

Karlsruhe Institute of Technology, Karlsruhe, Germany
{bingmann,sanders}@kit.edu

Abstract. We propose the design and an implementation of a bulk-parallel external memory priority queue to take advantage of both shared-memory parallelism and high external memory transfer speeds to parallel disks. To achieve higher performance by decoupling item insertions and extractions, we offer two parallelization interfaces: one using "bulk" sequences, the other by defining "limit" items. In the design, we discuss how to parallelize insertions using multiple heaps, and how to calculate a dynamic prediction sequence to prefetch blocks and apply parallel multiway merge for extraction. Our experimental results show that in the selected benchmarks the priority queue reaches 64% of the full parallel I/O bandwidth of SSDs and 49% of rotational disks, or the speed of sorting in external memory when bounded by computation.

1 Introduction

Priority queues (PQs) are fundamental data structures which have numerous applications like job scheduling, graph algorithms, time forward processing [8], discrete event simulation, and many greedy algorithms or heuristics. They manage a dynamic set of items, and support operations for inserting new items (*push*), and reading and deleting (*top*/*pop*) the item smallest w.r.t. some order.

Since the performance of such applications usually heavily depends on the PQ, it is unavoidable to consider parallelized variants of PQs as parallelism is today the only way to get further performance out of Moore's law. However, even the basic semantics of a parallel priority queue (PPQ) are unclear, since PQ operations inherently sequentialize and synchronize algorithms. Researchers have previously focused on parallelizing *main memory* PQs which provide lock-free concurrent access, and/or relaxed operations delivering *some* small item.

In this work we propose a PPQ for applications where data *does not* fit into internal memory and thus requires efficient *external memory techniques*. Parallelizing external memory algorithms is one of the main algorithmic challenges termed as "Big Data". We propose a "bulk" and a "limit" parallelization interface for PQs, since the requirements of external memory applications are different from those working on smaller PQ instances. One application of these interfaces is *bulk-parallel time forward processing*, where one uses the graph's structure to

---

This paper is a short version of the technical report [6].

© Springer International Publishing Switzerland 2015
E. Bampis (Ed.): SEA 2015, LNCS 9125, pp. 28–40, 2015.
DOI: 10.1007/978-3-319-20086-6_3

identify layers of nodes that can be processed independently. For example, the inducing process of an external memory suffix sorting algorithm [5] follows this pattern. This paper continues work started in Thomas Keh's bachelor thesis [13].

We implemented our PPQ design in C++ with OpenMP and STXXL [9], and compare it using four benchmarks against the fastest EM priority queue implementations available. In our experiments we achieve 49% of the full I/O throughput of parallel rotational disks and 64% of four parallel solid-state-disks (SSDs) with about 2.0/1.6 GiB/s read/write performance. We reach these percentages in all experiments except when internal work is clearly the limitation, where our PPQ performs equally well as a highly tuned sorter. For smaller bulk sequences, the PPQ's performance gradually degrades, however, already for bulks larger than 20 K or 80 K 64-bit integers (depending on the platform) our PPQ outperforms the best existing parallelized external memory PQ.

After preliminaries and related work, we discuss our parallelization interfaces in Section 2. Central is our PPQ design in Section 3 where we deal with parallel insertion and extraction. Details of our implementation, the rationale of our experiments, and their results are discussed in Section 4.

1.1 Preliminaries

A PQ is a data structure holding a set of items, which can be ordered w.r.t some relation. All PQs support two operations: *insert* or *push* to add an item, and *deleteMin* or *top* and *pop* to retrieve and remove the smallest item from the set. In this paper we use the *push*, *top*, *pop* notation, since our implementation's interface aims to be compatible to the C++ Standard Template Library (STL). Addressable PQs additionally provide a *decreaseKey* operation, but we omit this function since it is difficult to provide efficiently in external memory.

We use the external memory (EM) model [21], which assumes an internal memory (called RAM) containing up to M items, and D disks containing space for N items, used for input, output and temporary data. Transfer of B items between disks and internal memory costs one I/O operation, whereas internal computation is free. While the EM model is good to describe asymptotically optimal I/O efficient algorithms, omitting computation time makes the model less and less practical as I/O throughput increases. For example, data transfer to a single modern SSD reaches more than 450 MiB/s (MiB = 2^{20} bytes), while sorting 1 GiB of random 64-bit integers sequentially reaches only about 85 MiB/s on a current machine. Thus exploiting parallelism in modern machines is unavoidable to achieve good performance with I/O efficient algorithms. For this experimental paper, we assume a shared memory system with p processors or threads, which have a simple set of explicit synchronization primitives. In future, one could consider a detailed theoretical analysis using the parallel external memory model [3].

1.2 Related Work

There has been significant work on bulk-parallel PQs in an internal memory setting [11,15,17]. We owe to the earlier of these results [11,15] the idea to replace

Listing 1. Bulk Pop/Push Loop

```
vector<item> work;
while (!ppq.empty()) {
  ppq.bulk_pop(work, max_size);
  ppq.bulk_push_begin(approx_bulk_size);
#pragma omp parallel for
  for (i = 0; i < work.size(); ++i) {
    // process work[i], maybe bulk_push()
  }
  ppq.bulk_push_end();
}
```

Listing 2. Bulk-Limit Loop

```
for (...) {
  ppq.limit_begin(L, bulk_size);
  while (ppq.limit_top() < L) {
    top = ppq.limit_top();
    ppq.limit_pop();
    // maybe use limit_push()
  }
  ppq.limit_end();
}
```

elements in a heap by sorted sequence and the basic operations on heap nodes by sorting and merging. Previous work on external sequential PQs [2,7,18] reduced the number of times an element moves between heap nodes from $\mathcal{O}(\log_2 N)$ to $\mathcal{O}(\log_{M/B} N/M)$ by increasing the degree of the involved tree structures.

There has also been a lot of work on concurrent PQs that allow asynchronous insertion and deletion by independent threads. Since this is not scalable with strict PQ semantics, there has recently been interest in concurrent PQs with relaxed semantics. Bulk-parallel PQs can be viewed as synchronous relaxed PQs with simple and clear semantics. We refer to recent work for details [1,16].

We parallelize the external *sequence heap* [18]. At the bottom level, a sequence heap consists of R groups of $k = \mathcal{O}(M/B)$ sorted external arrays. This PQ design was implemented for external memory in STXXL [9], and later also in TPIE [14], so it is probably the most widely used today. Beckmann, Dementiev and Singler [4] have partially parallelized sequence heaps without touching the sequential semantics. However, this gives only little opportunity for parallelization – mostly for merging in groups with large external arrays.

The most sophisticated parallelization tool we use in our PPQ is the parallel k-way merge algorithm first proposed by Varman et al. [20], and engineered by Singler et al. in the MCSTL [19] and later the GNU Parallel Mode library. Since this algorithm's details and implementation are important for our PPQ design, we briefly describe it: given p processors and k sorted arrays with in total n items and of maximum length m, each array is split into p range-disjoint parts where the sum of each processor's parts are of equal size. The partition is calculated by running p intertwined multisequence selection algorithms, which take $\mathcal{O}(k \cdot \log k \cdot \log m)$. After partitioning, the work of merging the p disjoint areas can be done independently by the processors, e.g., using a k-way tournament tree in time $\mathcal{O}(\frac{n}{p} \log k)$. For our EM setting it is important that the output is generated as p equal-sized parts in parallel, with each part being written in sequence. We also note that the multisequence selection is implemented sequentially.

2   Bulk-Parallel Interface and Limit Items

Before we discuss our PPQ design, we focus on the proposed application interface. As suggested by the related work on PQs, substantial performance

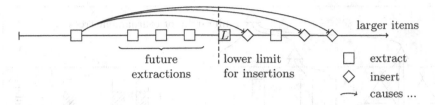

Fig. 1. Decoupling insertion and extraction operations with a limit item L

gains from parallelization are only achievable when loosening some semantics of the PQ. Put plainly, an alternating sequence of dependent *push/pop* s is inherently sequential. Since we focus on large amounts of data, the more natural relaxation of a PQ is to require insertion and extraction of multiple items, or "bulks" of items. This looser semantic *decouples* insert and delete operations both among themselves (i.e., items within a bulk) as well as the operation phases from another. This enables us to apply parallel algorithms on larger amounts of items, and our experiments in Section 4 show how speedup depends on the bulk sizes.

Thus the primary interface of our EM PPQ is bulk insertion and extraction (see Listing 1). A bulk insertion phase is started with *bulk_push_begin*(k), where k is an estimate of the bulk size. Thereafter, the application may insert a bulk of items using *bulk_push*, possibly concurrently from multiple threads, and terminate the sequence with *bulk_push_end*. There are two bulk extraction primitives: *bulk_pop*(v, k) which extracts up to k items into v, and *bulk_pop_limit*(v, L, k), which extracts at most k items strictly smaller than a limit item L. The limit extraction also indicates whether more items smaller than L are available.

Beyond the primary bulk interface, we also propose a second interface (see Listing 2), which is geared towards the canonical processing loop found in most sequential applications using a PQ: extract an item, inspect it, and reinsert zero or more items into the PQ. To decouple insertions and extractions in this loop, we let the application define a "limit item" L, and require that all insertions thereafter are larger or equal to L (see Figure 1). By defining this limit, all extractions of items less than L become decoupled from insertions. The drawback of this second interface is that the application does not process items in parallel, however, this can easily be accomplished by using *bulk_pop_limit*.

3   Design of a Bulk-Parallel Priority Queue

Our PPQ design (see Figure 2) is based on Sanders' sequence heap [18], but we have to reevaluate the implicit assumptions, duplicate data structures for independent parallel operations and apply parallel algorithms where possible. After briefly following the lifetime of an item in the PPQ, we first discuss separately how insertions and extractions can be processed in parallel, and then focus on the difficulty of balancing both.

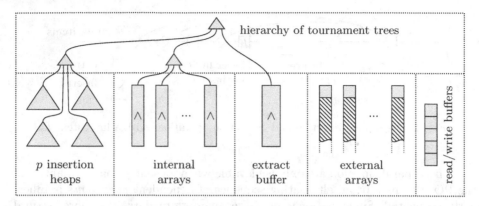

Fig. 2. Components of our PPQ. All lightly shaded parts are in internal memory

An item is first inserted into an *insertion heap*, which is kept in heap order. As simple binary heaps are not particularly cache-efficient, they are given a fixed maximum size. When full, an insertion heap is sorted and transformed into an *internal array*. To limit the number of internal arrays, they may be merged with others to form longer internal arrays. When memory is exhausted, all internal arrays and the extract buffer are merged into one sorted *external array* which is written to disk. Again, shorter external arrays may also be merged together. Extracts from the set of external arrays are amortized using the extract buffer.

Insertion, Multilevel Merging, and External Writing. To accelerate parallel push operations, the first obvious step is to have p insertion heaps, one for each processor. This decouples insertions on different processors and parallelizes the work of maintaining the heaps. Once a heap is full, the processor can independently sort the heap using a general sorter. Remarkably, these initial steps are among the most time consuming in a sequence heap, and can be parallelized well. In our PPQ design, we then use a critical section primitive to synchronize adding the new internal array to the common list. This was never a bottleneck, since such operations happen only when an insertion heap is full.

In bulk push sequences, we can accelerate individual push operations much further. While pushing, no items from the insertion heap can be extracted, thus we can postpone reestablishing heap order to *bulk_push_end*; a *bulk_push* just appends to the insertion heap's array. If the heap overflows, then the array is sorted anyway. In our experiments, this turned out to be the best option, probably because the loop sifting items up the heap becomes very tight and cache efficient. For larger bulk operations (as indicated by the user's estimation) we even let the insertion heap's array grow beyond the usual limit to fill up the available RAM, since sorting is more cache efficient than keeping a heap.

Instead of separating internal arrays into groups, as in a sequence heap, we label them using a *level number* starting at zero. If the number of internal arrays on one level grows larger than a tuning parameter (about 64) and there is enough RAM available, then all internal arrays of one level are merged together and added to the next higher level. The decisive difference of parallel multiway merge over sequentially merging sorted arrays is that *no state* is kept to amortize operations. Hence, in our PPQ design the indicated tournament trees over the insertion heaps and arrays are useless for parallel operations. When applying parallel multiway merge, we want to have the total number of items as large as possible, however, at the same time the number of sequences should be kept as small as possible.

When the PPQ's alloted memory is exhausted, one large parallel multiway merge is performed directly into EM. This is possible without an extra copy buffer, by using just $\Theta(p)$ write buffers and overlapping I/O and computation, since parallel multiway merge outputs p sorted sub-sequences. We use $\geq 2p$ write buffer blocks to keep the merge boundaries in memory; thus avoiding any rereading of blocks from disk during the merge.

An item may travel multiple times to disk and back, since the extract buffer is included while merging into EM. However, as in the sequence heap structure, this only occurs when internal memory is exhausted and all items are written to disk; thus we can amortize the extra I/Os for the extract buffer with the $\Theta(M/B)$ I/Os needed to flush main memory.

Extraction, Prediction, and Minimum of Minima. To support fast non-bulk *pop* operations, we keep a hierarchy of tournament trees to save results of pairwise comparisons of items. The trees are built over the insertion heaps, internal arrays, and extract buffer. External arrays need not be included, since extraction from them is buffered using the extract buffer. The tournament trees need to be updated each time an insertion heap's minimum element changes, or a heap is flushed into an internal array. In bulk push operations these actions are obviously postponed until the bulk's end.

When merging external arrays with parallel multiway merge we are posed (again) with the discrepancy between parallelism, which requires large item counts for efficiency, and relatively small disk blocks (by default 2–8 MiB). To alleviate the problem, we increase the number of read buffers and calculate an *optimal block prediction sequence*, as also done for sorting [12], which contains the order in which the EM blocks are needed during merging and fetch as many as fit into RAM. In sorting, the prediction sequence is fixed and can be determined by sorting the smallest items of each block as a representative (also called "trigger" element). In the parallel disk model, the independent disks need to be considered as well. In our PPQ setting, the prediction sequence becomes a *dynamic problem*, since external arrays may be added. We define four states for an external block: in external memory, hinted for prefetching, loaded in RAM, and finished (see Figure 3). To limit the main memory usage of the PPQ, the number of prefetched and blocks loaded in RAM must be restricted.

Since the next k external blocks needed for merging are determined by the k smallest block minima, we keep track of these items in a tournament tree over the block minima sequences of the external arrays (see items h_i in Figure 3). This allows fast calculation of the next block when another can be prefetched. However, when a new external array is added, the dynamic prediction sequence

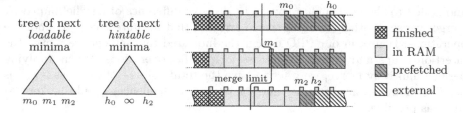

Fig. 3. Establishing the dynamic prefetching sequence and upper merge limit

changes, and we may have to cancel prefetch hints. This is done by resetting the tournament tree back to the first block minima merely hinted for prefetching, but not loaded in RAM, and replaying it till the new k smallest block minima are determined. This costs less than $k + k \log S$ comparisons, where S is the number of sequences. We then compare the new predictions with the old by checking how many blocks are to be prefetched in each array, and cancel or add hints.

For parallel merging, we need to solve another problem: the merge ranges within the blocks in RAM must be limited to items smaller than the smallest item still in EM, since otherwise the PQ invariant may be violated. To determine the smallest item in EM we reuse the block minima sequences, and build a second tournament tree over them containing the smallest items of the next "loadable" block, not guaranteed to be in RAM (items m_i in Figure 3). When performing a parallel multiway merge into the extract buffer, all hinted external blocks are first checked (in order) whether the prefetch is complete, and the tournament tree containing the smallest external items is updated. The tip then contains $\overline{m} = \min_i m_i$, the overall smallest external item, which serves as *merge limit*. We then use binary search within the loaded blocks of each array and find the largest items smaller than $\overline{m}$. We thus limit the multisequence selection and merge range on each array by $\overline{m}$. Additionally, by using smaller selection ranks during parallel multiway merging one can adapt the total number of elements merged. These rank limits enable us to efficiently limit the extract buffer's size and the output size of $bulk\_pop(v, k)$ and $bulk\_pop\_limit(v, L, k)$ operations. To limit extraction up to L, we simply use $\min(L, \overline{m})$ as merge limit.

As with internal arrays, the number of external arrays should be kept small for multiway merge to be efficient. One may suspect that merging from EM is I/O bound, however, if the merge output buffers are smaller than the read buffers, then this is obviously not the case. Thus, the parallelization bottleneck of refilling the extract buffer or of $bulk\_pop$ operations largely depends on the number of arrays. We also adapt the number of read buffers (both for prefetching and holding blocks in RAM) dynamically to the number of external arrays. Each newly added external array requires at least one additional read buffer.

As with internal arrays, instead of keeping external arrays in separate groups, we label them with a level number, and merge levels when the contained number grows too large. This enables more dynamic memory pooling than in the rigid sequence heap data structure, while maintaining the optimal I/O complexity.

Trade-Offs between Insertion and Extraction. As already discussed, to enable non-bulk *pop* operations we keep a hierarchy of tournament trees. Using this hierarchy instead of one large tournament tree skews the depth of nodes in the tree, making replays after *pop*s from the extract buffer and the insertion heaps cheaper than from internal arrays.

When a new external array is created, then the read prediction sequence may change and previous prefetch requests need to be canceled and new ones issued. In long bulk push sequences (as the ascending sequence in our experiments), this can amount to many superfluous prefetch reads of blocks. Thus we disable prefetching during *bulk_push* operations and issue all hints at the end. This suggests that bulk push sequences should be as long as possible, and that they are interleaved with *bulk_pop* operations.

4   Implementation in STXXL and Experimental Results

We implemented our PPQ design in C++ with OpenMP and the STXXL library [9], since it provides a well-designed interface to asynchronous I/O, and allowing easy overlapping of I/O and computation. It also contains two other PQ implementations that we compare our implementation to. Our implementation will be available as part of the next STXXL release 1.4.2, which will be publicly available under the liberal Boost software license. At the time of submission it is available in the public development repository.

Other PQ Implementations. In these experiments we compare our PPQ implementation (**PPQ**) with the sequential sequence heap [18] (**SPQS**) contained in the STXXL, a partially parallelized version [4] of it (**SPQP**), which uses parallel multiway merging only when merging external arrays, and with the STXXL's highly tuned stream sorting implementation [10] (**Sorter**) as a baseline.

Experimental Platforms. We run the experiments on two platforms. Platform **A-Rot** is an Intel Xeon X5550 from 2009 with 2 sockets, 4 cores and 4 Hyperthreading cores per socket at 2.66 GHz clock speed and 48 GiB RAM, and eight rotational Western Digital Blue disks with 1 TB capacity and about 127 MiB/s transfer speed each, which are attached via an Adaptec ASR-5805 RAID controller. Platform **B-SSD** is an Intel Xenon E5-2650 v2 from 2014 with 2 sockets, 8 cores and 8 Hyperthreading cores per socket at 2.6 GHz clock speed with 128 GiB RAM. There are four Samsung SSD 840 EVO disks with 1 TB each attached via an Adaptec ASA-7805H Host adapter, yielding together 2 GiB/s read and 1.6 GiB/s write transfer speed to/from EM. The platforms run Ubuntu Linux 12.04 and 14.04, respectively, and all our programs were compiled with gcc 4.6.4 and 4.8.2 in *Release* performance mode using STXXL's CMake build system.

Experiments and Parameters. To compare the three PQs we report results of four sets of experiments. In all experiments the PQ's items are plain 64-bit integer keys (8 bytes), which places the spotlight on internal comparison work as payload only increases I/O volume. (See our report [6] for additional results with 24 byte items.) The PQs are allotted 16 GiB of RAM on both platforms, since in a real EM application multiple data structures exist simultaneously and thus have to share RAM.

In the first two experiments, called a) *push-rand-pop* and b) *push-asc-pop*, the PQ is filled a) with n uniformly random generated integer items, or b) with n ascending integers, and then the n items are extracted again. In these canonical benchmarks, the PQ is used to just sort the items, but it enables us to compare the PQs against the highly optimized sorting implementation, which also employs parallelism where possible. In the ascending sequence, the first items inserted are removed first, forcing the PQs to cycle items. Considering the amount of internal work, the *push-asc-pop* benchmark is an easy case, since all buffers are sorted and merging is skewed. Thus the focus of this benchmark is on I/O overlapping. On the other hand, in the *push-rand-pop* benchmark the internal work to sort and merge the random numbers is very high, which makes it a test of internal processing speed. We ran the experiments for $n = 2^{27}, \ldots, 2^{35}$, which is an item volume of 1 GiB, . . . , 256 GiB.

The third and forth experiments, *asc-rbulk-rewrite* and *bulk-rewrite*, fully rewrite the PQ in bulks: the PQ is filled with n ascending items, then the n items are extracted in bulks of random or fixed size v, and after each bulk extraction v items are pushed again. During the rewrite, in total n items are extracted and n items inserted with higher ids. We measure only the bulk pop/push cycles as these experiments are designed to emulate traversing a graph for time forward processing. We use bulk rewriting in two different experiment scenarios: in the first, we select the bulk size uniformly at random from 0 to 640 000, and let n increase as in the first two experiments. For the second, $n = 4 \cdot 2^{30}$ items (32 GiB) is fixed and the bulk size v is varied from 5 000 to 5 120 000.

All experiments were run only once due to long execution times and little variation in the results over large ranges of input size. During the runs we pinned the OpenMP threads to cores, which is important since it keeps the insertion heaps local. Due to the large I/O bandwidth of the SSDs, we increased the number of write buffers of the PPQ to 2 GiB on B-SDD to better overlap I/O and computation. Likewise, we allotted 128 MiB read buffers per external array. On A-Rot we set only 256 MiB write buffers and 32 MiB read buffers per array. For the STXXL PQ, of the 16 GiB of RAM one fourth is allocated for read and one fourth for write buffers. We used in all experiments the new "linuxaio" I/O

Table 1. Speedup of PPQ over parallelized STXXL PQ (SPQP), sequential STXXL PQ (SPQS), and STXXL Sorter for 64-bit integers, averaged for experiments $n \geq 2^{30.5}$

	Platform A-Rot			Platform B-SSD		
Experiment	SPQP	SPQS	Sorter	SPQP	SPQS	Sorter
push-rand-pop	1.39	3.58	0.87	2.25	4.83	0.83
push-asc-pop	1.81	3.40	1.37	4.29	6.71	1.20
asc-rbulk-rewrite	1.89	4.70		2.91	3.43	

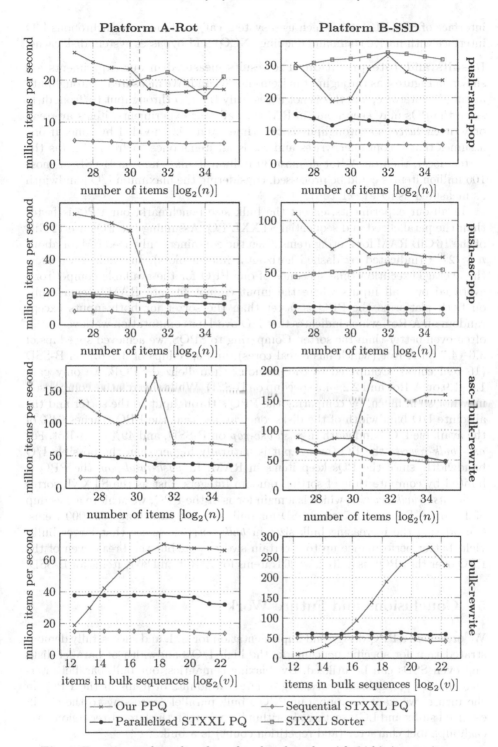

Fig. 4. Experimental results of our four benchmarks with 64-bit integer items

interface of STXXL 1.4.1, which uses system calls to Linux's asynchronous I/O interface with native command queuing (NCQ) and bypasses system disk cache.

Results and Interpretation. The results measured in our experiments are shown in Figure 4 as throughput in items per second. We measured "throughput" at the PQ interface, and this is not necessarily the I/O throughput to/from disk, since the PQs may keep items in RAM. In all four experiments, items are read or written *twice*, so throughput is two times item size divided by time. If one assumes that a container writes and reads all items once to/from disk (as the sorter does), then on A-Rot at most 39 million items/s and on B-SSD at most 106 million item/s could be processed, considering the maximum I/O bandwidth as measured using `stxxl_tool`.

In all our experiments, except the bulk size benchmark, our PPQ is faster than the parallelized and sequential STXXL PQ. Assuming the PQs use 12 GiB of the 16 GiB RAM for storing items, then the containers only need EM for about $n \geq 2^{30.5}$ (indicated by dashed horizontal line in plots). In Table 1 we show the average execution time speedups of our PPQ for the available competitors, averaged over all inputs where the input cannot fit into RAM. Remarkably, on both platforms the PPQ is faster than the sorter for both inputs except random on A-Rot, which indicates that I/O overlaps computation work very well, often even better than the sorter. Comparing to SPQS, we achieved speedups of $3.6-4.7$ on A-Rot (which has 8 real cores), and speedups of $3.4-6.7$ on B-SSD (16 real cores). Compared to the previously parallelized SPQP, we only gain $1.4-1.9$ on A-Rot and $2.2-4.3$ speedup on B-SSD. While this relative comparison may not seem much, by comparing the PPQ's throughput to the sorter and the absolute I/O bandwidth of the disks, one can see that the PPQ reaches 64% of the available I/O bandwidth in *push-asc-pop* on B-SSD, and 49% on A-Rot. For *asc-rbulk-rewrite* the PQ-throughput is naturally higher than the possible I/O bandwidth, since the PQs keep items in RAM. In *push-rand-pop*, the PPQ is limited by compute time of sorting random integers, just as the STXXL sorter is. For *asc-rbulk-rewrite*, which is a main focus of the PPQ, we achieve a speedup of 1.9 on A-Rot and 2.7 on B-SSD for bulk sizes of on average 320 000 items. Considering the increasing bulk sizes in *bulk-rewrite*, we see that larger bulks yield better performance up to a certain sweet spot, but the break even of the PPQ over the SPQP is quite low: 20 K items for A-Rot and 80 K items for B-SSD.

5   Conclusions and Future Work

We presented a PPQ design and implementation for EM, and successfully demonstrated that for specific benchmarks the high I/O bandwidth of parallel disks and even SSDs can be utilized. By relaxing semantics, our bulk-parallel interface enables parallelized processing of larger amounts of items in the PPQ. In the future, we want to apply our PPQ's bulk-parallel processing to the eSAIS external suffix and LCP sorting algorithm [5], where in the largest recursion level each alphabet character (and repetition count) is a bulk.

During our work on the PPQ two important issues remained untouched: how does one balance work in an EM algorithm library when the user application, the EM containers, and I/O overlapping require parallel work? We left this to the operating system scheduler and block the user application during parallel merging, which is not desirable. As indicated by theory and experiments, *bulk_pop_limit* requires large bulks to work efficiently, however, the PPQ cannot know the resulting bulk sizes without performing a costly multisequence selection. One could require the user application to provide an estimate of the resulting size, or develop an online oracle. Finally, experiments with other internal memory PPQs and d-ary heaps may improve performance by using larger insertion heaps.

References

1. Alistarh, D., Kopinsky, J., Li, J., Shavit, N.: The SprayList: A scalable relaxed priority queue. Tech. Rep. MSR-TR-2014-16, Microsoft Research, September 2014
2. Arge, L.: The buffer tree: A technique for designing batched external data structures. Algorithmica **37**(1), 1–24 (2003)
3. Arge, L., Goodrich, M.T., Nelson, M., Sitchinava, N.: Fundamental parallel algorithms for private-cache chip multiprocessors. In: SPAA, pp. 197–206. ACM (2008)
4. Beckmann, A., Dementiev, R., Singler, J.: Building a parallel pipelined external memory algorithm library. In: IPDPS 2009, pp. 1–10. IEEE (2009)
5. Bingmann, T., Fischer, J., Osipov, V.: Inducing suffix and LCP arrays in external memory. In: ALENEX 2013, pp. 88–102. SIAM (2013)
6. Bingmann, T., Keh, T., Sanders, P.: A bulk-parallel priority queue in external memory with STXXL, April 2015. see ArXiv e-print arXiv:1504.00545
7. Brodal, G.S., Katajainen, J.: Worst-case efficient external-memory priority queues. In: Arnborg, S. (ed.) SWAT 1998. LNCS, vol. 1432, pp. 107–118. Springer, Heidelberg (1998)
8. Chiang, Y.J., Goodrich, M.T., Grove, E.F., Tamassia, R., Vengroff, D.E., Vitter, J.S.: External-memory graph algorithms. In: SODA 1995, pp. 139–149. SIAM (1995)
9. Dementiev, R., Kettner, L., Sanders, P.: STXXL: Standard template library for XXL data sets. Software & Practice and Experience **38**(6), 589–637 (2008)
10. Dementiev, R., Sanders, P.: Asynchronous parallel disk sorting. In: SPAA 2003, pp. 138–148. ACM (2003)
11. Deo, N., Prasad, S.: Parallel heap: An optimal parallel priority queue. The Journal of Supercomputing **6**(1), 87–98 (1992)
12. Hutchinson, D.A., Sanders, P., Vitter, J.S.: Duality between prefetching and queued writing with parallel disks. SIAM Journal on Computing **34**(6) (2005)
13. Keh, T.: Bulk-parallel priority queue in external memory, Bachelor Thesis, Karlsruhe Institute of Technology, Germany (2014)
14. Petersen, L.H.: External Priority Queues in Practice. Master's thesis, Aarhus Universitet, Datalogisk Institut, Denmark (2007)
15. Pinotti, M.C., Pucci, G.: Parallel priority queues. IPL **40**(1), 33–40 (1991)
16. Rihani, H., Sanders, P., Dementiev, R.: Multiqueues: Simpler, faster, and better relaxed concurrent priority queues. arXiv preprint arXiv:1411.1209 (2014)
17. Sanders, P.: Randomized priority queues for fast parallel access. Journal of Parallel and Distributed Computing **49**(1), 86–97 (1998)

18. Sanders, P.: Fast priority queues for cached memory. JEA **5**, 7 (2000)
19. Singler, J., Sanders, P., Putze, F.: MCSTL: the multi-core standard template library. In: Kermarrec, A.-M., Bougé, L., Priol, T. (eds.) Euro-Par 2007. LNCS, vol. 4641, pp. 682–694. Springer, Heidelberg (2007)
20. Varman, P.J., Scheufler, S.D., Iyer, B.R., Ricard, G.R.: Merging multiple lists on hierarchical-memory multiprocessors. Journal of Parallel and Distributed Computing **12**(2), 171–177 (1991)
21. Vitter, J.S., Shriver, E.A.: Algorithms for parallel memory, i: Two-level memories. Algorithmica **12**(2–3), 110–147 (1994)

Graph Problems I

Greedily Improving Our Own Centrality in a Network

Pierluigi Crescenzi[1], Gianlorenzo D'Angelo[2](✉), Lorenzo Severini[2], and Yllka Velaj[2]

[1] Department of Information Engineering, University of Florence, Viale Morgagni, 65, 50134 Florence, Italy
pierluigi.crescenzi@unifi.it
[2] Gran Sasso Science Institute (GSSI), Viale F. Crispi, 7, 67100 L'Aquila, Italy
{gianlorenzo.dangelo,lorenzo.severini,yllka.velaj}@gssi.infn.it

Abstract. The closeness and the betweenness centralities are two well-known measures of importance of a vertex within a given complex network. Having high closeness or betweenness centrality can have positive impact on the vertex itself: hence, in this paper we consider the problem of determining how much a vertex can increase its centrality by creating a limited amount of new edges incident to it. We first prove that this problem does not admit a polynomial-time approximation scheme (unless $P = NP$), and we then propose a simple greedy approximation algorithm (with an almost tight approximation ratio), whose performance is then tested on synthetic graphs and real world networks.

1 Introduction

Looking for the most important vertices within a given complex network has always been one of the main goals in the field of real-world network analysis. Different measures of importance have been introduced in the literature, and several of them are related to the notion of "centrality" of a vertex. This latter notion, in turn, has been explicitly formalized in different ways: two of the most popular ways are closeness centrality and betweenness centrality (see, for example, [5]). The first one somehow measures the efficiency of a vertex while spreading information to all other vertices in its connected component, while the second one intuitively quantifies how much a vertex controls the information flow between all pairs of vertices in a graph. More formally, the closeness centrality of v is equal to the sum of the reciprocal of the distances to v from all other vertices, while the betweenness centrality of a given vertex v is the portion of the shortest paths between all pairs of vertices that pass through v.

Both closeness and betweenness centrality, however, are computationally expensive, since they require $O(nm)$ time [7] (in order to be computed for each vertex) which is clearly infeasible for networks with millions of vertices and edges (which is the "normal" size of many interesting real-world networks).

© Springer International Publishing Switzerland 2015
E. Bampis (Ed.): SEA 2015, LNCS 9125, pp. 43–55, 2015.
DOI: 10.1007/978-3-319-20086-6_4

For this reason, several randomized and/or approximation algorithms have been proposed for the computation of these two centrality measures [9,20].

In this paper, instead, we consider a different problem related to the closeness and betweenness centrality, that is, the problem of identifying which "strategy" a vertex should adopt in order to increase its own centrality value. Indeed, increasing its own ranking in terms of centrality, can have positive consequences for the vertex. For example, in the field of author citation networks both closeness and betweenness centrality seem to be significantly correlated with citation counts (as it has been already observed in the case of collaboration networks) [23], while, in the field of transportation network analysis, the betweenness centrality seems to be positively related to the efficiency of an airport, as observed in [16] where a network of 57 European airports has been analyzed.

More specifically, we consider the problem of efficiently determining, for a given vertex v, the set of k edges entering v that, when added to the original directed graph, allows v to increase as much as possible its closeness (respectively, betweenness) centrality and its ranking according to this measure. We first prove that this problem is hard to be approximated within an approximation factor greater than $1 - \frac{1}{3e}$ (respectively, $1 - \frac{1}{2e}$), and we then show that a greedy approach yields an $(1 - \frac{1}{e})$-approximation algorithm (for both closeness and betweenness). Successively, we present several experiments that we have performed (i) in order to evaluate how good is the approximation factor in the case of relatively small randomly generated graphs, and (ii) in order to apply the greedy approach to real-world citation and transportation networks. As a result of the first set of experiments, we have that the greedy algorithm seems to perform much better than the theoretical results, since it often computes an optimal solution and, in any case, it achieves an approximation factor significantly larger than $1 - \frac{1}{e}$. By applying the greedy algorithm to real-world networks, instead, we observe that by adding very few edges a vertex can drastically increase its centrality measure and, hence, its ranking. For example, the first (respectively, second) author of this paper could pass from ranking 2540 to ranking 346 (respectively, from ranking 6398 to 380), with just three citations. However, he has to convince Robert Tarjan, Christos Papadimitriou and Leslie Valiant (respectively, Richard Karp) to cite one of his papers. In the field of transportation networks, instead, the Paris Orly airport could increase its betweenness centrality (in the Easyjet connection network) by 218%, and pass from ranking 22 to ranking 15, with just three new connections from Ljubljana, Newquay, and Ponta Dalgada.

As far as we know, the problem analyzed in this paper has never been attacked before, even though similar problems have been studied for other centrality measures, i.e. page-rank [4,19], eccentricity [10], average distance [17], and some measures related to the number of paths passing through a given node [13]. Hence, we had no other algorithms to compare with. However, we also consider the naive approach of connecting the vertex with the top-k vertices in the centrality ranking and we experimentally show that the greedy algorithm significantly outperforms this simple heuristic, whenever $k > 1$.

1.1   Preliminary Definitions and Results

Let $G = (V, E)$ be a directed graph. For each node v, N_v denotes the set of in-neighbors of v, i.e. $N_v = \{u \mid (u, v) \in E\}$. Given two vertices s and t, we denote by d_{st}, σ_{st}, and σ_{stv} the distance from s to t in G, the number of shortest paths from s to t in G, and the number of shortest paths from s to t in G that contain v, respectively. Given a set S of edges not in E, we denote by $G(S)$ the graph augmented by adding the edges in S to G, i.e. $G(S) = (V, E \cup S)$. For a parameter x of G, we denote by $x(S)$ the same parameter in graph $G(S)$, e.g. the distance from s to t in $G(S)$ is denoted as $d_{st}(S)$. For each node v, the *closeness centrality* (also called *harmonic centrality* [5]) of v is defined as follows

$$c_v = \sum_{\substack{s \in V \setminus \{v\} \\ d_{sv} < \infty}} \frac{1}{d_{sv}},$$

while the *betweenness centrality* [5] of v is defined as

$$b_v = \sum_{\substack{s,t \in V \\ s \neq t; s, t \neq v \\ \sigma_{st} \neq 0}} \frac{\sigma_{stv}}{\sigma_{st}}.$$

The closeness and the betweenness centralities of a vertex clearly depend on the graph structure. if we augment a graph by adding a set of edges S, then the centrality of a vertex might change. Generally speaking, adding edges incident to some vertex v can only increase the centrality of v. We are interested in finding the set S of edges incident to a particular vertex v that maximizes such an increment. Therefore, we define the following optimization problem.

Maximum Closeness Improvement (MCI)

Given: A directed graph $G = (V, E)$; a vertex $v \in V$; and an integer $k \in \mathbb{N}$

Solution: A set S of edges incident to v, $S = \{(u, v) \mid u \in V \setminus N(v)\}$, such that $|S| \leq k$

Goal: Maximize $c_v(S)$

Analogously, we can define the **Maximum Betweenness Improvement** (in short, **MBI**), by referring to the betweeness centrality measure.

In this paper, we will use the the *maximum set coverage problem* [12] to derive approximation hardness results. Such problem is defined as follows.

Maximum Set Coverage (MSC)

Given: A set X; a family of subsets of X, $\mathcal{F} = \{S_1, S_2, \ldots S_{|\mathcal{F}|}\}$; and an integer k'

Solution: A family $\mathcal{F}' \subseteq \mathcal{F}$ such that $|\mathcal{F}'| \leq k'$

Goal: Maximize $s(\mathcal{F}') = |\cup_{S_i \in \mathcal{F}'} S_i|$

It has been shown [12] that MSC cannot be approximated within a factor greater than $1 - \frac{1}{e}$, unless $P = NP$. Moreover, the following greedy algorithm matches

---

Algorithm: GREEDYIMPROVEMENT

Input : A directed graph $G = (V, E)$; a vertex $v \in V$; and an integer $k \in \mathbb{N}$

Output: Set of edges $S \subseteq \{(u, v) \mid u \in V \setminus N_v\}$ such that $|S| \leq k$

1 $S := \emptyset$;

2 **for** $i = 1, 2, \ldots, k$ **do**

3   **foreach** $u \in V \setminus (N_v \cup S)$ **do** Compute $f_v(S \cup \{(u, v)\})$;

4      $u_{\max} := \arg\max\{f_v(S \cup \{(u, v)\}) \mid u \in V \setminus (N_v \cup S)\}$;

5      $S := S \cup \{(u_{\max}, v)\}$;

6 **return** S;

Fig. 1. The greedy centrality improvement algorithm (f_v denotes c_v or b_v)

---

such upper bound [18]: start with the empty set, and repeatedly add an element that gives the maximal marginal gain. The greedy algorithm can be extended to any *monotone submodular*[1] objective function defined on $\mathcal{F}$ thanks to the following result.

Theorem 1 ([18]). *For a non-negative, monotone submodular function f, let S be a set of size k obtained by selecting elements one at a time, each time choosing an element that provides the largest marginal increase in the value of f. Then S provides a $\left(1 - \frac{1}{e}\right)$-approximation.*

In this paper, we exploit this result by showing that c_v and b_v are monotone and submodular w.r.t. the possible set of edges incident to v. Hence, the greedy algorithm reported in Fig. 1 (where f_v denotes either c_v or b_v) provides a $\left(1 - \frac{1}{e}\right)$-approximation. Note that the computational complexity of such algorithm is $O(k \cdot n \cdot g(n, m))$, where $g(n, m)$ is the complexity of computing either c_v or b_v.

2   Improving Closeness Centrality

We first prove that the problem of improving the closeness centrality of a vertex does not admit a polynomial-time approximation scheme.

Theorem 2. *Problem MCI cannot be approximated within a factor greater than $1 - \frac{1}{3e}$, unless $P = NP$.*

Proof. We give an *L*-reduction with parameters a and b [22]. In detail, we will give a polynomial-time algorithm that transforms any instance I_{MSC} of MSC into an instance I_{MCI} of MCI and a polynomial-time algorithm that transforms any solution S for I_{MCI} into a solution $\mathcal{F}'$ for I_{MSC} such that the following two

---

[1] For a ground set X, a function $f : 2^X \to \mathbb{N}$ is submodular if for any pair of sets $S \subseteq T \subseteq X$ and for any element $e \in X \setminus T$, $f(S \cup \{e\}) - f(S) \geq f(T \cup \{e\}) - f(T)$ [18].

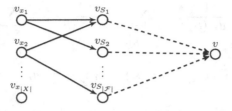

Fig. 2. The reduction used in Theorem 2 (in this example, $x_1 \in S_1$, $x_1 \in S_2$, $x_2 \in S_1$, and $x_2 \in S_{|\mathcal{F}|}$). The dashed edges denote those added in a solution.

conditions are satisfied for some values a and b:

$$OPT(I_{\mathrm{MCI}}) \leq a OPT(I_{\mathrm{MSC}}) \tag{1}$$

$$OPT(I_{\mathrm{MSC}}) - s(\mathcal{F}') \leq b\left(OPT(I_{\mathrm{MCI}}) - c_v(S)\right). \tag{2}$$

where OPT denotes the optimal value of an instance of an optimization problem. If the above conditions are satisfied and there exists a α-approximation algorithm for MCI, then there exists a $(1 - ab(1 - \alpha))$-approximation algorithm for MSC [22]. Since MSC is hard to approximate within a factor greater than $1 - \frac{1}{e}$, then $1 - ab(1 - \alpha) < 1 - \frac{1}{e}$, unless $P = NP$. This implies that $\alpha < 1 - \frac{1}{abe}$.

Given an instance $I_{\mathrm{MSC}} = (X, \mathcal{F}, k')$ of MSC, we define an instance $I_{\mathrm{MCI}} = (G, v, k)$ of MCI as follows (see Fig. 2): $k = k'$ and $G = (V, E)$, where $V = \{v\} \cup \{v_{x_i} \mid x_i \in X\} \cup \{v_{S_j} \mid S_j \in \mathcal{F}\}$ and $E = \{(v_{x_i}, v_{S_j}) \mid x_i \in S_j\}$.

Without loss of generality, we can assume that any solution S of MCI contains only edges (v_{S_j}, v) for some $S_j \in \mathcal{F}$. In fact, if a solution does not satisfy this property, then we can improve it in polynomial time by repeatedly applying the following rule: if S contains an edge (v_{x_i}, v), for some $x_i \in X$, then exchange such edge with an edge (v_{S_j}, v) such that $(v_{S_j}, v) \notin S$ (note that such an edge must exist, since otherwise $|\mathcal{F}| \leq k = k'$ and I_{MSC} could be easily solved). The above rule does not decrease the value of $c_v(S)$: indeed, if we exchange an edge (v_{x_i}, v) with an edge (v_{S_j}, v) such that $(v_{S_j}, v) \notin S$, then the closeness centrality of v decreases by either 1 or $\frac{1}{2}$ (because of the deletion of (v_{x_i}, v)) but certainly increases by 1 (because of the insertion of (v_{S_j}, v)).

Given a solution S of MCI, let $\mathcal{F}'$ be the solution of MSC such that $S_j \in \mathcal{F}'$ if and only if $(v_{S_j}, v) \in S$. We now show that $c_v(S) = \frac{1}{2}s(\mathcal{F}') + k$. To this aim, let us note that the distance from a vertex v_{x_i} to v is equal to 2 if an edge (x_{S_j}, v) such that $x_i \in S_j$ belongs to S, and it is ∞ otherwise. Similarly, the distance from a vertex v_{S_j} to v is equal to 1 if $(x_{S_j}, v) \in S$, and it is ∞ otherwise. Moreover, the set of elements x_i of X such that $d_{v_{x_i} v}(S) < \infty$ is equal to $\{x_i \mid x_i \in S_j \wedge (v_{S_j}, v) \in S\} = \bigcup_{S_j \in \mathcal{F}'} S_j$. Therefore,

$$c_v(S) = \sum_{\substack{s \in V \setminus \{v\} \\ d_{sv}(S) < \infty}} \frac{1}{d_{sv}(S)} = \sum_{\substack{x_i \in X \\ d_{v_{x_i} v}(S) < \infty}} \frac{1}{d_{v_{x_i} v}(S)} + \sum_{\substack{S_j \in \mathcal{F} \\ d_{v_{S_j} v}(S) < \infty}} \frac{1}{d_{v_{S_j} v}(S)}$$

$$= \frac{1}{2} |\{x_i \in X \mid d_{v_{x_i} v}(S) < \infty\}| + |\{S_j \in \mathcal{F} \mid d_{v_{S_j} v}(S) < \infty\}|$$

$$= \frac{1}{2} \left| \bigcup_{S_j \in \mathcal{F}'} S_j \right| + |\{S_j \mid (v_{S_j}, v) \in S\}| = \frac{1}{2} s(\mathcal{F}') + k' = \frac{1}{2} s(\mathcal{F}') + k.$$

It follows that Conditions (1) and (2) are satisfied for $a = \frac{3}{2}$ and $b = 2$. Indeed, $OPT(I_{\mathrm{MCI}}) = \frac{1}{2} OPT(I_{\mathrm{MSC}}) + k \le \frac{3}{2} OPT(I_{\mathrm{MSC}})$, where the inequality is due to the fact that $OPT(I_{\mathrm{MSC}}) \ge k$, since otherwise the greedy algorithm would find an optimal solution for I_{MSC}. Moreover, $OPT(I_{\mathrm{MSC}}) - s(\mathcal{F}') = 2(OPT(I_{\mathrm{MCI}}) - k) - 2(c_v(S) - k) = 2(OPT(I_{\mathrm{MCI}}) - c_v(S))$. The theorem follows by plugging the values of a and b into $\alpha < 1 - \frac{1}{abe}$.   □

2.1 Greedy Algorithm and Submodularity

We now prove that the GREEDYIMPROVEMENT algorithm provides a $(1 - \frac{1}{e})$-approximation for the MCI problem. To this aim, because of Theorem 1, it suffices to prove that the closeness centrality measure is monotone and submodular.

Theorem 3. *For each vertex v, function c_v is monotone and submodular with respect to any feasible solution for MCI.*

Proof. To show that c_v is monotone increasing, it is enough to observe that for each solution S to MCI, each vertex u such that $(u,v) \notin E \cup S$, and each $s \in V \setminus \{v\}$ such that $d_{sv}(S \cup \{(u,v)\}) \ne \infty$, then $d_{sv}(S \cup \{(u,v)\}) \le d_{sv}(S)$ and therefore $\frac{1}{d_{sv}(S \cup \{(u,v)\})} \ge \frac{1}{d_{sv}(S)}$. We now show that for each pair S and T of solutions to MCI such that $S \subseteq T$ and for each vertex u such that $(u,v) \notin T \cup E$,

$$c_v(S \cup \{(u,v)\}) - c_v(S) \ge c_v(T \cup \{(u,v)\}) - c_v(T).$$

To simplify notation, we assume that $\frac{1}{d_{st}(X)} = 0$ whenever $d_{st}(X) = \infty$, for any solution X to MCI. We prove that each term of c_v is submodular, that is, that, for each vertex $s \in V \setminus \{v\}$ such that $d_{sv}(T \cup \{(u,v)\}) \ne \infty$, we show that

$$\frac{1}{d_{sv}(S \cup \{(u,v)\})} - \frac{1}{d_{sv}(S)} \ge \frac{1}{d_{sv}(T \cup \{(u,v)\})} - \frac{1}{d_{sv}(T)}. \tag{3}$$

Let us consider the shortest paths from s to v in $G(T \cup \{(u,v)\})$. The following two cases can arise:

1. The last edge of a shortest path from s to v in $G(T \cup \{(u,v)\})$ is (u,v) or belongs to $S \cup E$. In this case, such a path is a shortest path also in $G(S \cup \{(u,v)\})$, as it cannot contain edges in $T \setminus S$. Then, $d_{sv}(S \cup \{(u,v)\}) = d_{sv}(T \cup \{(u,v)\})$ and $\frac{1}{d_{sv}(S \cup \{(u,v)\})} = \frac{1}{d_{sv}(T \cup \{(u,v)\})}$. Moreover, $d_{sv}(S) \ge d_{sv}(T)$ and, therefore, $-\frac{1}{d_{sv}(S)} \ge -\frac{1}{d_{sv}(T)}$.

2. The last edge of all shortest paths from s to v in $G(T \cup \{(u,v)\})$ belongs to $T \setminus S$. In this case, $d_{sv}(T) = d_{sv}(T \cup \{(u,v)\})$ and, therefore, $\frac{1}{d_{sv}(T \cup \{(u,v)\})} - \frac{1}{d_{sv}(T)} = 0$. As $\frac{1}{d_{sv}(S)}$ is monotone increasing, then $\frac{1}{d_{sv}(S \cup \{(u,v)\})} - \frac{1}{d_{sv}(S)} \geq 0$.

In both cases, we have that the inequality (3) is satisfied and, hence, the theorem follows. $\qquad\square$

2.2 Experimental Evaluation

We conducted two types of experiments: in the first type we evaluate the quality of the solution produced by the greedy algorithm by measuring the approximation ratio on several randomly generated networks; in the second type we measured the improvement in the value of closeness of v and in the closeness ranking of v within the network (these latter experiments are conducted on three real-world networks). All our experiments have been performed on a computer equipped with an AMD Opteron 6376 CPU with 16 cores clocked at 2.30GHz and 64GB of main memory, and our programs have been implemented in C++ (gcc compiler v4.8.2 with optimization level O3).

We measured the approximation ratio of the greedy algorithm on four types of randomly generated networks, namely directed Preferential Attachment (in short, PA) [6], Erdős-Rényi (in short, ER) [11], Copying (in short, COPY) [14], and Compressible Web (in short, COMP) [8]. The size of the graphs is reported in Table 1. For each combination (n, m), we generated five random graphs and used five vertices as v. These vertices have been chosen on the basis of their original closeness ranking: in particular, we divided the list of vertices sorted by their original ranking in five parts and choose the vertices in the boundaries. We denote by $v_{X\%}$ the vertex on the boundary of the top Xth percentile (e.g. $v_{25\%}$ is a vertices on the boundary of the top 25th percentile).

In the experiments, we measured the ratio between the value of the solution found by the greedy algorithm and the optimal value computed by using an Integer Program (in short, IP). We solved IP by using the GLPK solver [3]. However, since solving IP requires long time on large instances, in some cases we used the solution to the Linear Relaxation (in short, LP) of IP as an upper bound to the optimal value. In these cases, the ratio is obtained by using the LP upper bound as a denominator, and therefore it represents a lower bound to the actual approximation ratio.

The results are reported in Table 1, where we show the number of times that the approximation ratio is equal to one and the minimum ratio obtained. The experiments clearly show that the measured approximation ratio is by far better than the theoretical one proven in the previous section. In fact, in more than 91% cases, the greedy algorithm found an optimal solution, and in the worst case the ratio is 0.9694.

For the second type of experiments we used real-world citation networks obtained by the Arnetminer database [1]. In such networks, there is a vertex for each author and an edge from vertex x to vertex y if the author corresponding to vertex x cited in his paper one paper written by the author corresponding to y.

Table 1. Closeness centrality: comparison between the greedy algorithm and the optimum (or an upper bound to the optimum). The first three columns report the type and size of the graphs; the fourth column reports the relative number of times that the greedy algorithm finds an optimal solution. The fifth column reports the minimum measured approximation ratio. The last column indicates whether the optimum has been found by using the integer program (IP) or its linear relaxation (LP), in the latter case it is an upper bound to the optimum.

Network	$n = |V|$	$m = |E|$	OPT%	Min Approx. Ratio	IP/LP
PA	100,500,1000	$\approx 1.3 \times n$	93.25	0.9831	IP
ER	100	200, 500, 1000	88.60	0.9788	IP
ER	500	5000, 12500, 25000	74.28	0.9694	LP
COMP	100	200, 500, 1000	99.88	0.9764	LP
COMP	500	5000, 12500, 25000	99.47	0.9854	LP
COPY	100	200, 500, 1000	97.48	0.9885	LP
COPY	500	5000, 12500, 25000	89.65	0.9697	LP

We parsed the Arnetminer database in order to select three sub-networks induced by the authors that published at least a paper in one of the main conferences or a journals in (i) algorithms (THEORY network, with 9274 vertices and 130419 edges), social network analysis (SN network, with 3666 vertices and 32413 edges), and computer science education (CSE network, with 3680 vertices and 35691 edges). As in the previous experiment, for each graph, we used five vertices as v. The value of k ranges from 1 to 100.

The results for THEORY are plotted in Fig. 3 (the results for the other networks are similar). In the two top charts we plot the closeness centrality and the ranking of vertex v as a function of k. We observe that any vertex become central by adding just few edges. For example a vertex with the smallest closeness centrality which initially has closeness 0 and is ranked 6398, improves its closeness and ranking to 2705.97 and 215, respectively, by adding only 10 edges, and it is among the top 10 vertices by adding 57 edges. On average, the algorithm required 3.68 seconds of computational time for each iteration of the algorithm (i.e. for each added edge).

In the charts on the bottom, we compare the greedy algorithm with a naive algorithm that adds the edges from the k vertices with the highest closeness centrality to v. In this case we choose 10 vertices for v instead of 5. We report the ratio between the closeness value (respectively, ranking) obtained by the naive algorithm and that obtained by the greedy one. It is easy to prove that the two algorithms find the same solution for $k = 1$, while in any other case the experiments show that the greedy algorithm outperforms the naive approach. In fact, the solution computed by this latter is up to 12 times worse in terms of ranking.

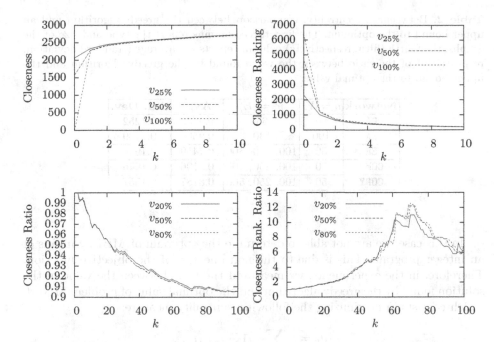

Fig. 3. Closeness centrality: (Top) performance of the greedy algorithm on network THEORY. (Bottom) comparison of the greedy algorithm with the naive method on network THEORY.

3   Improving Betweenness Centrality

Similarly to the case of the closeness centrality, we can prove, in the case of the betweenness centrality, the following two results.

Theorem 4. *Problem* MBI *cannot be approximated within a factor greater than* $1 - \frac{1}{2e}$, *unless* $P = NP$.

Theorem 5. *For each node* v, *function* b_v *is monotone and submodular with respect to any feasible solution for* MBI.

As a consequence of the previous theorem and of Theorem 1, we have that the GREEDYIMPROVEMENT algorithm is a $(1 - \frac{1}{e})$-approximation algorithm for the MBI problem. We now report the results of our experimental study on this algorithm. We used the same platform used for closeness and the parallel implementation of betweenness centrality of the NetworKit library [21]. First, we measured the approximation ratio of the greedy algorithm on the four types of randomly generated networks used for closeness centrality. The size of the graphs is reported in Table 2. For each combination (n, m), we generated five random graphs and used five vertices as v chosen like in the case of closeness centrality. The value of k ranges from 1 to 100.

Table 2. Betweenness centrality: comparison between the greedy algorithm and an upper bound to the optimum. The first three columns report the type and size of the graphs; the fourth (fifth, respectively) column reports the average (standard deviation, respectively) of the ratio between the value found by the greedy algorithm and the upper bound to the optimal value.

Network	$n = |V|$	$m = |E|$	Avg.	Std. Dev.
PA	50	65	0.9586	0.1252
PA	100	130	0.9500	0.1739
ER	50	100, 250, 500	0.7459	0.1946
COMP	50	100, 250, 500	0.8196	0.1946
COPY	50	100, 250, 500	0.8187	0.1557

In this case, we are not able to determine the optimum of MBI by means of an integer program. This is due to the non-linearity of the objective function. Therefore, in the experiments, we measured the ratio between the value of the solution found by the greedy algorithm and the optimal value of problem MBI-d, which consists in maximizing the following centrality measure:

$$d_v = \sum_{\substack{s,t \in V \\ s \neq t; s, t \neq v}} \mathbf{1}_{SP(s,t)}(v).$$

In the above formula $SP(s,t)$ denotes the set of all the vertices that belong to a shortest path from s to t and $\mathbf{1}_A(x)$ is the indicator function (i.e. $\mathbf{1}_A(x) = 1$ if $x \in A$ and $\mathbf{1}_A(x) = 0$, otherwise). It is easy to show that the optimal value of any instance of MBI-d is an upper bound for the optimal value of the corresponding MBI instance. The optimal value for MBI-d is computed by using an integer program. We solved the linear relaxation of such integer program by using GLPK. The results are reported in Table 2, where we show the average value and the standard deviation of the measured lower bound to the approximation ratio. Also in this case, the experiments show that the measured approximation ratio is by far better than the theoretical one.

For the second type of experiments we used real-world networks representing flight connection. Vertices in these networks represent airports and edges represent a connection from one airport to another. In detail, we used three networks: (i) a network obtained by crawling the EasyJet website [2] (EasyJet network, with 136 vertices and 1510 edges), (ii) the directed network of flights between US airports in 2010 (USAirports network, with 501 vertices and 5960 edges), and (iii) a network constructed from the USA Federal Aviation Administration (USA Traffic Control network, with 1227 vertices and 2615 edges). The last two networks are available from Konect [15]. As in the previous experiments, for each graph we used five vertices as v and we let k range from 1 to 100. The results for EasyJet are plotted in Fig. 4 (the results for the other networks are similar). As in the case of closeness, in the two top charts we plot the betweenness centrality and the ranking of node v as a function of k, in the two bottom charts,

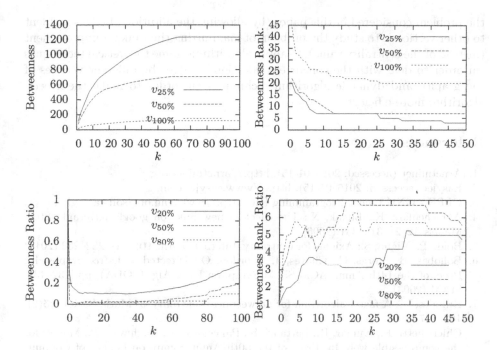

Fig. 4. Betweenness centrality: (Top) Performance of the greedy algorithm on network EasyJet; (Bottom) Comparison of the greedy algorithm with the naive method on network EasyJet

we compare the greedy algorithm with the naive algorithm. Similar results as for closeness can be observed. However, in this case, the improvement in value and in ranking is smaller than in the case of closeness. This is due to the fact that we only add incoming edges while the number of shortest paths passing through v also depends on the edges outgoing from v. We leave the problem of adding both incoming and outgoing edges as an open problem. Also in this case our algorithm outperforms the naive approach by computing solutions that are up to 7 times better in terms of ranking. On average, the algorithm required 0.33 seconds of computational time for each iteration of the algorithm (i.e. for each added edge).

4 Conclusion and Future Research

In this paper, we have proposed a greedy approximation algorithm for efficiently computing a set of edges that a node can decide to add to a graph in order to increase its betweenness or closeness centrality. The algorithm has been tested on several relatively small random graphs and, then, applied to several real-world collaboration networks. As future works, we plan to extend our approach to weighted graphs and to other centrality measures, to analyze a generalization of the problem considered in this paper (by allowing the addition of edges incident to other vertices), to study the problem of maximizing the ranking improvement (instead of the centrality value), to apply algorithmic game theoretical techniques (in order to deal with the concurrent addition of edges by different vertices of the graph), and dynamic algorithm techniques (in order to make the greedy algorithm more efficient).

References

1. Arnetminer (accessed: 2015–01-15). http://arnetminer.org
2. Easyjet (accessed: 2015–01-15). http://www.easyjet.com
3. GLPK - GNU Linear Programming Kit. http://www.gnu.org/software/glpk
4. Avrachenkov, K., Litvak, N.: The effect of new links on google pagerank. Stoc. Models **22**(2), 319–331 (2006)
5. Boldi, P., Vigna, S.: Axioms for centrality. Internet Math. **10**(3–4), 222–262 (2014)
6. Bollobás, B., Borgs, C., Chayes, J., Riordan, O.: Directed scale-free graphs. In: Proc. of the 14th Annu. ACM-SIAM Symp. on Disc. Alg. (SODA), pp. 132–139. SIAM (2003)
7. Brandes, U.: A faster algorithm for betweenness centrality. J. Math. Sociol. **25**(2), 163–177 (2001)
8. Chierichetti, F., Kumar, R., Lattanzi, S., Panconesi, A., Raghavan, P.: Models for the compressible web. In: Proc. of the 50th Annu. Symp. on Found. of Comput. Sci. (FOCS), pp. 331–340. IEEE (2009)
9. Cohen, E., Delling, D., Pajor, T., Werneck, R.F.: Computing classic closeness centrality, at scale. Technical Report MSR-TR-2014-71 (2014)
10. Demaine, E.D., Zadimoghaddam, M.: Minimizing the diameter of a network using shortcut edges. In: Kaplan, H. (ed.) SWAT 2010. LNCS, vol. 6139, pp. 420–431. Springer, Heidelberg (2010)
11. Erdős, P., Rényi, A.: On random graphs I. Publ. Math. **6**, 290–297 (1959)
12. Feige, U.: A threshold of ln n for approximating set cover. J. ACM **45**(4) (1998)
13. Ishakian, V., Erdös, D., Terzi, E., Bestavros, A.: A framework for the evaluation and management of network centrality. In: Proc. of the 12th SIAM Int. Conf. on Data Mining (SDM), pp. 427–438. SIAM (2012)
14. Kumar, R., Raghavan, P., Rajagopalan, S., Sivakumar, D., Tomkins, A., Upfal, E.: Stochastic models for the web graph. In: Proc. of the 41st Annu. Symp. on Found. of Comput. Sci. (FOCS), pp. 57–65. IEEE (2000)
15. Kunegis, J.: KONECT - The Koblenz network collection. In: Proc. of the 1st Int. Web Observatory Work. (WOW), pp. 1343–1350 (2013)
16. Malighetti, P., Martini, G., Paleari, S., Redondi, R.: The impacts of airport centrality in the EU network and inter-airport competition on airport efficiency. Technical Report MPRA-7673 (2009)
17. Meyerson, A., Tagiku, B.: Minimizing average shortest path distances via shortcut edge addition. In: Dinur, I., Jansen, K., Naor, J., Rolim, J. (eds.) Approximation, Randomization, and Combinatorial Optimization. LNCS, vol. 5687, pp. 272–285. Springer, Heidelberg (2009)
18. Nemhauser, G., Wolsey, L., Fisher, M.: An analysis of approximations for maximizing submodular set functions-I. Math. Program. **14**(1), 265–294 (1978)
19. Olsen, M., Viglas, A.: On the approximability of the link building problem. Theor. Comput. Sci. **518**, 96–116 (2014)

20. Riondato, M., Kornaropoulos, E.M.: Fast approximation of betweenness centrality through sampling. In: Proc. of the 7th ACM Int. Conf. on Web Search and Data Mining, (WSDM), pp. 413–422. ACM (2014)
21. Staudt, C.L., Sazonovs, A., Meyerhenke, H.: Networkit: An interactive tool suite for high-performance network analysis. arXiv preprint arXiv:1403.3005 (2014)
22. Williamson, D., Shmoys, D.: The Design of Approximation Algorithms. Cambridge University Press (2011)
23. Yan, E., Ding, Y.: Applying centrality measures to impact analysis: A coauthorship network analysis. J. Ass. Inf. Sci. Tech. **60**(10), 2107–2118 (2009)

An Exact Algorithm for Diameters
of Large Real Directed Graphs

Takuya Akiba, Yoichi Iwata, and Yuki Kawata[(⊠)]

The University of Tokyo, Bunkyo 113-0033, Japan
{t.akiba,y.iwata}@is.s.u-tokyo.ac.jp,
kawatea@lager.is.s.u-tokyo.ac.jp

Abstract. We propose a new algorithm to compute the diameters of
large real directed graphs. In contrast to recent algorithms, the proposed
algorithm is designed for general directed graphs, i.e., it does not assume
that given graphs are undirected or strongly connected. Experimental
results on large real graphs show that the proposed algorithm is several
orders of magnitude faster than the naive approach, and it reveals the
exact diameters of large real directed graphs, for which only lower bounds
have been known.

1 Introduction

The *diameter* of a graph is the distance between the most distant pairs of
vertices. Because of the connection to the *small-world phenomenon*, the diameters of real-world graphs have been of interest in sociology and network science [1,3,11,14].

However, even with the recent availability of large real graph data, the diameters of such graphs have remained unclear because computing diameters has
been too computationally expensive. In theory, currently no algorithm has better time complexity than all-pairs shortest path algorithms. Our target graphs
are sparse unweighted graphs; thus, this approach corresponds to a naive algorithm that conducts a breadth first search (BFS) from every vertex, which is
obviously too slow for large graphs with millions of vertices.

Therefore, practical heuristic algorithms that are tailored to real graphs have
been studied [2,5–7,10,12]. Table 1 summarizes such previous methods. Experiments on real-world networks have shown that these methods work surprisingly
well; they can successfully give accurate bounds or exact values of diameters,
especially for undirected or strongly connected graphs. Nevertheless, the table
highlights a lack of practical exact algorithms for general (i.e., not necessarily
strongly connected) directed graphs. On the other hand, from a practical perspective, it is of great interest to reveal the exact values of the diameters of large
real directed graphs, such as web graphs and some online social networks.

In this study, we propose a new algorithm to compute the diameters of large
real directed graphs. The proposed algorithm is designed for general directed
graphs, i.e., it does not assume that given graphs are undirected or strongly

© Springer International Publishing Switzerland 2015
E. Bampis (Ed.): SEA 2015, LNCS 9125, pp. 56–67, 2015.
DOI: 10.1007/978-3-319-20086-6_5

Table 1. Previous empirical algorithms for graph diameters

Method	Graph	Output
Classic Methods:		
Naive (all-pairs)	any	exact value
Random sampling	any	lower bound
Double sweep [4,8]	any	lower bound
Recent Methods:		
Magnien, et al., 2009[10]	undirected	upper and lower bound
Crescenzi, et al., 2010 [6]	undirected	upper and lower bound
Takes and Kosters, 2011 [12]	undirected	exact value
Crescenzi, et al., 2012 [7]	strongly connected	exact value
Crescenzi, et al., 2013 [5]	undirected	exact value
Borassi, et al., 2014 [2]	strongly connected	exact value
This work	any	exact value

connected. The main technical challenge is to efficiently deal with unreachable
vertex pairs, which do not exist in undirected or strongly connected graphs. As
demonstrated by our experimental results, the proposed algorithm works quite
well on large real directed graphs, and, to the best of our knowledge, it has
yielded a list of precise diameters of famous large real graph datasets for the
first time.

Contributions. Our technical contributions are as follows:

1. We propose the first practical algorithm that can compute the exact diameters of large real directed graphs with millions of vertices and edges.

2. Using the proposed algorithm, we present the first precise list of the diameters of famous large real directed graph datasets from SNAP [9].

3. Using the exact diameter values, the accuracy of classic approximation
   heuristics is evaluated on large real directed graphs for the first time.

4. As mentioned in Section 7, our implementation is publicly available online.

Organization. In Section 2, we discuss the definitions and notations used in
this paper. In Section 3, we examine inequalities on vertex eccentricities, which
are key components of the proposed algorithm. Section 4 details the proposed
algorithm. We present experimental results in Section 6 and conclude the paper
in Section 7.

2 Preliminaries

Let $G = (V, E)$ be a graph, where V is the vertex set and E is the edge set.
The number of vertices and number of edges are denoted n and m, respectively.
We assume that graph G is weakly connected and unweighted. We define $N_{in}(v)$
and $N_{out}(v)$ as the in- and out-neighbors of vertex v. The subgraph induced

by a vertex set S is denoted $G[S]$, and G^T describes the transpose graph of a graph G. In addition, let $d(v, w)$ denote the distance from a vertex v to a vertex w. When w is not reachable from v, for simplicity, we define $d(v, w) = \infty$, where ∞ is considered a sufficiently large number. The eccentricity $ecc(v)$ of vertex v is defined as the distance from v to the farthest reachable vertex, i.e., $ecc(v) = \max\{d(v, w) \mid w \in V, d(v, w) \neq \infty\}$. The diameter D of a graph G is the maximum eccentricity, i.e., $D = \max\{ecc(v) \mid v \in V\}$.

Strongly Connected Components. A strongly connected component (SCC) of a graph is defined as a strongly connected maximal subgraph. Each SCC does not overlap; therefore, we can partition the vertex set V into SCCs. The set of SCCs is denoted V_{SCC}, and the vertex set of the SCC that a vertex v belongs to is denoted $V_{SCC}(v)$. Note that this decomposition can be obtained in linear time [13].

Double Sweep Algorithm. The double sweep algorithm [4,8] works as follows. First, we randomly select a vertex v and conduct a breadth first search (BFS) from this vertex. Then, we select a vertex w such that $d(v, w) = ecc(v)$ and conduct a backward BFS from w. Finally, the double sweep algorithm outputs $\max\{d(u, w) \mid u \in V, d(u, w) \neq \infty\}$ as the lower bound of the diameter. Note that this runs in $O(n + m)$ time. The precision can be improved by conducting the algorithm repeatedly with a different initial vertex. Typically, a few iterations of double sweep give quite precise approximation for real graphs [6,7,10]. In the proposed algorithm, we use the double sweep algorithm to initialize the diameter lower bound.

3   Upper Bounds

In this section, we study the upper bounds on the eccentricities. In previous work that focused on undirected or strongly connected graphs, upper bounds based on the triangle inequality turned out to be quite effective [2,12]. However, as discussed in Section 3.1, this does not hold between vertices in different SCCs. Therefore, we propose new inequalities (Sections 3.2 and 3.3) that can propagate the upper bounds of vertices in other SCCs.

3.1   Triangle Inequality

Due to its effectiveness, the following inequality is often solely used as the upper bound of the eccentricity in undirected or strongly connected graphs [2,12]. This inequality can be obtained easily from the triangle inequality.

Lemma 1 (Borassi, et al. [2]). *If the graph is strongly connected, for any pair of vertices v and u, $ecc(v) \leq d(v, u) + ecc(u)$.*

Unfortunately, on general directed graphs, this inequality does not hold between any pair of vertices. Here, we illustrate this point with an example.

In Figure 1, $ecc(2)$ is 2. However, the sum of the distance $d(2,1)$ and $ecc(1)$ is $1+0 = 1$, which is less than $ecc(2)$. This is because vertex 1 cannot reach vertices 3 and 4.

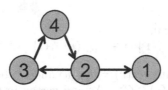

Fig. 1. An example of a general directed graph, on which the upper bound based on the triangle inequality does not necessarily hold

Therefore, we use this upper bound only inside an SCC to form the following corollary.

Corollary 1. *If vertices v and u are in the same SCC, $ecc(v) \leq d(v,u)+ecc(u)$.*

3.2   Inequality for Propagation

For vertices in large SCCs, similar to those in undirected or strongly connected graphs, good eccentricity upper bounds can be obtained using the above inequality. However, as there are many small SCCs in real-world networks, using the aforementioned inequality results in the eccentricities of vertices in those small SCCs remaining unbounded.

Therefore, we propose new inequalities that can propagate the eccentricity bounds of vertices in different SCCs. We begin with the following simple inequalities.

Lemma 2. *For any vertex v, $ecc(v) \leq \max\{ecc(w) + 1 \mid w \in N_{out}(v)\}$.*

Proof. Let u be a vertex with $ecc(v) = d(v,u)$. Since there exists a vertex $w^* \in N_{out}(v)$ such that $d(v,u) = d(w^*,u)+1$, we obtain $ecc(v) = d(v,u) = d(w^*,u)+ 1 \leq ecc(w^*) + 1 \leq \max\{ecc(w) + 1 \mid w \in N_{out}(v)\}$.

Although this inequality is quite simple, it enables us to obtain the upper bounds of vertices in small SCCs effectively. Moreover, it becomes significantly more effective when combined with vertex ordering strategies that consider the topological order of SCCs. Note that if we contract each SCC to a single vertex, we obtain a directed acyclic graph, which defines the topological order of SCCs. Therefore, by visiting each SCC in reverse topological order, this inequality works more effectively because, when we examine a vertex, we have already visited all of its out-neighbors. Thus, their eccentricity bounds are already tightened with the inequality using the bounds of their out-neighbors. In other words, we can propagate the upper bounds from lower SCCs to upper SCCs efficiently.

3.3 Tight Inequality for Propagation

By considering SCCs, we can obtain the following inequality, which is tighter than the above simple inequality. Note that we use this inequality in the proposed algorithm.

Lemma 3. *For any vertex* v,
$$ecc(v) \leq \max \left\{ \min \left\{ ecc(w) + 1 \mid w \in N_{out}(v) \cap C \right\} \mid C \in \mathcal{V}_{SCC} \right\}.$$

Proof. Let u be a vertex with $ecc(v) = d(v, u)$ and C^* be an SCC with $d(v, u) = \min(d(w, u) + 1 \mid w \in N_{out}(v) \cap C^*)$. As any vertex in C^* can reach u, it holds that $d(w, u) \leq ecc(w)$ for any vertex $w \in C^*$. Thus, we obtain $ecc(v) = d(v, u) = \min(d(w, u) + 1 \mid w \in N_{out}(v) \cap C^*) \leq \min(ecc(w) + 1 \mid w \in N_{out}(v) \cap C^*) \leq \max(\min(ecc(w) + 1 \mid w \in N_{out}(v) \cap C) \mid C \in \mathcal{V}_{SCC})$

4   Algorithm Description

Here, we present the proposed algorithm for computing diameters. The proposed algorithm and some of the previous algorithms share a common approach that maintains and gradually tightens a diameter lower bound and eccentricity upper bounds [2,12]. The main technical challenge in designing the proposed algorithm is handling pairs in different SCCs efficiently using SCC decomposition and the new upper bound inequalities.

4.1   Overall Algorithm

Algorithm 1 shows an overview of the proposed algorithm. The variable $\overline{ecc}[v]$ maintains an upper bound of the eccentricity of each vertex v (i.e., $\overline{ecc}[v] \geq ecc(v)$), and variable $\underline{D}$ maintains a lower bound of the diameter of the given graph (i.e., $\underline{D} \leq D$). At the end of Algorithm 1, $\overline{ecc}[v]$ becomes less than or equal to $\underline{D}$ for any vertex v. From the inequality $\underline{D} \leq D = \max(ecc(v) \mid v \in V) \leq \max(\overline{ecc}(v) \mid v \in V) \leq \underline{D}$, $\underline{D}$ matches the exact value of the diameter D.

Algorithm 1. Proposed algorithm

1: **procedure** DIAMETER(G)
2:      $\mathcal{V}_{SCC} \leftarrow$ STRONGLYCONNECTEDCOMPONENTS(G)
3:      $\overline{ecc}[v] \leftarrow \infty$ **for all** $v \in V$
4:      $\underline{D} \leftarrow$ DOUBLESWEEP(G)
5:      **for all** $v \in V$ **do**
6:          $e' \leftarrow \max \left\{ \min \left\{ \overline{ecc}[w] + 1 \mid w \in N_{out}(v) \cap C \right\} \mid C \in \mathcal{V}_{SCC} \right\}$
7:          $\overline{ecc}[v] \leftarrow \min(\overline{ecc}[v], e')$
8:          **if** $\overline{ecc}[v] > \underline{D}$ **then**
9:              SEARCHANDBOUND(G, v)
10:     **return** $\underline{D}$

First, we decompose the graph into SCCs. The SCCs are later used for vertex ordering and eccentricity upper bounds. Next, we initialize the diameter lower bound $\underline{D}$ using the double sweep algorithm. We also set the initial eccentricity upper bound $\overline{ecc}[v]$ to ∞ for each vertex v.

We then examine every vertex $v \in V$ in sequence. First, we refine the eccentricity upper bound $\overline{ecc}[v]$ by Lemma 3. We then compare $\overline{ecc}[v]$ and $\underline{D}$, and skip vertex v if $\overline{ecc}[v] \leq \underline{D}$ because the lengths of paths from v are bound by $\overline{ecc}[v]$; thus, they never update the diameter lower bound $\underline{D}$. Otherwise, we conduct breadth first searches (BFSs) from v to update the bounds.

4.2   Updating Bounds by BFSs

The process of updating bounds by BFSs from vertex u is described in Algorithm 2. We conduct two BFSs, a forward BFS from u to the whole graph (lines 1–4) and a backward BFS from u only to the SCC to which u belongs (lines 5–7). In Algorithm 2, we assume that the BREADTHFIRSTSEARCH function performs a BFS and returns an array that represents distances, i.e., $d_f[v] = d(u,v)$ for any $v \in V$ and $d_b[v] = d(v,u)$ for any $v \in V_{\text{SCC}}(u)$.

Algorithm 2. Breadth first searches from vertex u to update bounds

1: **procedure** SEARCHANDBOUND(G, u)
2:    $d_f \leftarrow$ BREADTHFIRSTSEARCH(G, u)
3:    $ecc[u] \leftarrow \max\{d_f[v] \mid v \in V, d_f[v] \neq \infty\}$
4:    $\overline{ecc}[u] \leftarrow ecc[u]$
5:    $\underline{D} \leftarrow \max(\underline{D}, ecc[u])$
6:    $d_b \leftarrow$ BREADTHFIRSTSEARCH$(G[V_{\text{SCC}}(u)]^T, u)$
7:    $\overline{ecc}[v] \leftarrow \min(\overline{ecc}[v], d_b[v] + ecc[u])$ **for all** $v \in V_{\text{SCC}}(u)$

The forward BFS computes the exact eccentricity of u. Once the eccentricity is obtained, the diameter lower bound is compared to the eccentricity and updated if necessary. We then perform a reverse BFS to compute the distances to vertex u. This reverse BFS updates the eccentricity upper bounds of other vertices. Here we only visit vertices in the SCC to which u belongs because we are dealing with general (not necessarily strongly connected) directed graphs. As discussed in Section 3, this upper bound holds only within an SCC (Corollary 1).

5   Vertex Ordering Strategies

Here, we discuss vertex ordering strategies. The proposed algorithm examines each vertex sequentially and conducts BFSs if necessary. While its correctness is not dependent on the order of vertices, the order has considerable effect on the performance of the proposed algorithm.

Ordering SCCs. First, we consider the overall graph. We update the upper bound of the eccentricity by checking all out-neighbors and using the upper bound for the out-neighbors. Therefore, we check each SCC in reverse topological order. This enables us to obtain a better upper bound because, when we check a neighbor in a different SCC, the neighbor has been visited and the upper bound of the neighbor has been tightened.

Ordering inside an SCC. Next, we consider each SCC. In each SCC, we use the triangle inequality to update the upper bound of the eccentricity. When we examine a central vertex, distances to the vertex are small; therefore, by examining such vertices first, we can tighten the upper bounds. We order vertices based on the degree of vertices to select central vertices. However, we can consider various strategies based on the degree in directed graphs. We propose and empirically compare the following five vertex ordering strategies.

- RANDOM: A random permutation of vertices is used as the vertex order.
- IN-DEGREE: We order vertices based on the in-degree of vertices, i.e., we use $|N_{in}(v)|$ as the index.
- OUT-DEGREE: We order vertices based on the out-degree of vertices, i.e., we use $|N_{out}(v)|$ as the index.
- DEGREE-PRODUCT: We order vertices based on the product of the in-degree and out-degree of vertices, i.e., we use $|N_{in}(v)| \cdot |N_{out}(v)|$ as the index.
- DEGREE-COMPONENT-PRODUCT: We order vertices based on the product of the in-degree and out-degree for vertices in the same SCC, i.e., we use $|N_{in}(v) \cap V_{SCC}(v)| \cdot |N_{out}(v) \cap V_{SCC}(v)|$ as the index. We consider that vertices in different SCCs have little to do with the centrality in the SCC.

6   Experiments

Here, we present our experimental results. In Section 6.1, we discuss an experimental evaluation of the practicality of the proposed method and compare it to previous methods. In Section 6.3, we analyze the behavior of the proposed method empirically.

Setup. We conducted experiments on a Linux server with an Intel Xeon X5675 processor (3.06 GHz) and 288 GB of main memory. All algorithms were implemented in C++ and compiled with gcc 4.1.2 using optimization level 3. We did not parallelize the proposed algorithm and used only a single core. We used various real-world directed networks from the Stanford Large Network Dataset Collection [9]. We treated all graphs as unweighted. In addition, we used the largest weakly connected component of the graphs. Unless otherwise stated, we used the DEGREE-COMPONENT-PRODUCT strategy for vertex ordering, and the initial lower bound was obtained by performing the double sweep algorithm from 10 randomly selected vertices.

6.1 Evaluating the Proposed Method

We evaluate the performance of the proposed method with real graphs and virtually compare performance to the naive method. Table 2 shows the results. We counted each call of function SEARCHANDBOUND (Algorithm 2) as a single BFS; SEARCHANDBOUND conducts one whole BFS and one small BFS that is limited

Table 2. Performance of the proposed algorithm on real graphs. #BFS denotes the number of BFSs to compute the diameter and Speed-up denotes the improvement factor of the number of BFSs against the naive approach.

| Dataset | | | | Proposed method | |
Name	n	m	D	#BFS	Speed-up
Social Networks:					
ego-Twitter	81,306	1,768,149	15	353	230.3
soc-Epinions1	75,878	508,837	16	1,066	71.2
soc-LiveJournal1	4,843,953	68,983,820	22	53,819	90.0
soc-Pokec	1,632,803	30,622,564	18	2,110	773.8
soc-Slashdot0811	77,361	905,469	12	6,744	11.5
soc-Slashdot0922	82,169	948,465	13	6,891	11.9
soc-sign-Slashdot081106	77,350	516,575	15	970	79.7
soc-sign-Slashdot090216	81,867	545,671	15	983	83.3
soc-sign-Slashdot092221	82,140	549,202	15	975	84.2
soc-sign-epinions	119,130	833,390	16	1,269	93.9
wiki-Vote	7,067	103,664	10	2	3,533.5
Communication Networks:					
email-EuAll	224,833	395,271	11	206	1,091.4
wiki-Talk	2,388,954	5,018,446	11	935	2,555.0
Citation Networks:					
cit-HepPh	34,401	421,485	49	915	37.6
cit-HepTh	27,400	352,542	37	2,412	11.4
Web Graphs:					
web-BerkStan	654,783	7,499,426	694	5,977	109.6
web-Google	855,803	5,066,843	51	12,347	69.3
web-NotreDame	325,730	1,497,135	93	3,626	89.8
web-Stanford	255,266	2,234,573	580	3,027	84.3
Product Co-Purchasing Networks:					
amazon0302	262,112	1,234,878	88	1,666	157.3
amazon0312	400,728	3,200,441	53	593	675.8
amazon0505	410,237	3,356,825	55	388	1,057.3
amazon0601	403,365	3,387,225	54	450	896.4
Internet Peer-to-Peer Networks:					
p2p-Gnutella04	10,877	39,995	26	30	362.6
p2p-Gnutella05	8,843	31,838	22	24	368.5
p2p-Gnutella06	8,718	31,526	21	50	174.4
p2p-Gnutella08	6,300	20,777	20	51	123.5
p2p-Gnutella09	8,105	26,009	20	110	73.7
p2p-Gnutella24	26,499	65,360	29	45	588.9
p2p-Gnutella25	22,664	54,694	22	138	164.2
p2p-Gnutella30	36,647	88,304	24	165	222.1
p2p-Gnutella31	62,562	147,879	31	210	297.9

to an SCC. Speed-up represents the improvement factor of the number of BFSs against the naive approach, which requires n BFSs (i.e., n divided by #BFS).

We emphasize that the proposed algorithm can compute the diameters of large real directed graphs with up to tens of millions of edges. Obviously, this is too large for the naive method. However, the proposed algorithm can compute diameters several tens to several thousand times faster than the naive method. In addition, we do not necessarily visit all vertices in a BFS. Thus, the proposed algorithm can compute diameters faster than it appears.

6.2   Evaluating the Previous Approximation Heuristics

We then evaluate the classic lower bound heuristics. Note that the evaluation of these methods on large directed networks became possible for the first time using the proposed algorithm. We evaluated both random sampling and the double sweep algorithm. For random sampling, we sampled 10, 100, or 1,000 vertices and conducted BFSs from these vertices. We ran the double sweep algorithm from 10 randomly selected vertices. The results are given in Table 3.

Note that the lower bounds obtained by random sampling are far from exact diameters. For any graph, we could not compute exact diameters by sampling less than 100 vertices. Even if we sampled 1,000 vertices, we could compute exact diameters in only eight graphs. On the other hand, the lower bounds obtained by the double sweep algorithm were correct for many graphs. However, these bounds varied greatly from the exact diameters in a few graphs.

6.3   Analysis

Upper Bounds. Table 4 compares the effectiveness of inequalities for eccentricity upper bounds. In Table 4, (1), (2), and (3) represent Corollary 1, Lemma 2, and Lemma 3, respectively. We reduced the number of BFSs by approximately five percent on average using Lemma 3 rather than Lemma 2. This indicates that Lemma 3 is effective for obtaining tighter upper bounds. In addition, if we did not use Lemma 3, the number of BFSs increased drastically due to many small SCCs. On the other hand, we could compute diameters efficiently for many graphs using only Lemma 3. These results confirm the effectiveness of our new upper bounds by propagation.

Vertex Ordering Strategies. Table 5 shows the number of BFSs for each vertex ordering strategy. Essentially, we used the DEGREE-COMPONENT-PRODUCT strategy. However, all strategies demonstrated nearly the same performance, with exception of the RANDOM strategy. The reason is as follows. We use two inequalities with Corollary 1 and Lemma 3. As mentioned previously, Lemma 3 is quite powerful. In addition, the ordering inside an SCC affects only the triangle inequality. Therefore, performance is insulated from the influence of ordering inside an SCC. As can be seen, the RANDOM strategy is worse. However, the difference between the RANDOM strategy and other strategies is small in many cases. This confirms that the order of vertices in each SCC has less effect on performance.

Table 3. Comparison between exact diameters and lower bounds using heuristics. Sampling denotes the lower bound computed by 10, 100, or 1,000 random samplings. Double Sweep denotes the lower bound computed by performing the double sweep algorithm 10 times.

Dataset Name	n	m	D	Sampling 10	100	1,000	Double Sweep
Social Networks:							
ego-Twitter	81,306	1,768,149	15	10	10	11	**15**
soc-Epinions1	75,878	508,837	16	12	12	14	**16**
soc-LiveJournal1	4,843,953	68,983,820	22	15	17	18	**22**
soc-Pokec	1,632,803	30,622,564	18	12	14	14	**18**
soc-Slashdot0811	77,361	905,469	12	9	9	**12**	**12**
soc-Slashdot0922	82,169	948,465	13	10	10	**13**	**13**
soc-sign-Slashdot081106	77,350	516,575	15	13	13	13	**15**
soc-sign-Slashdot090216	81,867	545,671	15	11	12	13	**15**
soc-sign-Slashdot092221	82,140	549,202	15	10	13	13	**15**
soc-sign-epinions	119,130	833,390	16	12	12	13	**16**
wiki-Vote	7,067	103,664	10	6	8	8	**10**
Communication Networks:							
email-EuAll	224,833	395,271	11	9	9	**11**	**11**
wiki-Talk	2,388,954	5,018,446	11	7	7	8	**11**
Citation Networks:							
cit-HepPh	34,401	421,485	49	32	47	**49**	**49**
cit-HepTh	27,400	352,542	37	29	32	34	**37**
Web Graphs:							
web-BerkStan	654,783	7,499,420	694	580	585	645	**694**
web-Google	855,803	5,066,843	51	36	39	41	**51**
web-NotreDame	325,730	1,497,135	93	8	54	66	64
web-Stanford	255,266	2,234,573	580	148	149	557	244
Product Co-Purchasing Networks:							
amazon0302	262,112	1,234,878	88	69	75	85	**88**
amazon0312	400,728	3,200,441	53	38	40	48	**53**
amazon0505	410,237	3,356,825	55	37	41	**55**	**55**
amazon0601	403,365	3,387,225	54	36	48	48	**54**
Internet Peer-to-Peer Networks:							
p2p-Gnutella04	10,877	39,995	26	19	22	23	**26**
p2p-Gnutella05	8,843	31,838	22	18	20	**22**	**22**
p2p-Gnutella06	8,718	31,526	21	16	19	20	**21**
p2p-Gnutella08	6,300	20,777	20	16	17	**20**	**20**
p2p-Gnutella09	8,105	26,009	20	18	19	19	**20**
p2p-Gnutella24	26,499	65,360	29	25	25	27	**29**
p2p-Gnutella25	22,664	54,694	22	17	18	19	21
p2p-Gnutella30	36,647	88,304	24	20	22	22	**24**
p2p-Gnutella31	62,562	147,879	31	24	28	**31**	**31**

Table 4. Comparison of the number of BFSs between different sets of eccentricity upper bound inequalities

Dataset			Upper Bounds			
Name	n	m	(1)+(3)	(1)+(2)	(1)	(3)
ego-Twitter	81,306	1,768,149	353	350	11,704	540
soc-Pokec	1,632,803	30,622,564	2,110	2,118	307,827	5,476
soc-sign-epinions	119,130	833,390	1,269	1,022	53,346	1,254
wiki-Talk	2,388,954	5,018,446	935	515	2,257,197	750
cit-HepPh	34,401	421,485	915	2,023	19,265	1,315
web-Google	855,803	5,066,843	12,347	12,550	304,516	25,302
amazon0601	403,365	3,387,225	450	450	1,573	5,637
p2p-Gnutella30	36,647	88,304	165	226	27,969	892
p2p-Gnutella31	62,562	147,879	210	256	47,906	1,430

Table 5. Comparison of the number of BFSs between different vertex ordering strategies

Dataset	DEGREE				RANDOM
	IN	OUT	PRODUCT	COMPONENT PRODUCT	
ego-Twitter	364	**350**	355	353	406
soc-Pokec	**2,110**	2,111	**2,110**	**2,110**	3,799
soc-sign-epinions	**1,027**	1,040	1,103	1,269	1,050
wiki-Talk	**925**	1,272	978	935	983
cit-HepPh	**856**	1,573	859	915	977
web-Google	12,071	12,085	**12,064**	12,347	12,118
amazon0601	**450**	466	**450**	**450**	508
p2p-Gnutella30	213	269	244	**165**	327
p2p-Gnutella31	**187**	273	264	210	241

7   Conclusion

In this paper, we have proposed an algorithm to exactly compute the diameters of large real graphs. The proposed algorithm gradually tightens the diameter lower bound and eccentricity upper bound, and they are guaranteed to match when the algorithm terminates. Note that the worst-case time complexity is still $\Theta(nm)$. However, as demonstrated by our experimental results, the proposed algorithm works surprisingly well in practice; it yielded the exact diameters of real directed graphs with millions of vertices and edges.

Software Available Publicly. Our implementation of the proposed algorithm is publicly available online at http://git.io/graph-diameter. It is our hope that further scientific findings will be enabled by our public code.

Acknowledgments. Takuya Akiba and Yoichi Iwata are supported by Grant-in-Aid for JSPS Fellows 256563 and 256487, respectively. Yuki Kawata is supported by JST, ERATO, Kawarabayashi Project.

References

1. Albert, R., Jeong, H., Barabási, A.-L.: Diameter of the World-Wide web. Nature **401**(6749), 130–131 (1999)
2. Borassi, M., Crescenzi, P., Habib, M., Kosters, W., Marino, A., Takes, F.: On the solvability of the six degrees of kevin bacon game. In: Ferro, A., Luccio, F., Widmayer, P. (eds.) FUN 2014. LNCS, vol. 8496, pp. 52–63. Springer, Heidelberg (2014)
3. Broder, A., Kumar, R., Maghoul, F., Raghavan, P., Rajagopalan, S., Stata, R., Tomkins, A., Wiener, J.: Graph Structure in the Web. Computer Networks: The International Journal of Computer and Telecommunications Networking **33**(1–6), 309–320 (2000)
4. Corneil, D.G., Dragan, F.F., Habib, M., Paul, C.: Diameter Determination on Restricted Graph Families. Discrete Applied Mathematics **113**(2), 143–166 (2001)
5. Crescenzi, P., Grossi, R., Habib, M., Lanzi, L., Marino, A.: On Computing the Diameter of Real-World Undirected Graphs. Theoretical Computer Science **514**, 84–95 (2013)
6. Crescenzi, P., Grossi, R., Imbrenda, C., Lanzi, L., Marino, A.: Finding the diameter in real-world graphs. In: de Berg, M., Meyer, U. (eds.) ESA 2010, Part I. LNCS, vol. 6346, pp. 302–313. Springer, Heidelberg (2010)
7. Crescenzi, P., Grossi, R., Lanzi, L., Marino, A.: On computing the diameter of real-world directed (weighted) graphs. In: Klasing, R. (ed.) SEA 2012. LNCS, vol. 7276, pp. 99–110. Springer, Heidelberg (2012)
8. Handler, G.Y.: Minimax Location of a Facility in an Undirected Tree Graph. Transportation Science **7**(3), 287–293 (1973)
9. Leskovec, J., Krevl, A.: SNAP Datasets: Stanford Large Network Dataset Collection. http://snap.stanford.edu/data
10. Magnien, C., Latapy, M., Habib, M.: Fast Computation of Empirically Tight Bounds for the Diameter of Massive Graphs. Journal of Experimental Algorithmics **13**(10), 1.10–1.9 (2009)
11. Milgram, S.: The Small-World Problem. Psychology Today **2**(1), 60–67 (1967)
12. Takes, F.W., Kosters, W.A.: Determining the Diameter of Small World Networks. In: CIKM, pp. 1191–1196 (2011)
13. Tarjan, R.: Depth-First Search and Linear Graph Algorithms. SIAM Journal on Computing **1**(2), 146–160 (1972)
14. Watts, D.J., Strogatz, S.H.: Collective Dynamics of 'Small-World' Networks. Nature **393**(6684), 440–442 (1998)

Graph Partitioning for Independent Sets

Sebastian Lamm, Peter Sanders, and Christian Schulz[✉]

Karlsruhe Institute of Technology, Karlsruhe, Germany
lamm@ira.uka.de, {sanders,christian.schulz}@kit.edu

Abstract. Computing maximum independent sets in graphs is an important problem in computer science. In this paper, we develop an evolutionary algorithm to tackle the problem. The core innovations of the algorithm are very natural combine operations based on graph partitioning and local search algorithms. More precisely, we employ a state-of-the-art graph partitioner to derive operations that enable us to quickly exchange whole blocks of given independent sets. To enhance newly computed offsprings we combine our operators with a local search algorithm. Our experimental evaluation indicates that we are able to outperform state-of-the-art algorithms on a variety of instances.

1 Introduction

In a simple and connected graph, an *independent set* is a subset of the nodes such that every pair of nodes that can be formed from the set is not adjacent. The *maximum independent set* problem is then to find the independent set in the graph with the largest possible cardinality. There are lots of applications that benefit from large independent sets such as information retrieval, signal transmission analysis, classification theory, economics, scheduling or computer vision [9]. As a more specific example, finding large independent sets is useful in map labeling [11] where one wants to maximize the number of visible non-overlapping labels on a map. Here, a graph model is built such that labels correspond to nodes and there is an edge between two nodes if the associated labels are overlapping. It is easy to see that a maximum independent set in the model yields a maximum number of non-overlapping labels.

The maximum independent set problem is closely related to the maximum clique problem and the minimum vertex cover problem. More precisely, the complement of an independent set $\mathcal{I}$ results in a vertex cover $V \backslash \mathcal{I}$ and an independent set is a clique in the complement graph $\overline{G}$. However, note that results from the maximum clique problem are usually only partially transferable to practical algorithms for the maximum independent set problem since building the complement of sparse graphs yields dense graphs. It is well known that all of these problems are NP-hard [10]. Thus, one relies on heuristic algorithms to find good solutions on large graphs. Most of the work in literature considers heuristics

Graph Partitioning for Independent Sets—Partially supported by DFG Gottfried Wilhelm Leibniz Prize 2012 for Peter Sanders.

ⓒ Springer International Publishing Switzerland 2015
E. Bampis (Ed.): SEA 2015, LNCS 9125, pp. 68–81, 2015.
DOI: 10.1007/978-3-319-20086-6_6

and local search algorithms for the maximum clique problem (see for example [5,13–16,21]). These algorithms keep a single solution and try to improve it by using node deletions, insertions, swaps as well as the concept of plateau search. In this context, plateau search only accepts moves that do not change the objective function of the optimization problem. Heuristics usually employ node swaps to achieve that. A node swap refers to the replacement of a node by one of its neighbors; Hence, a node swap cannot directly increase the size of the independent set but can yield a situation where an additional node may get inserted to the solution. A very successful approach for the maximum clique problem has been presented by Grosso et al. [14]. In addition to the plateau search approach, different diversification operations are performed and restart rules are added. In the independent set context, Andrade et al. [1] extended the notion of swaps to (j, k)-swaps. A (j, k)-swap removes j nodes from the current solution and inserts k nodes. The authors present a fast linear-time implementation that, given a maximal solution, can find a $(1, 2)$-swap or prove that none exists. We implemented the algorithm and use it within our evolutionary algorithm to improve newly computed offsprings.

There are very few papers considering evolutionary algorithms for the maximum independent set problem. The general idea behind evolutionary algorithms is to use mechanisms which are highly inspired by biological evolution such as selection, mutation, recombination and survival of the fittest. An evolutionary algorithm starts with a population of individuals (in our case independent sets of the graph) and evolves the population into different populations over several rounds. In each round, the evolutionary algorithm uses a selection rule based on the fitness of the individuals of the population to select good individuals and combine them to obtain improved offspring [12].

Bäck and Khuri [3] and Borisovsky and Zavolovskaya [6] use fairly similar approaches. They encode solutions as bitstrings such that the value at position i equals one if and only if node i is in the current solution. In both cases a classic two-point crossover is used which randomly selects two crossover points p_1, p_2. Then all bits in between these positions are exchanged between both input individuals. Note that this likely results in invalid solutions. To guide the search towards valid solutions a penalty approach is used. A major drawback of the work by Bäck and Khuri [3] is that the authors only test their algorithm on synthetic instances. Moreover, in both cases the graphs under consideration are very small.

The *main contribution* of our paper is a very natural evolutionary framework for the computation of large maximal independent sets. The core innovations of the algorithm are combine operations based on graph partitioning and local search algorithms. More precisely, we employ the state-of-the-art graph partitioner KaHIP [22] to derive operations that enable us to quickly *exchange whole blocks* of given individuals. The newly computed offsprings are then improved using a local search algorithm. In *contrast* to previous evolutionary algorithms, each computed offspring is valid. Hence, we only allow valid solutions in our population and thus are able to use the cardinality of the independent set as a fitness function. The rest of paper is organized as follows. We begin in Section 2 by introducing basic concepts and related work. We describe the core components of our evolutionary algorithm in Section 3. This includes a number of partitioning based combine operators that take two individuals as input as well as combine operators that can take *multiple* individuals as input. A summary of extensive experiments done to tune the algorithm and evaluate its performance is presented in Section 4. Experiments indicate that our algorithm computes very good independent sets and outperforms state-of-the-art algorithms on a large variety of instances. Finally, we conclude with Section 5.

2 Prelimiaries

2.1 Basic Concepts

Let $G = (V = \{0, \ldots, n-1\}, E)$ be an undirected graph with $n = |V|$ and $m = |E|$. The set $N(v) := \{u : \{v, u\} \in E\}$ denotes the neighbors of v. The *complement* of a graph is defined as $\overline{G} = (V, \overline{E})$ with $\overline{E}$ being the complement of E. An *independent set* is a subset $\mathcal{I} \subseteq V$, such that there are no adjacent nodes in $\mathcal{I}$. It is *maximal*, if it is not a subset of any larger independent set. The *independent set problem* is that of finding the maximum cardinality set among all possible independent sets. A *vertex cover* is a subset of nodes $C \subseteq V$, such that every edge $e \in E$ is at least incident to one node within the set. The *minimum vertex cover problem* asks for the vertex cover with the minimum number of nodes. It is worth mentioning that the complement of a vertex cover $V \setminus C$ always is an independent set by definition. A *clique* is a subset of the nodes $Q \subseteq V$ such that there is an edge between all pairs of nodes from Q.

A k-way partition of a graph is a division of V into *blocks* of nodes $V_1, \ldots, V_k$, i.e. $V_1 \cup \cdots \cup V_k = V$ and $V_i \cap V_j = \emptyset$ for $i \neq j$. A *balancing constraint* demands that $\forall i \in \{1..k\} : |V_i| \leq L_{\max} := (1 + \epsilon)\lceil \frac{|V|}{k} \rceil$ for some imbalance parameter ϵ. The objective is to minimize the total *cut* $\sum_{i<j} w(E_{ij})$ where $E_{ij} := \{\{u, v\} \in E : u \in V_i, v \in V_j\}$. The set of cut edges is also called *edge separator*. The *k-node separator problem* asks to find $k + 1$ blocks, $V_1, V_2, \ldots, V_k$ and a separator S, that partition V, such that there are no edges between the blocks. Again, a balancing constraint demands $|V_i| \leq (1 + \epsilon)\lceil |V|/k \rceil$. However, there is no balancing constraint on the separator S. The objective is to minimize the size of the separator $|S|$. Note that removing the set S from the graph results in at least k connected components and that the blocks V_i itself do not need to be connected components. By default, our initial inputs will have unit edge and node weights.

2.2 Detailed Related Work

We now discuss algorithmical details of the algorithm by Andrade et al. [1]. We call the algorithm ARW as an abbreviation for Andrade, Resende and Werneck. While we compare our algorithm against ARW, we also use it within our algorithm to improve newly created offsprings. Moreover, we shortly present the KaHIP graph partitioning framework since we use it to compute partitions and node separators.

ARW. One *iteration* of the ARW algorithm consists of a perturbation and a local search step. The ARW *local search* algorithm uses simple 2-improvements or $(1, 2)$-swaps to gradually improve a single current solution. A (j, k)-swap removes j nodes from the solution and then inserts k new nodes into it. A $(1, 2)$-swap in particular removes a single node from the solution and adds two other free nodes. A node is called *free*, if none of its neighbouring nodes can be found in the current solution. The *tightness* of a node $\tau(v)$ is the number of neighbouring solution nodes. Hence, free nodes have zero tightness. The simple version of the local search algorithm then iterates over all solution nodes of the graph and looks for a $(1, 2)$-swap. It is shown, that this procedure can find a valid $(1, 2)$-swap in linear time $\mathcal{O}(m)$, if it exists. This is achieved by using a data structure that allows insertion and removal operations on nodes in time proportional to their degree. The data structure basically divides the nodes into solution nodes, free nodes and non-free non-solution nodes. The *perturbation step* used for diversification, forces nodes into the solution and removes neighboring nodes as necessary. In most cases, one node is forced into the solution per iteration. With a small probability the number of forced nodes f is set to higher value: f is set to $i + 1$ with probability $1/2^i$. Moreover, the current node to be forced into a solution is picked from a number of random candidates. Among those candidates the vertex that has been outside the solution for the longest time is picked. We refer the reader to original paper for more details about the ARW algorithm [1]. There is also an even faster incremental version of the algorithm that maintains a list of candidates. We use this version of the algorithm in our framework.

KaHIP. Karlsruhe High Quality Partitioning – is a family of graph partitioning programs that tackle the balanced graph partitioning problem [22,23]. The algorithms in KaHIP have been able to compute the best results in various benchmarks. It implements different sequential and parallel algorithms to compute k-way partitions and node separators. In this work, we use the sequential multilevel graph partitioner KaFFPa (Karlsruhe Fast Flow Partitioner) to obtain partitions and separators for the graphs. In particular, we use specialized partitioning techniques based on multi-level size-constrained label propagation [19].

3   Evolutionary Components

We now discuss the main contributions of the paper. We begin by outlining the general structure of our evolutionary algorithm and then explain how we build the initial population. Finally, we present our new combine operations and the methods we use formutation.

Algorithm 1. Steady State Evolutionary Algorithm with Local Search

> create initial population P
> **while** stopping criterion not fulfilled
>   *select* parents $\mathcal{I}_1, \mathcal{I}_2$ from P
>   *combine* $\mathcal{I}_1$ with $\mathcal{I}_2$ to create offspring O
>   *ARW local search+mutation* on offspring O
>   *evict* individual in population using O
> **return** the fittest individual that occurred

3.1 General Structure

As previous work [3,6] we use bitstrings as a natural way to represent solutions in our population. More precisely, an independent set $\mathcal{I}$ is represented as an array $s = \{0,1\}^n$ where $s[v] = 1$ if and only if $v \in \mathcal{I}$. The general structure of our evolutionary algorithm is very simple. Our algorithm starts with the creation of a population of individuals (in our case independent sets in the graph) and evolves the population into different populations over several rounds until a stopping criterion is reached.

In each round, our evolutionary algorithm uses a selection rule that is based on the fitness of the individuals (in our case the size of the independent set) of the population to select good individuals and combine them to obtain improved offspring. In contrast to previous work [3,6], our combine and mutation operators always create valid independent sets. Hence, we use the size of the independent set as a fitness function. That means that there is no need to use a penalty function to ensure that the final individuals generated by our algorithm are independent sets. As we will see later when an offspring is generated it is possible that it is a non-maximal independent set. Hence, we apply one iteration of ARW local search without the perturbation step to ensure that it is locally maximal and apply a mutation operation to the offspring. We use mutation operations since it is of major importance to keep the diversity in the population high [2], i.e. the individuals should not become too similar, in order to avoid a premature convergence of the algorithm.

We then use an eviction rule to select a member of the population and replace it with the new offspring. In general one has to take both into consideration, the fitness of an individual and the distance between individuals in the population [2]. Our algorithm evicts the solution that is *most similar* to the newly computed offspring among those individuals of the population that have a smaller or equal objective than the offspring itself. Once an individual has been accepted into the population we further refine it using additional iterations of the ARW algorithm. The general structure of our evolutionary algorithm follows the steady-state approach [8] which generates only one offspring per generation. We give an outline in Algorithm 1.

3.2   Initial Solutions

We use three different approaches to create initial solutions. Each time we create an individual for the population we pick one of the approaches uniformly at random. The first and most simplistic way is to start from an empty independent set and add nodes at random until no further nodes can be added. To ensure that adding a node results in a valid independent set we have to check if the node is free. We do this by simply checking if any of the surrounding nodes is already in the set. The method adds a decent amount of diversity during the construction phase, which over an extended period of time can lead to good solutions.

Secondly, we use a greedy approach similar to Andrade et al. [1]. Starting from an empty solution, we always add the node with the least residual degree which is the number of free neighbors. After a node is added to the solution, we remove all its neighbouring nodes from the graph and update the residual degree of their neighbors. We repeat the procedure until no further node can be added. The implementation is done using a simple bucket priority queue which groups nodes into buckets based on their residual degree. This allows us to pick a random node each time multiple nodes share the same residual degree.

The last approach that we use to create initial solutions is also a greedy one. Here, we take a detour and generate an independent set by computing a vertex cover. We first create a vertex cover and then compute its complement to get an independent set. The algorithms also starts with an empty solution and then always adds the node that will cover the most currently uncovered edges. We repeat this until all edges are covered and then return the corresponding independent set. Note that the two greedy algorithms compute different independent sets (e.g. consider a path with five nodes). While the first approach always maintains an independent set and tries to improve it, the second approach can only return an independent set once the algorithm has terminated.

3.3   Combine Operations

We perform different kinds of combine operations which are all based on graph partitioning. The main idea of our operators is to use a partition of the graph to exchange whole blocks of solution nodes. In *general* our combination operators try to generate new independent sets that are not necessarily maximal. We then perform a maximization step that adds as many free nodes as possible. Afterwards, we apply a single iteration of the ARW local search algorithm to ensure that our solution is locally optimal. Depending on the type of the operator, we use a node separator or an edge separator of the graph that has been computed by the graph partitioning framework KaHIP. As a side note, small edge or node separators are vital for our combine operations to work well. This is due to the fact that large separators in the combine operations yield offsprings that are far from being maximal. Hence, the maximization step performs lots of fixing and the computed offspring is not of high quality. This is supported by experiments presented in Section 4.1.

The first and the second operator need precisely two input solutions while our third operator is a multi-point combine operator – it can take multiple input solutions. In the first case, we use a simple tournament selection rule [20] to determine the inputs, i.e. $\mathcal{I}_1$ is the fittest out of two random individuals r_1, r_2 from the population. The same is done to select $\mathcal{I}_2$. Note that due to the fact that our algorithms are randomized, a combine operation performed twice using the same parents can yield a different offspring.

Node Separator Combination. In its simplest form, the operator starts by computing a node separator $V = V_1 \cup V_2 \cup S$ of the input graph. We then use S as a crossover point for our operation. The operator generates two offsprings. More precisely, we set $O_1 = (V_1 \cap \mathcal{I}_1) \cup (V_2 \cap \mathcal{I}_2)$ and $O_2 = (V_1 \cap \mathcal{I}_2) \cup (V_2 \cap \mathcal{I}_1)$. In other words, we exchange whole parts of independent sets from the blocks V_1 and V_2 of the node separator. Note that the exchange can be implemented in time linear in the number of nodes. Recall that the definition of a node separator implies that there are no edges running between V_1 and V_2. Hence, the computed offsprings are independent sets, but may not be maximal since separator nodes have been ignored and potentially some of them can be added to the solution. We maximize the offsprings by using the greedy independent set algorithm from Section 3.2. The operator finishes with one iteration of the ARW algorithm to ensure that we reached a local optimum and to add some diversification. An example illustrating the combine operation is shown in Figure 1.

Edge Separator Combination. This operator computes offsprings by taking a detour over vertex covers. It starts by computing a bipartition $V = V_1 \cup V_2$ of the graph. Let C_i be the vertex cover $V \backslash \mathcal{I}_i$. We define temporary vertex cover offsprings similar to before: $D_1 = (C_1 \cap V_1) \cup (C_2 \cap V_2)$ and $D_2 = (C_1 \cap V_2) \cup (C_2 \cap V_1)$. Unfortunately, it is possible that an offspring created this way contains some non-covered edges. These edges can only be a subset of the cut edges of the partition. We want to add as little nodes as possible to our solution to fix this. Hence, we add a minimum vertex cover of the bipartite graph induced by the non-covered cut edges to our vertex cover offspring. The minimum vertex cover in a bipartite graph can be computed using the Hopcroft-Karp algorithm. Afterwards, we transform the vertex cover back to an independent set, and follow our general approach by applying ARW local search to reach a local optimum.

Multi-way Combination. Our last two operators are multi-point crossover operators that extend the previous two operators. Both of them divide the graph into a number of blocks k. Depending on the type of the operator, a node or edge separator is used. We start with the description of the node separator approach where $V = V_1 \cup \ldots \cup V_k \cup S$. The operator selects a number of parents. We then calculate the score for *every* possible pair of a parent $\mathcal{I}_i$ and a block V_j. The score of a pair is the number of the parents solution nodes inside the given block. We then select the parent with the *highest score* for each of the blocks to compute the offspring. As before, since we left out the separator nodes we use

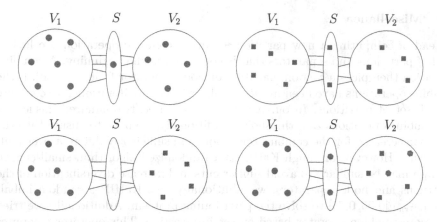

Fig. 1. An example combine operation using a node separator $V = V_1 \cup V_2 \cup S$. On top two input individuals/independent sets, $\mathcal{I}_1$ and $\mathcal{I}_2$, are shown. Bottom left: a possible offspring that uses the independent set of $\mathcal{I}_1$ in block V_1 and the independent set of $\mathcal{I}_2$ in block V_2. Bottom right: the improved offspring after ARW local search has been applied to improve the given solution and to add nodes from the separator to the independent set.

a maximization step to make the solution maximal and afterwards apply ARW local search to ensure that our solution is a local optimum.

If we use an edge separator for the combination, we start with a k-way partition of the nodes $V = V_1 \cup \ldots \cup V_k$. This approach also computes scores for each pair of parent and block. This time the score of a pair is defined as the number of the vertex cover nodes of the complement of an independent set inside the given block. We then select the parent with the *lowest score* for each of the blocks to compute the offspring. As in the simple vertex cover combine operator, it is possible that some cut edges are not covered. We use the simple greedy vertex cover algorithm to fix the offspring since the graph induced by the non-covered cut edges is not bipartite anymore. We then once again complement our vertex cover to get our final offspring.

3.4 Mutation Operations

After we performed a combine operation, we apply a mutation operator to introduce further diversification. Previous work [3,6] uses bit-flipping for mutation, i.e. every bit in the representation of a solution has a certain probability of being flipped. We can not use this approach since our population only allows valid solutions. Instead we perform forced insertions of new nodes into the solution and remove adjacent solution nodes if necessary as in the perturbation routine of the ARW algorithm. Afterwards we perform ARW local search to improve the perturbed solution.

3.5 Miscellanea

Instead of computing a new partition for every combine operation, we hold a pool of partitions and separators that is computed in the beginning. A combine operation then picks a random partition or node separator from the pool. If the combine operations have been unsuccessful for too many iterations, we compute a fresh set of partitions. In our experiments we used two-hundred unsuccessful combine operations as a threshold. Additionally, we have to ensure that the partitions created for the combine operations are sufficiently different over multiple runs. However, although KaHIP is a randomized algorithm, small cuts in a graph may be similar. To avoid similar cuts and increase diversification of the partitions and node separators, we additionally give KaHIP a random imbalance $\epsilon \in_{\mathrm{rnd}} [0.05, 0.75]$ to solve the partitioning problem. Additionally, we tried one more combine operator based on set intersection. This operator computes an offspring by keeping the nodes that are in both inputs which is by definition an independent set. However, our experiments with the operator did not yield good results so that we omit further investigations here.

4 Experimental Evaluation

Methodology. We have implemented the algorithm described above (EvoMIS) using C++ and compiled all algorithms using gcc 4.63 with full optimization's turned on (-O3 flag). We mainly compare our algorithm against the ARW algorithm since it has a relatively clear advantage in Resende et al. [1]. The algorithm by Grosso et al. [14] has originally been formulated for the maximum clique problem. Andrade et al. [1] used an implementation of the algorithm for the maximum independent set problem. Hence, we also compare against the results of the algorithm by Grosso et al. presented in the paper of Andrade et al. [1]. Additionally, we compare ourselves with our implementation of the evolutionary algorithm presented by Bäck and Khuri [3].

Unless otherwise mentioned, we perform five repetitions where each algorithm that we run gets ten hours of running time to compute a solution. Each run was made on a machine that is equipped with two Quad-core Intel Xeon processors (X5355) which run at a clock speed of 2.667 GHz. It has 2x4 MB of level 2 cache each, 64 GB main memory and runs Suse Linux Enterprise 10 SP 1. We used the fastsocial configuration of the KaHIP v0.6 graph partitioning package [22] to obtain graph partitions and node separators. The test results for the ARW algorithm were obtained by using the original algorithm from Andrade et al. [1]. Within the evolutionary algorithm we used our own implementation of the ARW algorithm.

We mostly present two kinds of data: maximum values, average values, minimum values as well as plots that show the evolution of solution quality. We now explain how we compute the convergence plots. Whenever an algorithm creates a new best independent set S it reports a tuple $(t, |S|)$, where the time stamp t is the currently elapsed time and $|S|$ refers to the size of the independent set that has been created. Since we perform multiple repetitions, the final plots correspond to average values over these repetitions. To compute these we take the time stamps of all repetitions and sort them in ascending order. For each time stamp in this series, we report the average value of the best solution size of each repetition at that time.

Algorithm Configuration. After an extensive evaluation of the parameters [17], we fixed the population size to two hundred fifty, the partition pool size to thirty, the number of ARW iterations to 15 000 as well as the number of blocks used for the multi-way combine operations to sixty-four. In each iteration, one of our three combine operations is picked uniformly at random. However, our experiments indicate that our algorithm is not too sensitive about the precise choice of the parameters. We mark the instances that have also been used for the parameter tuning in [17] in the TR [18] with a *.

Instances. We use graphs from various sources to test our algorithm. We divide them into five categories: social networks, meshes, road networks, networks from finite element computations as well as networks stemming from matrices. Social networks include citation networks, autonomous systems graphs or web graphs taken from the 10th DIMACS Implementation Challenge benchmark set [4]. Road networks and meshes are taken from Andrade et al. [1] and have been kindly provided by Renato Werneck. Meshes are dual graphs of triangular meshes. Networks stemming from finite element computations have been taken from Chris Walshaw's benchmark archive [24]. Graphs stemming from matrices have been taken from the Florida Sparse Matrix Collection [7]. We randomly selected one from each group of all real, symmetric matrices having between 10K and 65K columns. A graph is derived by inserting a node for each column and creating an edge between two nodes u, v if the corresponding matrix entry is non-zero. Singletons and self-loops are removed from the graphs.

4.1   Main Results

We now shortly summarize the main results of our experiments. First of all, in 50 out of the 67 instances, we either improve or reproduce the maximum result computed by the ARW algorithm. Our algorithm computes a maximum solution that is strictly larger than the maximum solution computed by the ARW algorithm in 21 cases. Contrarily, in 17 cases the maximum result of the ARW algorithm is larger then the maximum result of our algorithm. When looking at average values, we get 23 cases in which our algorithm strictly outperforms the ARW algorithm, and 17 cases for the opposite direction. Remarkably, when looking at the graphs obtained from the Florida Sparse Matrix collection, the average value of the ARW algorithm only outperforms our algorithm on one instance. The mesh family that we use in this paper has also been used in the original ARW paper [1]. We like to stress that most of the maximum results of the ARW algorithm are strictly larger than the maximum values originally reported by Andrade et al. [1] (including the maximum values presented there of the algorithm by Grosso et al. [14]). Except for four instances the same holds for our algorithm. On these four instances, our algorithm is worse than the original maximum value of the ARW algorithm. On the mesh family, in 8 out of 14 cases our algorithm computes the best result ever reported in literature. On road networks and the largest graphs from the mesh family as well as Walshaw family the ARW algorithm outperforms our algorithm. We tried to give both algorithms more time, i.e. a whole day of computation, but did not see much different results. Lastly, there is an interesting observation on social networks, that is in 5 out of 9 cases the minimum, average and maximum result produced by both algorithms are precisely the same. We suspect that these instances are in a sense easy and that both algorithms compute the optimal result or are very close to the optimum. We provide detailed per instance results in the TR [18].

Figure 2 shows how solution quality evolves over time on four example instances from the mesh family for both algorithms. As one would suspect, our algorithm almost keeps its level of solution quality in the beginning since it has to build the full population before it can start with combine and mutation operations. Contrarily, the ARW algorithm can directly start with local search and improve its solution. Hence, the solution quality of the ARW algorithm rises above the solution quality of our algorithm. As soon as our algorithm finished to

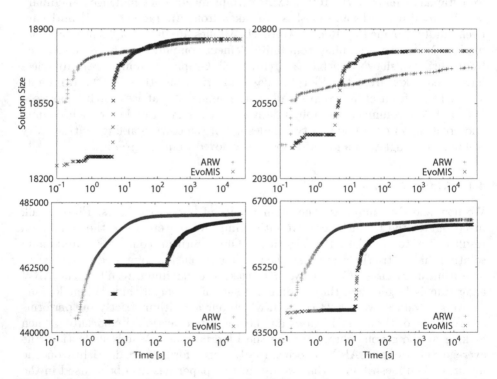

Fig. 2. Solution size evolving over time for four instances from the mesh family: `feline` and `gameguy` [top] as well as `buddha` and `dragon` [bottom].

compute the population, solution quality starts to improve and eventually the size of the computed independent sets becomes better than the solution quality of the ARW algorithm.

We also implemented the algorithm presented by Bäck and Khuri [3]. The algorithm uses a two-point crossover as a combine operation, as well as a bit-flip approach for mutation. Solutions created by the combine and mutation operations can be invalid. Hence, a penalty approach is used to deal with invalid solution candidates. In the original paper, the algorithm is only tested on small synthetic or random instances (≤ 200 nodes). We tested the algorithm on the four smallest graphs from the mesh family and gave the algorithm ten hours of time to compute a solution. However, the best valid solution created during the course of the algorithm *never* exceeded the size of the best solution after the initial population has been created. This is due to the fact that the two-point crossover and the mutation operations found valid solutions very rarely so that the average solution quality of the population degrades over time. On average, final solution quality of the algorithm has been more than 20% worse than the final result of our algorithm. Due to the bad solution quality observed, we did not perform additional experiments with this algorithm.

The Role of Graph Partitioning. To estimate the influence of good partitionings in this context, we performed an experiment in which partitions of the graph have been obtained by simple breadth first searches. More precisely, we obtain a two-way partition of the graph using a breadth first search starting from a random node. Every node touched by the breadth first search is added to the first block, and every node not touched by the breadth first search is added to the second block. The breadth first search is stopped as soon as a specified number of nodes has been touched. In our experiments, using this approach instead of the approach that uses a graph partitioner to compute a partition yields significantly worse results. The influence of all the different combine operators that we use here is presented in the thesis [17].

5   Conclusion

We presented a very natural evolutionary framework for the computation of large maximal independent sets. Our core innovations are combine operations that are based on graph partitioning and local search algorithms. More precisely, our combine operations enable us to quickly *exchange whole blocks* of given individuals. In contrast to previous evolutionary algorithms for the problem, our operators are able to guarantee that the created offspring is valid. Experiments indicate that our algorithms outperforms state-of-the-art algorithms on a large variety of instances – some of which are better than ever reported in literature. Important future work includes a coarse-grained parallelization of our approach which can be done by using an island-based approach. Moreover, it would be interesting to improve the solution quality of our approach on road networks and to compare our algorithms with exact approaches. Additionally, it would be interesting to overcome the slow start of our algorithm due to the initialization of the population. For example, one could try to adjust the size of the population dynamically.

Acknowledgments. We would like to thank Renato Werneck for providing us the source code of the local search algorithms presented in Andrade et al. [1]. Moreover, we thank the Steinbuch Centre of Computing for giving us access to the IC2 machine.

References

1. Andrade, D.V., Resende, M.G.C., Werneck, R.F.: Fast Local Search for the Maximum Independent Set Problem. J. Heuristics **18**(4), 525–547 (2012)
2. Bäck, T.: Evolutionary Algorithms in Theory and Practice: Evolution Strategies, Evolutionary Programming, Genetic Algorithms. Ph.D thesis (1996)
3. Bäck, T., Khuri, S.: An evolutionary heuristic for the maximum independent set problem. In: Proc. 1st IEEE Conf. on Evolutionary Computation, pp. 531–535. IEEE (1994)
4. Bader, D., Kappes, A., Meyerhenke, H., Sanders, P., Schulz, C., Wagner, D.: Benchmarking for graph clustering and partitioning. In: Encyclopedia of Social Network Analysis and Mining. Springer (2014)
5. Battiti, R., Protasi, M.: Reactive Local Search for the Maximum Clique Problem. Algorithmica **29**(4), 610–637 (2001)
6. Borisovsky, P.A., Zavolovskaya, M.S.: Experimental comparison of two evolutionary algorithms for the independent set problem. In: Raidl, G.R., Cagnoni, S., Cardalda, J.J.R., Corne, D.W., Gottlieb, J., Guillot, A., Hart, E., Johnson, C.G., Marchiori, E., Meyer, J.-A., Middendorf, M. (eds.) EvoIASP 2003, EvoWorkshops 2003, EvoSTIM 2003, EvoROB/EvoRobot 2003, EvoCOP 2003, EvoBIO 2003, and EvoMUSART 2003. LNCS, vol. 2611, pp. 154–164. Springer, Heidelberg (2003)
7. Davis, T.: The University of Florida Sparse Matrix Collection
8. De Jong, K.A.: Evolutionary Computation: A Unified Approach. MIT Press (2006)
9. Feo, T.A., Resende, M.G.C., Smith, S.H.: A Greedy Randomized Adaptive Search Procedure for Maximum Independent Set. Operations Research **42**(5), 860–878 (1994)
10. Garey, M.R., Johnson, D.S.: Computers and Intractability: A Guide to the Theory of NP-Completeness. W. H. Freeman (1979)
11. Gemsa, A., Niedermann, B., Nöllenburg, M.: Trajectory-based dynamic map labeling. In: Cai, L., Cheng, S.-W., Lam, T.-W. (eds.) Algorithms and Computation. LNCS, vol. 8283, pp. 413–423. Springer, Heidelberg (2013)
12. Goldberg, D.E.: Genetic Algorithms in Search, Optimization, and Machine Learning. Addison-Wesley (1989)
13. Grosso, A., Locatelli, M., Della Croce, F.: Combining Swaps and Node Weights in an Adaptive Greedy Approach for the Maximum Clique Problem. J. Heuristics **10**(2), 135–152 (2004)
14. Grosso, A., Locatelli, M., Pullan, W.: Simple Ingredients Leading to Very Efficient Heuristics for the Maximum Clique Problem. J. Heuristics **14**(6), 587–612 (2008)
15. Hansen, P., Mladenović, N., Urošević, D.: Variable Neighborhood Search for the Maximum Clique. Discrete Applied Mathematics **145**(1), 117–125 (2004)
16. Katayama, K., Hamamoto, A., Narihisa, H.: An Effective Local Search for the Maximum Clique Problem. Inf. Proc. Letters **95**(5), 503–511 (2005)
17. Lamm, S.: Evolutionary Algorithms for Independent Sets. Bachelor's Thesis, KIT (2014)
18. Lamm, S., Sanders, P., Schulz, C.: Graph Partitioning for Independent Sets. Technical Report arxiv:1502.01687 (2015)

19. Meyerhenke, H., Sanders, P., Schulz, C.: Partitioning complex networks via size-constrained clustering. In: Gudmundsson, J., Katajainen, J. (eds.) SEA 2014. LNCS, vol. 8504, pp. 351–363. Springer, Heidelberg (2014)
20. Miller, B.L., Goldberg, D.E.: Genetic Algorithms, Tournament Selection, and the Effects of Noise. Evolutionary Computation 4(2), 113–131 (1996)
21. Pullan, W.J., Hoos, H.H.: Dynamic Local Search for the Maximum Clique Problem. J. Artif. Intell. Res. (JAIR) 25, 159–185 (2006)
22. Sanders, P., Schulz, C.: KaHIP - Karlsruhe High Qualtity Partitioning Homepage. http://algo2.iti.kit.edu/documents/kahip/index.html
23. Sanders, P., Schulz, C.: Think locally, act globally: highly balanced graph partitioning. In: Demetrescu, C., Marchetti-Spaccamela, A., Bonifaci, V. (eds.) SEA 2013. LNCS, vol. 7933, pp. 164–175. Springer, Heidelberg (2013)
24. Soper, A.J., Walshaw, C., Cross, M.: A Combined Evolutionary Search and Multilevel Optimisation Approach to Graph-Partitioning. J. of Global Optimization 29(2), 225–241 (2004)

Finding Connected Subgraphs of Fixed Minimum Density: Implementation and Experiments

Christian Komusiewicz, Manuel Sorge$^{(\boxtimes)}$, and Kolja Stahl

Institut für Softwaretechnik und Theoretische Informatik, TU, Berlin, Germany
{christian.komusiewicz,manuel.sorge}@tu-berlin.de,
kolja.stahl@campus.tu-berlin.de

Abstract. We consider the following problem. Given a graph and a rational number μ, $0 < \mu \leq 1$, find a connected subgraph of density at least μ with the largest number of vertices. Here, the density of an n-vertex graph with m edges is $m/\binom{n}{2}$. This problem arises in many application contexts such as community detection in social networks. We implement a branch and bound algorithm and tune it for efficiency on sparse real-world graphs for the case $\mu \geq 1/2$. Central issues for the implementation are the choice of branching candidates, two new upper bounding procedures, and several data reduction and early termination rules.

1 Introduction

Identifying dense subgraphs is a problem arising in the analysis of social [4], financial [5], and biological networks [3]. In most applications, the desired dense subgraphs do not contain an edge between each vertex pair but rather adhere to a more relaxed notion of density. Many different, mathematically precise definitions of such desired subgraphs have been proposed [4,9]. We consider the concept of μ-cliques, used for example by Abello et al. [1,2]. It is defined as follows.

Definition 1. *The* density *of an n-vertex graph with m edges is $m/\binom{n}{2}$. A graph is a μ-clique if its density is at least μ.*

In general, μ-cliques need not be connected. However, this is an important property expected from a community. Hence, we impose connectivity as a further constraint on the μ-cliques we are looking for. As observed previously, demanding connectivity also allows for a simple solving algorithm [8].

Our goal in this work is to develop an implementation for finding large connected μ-cliques in a given graph for some fixed $\mu \geq 1/2$. Most input graphs in the mentioned applications are sparse with few high-degree vertices [5,13]. We thus aim to tune the implementation to perform well on graphs with this structure. Our implementation is based on an *exact* algorithm which, given k, either finds a μ-clique with k vertices or determines correctly that no such subgraph exists. Exact algorithms are desirable because they yield reference points for the performance of heuristics and because surprising results can be attributed to the model (here: connected μ-cliques) rather than to deficiencies of the algorithm.

© Springer International Publishing Switzerland 2015
E. Bampis (Ed.): SEA 2015, LNCS 9125, pp. 82–93, 2015.
DOI: 10.1007/978-3-319-20086-6_7

Contribution. Our implementation follows the branch and bound paradigm and is based on an algorithm proposed by a subset of the authors [8]. The input is a graph G, the density threshold μ, and the minimum required number k of vertices in the desired μ-clique. The algorithm proceeds roughly as follows. In each step, we maintain a set P of vertices which we aim to extend to a μ-clique. To do this, we maintain also an *active vertex* v whose neighbors we will consider to add to P first. That is, given P and v we branch into all possibilities of adding a neighbor of v to P, and into the possibility of making v permanently inactive and consequently choosing a new active vertex in P. We terminate this process if P has size k and report $G[P]$ if it is a connected μ-clique.

This algorithm is called for increasing values of k. If for some value of k no connected μ-clique is found, then it stops and returns the largest μ-clique computed so far. This approach is only correct if the nonexistence of a μ-clique of order k implies that there is also no μ-clique of order $k + 1$. In a first step, we thus examine whether connected μ-cliques fulfill a nestedness property (which is called *quasi-heredity* [11]). We obtain that for $\mu \geq 1/2$, connected μ-cliques are quasi-hereditary, but that for $\mu < 1/2$, they cannot be assumed to be quasi-hereditary. Accordingly, we focus on the case $\mu \geq 1/2$ in our experiments.

We develop several approaches to improve the running time of the above algorithm and we detail them in Sections 3 and 4: First, we consider upper bounds on the density we can achieve when we are given P. If the upper bound is smaller than the given μ, then we can terminate branching early. We modify a known upper bound [10], obtaining two new variants. Second, we develop scoring functions to determine which vertex should be chosen as active vertex and which of its neighbors should be included into P first, so to quickly find solutions. Finally, we also employ several further "early termination" rules (either finding a connected μ-clique of the desired order or deciding that there is none), improved branching rules, as well as several heuristic tricks that speed up the computation of the upper bounds, for example.

In Section 5 we report our experimental findings. Briefly, we find that our branching approach for connected μ-cliques is competitive with the state of the art algorithm for possibly disconnected μ-cliques. The upper bound k imposed on the solution order (which is incremented until we face a no-instance) seems to be crucial to limit the search space. Using this approach we find optimal connected μ-cliques for several real-world instances for the first time. Furthermore, we find that a very simple bound performs best, since the upper bounds are often applied without avail. Due to lack of space, we defer proofs to a full version of this article.

Related Work. Finding a μ-clique of order k in a given graph is a decision version of DENSEST k-SUBGRAPH, where we seek to find a k-vertex subgraph with the maximum number of edges. This problem is NP-hard even on graphs with maximum degree three [7]. Moreover, it is W[1]-hard with respect to k [6] and thus unlikely to be solvable in $f(k) \cdot n^{O(1)}$ time. Under the Unique Games Conjecture there is no polynomial-time constant-factor approximation algorithm [12]. Finding μ-cliques with k vertices remains NP-hard for every rational μ [11]. On the positive side, finding μ-cliques of maximum order is tractable on graphs with small maximum degree and on graphs with few high-degree vertices [8].

We are aware of two experimental studies for finding large μ-cliques via exact algorithms. Pattillo et al. [11] develop two mixed-integer programming (MIP) formulations for this problem, which were used to solve several real-world instances with up to 154 vertices with CPLEX. Pajouh et al. [10] instead implemented an algorithm which implicitly enumerates vertex subsets. They developed an easy-to-compute upper bound for the number of edges induced by any extension of a vertex set P to one with k vertices. Their algorithm seems to be the state of the art, improving on the MIP formulation in almost all test instances. Hence, we use it as a main reference point here. Note that, in contrast to our algorithm, both algorithms may report *disconnected* μ-cliques.

There is also a large body of work on heuristic algorithms for finding μ-cliques (see [1,2,14], for example) as well as heuristics and exact algorithms for other concepts of dense subgraphs (see Balasundaram and Pajouh [4] for a survey).

Preliminaries. We consider only undirected and simple graphs $G = (V, E)$ where $V = V(G)$ denotes the vertex set and $E = E(G)$ denotes the edge set. Unless stated otherwise, n denotes the number of vertices, also called *order* of the graph, and m the number of edges of G. The open neighborhood of a vertex v is denoted by $N(v)$. The degree of a vertex v is denoted by $\deg(v) := |N(v)|$. For a vertex set $S \subseteq V$, we use $N_S(v) := N(v) \cap S$ and $\deg_S(v) := |N_S(v)|$ to denote the neighborhood and degree restricted to S. Furthermore, we use $G[S] := (S, \{\{u,v\} \in E \mid \{u,v\} \subseteq S\})$ to denote the subgraph of G *induced by* S. The *degeneracy* of a graph G is the smallest integer d such that every subgraph of G has a vertex of degree at most d.

2   Connected μ-Cliques and Quasi-Heredity

We now study some properties of connected μ-cliques. The property of being a μ-clique, without the connectivity constraint, is not hereditary [9,11]. That is, there are μ-cliques G such that some induced subgraph of G is not a μ-clique. Being a μ-clique is, however, quasi-hereditary, that is, every μ-clique G or order n has an induced subgraph of order $n-1$ which is a μ-clique. This is implied by the following which slightly extends [9, Proposition 6.3.2] and [11, Proposition 2].

Lemma 1. *Let* $G = (V, E)$ *be a graph with density exactly* μ *and let* v *be a vertex in* G. *Then,* $G[V \setminus \{v\}]$ *has density at least* μ *if and only if* $\deg(v) \leq 2m/n$.

Thus, removing a vertex of minimum degree in a μ-clique yields a μ-clique, implying the quasi-heredity of μ-cliques.

The argument for μ-cliques does not extend easily to *connected* μ-cliques: it could be the case that all vertices v with $\deg(v) \leq 2m/n$ are cut-vertices. Moreover, it is not hard to check that additionally demanding connectedness does not yield a hereditary graph property (consider a clique with a degree-one vertex attached to it). Thus, it is interesting to know whether connected

μ-cliques are at least quasi-hereditary. Somewhat surprisingly, this depends on μ: for large μ we observe quasi-heredity whereas for small μ this is impossible.

Theorem 1. *If $\mu \geq 1/2$, then "being a connected μ-clique" is quasi-hereditary.*

In contrast, for $\mu < 1/2$, we obtain a family of counterexamples, showing that we can use quasi-heredity safely only when $\mu \geq 1/2$.

Theorem 2. *For any fixed rational $\mu = a/b$ such that $0 < \mu < 1/2$ and b is odd, "being a connected μ-clique" is not quasi-hereditary.*

3   Upper Bounds

In this section we detail several upper bounds that are used in the algorithm. We start with a previously known upper bound on the order of the μ-clique that depends on the number of edges m and number of vertices n in the graph G.

Proposition 1 (Edge bound [11]). *If $G[S]$ is a μ-clique in a connected graph G, then $|S| \leq \left(\mu + 2\sqrt{(\mu + 2)^2 + 8(m - n)\mu}\right)/2\mu$.*

This upper bound obviously also applies to connected μ-cliques. In the course of the algorithm, some vertices of the input graph G are discarded in some recursive branches. Thus m and n decrease in these branches and the bound may then show that no μ-clique of order k exists. While the bound is easy to compute, it rarely leads to early termination.

The following bounds are based on the strategy to gradually extend the "pivot" set P. The aim is to decide whether it is still possible to extend P to a μ-clique of order k. In the following, let $\ell := k - |P|$ denote the number of vertices that we still need to add. Moreover, for a vertex set $S \subseteq V$ let $\mathrm{m}(S)$ denote the number of edges in $G[S]$. Pajouh et al. [10] proved the following.

Proposition 2 (Inner P-bound [10]). *Let $G = (V, E)$ be a graph and $P \subseteq V$ a vertex subset. Then, for any $S \supseteq P$ with $|S| - |P| = \ell$, we have*

$$\mathrm{m}(S) \leq \mathrm{m}(P) + \frac{1}{2} \sum_{v \in P} \min\{\deg_{V \setminus P}(v), \ell\}$$

$$+ \frac{1}{2} \sum_{i=1}^{\ell} \deg_P(v_i) + \min\{\deg_{V \setminus P}(v_i), \ell - 1\},$$

where $v_1, \ldots, v_\ell \in V \setminus P$ exhibit the largest values of

$$(\deg_P(v_i) + \min\{\deg_{V \setminus P}(v), \ell - 1\})/2.$$

Note that the degree of the vertices in P can be large, and hence, the sum over all $v \in P$ does not make a good estimate on the number of edges between P and $S \setminus P$ in this case. We now aim to make this estimate from "outside" of P instead. This often yields a better bound because $|P|$ is usually relatively small in the course of the algorithm.

Proposition 3 (Outer P-bound). *Let $P \subseteq V$ be a vertex set in $G = (V, E)$. Then for any $S \supseteq P$ with $|S| - |P| = \ell$, we have*

$$m(S) \le m(P) + \sum_{i=1}^{\ell} \left(\deg_P(v_i) + \min\{\deg_{V\setminus P}(v_i), \ell - 1\}/2 \right),$$

where $v_1, \ldots, v_\ell \in V \setminus P$ exhibit the largest values of

$$\deg_P(v_i) + \min\{\deg_{V\setminus P}(v_i), \ell - 1\}/2.$$

By replacing the estimate of the edges contained in $G[S \setminus P]$ by the trivial upper bound $\binom{\ell}{2}$, we get the following.

Proposition 4 (Simple P-bound). *Let $P \subseteq V$ be a vertex set in $G = (V, E)$. Then for any $S \supseteq P$ with $|S| - |P| = \ell$, we have*

$$m(S) \le m(P) + \binom{\ell}{2} + \sum_{i=1}^{\ell} \deg_P(v_i)/2,$$

where $v_1, \ldots, v_\ell \in V \setminus P$ exhibit the largest values of $\deg_P(v_i)/2$.

While the simple P-bound is the least tight of these three P-bound variants, it is also the one with the least computational overhead. It thus is a crucial feature of our algorithm (see Section 5).

4 Algorithm and Heuristic Improvements

We now describe our algorithm in detail, including several heuristic speed-ups; a pseudocode is shown in Algorithm 1. As outlined in the introduction, we maintain a partial solution P throughout the execution of the algorithm as well as an active vertex v. Initially, P contains a single vertex (we try all possibilities). We successively either add a neighbor of v to P (trying all possibilities) or make v inactive, meaning that no further neighbors of v should be added to P. Inactive vertices are maintained in a set $I \subseteq P$. The procedure is terminated if P reaches size k or all vertices are inactive. As previously shown, this strategy finds a connected μ-clique with k vertices if there is one [8]. After each step of either adding a vertex to P or making a vertex inactive, we check whether the bounds from Section 3 imply that no k-vertex μ-clique containing P exists.

Our general strategy to find the largest μ-clique is to apply Algorithm 1 with successively increasing k. Due to quasi-heredity, once the algorithm asserts that there is no k-vertex μ-clique subgraph, then there is also none with more than k vertices. Next, we describe several speed-up tricks.

4.1 Simple Early Termination Rules and Improved Branching

The goal of the following modifications is to avoid branching (Line 9, Algorithm 1) if a solution can be obtained greedily or if some branches are symmetric to others that have been already explored.

Algorithm 1. Find μ-clique

Input: A graph G, $k \in \mathbb{N}$, $1/2 \leq \mu \leq 1$
Output: A connected μ-clique in G of order k if there is one, otherwise $\bot$.

1 **foreach** $v \in V(G)$ **do**
2 | Recurse$(G, \{v\}, \emptyset, v)$
3 | Remove v from G

4 **return** $\bot$

5 **Procedure** Recurse(G, P, I, a)
6 | **if** $|P| = k$ and $G[P]$ is a μ-clique **then return** P
7 | **if** $|P| = k$ and $G[P]$ is not a μ-clique **then** break
8 | **if** edge bound, simple P-bound, or outer P-bound are violated **then** break
9 | **foreach** $u \in N(a) \setminus P$ **do**
10 | Recurse$(G, P \cup \{u\}, I, a)$
11 | Remove u from G
12 | $I \leftarrow I \cup \{a\}$
13 | **if** $P = I$ **then** break
14 | Remove all vertices in $N(a) \setminus P$ from G, choose $w \in P \setminus I$, and set $a \leftarrow w$
15 | Recurse$(G, P, I \cup \{u\}, a)$

Simple Rules. We use two greedy termination rules. First, if at some time in Algorithm 1 the graph G is a connected μ-clique, then we can obtain a k-vertex μ-clique using Theorem 1 by greedily deleting a non-cut vertex of minimum degree. Second, if adding $k - |P|$ edges to $G[P]$ would yield a μ-clique, then it suffices to simply check whether the connected component containing P is large enough.

Pending Trees. The latter observation can be extended to any *pending tree* on P, that is, an induced tree T in G containing exactly one vertex v of P such that deleting v cuts T from the rest of the graph. We avoid branching on vertices in pending trees as follows. Adding ℓ vertices from such a tree to P adds exactly ℓ edges. Hence, any solution containing pending tree vertices is found by first branching on the vertices that are not in pending trees and then applying the simple check described above. Hence, after computing the set of all pending trees, we can restrict the branching step in Line 9 of Algorithm 1 to vertices *not* contained in any pending tree.

Twins. We call two vertices u and v *twins* if $N(u) \setminus \{v\} = N(v) \setminus \{u\}$. While we cannot assume that, if a vertex is in a solution, then also all its twins are, we do have the following property. Given $P \subseteq V(G)$, if there is no k-vertex μ-clique that contains P and a vertex $v \in V(G) \setminus P$, then there is no μ-clique containing P and any of the twins of v in $V(G) \setminus P$. Note that after Line 10 in Algorithm 1 we know that no k-vertex μ-clique containing u exists. Hence, we may not only remove u in Line 11, but also all its twins in $V(G) \setminus P$. In order to do this, we compute the set of twins for each vertex in the beginning. (Note that two twins in a graph remain twins after deleting any subset of vertices.)

Pre-evaluation of the Modifications. Since the simple rules above can be computed very quickly, we enabled them in all variants of Algorithm 1 we tested. Regarding pending trees and twins, we found that their benefits overlapped strongly in our benchmark instances of Section 5. That is, enabling both at the same time did not yield meaningful speed-up over the variants in which only one of them was enabled. Hence, we enabled only the twin modification, which showed a slightly greater reduction in calls to `Recurse`.

4.2   Order of Adding Vertices to P

We now consider the order in which vertices are added to the partial solution P in Lines 1 and 9 of Algorithm 1. Intuitively, for a yes-instance, we would like to order the vertices in such a way that a solution is discovered within only few branches. This approach is followed in our *optimistic* ordering. The optimistic ordering also serves as a greedy heuristic by determining which vertices to add in the first descent in the recursion. For a no-instance, however, it is better to add vertices to P that lead to sparse partial solutions, so that it can be easily determined that these vertices are not in a solution. Subsequently, these vertices will be removed in Lines 3 and 11, truncating the search space. This approach is followed in our *pessimistic* ordering.

Basic Optimistic Ordering. The optimistic ordering is based on two simple heuristics. The first heuristic, `MaxDegKeep`, starts with a highest-degree vertex and then recursively selects neighbors of already selected vertices with highest degree, until k vertices are selected. The second one, `MinDegDel`, instead removes vertices of minimum degree—omitting cut vertices —until only k vertices remain. In preliminary experiments we observed that the inequality $d \geq \Delta/10$ seems to be a good predictor on which of the two heuristics performed better. Here, Δ is the maximum degree of the input graph, and d is its degeneracy. If $d \geq \Delta/10$, then `MaxDegKeep` worked better and `MinDegDel` otherwise.

Based on the above observation we define score(v) for each vertex v and we first add vertices with the higher scores to P in Lines 1 and 9. If $d \geq \Delta/10$, then score(v) is simply the degree of v in the input graph. If $d < \Delta/10$, then score(v) is the largest degree encountered when deleting vertices of minimum degree from the input graph until v is deleted.

Breaking Ties. Most of our instances, and most of the instances we expect to be encountered in practice, fall into the "$d < \Delta/10$" category. Since often these graphs have thousands of vertices and small maximum score, many vertices receive the same score. Thus, we try to break ties by modifying the score. We tested two alternatives for tie-breaking: a) the number of neighbors with larger score, and b) the number of edges in the neighborhood of the vertex. Interestingly, pre-evaluation showed that a) performed worse than without tie breaking, increasing running times and calls to `Recurse`. Tie breaker b) showed improvements on some instances, so we opted to test only b) in Section 5.

Neighborhood-based Scoring. As the set P grows, it intuitively becomes more important to add many edges to $G[P]$ when adding vertices. Thus, in a variant of the vertex scoring, for each vertex $v \in V \setminus P$, we add $|N_P(v)|$ to score(v).

Pessimistic Ordering. The pessimistic ordering is obtained by essentially reversing the ordering given by the score of the vertices. That is, we first consider vertices, which we expect to not be in a μ-clique of order k. We break ties among them by considering first vertices with the fewest number of edges in the neighborhood. In the neighborhood-based scoring for the pessimistic variant, we score vertices with the fewest neighbors in P highest.

4.3 Application of the Upper Bounds

We now list several optimizations we employed for Line 8 of Algorithm 1.

– Since the edge bound and simple P-bound can be computed quickly, we determined in preliminary experiments that it is always better to enable both bounds. In particular, the simple P-bound has to be enabled in any good configuration of the algorithm. Thus, both bounds are always enabled in Section 5.

– The simple P-bound and outer P-bound rely on knowing the number of neighbors in P for each vertex outside of P. To amortize the corresponding computation cost, this information is kept and updated in each call to **Recurse**.

– The outer P-bound is based on certain values for each vertex. Then, from the ℓ largest of these values, it derives an upper bound on the density achievable in a k-vertex subgraph containing P. Compared to the trivial approach of computing all values, a considerable speed-up can be achieved by computing the values one-by-one, and only as long as the upper bound derived from the ℓ largest values computed so far still is below μ.

5 Implementation and Experiments

The algorithm described in Section 4 was implemented in Haskell and compiled using ghc version 7.4.1; the source code and test data is freely available, see http://fpt.akt.tu-berlin.de/connected-mu-clique. All experiments were run on an Intel Xeon E5-1620 computer with 4 cores at 3.6 GHz and 64 GB RAM. The operating system was Debian GNU/Linux 6.0. Our implementation does not use multiprocessing capabilities, however, up to four experiments were run on the machine at once (one on each core). Unless stated otherwise, the time limit was one hour.

We performed the following experiments. First, for $\mu = 0.7$, we compared all configuration variants of our algorithm in order to identify the best ones. The comparison is done on 25 real-world and benchmark instances. Then, we compare our algorithm to the one of Pajouh et al. [10] on a representative subset of the real-world instances for several values of μ. Finally, we perform experiments on random graphs to determine more precisely the limits of our algorithm and of the algorithm of Pajouh et al. [10].

5.1 Finding the Best Algorithm Variants

Our test bed consists of 25 networks overall. Of these networks, 12 are from the Second DIMACS Implementation Challenge, chosen to represent hard instances

Table 1. Reported μ-clique orders and running times (s) of the algorithm configurations across the test data set. The "# solved" column denotes the number of instances solved to optimality. Optimality is also indicated by a star on the μ-clique order. For any variant, the "# max k" column denotes the number of graphs where the largest μ-clique order was achieved among all variants. This is also indicated by a bold μ-clique order. A bold time means that this variant was the fastest among all variants that solved this instance.

	#max k	#solved	ERDOS-99-2	Human-all	GEOM-0	email-Enron	Acker-all	Acker-pc
(O)-(↑)-(B)	24	6	**15*** (2850.14)	25 (3600.0)	28 (3600.0)	58 (3600.0)	25 (3600.0)	**17*** (199.03)
(O)-(↑)-(B,N)	24	6	**15*** (2850.27)	25 (3600.0)	28 (3600.0)	58 (3600.0)	25 (3600.0)	**17*** (201.31)
(↑)-(B,N)	24	6	**15*** **(2809.09)**	25 (3600.0)	28 (3600.0)	58 (3600.0)	25 (3600.0)	**17*** (190.91)
(↓)-(B)	6	3	**15*** (3319.91)	16 (3600.0)	28 (3600.0)	20 (3600.0)	20 (3600.0)	**17*** **(64.36)**

for dense subgraph problems, and 13 are real-world social and biological networks, chosen from several applications to represent instances one might face in practice. Table 1 shows the performance of four algorithm variants (including the three best) on a subset of these instances. Each variant is represented by a string in which O denotes that the outer P-bound is enabled, B denotes that tie-breaking is enabled, N denotes that neighborhood-based scoring is enabled, ↑ denotes the optimistic ordering and ↓ denotes the pessimistic ordering.

Our observations are roughly as follows: For instances with larger maximum k, the optimistic ordering outperforms the pessimistic one. Those with small maximum k are solved slightly faster with pessimistic ordering. The outer P-bound usually does not reduce search tree size significantly but it runs fast enough to have only a small negative effect on running times. Tie-breaking allows to discover several μ-cliques in instances of medium difficulty which otherwise seem to be hard to find. The effect of neighborhood-based scoring is negligible.

5.2 Comparison with a Previous Approach

We compared our algorithms with an exact branch and bound algorithm for finding μ-cliques by Pajouh et al. [10]. In the following, we denote their algorithm by BB. (Recall that BB may report disconnected μ-cliques.) For the comparison, we chose several real-world instances from the test bed above and the three values of $\mu = 0.55, 0.7, 0.9$. The results are shown in Table 2. In terms of quickly finding large solutions, BB performs better than our algorithm but

Table 2. Largest μ-cliques found by the branch and bound algorithm (BB) by Pajouh et al. [10], and by our algorithm (O)-($\uparrow$)-(B,N), indicated by A1, and (O)-($\uparrow$)-(N), indicated by A2. Bold values represent *maximum* connected μ-clique orders as reported by the corresponding algorithm.

	$\mu = 0.55$			$\mu = 0.7$			$\mu = 0.9$		
	BB	A1	A2	BB	A1	A2	BB	A1	A2
Acker-all	32	32	32	25	25	25	15	**15**	**15**
Human-all	41	37	39	31	26	27	20	20	18
email-Enron	86	81	68	55	58	44	29	29	21
ERDOS-99-2	20	19	20	14	**14**	14	9	**9**	**9**
GEOM-0	39	32	32	30	28	28	23	**23**	**23**
wiki-Vote	104	84	103	65	62	61	31	28	26

the favor shifts towards ours for larger μ. Our algorithm could verify optimality for several instances with larger values of μ, whereas BB was never able to verify optimality within the time limit. Disabling the outer P-bound does not change the results or quality of running times. In some instances, enabling the outer P-bound reduces the number of calls to `Recurse`, but this is rare.

5.3   Evaluation on Random Instances

Erdős-Rényi Random Graphs. For each combination of $n = 10, 20, \ldots, 1200$ vertices and edge probability $p = 0.05, 0.1, 0.2$, we generated 15 Erdős-Rényi random graphs. The average running times of our algorithm variant ($\uparrow$)-(B,N) and algorithm BB are shown in Figure 1 for those n, where all 15 instances were solved to optimality within 20 minutes. For $p = 0.1$ and $p = 0.2$, the reported maximum μ-cliques of our algorithm were around ten at the cut-off points due to the time limit. Our algorithm clearly outperforms BB in terms of verifying

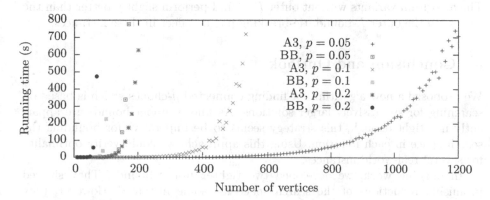

Fig. 1. Running times for varying order and edge probability p of Erdős-Rényi graphs. A3 denotes our algorithm in variant ($\uparrow$)-(B,N).

Table 3. Comparison of the retrieved μ-clique orders in random small-world networks. Here, k denotes the order of the planted μ-clique, n denotes the order of the input graph, and each * denotes an instance that was solved within the time limit.

k	n	(↑)-(B,N)	(↑)-(N)	(O)-(↑)-(N)	(O)-(↑)-(B,N)	BB
	500	11.0 (**)	11.0 (**)	11.0 (**)	11.0 (**)	10.0
10	1000	12.0 (**)	12.0 (**)	12.0 (**)	12.0 (**)	11.0
	2000	12.0 (**)	12.0 (**)	12.0 (**)	12.0 (**)	11.0
	500	21.5 (*)	21.5 (*)	21.0	21.0	21.5
20	1000	21.0	21.0	21.0	21.0	21.0
	2000	20.5	20.5	20.5	20.5	20.5
	500	26.0	26.0	25.5	25.5	30.5
30	1000	29.0	29.0	29.0	29.0	31.0
	2000	30.0	30.0	30.0	30.0	30.0

optimality on these instances. Furthermore, the differences get more pronounced as p gets smaller, that is, the graphs get sparser.

Random Small-World Graphs with Planted μ-cliques. In order to assess the order of the retrieved μ-cliques, we generated random networks with a planted μ-clique of order 10, 20, and 30. For each order, we created six networks, two networks with 500 vertices, two with 1000 vertices, and two with 2000 vertices. First, we sample a μ-clique of the appropriate order using the Erdős-Rényi model with edge probability $p = \mu$ and ensuring density at least μ. Then, we add vertices according to the Barabási-Albert model, making a new vertex adjacent to $\lfloor k/i \rfloor$ previous ones with probability proportional to their degrees. Herein, k is the μ-clique order and $i = 2$ for the first graph and $i = 4$ for the second one. Table 3 shows our results. If the planted μ-clique has order 10, our algorithm outperforms BB as it can exactly solve these instances. For planted μ-cliques of order 30, BB outperforms our algorithm. For order 20, they behave roughly the same. The algorithm variants without outer P-bound perform slightly better than the ones with the outer P-bound, tie-breaking has no effect in these instances.

6   Conclusion and Outlook

We proposed a new algorithm for finding connected μ-cliques which is based on searching for successively larger solutions. As known upper bounds are apparently not tight enough, this strategy seems to be imperative for bounding the search space in each iteration. Using this approach, we could verify optimality for several real-world instances.

In ongoing work, we developed two tighter upper bounds. They showed promising reductions of the search space for some instances. However, they require more computational overhead which increases the overall computation time. It is thus interesting to improve the corresponding implementations and to find easily checkable conditions on when the bounds might apply.

Acknowledgments. Manuel Sorge and Kolja Stahl gratefully acknowledge support by Deutsche Forschungsgemeinschaft (DFG), project DAPA, NI 369/12.

References

1. Abello, J., Pardalos, P.M., Resende, M.G.C.: On maximum clique problems in very large graphs. In: External Memory Algorithms and Visualization, vol. 50 of DIMACS, pp. 119–130. AMS (1999)
2. Abello, J., Resende, M.G.C., Sudarsky, S.: Massive quasi-clique detection. In: Rajsbaum, S. (ed.) LATIN 2002. LNCS, vol. 2286, pp. 598–612. Springer, Heidelberg (2002)
3. Bader, G.D., Hogue, C.W.: An automated method for finding molecular complexes in large protein interaction networks. BMC Bioinformatics **4**(1), 2 (2003)
4. Balasundaram, B., Pajouh, F.M.: Graph theoretic clique relaxations and applications. In: Handbook of Combinatorial Optimization, pp. 1559–1598. Springer (2013)
5. Boginski, V., Butenko, S., Pardalos, P.M.: On structural properties of the market graph. In: Innovations in Financial and Economic Networks. New Dimensions in Networks, pp. 29–45. Edward Elgar Publishing, Cheltenham, England (2003)
6. Downey, R.G., Fellows, M.R.: Fixed-parameter tractability and completeness II: On completeness for W[1]. Theoretical Computer Science **141**(1&2), 109–131 (1995)
7. Feige, U., Seltser, M.: On the densest k-subgraph problem. Technical report,The Weizmann Institute, Department of Applied Math and Computer Science (1997)
8. Komusiewicz, C., Sorge, M.: Finding dense subgraphs of sparse graphs. In: Thilikos, D.M., Woeginger, G.J. (eds.) IPEC 2012. LNCS, vol. 7535, pp. 242–251. Springer, Heidelberg (2012)
9. Kosub, S.: Local Density. In: Brandes, U., Erlebach, T. (eds.) Network Analysis. LNCS, vol. 3418, pp. 112–142. Springer, Heidelberg (2005)
10. Pajouh, F.M., Miao, Z., Balasundaram, B.: A branch-and-bound approach for maximum quasi-cliques. Annals of Operations Research **216**(1), 145–161 (2014)
11. Pattillo, J., Veremyev, A., Butenko, S., Boginski, V.: On the maximum quasi-clique problem. Discrete Applied Mathematics **161**(1–2), 244–257 (2013)
12. Raghavendra, P., Steurer, D.: Graph expansion and the unique games conjecture. In: Proc. 42nd STOC, pp. 755–764. ACM (2010)
13. Wagner, A., Fell, D.A.: The small world inside large metabolic networks. Proceedings of the Royal Society of London. Series B: Biological Sciences **268**(1478), 1803–1810 (2001)
14. Zhang, J., Chen, Y.: Monte Carlo algorithms for identifying densely connected subgraphs. Journal of Computational and Graphical Statistics (2014)

Combinatorial Optimization I

On the Generation of Cutting Planes
which Maximize the Bound Improvement

Stefano Coniglio$^{(\boxtimes)}$ and Martin Tieves

Lehrstuhl II für Mathematik, RWTH Aachen University,
Pontdriesch 14-16, 52062 Aachen, Germany
{coniglio,tieves}@math2.rwth-aachen.de

Abstract. We propose a new cutting plane algorithm for Integer Linear Programming, which we refer to as the *bound-optimal cutting plane method*. The algorithm amounts to simultaneously generating k cuts which, when added to the linear programming relaxation, yield the (provably) largest bound improvement. We show that, in the general case, the corresponding cut generating problem can be cast as a Quadratically Constrained Quadratic Program. We also show that, for a large family of cuts, the latter can be reformulated as a Mixed-Integer Linear Program. We present computational experiments on the generation of bound-optimal stable set and cover inequalities for the max clique and knapsack problems. They show that, with respect to standard algorithms, the bound-optimal cutting plane method allows for a substantial reduction in the number of cuts and iterations needed to achieve either a given bound or an optimal solution.

1 Introduction

Cutting planes are one of the key components of modern Integer and Mixed-Integer Linear Programming (ILP and MILP) solvers [BR07]. The *textbook* cutting plane method is fairly simple: it generates a single cut, adds it to the Linear Programming (LP) relaxation, performs a reoptimization, and iterates. Although practical methods are usually more involved, they typically share a feature with the textbook one: they are, primarily, driven by the maximization of the cut violation.

We speculate that looking for cuts which are maximally violated (as it is done in what we call the *standard* cutting plane method) might be not the most advisable option. Consider the correspondence between pivoting on a nonbasic column in the primal simplex method and generating a cut in a cutting plane method. In light of such relationship, adding a maximally violated cutting plane and reoptimizing amounts to, in the dual, pivoting on a column with the most negative reduced cost. Notably, this pivoting rule is usually considered, in the

---

This work is supported by the BMBF grant 05M13PAA and by the BMWi grant 03ET7528B.

E. Bampis (Ed.): SEA 2015, LNCS 9125, pp. 97–109, 2015.
DOI: 10.1007/978-3-319-20086-6_8

literature on the simplex method, a rule as poor as pivoting entirely at random [Bix09]. An alternative idea which, according to [Har73], dates back to Dantzig, is that of considering a "greatest change" criterion, corresponding to looking for the best improving solution within a pivoting operation. This idea has given rise to alternative pivoting rules which, although only providing an approximation of the actual objective function improvement, are, computationally, very effective [FG92]. In this work, we present an adaptation of the "greatest change" criterion to the context of cutting plane generation.

This paper[1] belongs to a larger stream of work, see, e.g., [ZFB11, ACG14], where alternative paradigms for cutting plane generation are sought. The aim is of, at least primarily, finding alternatives to the standard method which allow for a reduction in the number of cuts needed to achieve a given bound. This issue is of high practical relevance in branch-and-cut algorithms as, in their context, the number of cuts greatly affects the size of the LP relaxations that are solved at each node, with a large impact on the overall efficiency of the algorithm.

In this work, we propose and investigate a method which generates, at the same time, up to k cutting planes which simultaneously yield the (provably) largest bound improvement. We refer to such cuts as to *bound-optimal cuts*. We will show that, in the general case, they can be generated via a Quadratically Constrained Quadratic Program (QCQP) which, for many relevant families of cuts, can be cast as an MILP. Experiments on the generation of stable set and cover inequalities for the max clique and knapsack problems will show that, when compared to standard algorithms, the bound-optimal cutting plane method yields a substantial reduction in the number of cuts and cutting plane iterations.

2   Bound-Optimal Cutting Planes

Let P^I be the ILP $\max_{x \in \mathbb{Z}_+^n} \{cx : Ax \leq b\}$ with n variables and m constraints and let P be its LP relaxation. Consider a family Π of cutting planes $\pi x \leq \pi_0$, valid for P^I, with coefficients $(\pi, \pi_0) \in \Pi$. We assume that Π is finitely generated and that it can be expressed as a mixed-integer set.

Throughout the paper, the focus will be on optimizing over the closure of P under Π, i.e., over the set: $P_\Pi = \max_{x \in \mathbb{R}_+^n} \{cx : Ax \leq b, \pi x \leq \pi_0 \; \forall (\pi, \pi_0) \in \Pi\}$. W.l.o.g., we assume that P may subsume some of the inequalities in Π as already contained in $Ax \leq b$ (e.g., because added in a previous cutting plane iteration). We will denote rows and columns of an MILP by, resp., i and j and adopt the notation $[n]$ for index sets of type $\{1, \ldots, n\}$.

In this work, we look for k *bound-optimal cutting planes*, i.e., for k cuts $\pi^h x \leq \pi_0^h$, with $(\pi^h, \pi_0^h) \in \Pi$ for all $h \in [k]$, whose introduction into P yields the largest bound improvement. Let P' be the problem obtained after introducing the new cuts into P. Formally, by letting z and z' be the optimal values of P and P', we are thus looking for k cuts which maximize the quantity $z - z'$.

---

[1] A partial, preliminary version appeared in [Con13].

2.1 Generation of a Single Bound-Optimal Cut

Let us first consider the case of $k = 1$. We can cast the problem of generating a single bound-optimal cutting plane as the following nonlinear bilevel program:

$$
\max_{\substack{x \in \mathbb{R}^n_+ \\ (\pi, \pi_0) \in \Pi}} \left\{ \overbrace{z}^{const} - \overbrace{cx}^{z'} : x \in \underset{x \in \mathbb{R}^n_+}{\mathrm{argmax}} \overbrace{\left\{ cx : \begin{array}{c} Ax \leq b \\ \pi x \leq \pi_0 \end{array} \right\}}^{P'} \right\}, \tag{1}
$$

whose nonlinearity is a consequence of the bilinear terms in πx. Leveraging LP duality, we obtain the following characterization:

Proposition 1. *A bound-optimal cutting plane can be found by solving the following single level QCQP with $2n + m + 2$ variables and $n + m + 2$ constraints:*

$$
\min_{\substack{x \in \mathbb{R}^n_+ \\ y \in \mathbb{R}^{m+1}_+ \\ (\pi, \pi_0) \in \Pi}} \left\{ \begin{array}{ll} \sum_{j=1}^n c_j x_j : \sum_{j=1}^n a_{ij} x_j \leq b_i & \forall i \in [m] \\ \sum_{j=1}^h \pi_j x_j \leq \pi_0 & \\ \sum_{i=1}^m a_{ij} y_i + \pi_j y_{m+1} \geq c_j & \forall j \in [n] \\ \sum_{j=1}^n c_j x_j = \sum_{i=1}^m b_i y_i + \pi_0 y_{m+1} & \end{array} \right\}. \tag{2}
$$

Proof. For any $(\pi, \pi_0) \in \Pi$, we guarantee that $x \in \mathbb{R}^n_+$ be feasible for P' by imposing $\sum_{j=1}^n a_{ij} x_j \leq b_i$, for all $i \in [m]$, and $\sum_{j=1}^n \pi_j x_j \leq \pi_0$. Let $y \in \mathbb{R}^{m+1}$ be the dual variables of P', with y_{m+1} corresponding to the new inequality $\pi x \leq \pi_0$. We impose dual feasibility by introducing $\sum_{i=1}^m a_{ij} y_i + \pi_j y_{m+1} \geq c_j$, for all $j \in [n]$. By imposing $\sum_{j=1}^n c_j x_j - \sum_{i=1}^m b_i y_i + \pi_0 y_{m+1}$, we guarantee that (x, y) form an optimal primal-dual pair for any $(\pi, \pi_0) \in \Pi$. Since, this way, any feasible solution x is optimal for P', a bound-optimal cut is found by minimizing the objective function of P', rather than by maximizing it. □

By projecting the x variables out, we obtain a simpler construction:

Proposition 2. *A bound-optimal cutting plane can be found by solving the following single level QCQP with $n + m + 2$ variables and n constraints:*

$$
\min_{\substack{y \in \mathbb{R}^{m+1}_+ \\ (\pi, \pi_0) \in \Pi}} \left\{ \sum_{i=1}^m b_i y_i + \pi_0 y_{m+1} : \sum_{i=1}^m a_{ij} y_i + \pi_j y_{m+1} \geq c_j \ \forall j \in [n] \right\}. \tag{3}
$$

Problem (3) is a QCQP due to the bilinear products $\pi_j y_{m+1}$ and $\pi_0 y_{m+1}$. Under the following assumptions, the problem can be reformulated as an MILP:

Proposition 3. *Assume that $\pi \in \{0, 1\}^n$ and that π_0 is an affine function of π, i.e., $\pi_0 = \sum_{j=1}^n g_j \pi_j + g_0$, for some $(g, g_0) \in \mathbb{R}^{n+1}$. Let $z_j := \pi_j y_{m+1}$. Assuming $y_{m+1} \leq y^U$ for some $y^U \in \mathbb{R}_+$, Problem (3) can be cast as the MILP:*

$$
\min_{\substack{y \in \mathbb{R}^{m+1}_+ \\ z \in \mathbb{R}^n_+ \\ (\pi, \pi_0) \in \Pi}} \left\{ \begin{array}{ll} \sum_{i=1}^m b_i y_i + \sum_{j=1}^n g_j z_j + g_0 y_{m+1} : & \\ \sum_{i=1}^m a_{ij} y_i + z_j \geq c_j & \forall j \in [n] \\ z_j \geq y_{m+1} + \pi_j y^U - y^U & \forall j \in [n] \\ z_j \leq \pi_j y^U & \forall j \in [n] \\ z_j \leq y_{m+1} & \forall j \in [n] \end{array} \right\}. \tag{4}
$$

This result follows by linearizing the bilinear products $\pi_j y_{m+1}$ by means of the McCormick envelope [McC76]. We remark that Proposition 3 encompasses many families of "combinatorial" cutting planes, such as clique and stable set inequalities (where $\pi_0 = 1$), cut-set inequalities (where $\pi_0 = 1$ or 2), or knapsack cover inequalities and subtour elimination constraints (where $\pi_0 = \sum_{j=1}^n \pi_j - 1$).

2.2   Simultaneous Generation of k Bound-Optimal Cuts

One of the most interesting features of bound-optimal cutting planes is that Propositions 2 and 3 can be easily generalized so to generate *any* number k of cuts which, jointly, yield the largest bound improvement. The generalization of Proposition 2, from which that of Proposition 3 is straightforward, is as follows:

Proposition 4. *A set of k bound-optimal cutting planes can be found by solving the following QCQP with $m + kn + 2k$ variables and n constraints:*

$$\min_{\substack{y \in \mathbb{R}_+^{m+k} \\ (\pi^h, \pi_0^h) \in \Pi, h \in [k]}} \left\{ \begin{array}{l} \sum_{i=1}^m b_i y_i + \sum_{h=1}^k \pi_0^h y_{m+h} : \\ \sum_{i=1}^m a_{ij} y_i + \sum_{h=1}^k \pi_j^h y_{m+h} \geq c_j \ \forall j \in [n] \end{array} \right\}, \quad (5)$$

where, for each $h \in [k]$, y_{m+h} is the dual variable of $\pi^h x \leq \pi_0^h$.

2.3   Convergence of the Bound-Optimal Cutting Plane Method

The cutting plane method that is obtained when generating bound-optimal cuts, which we refer to as the *bound-optimal cutting plane method*, iterates over two steps: *i*) it solves Problem (3) for $k = 1$ or Problem (5) for $k > 1$ and *ii*) it adds the k new inequalities $\pi^h x \leq \pi_0^h$, for $h \in [k]$, to $Ax \leq b$.

Differently from the standard cutting plane algorithm, the bound-optimal cutting plane method does not require to reoptimize the LP relaxation, the dual of which is implicitly reoptimized in step *i*), nor it involves the concept of *separation* of an infeasible solution x^*. Notably, the method does not account for cut violation at all. Also note that, regardless of the number of inequalities in Π, it allows to solve any problem P_Π, at least in principle, by solving Problem (5) only once, for $k = n$, i.e., by generating $k = n$ bound-optimal cuts in a single iteration. This is because any vertex of P_Π is uniquely identified by at most n inequalities (exactly n in the full dimensional case).

Differently, for small values of k, an iteration of the bound-optimal cutting plane method may stall. As illustrated in Figure 1, this is the case when the current LP relaxation admits an optimal facet but Π does not contain any set of k cuts which, jointly, allow to cut the entire facet away. We remark that this issue can always be circumvented by increasing k whenever the method fails to improve the bound. Another option is to couple the method with a standard cutting plane algorithm whenever a stalling iteration occurs. We discuss it in Section 4.

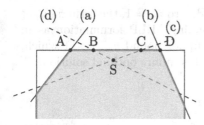

Fig. 1. Let Π contain four cuts (a), (b), (c), (d), with (a) and (b) already added to the relaxation. Assume that the objective function gradient is orthogonal to the facet AD. The bound (given by AD) cannot be improved by the introduction of a single cut, as a subset of AD will remain feasible in any case. Hence, for $k = 1$, the bound-optimal cutting plane method stalls while, for $k = 2$, it reaches the (unique) optimal solution S by generating (c) and (d) simultaneously.

3   Case Studies

In this section, we show an application of bound-optimal cutting planes to two classical combinatorial optimization problems: *max clique* and *binary knapsack*. For the sake of readability, we will report the bound-optimal cut generating problems as QCQPs for $k = 1$, as in Proposition 2. The derivation of the corresponding MILPs, as in Proposition 3, and the extension to $k > 1$ is left to the reader.

3.1   Application to the Max Clique Problem

Given an undirected graph $G = (V, E)$ with n vertices, consider the *max clique* problem calling for a clique of maximum cardinality. Let $\mathcal{S} := \{S_1, \ldots, S_{|\mathcal{S}|}\}$ be the set of all the (maximal) stable sets of G. For any $S_i \in \mathcal{S}$, the *stable set inequality* $\sum_{j \in S_i} x_j \leq 1$ is valid. The corresponding closure yields the following LP relaxation:

$$\max_{x \in \mathbb{R}^n_+} \left\{ \sum_{j \in V} x_j : \sum_{j \in S_i} x_j \leq 1 \, \forall S_i \in \mathcal{S} \right\}, \tag{6}$$

whose optimal value is the so-called *fractional clique number* of G. Its dual reads:

$$\min_{y \in \mathbb{R}^{|\mathcal{S}|}_+} \left\{ \sum_{S_i \in \mathcal{S}} y_i : \sum_{S_i \in \mathcal{S} : j \in S_i} y_i \geq 1 \, \forall j \in V \right\}. \tag{7}$$

Given a point $x^* \in \mathbb{R}^n_+$, the standard separation problem calling for a maximally violated stable set inequality can be cast as the ILP:

$$\max_{\pi \in \{0,1\}^n} \left\{ \sum_{j \in V} x^*_j \pi_j : \pi_i + \pi_j \leq 1 \, \forall \{i, j\} \in E \right\}. \tag{8}$$

Let us assume that m stable set inequalities have been generated. By virtue of Proposition 2, a bound-optimal stable set inequality is obtained by solving the following QCQP:

$$\min_{\substack{y \in \mathbb{R}^{m+1} \\ \pi \in \{0,1\}^n}} \left\{ \sum_{i=1}^m y_i + y_{m+1} : \begin{array}{ll} \sum_{i=1:j \in S_i}^m y_i + \pi_j y_{m+1} \geq 1 & \forall j \in V \\ \pi_i + \pi_j \leq 1 & \forall \{i, j\} \in E \end{array} \right\}. \tag{9}$$

Note that the only nonlinear term is $\pi_j y_{m+1}$ as, due to $\pi_0 = 1$, the term $\pi_0 y_{m+1}$ in Problem (3) reduces to y_{m+1}. The corresponding MILP formulation as in Proposition 3 is obtained after introducing the upper bound $y_{m+1} \leq 1$ which, due to the direction of the objective function, holds in any optimal solution.

3.2   Application to the Knapsack Problem

Consider the 0-1 *knapsack problem* with a set of items $[n]$, a weight function $a : [n] \rightarrow \mathbb{N}_+$, a profit function $c : [n] \rightarrow \mathbb{N}_+$, and a budget $b \in \mathbb{N}_+$. Let $\mathcal{C} = \{C_1, \ldots, C_{|\mathcal{C}|}\}$ be the set of *covers* of $[n]$, where $C_i \in \mathcal{C}$ if and only if $\sum_{j \in C_i} a_j \geq b + 1$. For any $C_i \in \mathcal{C}$, the *cover inequality* $\sum_{j \in C_i} x_j \leq |C_i| - 1$ is valid. When optimizing over the corresponding closure, we have the LP:

$$\max_{x \in [0,1]^n} \left\{ \sum_{j=1}^n c_j x_j : \begin{array}{l} \sum_{j=1}^n a_j x_j \leq b \\ \sum_{j \in C_i} x_j \leq |C_i| - 1 \ \forall C_i \in \mathcal{C} \end{array} \right\}. \tag{10}$$

By letting u be the dual variable of $\sum_{i=1}^n a_j x_j \leq b$, v_j that of $x_j \leq 1$, for all $j \in [n]$, and y_i that of each cover inequality of index i, the dual reads:

$$\min_{(u,v,y) \in \mathbb{R}_+^{1+n+|\mathcal{C}|}} \left\{ \begin{array}{l} ub + \sum_{j=1}^n v_j + \sum_{C_i \in \mathcal{C}} (|C_i| - 1) y_i : \\ a_j u + v_j + \sum_{C_i \in \mathcal{C}: j \in C_i} y_i \geq c_j \ \forall j \in [n] \end{array} \right\}. \tag{11}$$

Given $x^* \in \mathbb{R}_+^n$, the standard cutting plane generation problem is the ILP:

$$\min_{\pi \in \{0,1\}^n} \left\{ \sum_{j=1}^n (1 - x_j^*) \pi_j : \sum_{j=1}^n a_j \pi_j \geq b + 1 \right\}. \tag{12}$$

Let us assume that m cuts have been generated. Due to Proposition 2, a bound-optimal cover inequality is obtained by solving the following QCQP:

$$\min_{\substack{u \in \mathbb{R}_+, v \in \mathbb{R}_+^n \\ y \in \mathbb{R}_+^{m+1} \\ \pi \in \{0,1\}^n}} \left\{ \begin{array}{l} ub + \sum_{j=1}^n v_j + \sum_{i=1}^m (|C_i| - 1) y_i + (\sum_{j=1}^n \pi_j - 1) y_{m+1} : \\ a_j u + v_j + \sum_{i=1:j \in C_i}^m y_i + \pi_j y_{m+1} \geq c_j \ \forall j \in [n] \\ \sum_{j=1}^n a_j \pi_j \geq b + 1 \end{array} \right\}. \tag{13}$$

We can derive an MILP reformulation via Proposition 3 after the introduction of an upper bound $y_{m+1} \leq y^U$. As it is easy to see, if y^U is an upper bound on the value of an optimal solution to Problem (13), then $y_{m+1} \leq y^U$ holds in any of its optimal solutions. We can thus initialize $y^U = \max_{x \in [0,1]^n} \{\sum_{j=1}^n c_j x_j : \sum_{j=1}^n a_j x_j \leq b\}$ and update it at each iteration with the new, tighter bound that is found by generating a bound-optimal cutting plane.

3.3   Cut Domination

From a practical point of view, ensuring that we only generate nondominated cuts is paramount to a competitive cutting plane algorithm. Recall that a stable set (resp., cover) inequality is dominated if and only if the corresponding stable set (cover) is not inclusionwise maximal (minimal). To guarantee nondomination, we propose two families of inequalities.

For stable set inequalities, observe that a stable set S is maximal if and only if, for each vertex $j \in V \setminus S$, at least one vertex $i \in S$ shares an edge with j. Therefore, the following holds for any π denoting a maximal stable set:

$$\sum_{\{i,j\} \in E} \pi_i \geq 1 - \pi_j \quad \forall j \in V. \tag{14}$$

For cover inequalities, given a cover $C = \{j \in [n] : \pi_j = 1\}$, we must guarantee that, for all $\ell \in C$, $\sum_{j \in C \setminus \{\ell\}} a_j \leq b$. Thus, we introduce:

$$\sum_{j \in [n] \setminus \{\ell\}} a_j \pi_j \leq b + \left(\sum_{j \in [n] \setminus \{\ell\}} a_j - b \right)(1 - \pi_\ell) \quad \forall \ell \in [n]. \tag{15}$$

4   Computational Experiments

We evaluate the impact of the bound-optimal cutting plane method on the two problems discussed in the previous section. We address the following algorithms:

- BOC: generation of k *bound-optimal cuts* at time;
- STD: (*standard*) generation of a maximally violated inequality $\pi x \leq \pi_0$;
- COORD: generation of a *coordinated* cutting plane, as proposed in [ACG14]; at any iteration t, a cut $\pi x \leq \pi_0$ is *coordinated* if and only if, among all the cuts of maximum violation, it also maximizes the 1-norm difference between π and the average of the vectors $\pi^1, \ldots, \pi^{t-1}$ of the previous cuts,
- BOC+STD and BOC+COORD: extension of BOC where cut generation is switched to, resp., STD and COORD as soon as an iteration of BOC stalls.
- N-X: algorithm X where Constraints (14) or (15) are introduced to guarantee the generation of nondominated cuts.

We consider a set of 24 instances taken from the second DIMACS challenge on "cliques, coloring, and satisfiability" [JT96] for max clique and a set of 30 instances of type 2 (weakly correlated), 14 (bounded strongly correlated), and 15 (no small weights) generated via Pissinger's gen2.c generator [MPT99], with a range parameter of 10^5 and 125 items. We compiled the data sets so to have a pool of nontrivial instances for which the different cutting plane algorithms can be run within a reasonable amount of computing time.

All the experiments are carried out with CPLEX 12.4, adopting AMPL as modeling language, on a single threaded 3.40GHz Intel i7-3770 CPU with 32GB of RAM. For each method, we impose a time limit of 7200 seconds.

4.1   Comparisons for a Given Bound on Max Clique Instances

With these experiments, we illustrate an important feature of bound-optimal cutting planes, i.e., that, when compared to the other methods, the number of cuts and iterations that BOC needs to achieve a given bound (we consider that at which BOC halts due to a stalling iteration) is substantially smaller.

To better highlight this feature, we compare the different algorithms in their, arguably, best setting, i.e., when guaranteeing that they only generate nondominated inequalities by means of the introduction of Constraints (14). Hence, we compare thee methods: N-BOC, N-STD, and N-COORD. We consider three values of k, namely: $k = 1, 2, 3$, halting the execution of N-STD and N-COORD as soon as the target bound is reached.

Figure 2 reports a graphical representation of the typical evolution of the bounds for the different methods, highlighting that N-BOC provides a large improvement over N-STD and N-COORD. Indeed, it shows that, for all values of k and at every iteration, the bound provided by N-BOC is always tighter than that given by N-STD and N-COORD, and much tighter for $k = 2, 3$.

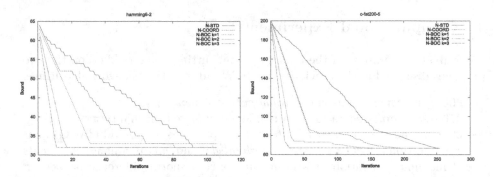

Fig. 2. Bound improvement vs number of iterations for N-BOC ($k = 1, 2, 3$), N-STD, and N-COORD on the instances **hamming-6-2** and **c-fat200-5**

The complete results are illustrated in Tables 1, 2, and 3, one per value of k. The tables not only confirm the improvement of N-BOC over the other two methods, but they also show that the difference in the number of cuts generated by N-BOC w.r.t. that for N-STD and N-COORD is very large for larger values of k. For $k = 1$, we register that N-STD and N-COORD generate, resp. and in geometric mean, 2.0 and 1.5 times the number of cuts generated by N-BOC. For $k = 2$, the factors increase to 2.3 and 1.8, reaching 2.4 and 1.8 for $k = 3$. The difference can be very large even for $k = 1$. Consider, as an example, the instance **hamming8-2**, where N-BOC generates only 127 cuts, as opposed to 367 for N-STD (2.9 times as much) and 412 for N-COORD (3.2 times as much).

Since N-BOC generates k cuts at a time, the comparison w.r.t. the number of iterations is even more favorable for it than that w.r.t. the number of cuts. Starting from $k = 1$, we have that N-STD and N-COORD require, resp., 2.0 and 1.5 times the iterations of N-BOC (same factors as for the number of cuts). The factors increase to, resp., 4.5 and 3.7 for $k = 2$, reaching 7.2 and 5.3 for $k = 3$.

We also note that the bound at which a stalling iteration of N-BOC occurs improves substantially for the different values of k. Neglecting the instances for which the time limit is hit, we have a gap, for $k = 1$, of 34.2% w.r.t. the value of

Table 1. Comparison w.r.t. the number of cuts on max clique instances between N-BOC, N-STD, and N-COORD, halting the latter two as they reach the bound at which N-BOC stalls, for $k = 1$. The best method per instance is highlighted in boldface.

Instance	$|V|$	$|E|$	Opt	Bnd	N-BOC Cuts	N-BOC Time	N-STD Cuts	N-STD Time	N-COORD Cuts	N-COORD Time
C-fat200-1	200	1533	12.0	14.0	**13**	0.5	38	0.8	14	0.3
C-fat200-2	200	3234	24.0	24.0	**22**	0.9	70	2.0	**22**	0.6
C-fat200-5	200	8472	66.7	83.0	59	4.4	171	8.9	69	3.6
Hamming6-2	64	1824	32.0	33.0	**31**	1.0	89	2.7	63	1.9
Hamming6-4	64	704	5.3	8.0	**8**	0.8	17	0.4	13	0.6
Hamming8-2	256	31616	128.0	129.0	**127**	222.4	367	583.2	412	657.5
Johnson8-4-4	70	1855	14.0	18.0	**19**	3.4	27	1.7	21	1.3
Johnson16-2-4	120	5460	8.0	14.0	**14**	13.0	16	3.2	16	3.8
MANN_a9	45	917	18.0	21.0	**12**	0.5	**12**	0.3	**12**	0.3
myciel4	23	70	3.2	5.0	**4**	0.1	5	0.0	5	0.1
myciel5	47	235	3.6	6.0	**5**	0.2	**5**	0.1	**5**	0.1
myciel6	95	755	3.8	7.0	**6**	0.4	**6**	0.1	**6**	0.2
myciel7	191	2359	4.1	8.0	**7**	1.8	**7**	0.4	**7**	0.5
queen6_6	36	289	7.0	9.0	**9**	0.2	14	0.2	12	0.2
queen7_7	49	475	7.0	7.0	**7**	0.2	54	0.8	31	0.5
queen8_12	96	1367	12.0	14.0	**13**	0.5	37	0.9	26	0.6
queen9_9	81	1055	9.0	13.0	**10**	0.6	12	0.3	12	0.3
queen10_10	100	1469	10.0	14.0	**12**	1.0	29	0.8	13	0.5
queen11_11	121	1979	11.0	14.0	**15**	2.0	38	1.3	27	1.1
queen12_12	144	2595	12.0	15.0	**15**	2.4	43	1.8	35	1.7
queen13_13	169	3327	13.0	17.0	**15**	3.8	36	1.8	33	2.1
queen14_14	196	4185	14.0	17.0	**17**	5.0	55	3.5	47	3.5
queen15_15	225	5179	15.0	18.0	**18**	11.1	63	4.6	48	4.3
queen8_8	64	627	8.4	11.0	**9**	0.5	10	0.2	13	0.3

Table 2. Comparison w.r.t. the number of cuts on max clique instances between N-BOC, N-STD, and N-COORD, halting the latter two as they reach the bound at which N-BOC stalls, for $k = 2$. The best method per instance is highlighted in boldface.

Instance	$|V|$	$|E|$	Opt	Bnd	Iters	N-BOC Cuts	N-BOC Time	N-STD Cuts	N-STD Time	N-COORD Cuts	N-COORD Time
C-fat200-1	200	1533	12.0	12.5	8	**16**	3.0	104	2.2	72	1.5
C-fat200-2	200	3234	24.0	24.0	11	**22**	1.7	70	2.0	**22**	0.6
C-fat200-5	200	8472	66.7	74.0	35	**70**	9.3	213	11.1	138	7.3
Hamming6-2	64	1824	32.0	32.0	17	**34**	9.1	91	2.8	80	2.5
Hamming6-4	64	704	5.3	7.0	6	**12**	6.6	24	0.5	15	0.6
Hamming8-2	256	31616	128.0	174.0	41	**82**	7200.0	204	324.2	230	364.8
Johnson8-4-4	70	1855	14.0	17.0	9	**18**	21.9	31	1.9	24	1.5
Johnson16-2-4	120	5460	8.0	14.0	7	**14**	212.2	16	3.2	16	3.8
MANN_a9	45	917	18.0	21.0	6	**12**	1.0	**12**	0.3	**12**	0.3
myciel4	23	70	3.2	5.0	2	**4**	0.2	5	0.0	5	0.1
myciel5	47	235	3.6	3.6	30	**60**	1007.9	77	5.3	69	5.1
myciel6	95	755	3.8	7.0	3	**6**	6.0	**6**	0.1	**6**	0.2
myciel7	191	2359	4.1	4.6	15	**30**	7200.0	81	30.7	39	8.6
queen6_6	36	289	7.0	8.0	6	**12**	0.7	16	0.2	17	0.2
queen7_7	49	475	7.0	7.0	5	**10**	1.2	54	0.8	31	0.5
queen8_12	96	1367	12.0	13.0	7	**14**	6.9	45	1.1	39	1.0
queen9_9	81	1055	9.0	12.0	6	**12**	5.9	28	0.6	18	0.4
queen10_10	100	1469	10.0	12.0	6	**12**	13.3	46	1.3	35	1.1
queen11_11	121	1979	11.0	13.0	7	**14**	24.0	55	1.9	40	1.6
queen12_12	144	2595	12.0	14.0	8	**16**	45.0	74	3.2	49	2.3
queen13_13	169	3327	13.0	15.0	8	**16**	40.5	79	4.2	59	3.5
queen14_14	196	4185	14.0	17.0	9	**18**	142.3	55	3.5	47	3.5
queen15_15	225	5179	15.0	142.0	3	**6**	7200.0	**6**	0.6	**6**	0.6
queen8_8	64	627	8.4	10.0	6	12	3.7	**10**	0.2	26	0.5

Table 3. Comparison w.r.t. the number of cuts on max clique instances between N-BOC, N-STD, and N-COORD, halting the latter two as they reach the bound at which N-BOC stalls, for $k = 3$. The best method per instance is highlighted in boldface.

Instance	$|V|$	$|E|$	Opt	Bnd	N-BOC Iters	N-BOC Cuts	N-BOC Time	N-STD Cuts	N-STD Time	N-COORD Cuts	N-COORD Time
C-fat200-1	200	1533	12.0	13.0	5	**15**	8.0	67	1.4	27	0.5
C-fat200-2	200	3234	24.0	24.0	8	24	5.0	70	2.0	**22**	0.6
C-fat200-5	200	8472	66.7	67.0	28	84	1771.7	248	13.1	189	10.1
Hamming6-2	64	1824	32.0	32.0	11	**33**	49.6	91	2.8	80	2.5
Hamming6-4	64	704	5.3	7.0	3	**9**	56.5	24	0.5	15	0.6
Hamming8-2	256	31616	128.0	250.0	2	**6**	7200.0	7	11.1	**6**	9.5
Johnson8-4-4	70	1855	14.0	14.0	7	**21**	144.0	56	3.3	56	3.4
Johnson16-2-4	120	5460	8.0	8.0	6	18	4941.3	**16**	3.2	**16**	3.8
MANN_a9	45	917	18.0	18.0	7	**21**	6.5	29	0.6	34	0.8
myciel4	23	70	3.2	3.5	3	9	2.2	14	0.1	**7**	0.1
myciel5	47	235	3.6	4.0	4	**12**	34.9	25	0.3	15	0.2
myciel6	95	755	3.8	4.3	7	**21**	7200.0	48	4.5	22	1.0
myciel7	191	2359	4.1	15.0	2	6	7200.0	**5**	0.3	**5**	0.3
queen6_6	36	289	7.0	8.0	3	**9**	2.3	16	0.2	17	0.2
queen7_7	49	475	7.0	7.0	3	**9**	4.5	54	0.8	31	0.5
queen8_12	96	1367	12.0	13.0	5	**15**	28.4	45	1.1	39	1.0
queen9_9	81	1055	9.0	11.0	4	**12**	84.8	38	0.8	28	0.6
queen10_10	100	1469	10.0	12.0	4	**12**	59.9	46	1.3	35	1.1
queen11_11	121	1979	11.0	13.0	5	**15**	1264.1	55	1.9	40	1.6
queen12_12	144	2595	12.0	14.0	6	**18**	2824.4	74	3.2	49	2.3
queen13_13	169	3327	13.0	15.0	6	**18**	7200.0	79	4.2	59	3.5
queen14_14	196	4185	14.0	16.0	6	**18**	7200.0	91	6.0	72	5.1
queen15_15	225	5179	15.0	30.0	5	**15**	7200.0	16	1.5	16	2.0
queen8_8	64	627	8.4	9.1	11	**33**	7200.0	56	1.0	49	1.0

an optimal solution to the fractional clique number problem (where all the stable set inequalities are introduced). For $k = 2$, the gap reduces to 21.5%, reaching 8.2% for $k = 3$. This corroborates that, as it can be expected, the chance of stalling decreases for larger values of k.

We remark that, in our experiments, the aim is of assessing the impact, as well as the potential, of bound-optimal cutting planes w.r.t. the reduction in number of cuts and iterations. Since, in N-BOC, we solve each separation problem to proven optimality (without any algorithmic sophistications), the computing times required to find bound-optimal cutting planes (which we report for completeness) become quite large for a large k. Starting from being, in geometric mean, 1.4 and 1.5 times those for, resp., N-STD and N-COORD for $k = 1$, they reach a factor of two and three orders of magnitude for, resp., $k = 2$ and $k = 3$. The number of instances for which the time limit is hit also increases from 0 for $k = 1$ to 3 for $k = 2$ to 7 for $k = 3$.

4.2   Comparisons to Optimal Solutions for the Knapsack Problem

We illustrate a set of experiments carried on the knapsack problem to highlight two features of bound-optimal cutting plane generation. First, that, if we switch to a standard cutting plane algorithm (such as STD or COORD) once an iteration of BOC stalls, the resulting method allows for a significant reduction in the number of cuts. Secondly, that the majority of the cuts produced by BOC are nondominated even if nondomination is not imposed. For this reason, in these experiments, we do not resort to Constraints (15).

We compare BOC+STD and BOC+COORD with $k = 1$ to STD and COORD. The results are reported in Table 4. All the instances are solved to optimality, except for 14_3, where BOC+STD and BOC+COORD are interrupted due to a numerical precision issue[2]. The column Cuts+ reports the number of "standard" cuts generated when switching to STD or COORD. The table shows that BOC+COORD and BOC+STD

Table 4. Comparison w.r.t. the number of cuts on knapsack instances for BOC+STD and BOC+COORD ($k = 1$), STD, and COORD. The best method per instance is highlighted in boldface.

Inst.	BOC+COORD			BOC+STD			COORD		STD	
	Cuts	Cuts+	Time	Cuts	Cuts+	Time	Cuts	Time	Cuts	Time
2_1	**128**	17	30.5	196	85	31.3	489	10.0	915	18.1
2_2	**127**	6	51.8	135	14	51.8	330	6.9	1942	51.8
2_3	**136**	8	60.1	147	19	60.2	329	7.0	1778	45.0
2_4	**122**	5	49.0	125	8	49.1	284	5.8	949	20.8
2_5	**128**	10	43.3	149	31	43.4	291	5.7	9071	649.4
2_6	**116**	7	36.9	143	34	37.0	242	4.5	1601	40.9
2_7	**107**	0	44.1	**107**	0	44.1	162	3.2	671	12.1
2_8	**129**	9	47.9	148	28	48.1	350	6.8	11100	1273.4
2_9	**118**	4	44.6	123	9	44.6	276	5.4	1109	24.1
2_10	**87**	0	17.8	**87**	0	17.8	153	2.9	626	10.4
14_1	117	22	73.6	116	21	73.5	**103**	1.7	301	4.3
14_2	103	6	41.7	105	8	41.6	**92**	1.6	330	4.8
14_3	173	0	680.8	173	0	680.8	**143**	2.8	492	9.1
14_4	**74**	5	12.9	75	6	12.8	90	1.6	267	3.4
14_5	132	35	74.3	143	46	74.3	**122**	2.4	350	5.5
14_6	**121**	1	227.0	**121**	1	226.9	132	2.8	365	5.9
14_7	87	0	13.5	87	0	13.4	**60**	0.9	284	3.6
14_8	**140**	23	785.9	143	26	785.9	162	3.4	533	9.9
14_9	100	4	31.8	100	4	31.8	**89**	1.5	270	3.7
14_10	93	16	26.3	94	17	26.2	**86**	1.5	292	3.9
15_1	**72**	0	10.1	**72**	0	10.2	118	2.2	464	7.1
15_2	**119**	0	64.0	**119**	0	63.8	172	3.4	613	10.9
15_3	**81**	1	15.8	82	2	15.8	109	1.9	362	4.8
15_4	**102**	13	26.9	107	18	27.0	130	2.5	378	5.4
15_5	**110**	0	62.7	**110**	0	62.7	144	2.4	445	7.1
15_6	**107**	0	123.7	**107**	0	123.7	123	2.4	350	5.5
15_7	**50**	0	5.4	**50**	0	5.5	81	1.6	303	3.8
15_8	**148**	0	93.3	**148**	0	93.4	198	3.7	706	13.4
15_9	**120**	0	94.8	**120**	0	94.8	151	2.9	428	7.0
15_10	**92**	0	27.8	**92**	0	27.7	151	2.9	420	6.3

manage to solve all the instances to optimality in a substantially smaller number of cutting planes (and iterations, as the two numbers coincide for $k = 1$). When compared to BOC+STD, STD and COORD generate, resp. and in geometric mean, 1.38 and 5.59 times the number of cuts generated by BOC+STD. When compared to BOC+COORD, the factors increase to, resp., 1.45 and 5.84.

---

[2] In all our experiments, validity is always checked before a cut is added to the relaxation, thus guaranteeing the correctness of the bounds that we produce.

As for the case of max clique, computing times for BOC are, in both variants, quite large, slightly more than an order of magnitude larger than those for COORD and slightly more than 4 times larger w.r.t. those for STD. Nevertheless, they still allow to assess the substantial reduction in the number of cutting planes provided by BOC, highlighting the potential of bound-optimal cutting plane generation.

We remark that the comparison between either BOC+STD or BOC+COORD and STD is quite impressive. Over all the instances, the latter produces a total of 37718 cuts, whereas BOC+COORD and BOC+STD only produce, resp., 3339 and 3524 cuts (equal to, resp., only 8.9% and 9.3% the number of cuts for STD). Also note that, for the cases where STD generates an extremely large number of cuts, such as for the instances 2_5 and 2_8, BOC+COORD and BOC+STD are much faster, exhibiting computing times that are, resp. for the two instances, more than one and two orders of magnitude smaller, in spite of the more involved cut generating problem that BOC entails. This better highlights that, even by generating a single cut at each iteration which maximizes the bound improvement, with BOC we are very likely to produce strong cutting planes, implicitly discarding the dominated ones. Indeed, 90% of the cuts generated by BOC+STD and BOC+COORD are, on arithmetic average, nondominated, as opposed to only 20% for STD. This feature might be very interesting for problems where nondomination cannot be easily imposed via the introduction of a compact set of inequalities, such as Constraints (14), (15).

5   Concluding Remarks

We have proposed the bound-optimal cutting plane method, a new paradigm for cutting plane generation which produces, at each iteration, up to k cuts which, jointly, yield the provably largest bound improvement.

We have compared our method to the standard separation of maximally violated inequalities and to the generation of coordinated cutting planes. Experiments on the fractional clique number and on the 0-1 knapsack problem show that, compared to other techniques, with bound-optimal cuts we can obtain a given bound within a substantially smaller number of cuts and iterations.

With this work, we have highlighted the potential of bound-optimal cutting plane generation, hopefully motivating further developments in this direction. Future studies include more efficient ways to solve the bound-optimal cut generating problem, possibly via heuristic approaches, as well as the investigation of its combinatorial nature for special classes of problems.

References

[ACG14] Amaldi, E., Coniglio, S., Gualandi, S.: Coordinated cutting plane generation via multi-objective separation. Math. Program. **143**, 87–110 (2014)

[Bix09] Bixby, R.: Advanced Mixed Integer Programming: Solving MIPs in practice. In: Combinatorial Optimization at Work 2, ZIB, Berlin (2009)

[BR07] Bixby, R., Rothberg, E.: Progress in computational mixed integer programminga look back from the other side of the tipping point. Annals of Operations Research **149**(1), 37–41 (2007)

[Con13] Coniglio, S.: Bound-optimal cutting planes. In: Proc. of Graph and Comb. Opt. (CTW 2013), pp. 59–62 (2013)

[FG92] Forrest, J., Goldfarb, D.: Steepest-edge simplex algorithms for linear programming. Math. Program. **57**(1), 341–374 (1992)

[Har73] Harris, P.: Pivot selection methods of the Devex LP code. Math. Program. **5**(1), 1–28 (1973)

[JT96] Johnson, D., Trick, M.: Cliques, Coloring, and Satisfiability, vol. 26 of DIMACS Series in Discrete Mathematics and Theoretical Computer Science. American Mathematical Society (1996)

[McC76] McCormick, G.: Computability of global solutions to factorable nonconvex programs: Part I - Convex underestimating problems. Math. Program. **10**(1), 147–175 (1976)

[MPT99] Martello, S., Pisinger, D., Toth, P.: Dynamic programming and strong bounds for the 0–1 knapsack problem. Manag. Sci. **45**(3), 414–424 (1999)

[ZFB11] Zanette, A., Fischetti, M., Balas, E.: Lexicography and degeneracy: can a pure cutting plane algorithm work? Math. Program. **130**(1), 153–176 (2011)

Separation of Generic Cutting Planes in Branch-and-Price Using a Basis

Marco E. Lübbecke and Jonas T. Witt[✉]

Operations Research, RWTH Aachen University,
Kackertstr. 7, 52072 Aachen, Germany
{marco.luebbecke,jonas.witt}@rwth-aachen.de

Abstract. Dantzig-Wolfe reformulation of a mixed integer program partially convexifies a subset of the constraints, i.e., it *implicitly* adds all valid inequalities for the associated integer hull. Projecting an optimal basic solution of the reformulation's LP relaxation to the original space does in general not yield a basic solution of the original LP relaxation. Cutting planes in the original problem that are separated using a basis like Gomory mixed integer cuts are therefore not directly applicable. Range [22] (and others) proposed as a remedy to heuristically compute a basic solution and separate this auxiliary solution also with cutting planes that stem from a basis. This might not only cut off the auxiliary solution, but also the solution we originally wanted to separate.

We discuss and extend Range's ideas to enhance the separation procedure. In particular, we present alternative heuristics and consider additional valid inequalities strengthening the original LP relaxation before separation. Our full implementation, which is the first of its kind, is done within the GCG framework. We evaluate the effects on several problem classes. Our experiments show that the separated cuts strengthen the formulation on instances where the integrality gap is not too small. This leads to a reduced number of nodes and reduced solution times.

1 Introduction

Branch-and-price has become a widely used technique for solving mixed integer programs (MIPs) with an embedded structure. The *original problem* is first reformulated using *Dantzig-Wolfe reformulation* and the reformulated problem is then solved with *branch-and-price* [12], where the *linear programming (LP) relaxation* is solved using *column generation* and specialized branching rules are applied. When additionally cutting planes are separated, the algorithm is called *branch-price-and-cut* [12].

Most often implementations are tailored for particular problems with known structure that can be exploited, but in the last decade also generic implementations were developed [15,20,21,25]. Bergner et al. [3] provide a computational proof-of-concept that the automatic detection of a suitable structure can be successful even when considering general problems.

Among others, cutting planes formulated with original variables were studied in the branch-price-and-cut literature. Adding these cuts to the problem does

© Springer International Publishing Switzerland 2015
E. Bampis (Ed.): SEA 2015, LNCS 9125, pp. 110–121, 2015.
DOI: 10.1007/978-3-319-20086-6_9

not change the structure of the pricing problem, whereas other types of cuts
do [11]. In several applications problem specific cuts formulated with original
variables are separated. Combinatorial cuts exploiting a particular substructure
can also be separated in a generic way [15]. Moreover, Range [22] introduced a
procedure to separate cuts in the original problem using a basis, but he only did
a preliminary computational study on elementary shortest path problems with
resource constraints, which was not successful (personal communication, 2013).

Our Contribution. A separation procedure that generates cuts in the original problem using a basis was mentioned by many authors [12,14], but only
Range [22] presented such a procedure without providing a computational study.
We discuss Range's ideas and present some extensions to enhance the separation
procedure. Furthermore, we implemented all ideas in the branch-price-and-cut
solver GCG [15] and tested the implementation on instances of several problem
classes. In particular, we computationally investigate the strength of the separated cutting planes and determine their influence on the overall solution process.

2   Dantzig-Wolfe Reformulation and Branch-and-Price

Let $n, m_1, m_2 \in \mathbb{Z}_{\geq 1}, q \in \mathbb{Z}_{\geq 0}$ be some integers, let $A \in \mathbb{Q}^{m_1 \times n}, D \in \mathbb{Q}^{m_2 \times n}$ be
some matrices, and let $b \in \mathbb{Q}^{m_1}, d \in \mathbb{Q}^{m_2}, c \in \mathbb{Q}^n$ be some vectors. Suppose we
are given the following *original problem*

$$\min\{c^T x : Ax \geq b, Dx \geq d, x \in \mathbb{Z}^{n-q} \times \mathbb{Q}^q\}$$

with mixed integer hull $P_{MIP} := \mathrm{conv}(\{x \in \mathbb{Z}^{n-q} \times \mathbb{Q}^q : Ax \geq b, Dx \geq d\})$,
where $\mathrm{conv}(S)$ denotes the convex hull of a set S. We will refer to its LP relax-
ation as *original LP relaxation* and denote the polyhedron of LP-feasible solutions by $P_{LP} := \{x \in \mathbb{Q}^n : Ax \geq b, Dx \geq d\}$.

When reformulating the original problem using Dantzig-Wolfe reformulation
for mixed integer programs [12], a part of the constraints, here $Dx \geq d$, is
convexified. Every solution $x \in X := \{x \in \mathbb{Z}^{n-q} \times \mathbb{Q}^q : Dx \geq d\}$ is reformulated
as a convex combination of extreme points $\{x^p\}_{p \in P}$ plus a non-negative linear
combination of extreme rays $\{x^r\}_{r \in R}$ of the associated convex hull $\mathrm{conv}(X)$:

$$\sum_{p \in P} x^p \lambda_p + \sum_{r \in R} x^r \lambda_r = x, \qquad \sum_{p \in P} \lambda_p = 1, \qquad \lambda_p \in \mathbb{Q}_{\geq 0} \quad \forall p \in P \cup R \ .$$

Replacing x by this combination while introducing new λ-variables results in
an *extended formulation* called the *master problem*. The corresponding LP
relaxation is called *linear master problem* and is solved with *column generation*, where a *pricing problem* over X is iteratively solved in order to generate
columns/variables having negative reduced cost. This procedure embedded in a
branch-and-bound tree is called *branch-and-price* [12].

It is known [26] that the optimal solution value of the linear master problem
is equal to $\min\{c^T x : Ax \geq b, x \in \mathrm{conv}(X)\}$, which corresponds to implicitly
adding all valid inequalities for $\mathrm{conv}(X)$ to the original LP relaxation.

3   Separation of Cutting Planes in Branch-and-Price

In each node of the branch-and-bound tree cutting planes can be added in order to strengthen the LP relaxation, which is then called *branch-price-and-cut* [12].

We can deal with valid inequalities formulated with original variables of the form $\pi^T x \geq \pi_0$, where $\pi \in \mathbb{Q}^n$ and $\pi_0 \in \mathbb{Q}$, in the same way as with constraints $Ax \geq b$. Therefore, adding these inequalities as cutting planes to the problem does not change the structure of the pricing problem, i.e., the set X is not affected. On the contrary, other types of cutting planes, e.g., cuts formulated with λ-variables that were introduced in the master problem, may change the structure of the pricing problem, which can hamper its computational tractability.

Let $\bar{\lambda}$ be an optimal *basic feasible solution* [4] of the linear master problem and suppose that the projection $\bar{x} := \sum_{p \in P} x^p \bar{\lambda}_p + \sum_{r \in R} x^r \bar{\lambda}_r$ onto the x-variables is not integer feasible for the original problem. We can try to separate $\bar{x}$ using problem specific cuts. Some of these cuts, e.g., knapsack or clique cuts, are implemented in state-of-the-art MIP solvers [1] and can automatically be applied in branch-price-and-cut algorithms if the original problem contains a particular substructure [15]. Additionally, there exist cuts that stem from a basis like Gomory mixed integer (GMI) cuts. These cuts are in general not directly applicable, because $\bar{x}$ does not have to be basic in the original LP relaxation [16], see Fig. 1.

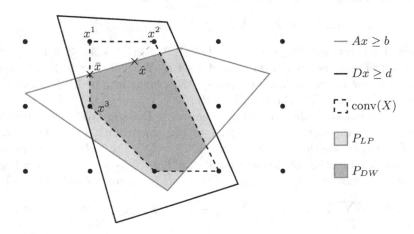

Fig. 1. Solution $\bar{x}$ is not a vertex of the polyhedron $P_{LP} = \{x : Ax \geq b, Dx \geq d\}$ and solution $\hat{x}$ is not even a vertex of $P_{DW} = \{x : Ax \geq b, x \in \text{conv}(X)\}$

In general we can check if the solution $\bar{x}$ is basic in the original LP relaxation by calculating the number of linear independent inequalities *active* at $\bar{x}$, i.e., satisfied with equality by $\bar{x}$. The solution $\bar{x}$ is basic if and only if this number is equal to the dimension n of the underlying vector space [4]. In case a description of $\text{conv}(X)$ is known explicitly, Goncalves' criterion [16] can be applied. Rios and Ross [23] proved that if the pricing problem consists of affinely independent extreme points, Goncalves' criterion is satisfied.

Motivated by the fact that cuts obtained from a basis have in general a larger impact than combinatorial cuts [5], we would like to apply these cuts in the context of branch-price-and-cut, too.

4 Basis Separation

We recall that cutting planes in the original problem stemming from a basis are not directly applicable, because the projected solution $\bar{x}$ is in general not basic in the original LP relaxation. An idea to overcome this issue is to calculate some basic feasible solution x^* and separate x^*. Since x^* is basic feasible, cuts stemming from a basis can be applied. The obtained cuts might not only cut off the basic feasible solution x^*, but also the solution $\bar{x}$ that we wanted to cut off initially. If the solution $\bar{x}$ is not cut off, we can strengthen the original LP relaxation by temporarily adding the obtained cuts to the problem formulation and repeat the procedure. Since the solution x^* is not feasible for the strengthened original LP relaxation, we will calculate a different basic feasible solution that can be used for separation in the following iteration. The resulting generic algorithm is described as Algorithm 1 and was initially proposed by Range [22].

Data: P_{MIP}, P_{LP}, $\bar{x}$, $p_{\min}$, and $i_{\max}$.
Result: Feasible solution $x^* \in P_{MIP}$ or set of coefficients $\bar{\Pi} \subseteq \mathbb{Q}^{n+1}$
            corresponding to cuts $\pi^T x \geq \pi_0$ with $(\pi, \pi_0) \in \bar{\Pi}$ separating $\bar{x}$.

$i := 0$, $\bar{\Pi} := \emptyset$, $\Pi^* := \emptyset$, $P'_{LP} := P_{LP}$;
while $|\bar{\Pi}| < p_{\min}$ *and* $i < i_{\max}$ **do**
  Calculate a vertex x^* of P'_{LP} (guided by $\bar{x}$);
  if $x^* \in P_{MIP}$ **then**
    | **return** x^*;
  Separate the solution x^* from P_{MIP} and let Π^* be the set of coefficients
  corresponding to the generated cuts;
  if $\Pi^* = \emptyset$ **then**
    | **break**;
  for $(\pi, \pi_0) \in \Pi^*$ **do**
    | **if** $\pi^T \bar{x} < \pi_0$ **then**
    | | $\bar{\Pi} := \bar{\Pi} \cup \{(\pi, \pi_0)\}$;
  end for
  $P'_{LP} := P'_{LP} \cap \{x : \pi^T x \geq \pi_0, (\pi, \pi_0) \in \Pi^*\}$;
  $i := i + 1$;
end while
return $\bar{\Pi}$;

Algorithm 1. The basis separation procedure

Note that we cannot guarantee to generate cuts that cut off the solution $\bar{x}$ when using Algorithm 1. Moreover, it highly depends on the types of cuts that are separated and how the basic feasible solutions are calculated.

4.1   Basis Heuristics

In this section we present approaches to cope with the crucial step in Algorithm 1 of calculating a basic feasible solution. Suppose we are given an optimal solution $\bar{\lambda}$ of the linear master problem. We want to find a basic feasible solution x^* of the original LP relaxation such that cuts separating x^* also tend to separate the solution $\bar{x} := \sum_{p \in P} x^p \bar{\lambda}_p + \sum_{r \in R} x^r \bar{\lambda}_r$. Approaches to obtain such a basic feasible solution are called *basis heuristics*. They were introduced by Range [22] in the context of branch-price-and-cut as well as by Dash and Goycoolea [10] in order to heuristically separate rank-1 GMI cuts. We will focus on basis heuristics based on solving linear programs in the following.

Original Objective. Probably the first idea that comes to mind is that we can obtain a basic feasible solution x^* of the original LP relaxation by solving the original LP relaxation. This basis heuristic will be called the *original objective*.

This approach is similar to *cut-first branch-and-price second* [6], where the original LP relaxation is solved first, cutting planes are added, and then the strengthened original problem is reformulated and solved using branch-and-price. A crucial difference is that cuts are added a priori to the problem in cut-first branch-and-price second without knowing if future solutions will ever violate these cutting planes. If we use the basis separation procedure instead, only cuts violated by the current solution of the master LP relaxation will be added, which is a clear advantage. A disadvantage of both approaches is that they are independent from the solution $\bar{x}$. They only depend on the original problem.

Range's Face Objective. In the following we present an alternative approach introduced by Range [22], where also the solution $\bar{x}$ is considered. Let

$$A' := \begin{pmatrix} A \\ D \end{pmatrix} \in \mathbb{Q}^{m \times n} \quad \text{and} \quad b' := \begin{pmatrix} b \\ d \end{pmatrix} \in \mathbb{Q}^m$$

with $m := m_1 + m_2$ be the constraint matrix and the left-hand side of the original problem. Furthermore, denote by A'_i the i-th row of the matrix A' for $i \in \{1, \ldots, m\}$ and let $I_0 := \{i \in \{1, \ldots, m\} : A'_i \bar{x} = b'_i\}$ be the set of indices corresponding to constraints of the original problem that are active at $\bar{x}$.

With the aim of obtaining a basic feasible solution of the original LP relaxation *near* $\bar{x}$, we solve the original LP relaxation using the *face objective* function

$$f(\bar{x}, x) := \sum_{i \in I_0} \frac{A'_i x - b'_i}{||A'_i||_2} \quad,$$

where $|| \cdot ||_2$ is the Euclidian norm.

The following proposition was initially proposed by Range [22].

Proposition 1 ([22]). *The solution $\bar{x}$ is an optimal feasible solution for the original LP relaxation with face objective and the optimal solution value is zero.*

Note that the *face* [4] $F := \{x \in \mathbb{Q}^n : A'x \geq b', A'_i x = b'_i \; \forall i \in I_0\}$ of the polyhedron P_{LP} is by definition of I_0 the face of smallest dimension containing $\bar{x}$. Since $x^* \in F$ holds, the solution x^* is at least contained in all faces the solution $\bar{x}$ is contained in. Hence, when solving the original LP relaxation with face objective using the simplex algorithm we obtain an optimal basic feasible solution x^* with $x^* \neq \bar{x}$ if and only if $\bar{x}$ is not basic feasible.

When the number of linearly independent rows A'_i with $i \in I_0$ is small in comparison to n, the information provided by the face objective is rather poor, because many linear independent inequalities active at $\bar{x}$ are missing to describe a basic solution.

Extended Face Objective. In the following we present an extension of the face objective taking also non-active constraints into account. For $k \in \mathbb{Z}_{\geq 0}$ we define the *k-activity* $g_k(\bar{x}, a, a_0)$ of an inequality $a^T x \geq a_0$ with $a \in \mathbb{Q}^n$ and $a_0 \in \mathbb{Q}$ at a given solution $\bar{x}$ as

$$g_k(\bar{x}, a, a_0) := \max\left(1 - \frac{a^T \bar{x} - a_0}{||a||_2}, 0\right)^k .$$

The k-activity $g_k(\bar{x}, a, a_0)$ describes how close to being active the inequality $a^T x \geq a_0$ is at $\bar{x}$. Note that $0 \leq g_k(\bar{x}, a, a_0) \leq 1$ holds and $g_k(\bar{x}, a, a_0) = 1$ if and only if the constraint $a^T x \geq a_0$ is active at $\bar{x}$. Furthermore, the greater the value k is chosen the smaller is the k-activity of a fixed non-active inequality.

We define the k-extended face objective, which is an extension of the face objective using the k-activity as a measure of the influence of a constraint:

$$f_k(\bar{x}, x) := \sum_{i=1}^{m} g_k(\bar{x}, A'_i, b'_i) \cdot \frac{A'_i x - b'_i}{||A'_i||_2} .$$

We additionally consider constraints that are almost active at $\bar{x}$, because if many of these constraints are active at a basic solution, this solution is intuitively a good approximation of the solution $\bar{x}$ we want to separate. Solving the original LP relaxation using the k-extended face objective yields such a basic solution.

Combination. We previously introduced three objectives that can be used as basis heuristics in combination with the original LP relaxation. The original objective is independent from the solution $\bar{x}$, whereas the face and the extended face objective are independent from the original objective function, they only depend on the solution $\bar{x}$ and the polyhedron P_{LP}. In the following we combine these objective functions in order to exploit as much information as possible.

We will combine the face and the original objective function by using a convex combination with coefficient $\alpha \in [0, 1]$

$$\min \alpha \cdot \frac{f(\bar{x}, x)}{|I_0|} + (1 - \alpha) \cdot \frac{c^T x}{||c||_2} .$$

Note that $|I_0|$ is the norm of the face objective. Analogously, we can combine the extended face objective and the original objective by using the norm $\sum_{i=1}^{m} g_k(\bar{x}, A'_i, b'_i)$ of the k-extended face objective. In the following we will present an approach to automatically choose a good value for α.

Remark that n linear independent inequalities are active at $\bar{x}$ if and only if the solution $\bar{x}$ is basic. Let $n(\bar{x})$ be the maximum number of linear independent inequalities active at $\bar{x}$ and define $\alpha(\bar{x}) := \frac{n(\bar{x})}{n} \in [0, 1]$, which can be used as a measure of how close $\bar{x}$ is to being basic. In the following we describe why $\alpha(\bar{x})$ is intuitively a suitable value for the convex combination coefficient α.

Obviously, $\alpha(\bar{x}) = 1$ if and only if $\bar{x}$ is basic. If $\alpha(\bar{x}) \approx 1$, then only few linear independent inequalities are missing to describe a basic solution. The influence of the face objective is increased, whereby the almost complete basis information of $\bar{x}$ will be exploited. On the contrary, if $\alpha(\bar{x}) \ll 1$, many linear independent inequalities are missing to describe a basic solution and the influence of the face objective, which contains only poor information, will be decreased.

5   Strengthening of the Original LP Relaxation Before Separation

In many applications the constraints $Dx \geq d$ are chosen in such a way that the LP relaxation of the master problem is much stronger than the one of the original problem. Thus, a basic solution of the original LP relaxation calculated during the basis separation procedure can only poorly approximate the solution $\bar{x}$ that was projected from the linear master problem. To counteract this and to enhance the basis separation procedure, we can try to imitate the convexification of the constraints $Dx \geq d$ by adding valid inequalities for $P_{DW} := \{x \in \text{conv}(X) : Ax \geq b\} \supseteq P_{MIP}$ before separation. In the following we present valid inequalities for P_{DW} that can be obtained while applying a branch-price-and-cut algorithm.

Range's Original Objective Cut. Range [22] suggests to add the *original objective cut* $c^T x \geq c^T \bar{x}$ to the problem in order to potentially strengthen the original LP relaxation. Note that this inequality holds for all $x \in P_{DW}$, because $\bar{x}$ is optimal for $\min\{c^T x : x \in P_{DW}\}$. If the objective function is known to be integral, e.g., $c \in \mathbb{Z}^n$ and $q = 0$, the inequality $c^T x \geq \lceil c^T \bar{x} \rceil$ can be added.

Reduced Cost Cuts. In each column generation iteration we solve a pricing problem over the set X in order to find negative reduced cost columns. Let $\pi^T x$ be the objective function of the pricing problem in some column generation iteration and let $\pi_0 := \min\{\pi^T x : x \in X\}$ be the optimal solution value of the corresponding pricing problem. Note that $\pi^T x \geq \pi_0$ is valid for $\text{conv}(X)$. Since $\text{conv}(X) \supseteq P_{DW}$, the inequality is also valid for P_{DW}. Inequalities of this type will be called *reduced cost cuts*, because they state that the reduced costs of all potential columns are greater than or equal to a specific value.

Pricing Cuts. Suppose the pricing problem is solved using branch-and-cut and in some column generation iteration a cutting plane $\pi^T x \geq \pi_0$ is separated in the pricing problem during separation at the root node. Since we optimize over X in the pricing problem, $\pi^T x \geq \pi_0$ is valid for $\text{conv}(X) \supseteq P_{DW}$. We will call such inequalities *pricing cuts*, because they are generated in the pricing problem.

6   Computational Setup and Results

We implemented the basis separation procedure including all presented features in GCG 2.0.1 [15] based on a development version of SCIP 3.1.0 [1] with CPLEX 12.5.0.0 as LP-solver. All computations were performed on Intel Core i7-2600 CPUs with 16GB of RAM on openSUSE 13.1 workstations running Linux kernel 3.11.10. We used a time limit of 3600 seconds in all our tests.

In GCG combinatorial cuts in the original problem are separated by default, but we will only report on the number of cuts that were separated by the basis separation procedure and were applied to the problem. Note that SCIP/GCG filters the separated cuts and only applies a subset of them. We used SCIP's separators with the aggressive setting to separate a basic feasible solution in Algorithm 1 and only separated cuts at the root node. In all our tests we used the values $p_{\min} = 1$ and $i_{\max} = 100$ for Algorithm 1. In order to compute $\alpha(\bar{x})$, we used the QR decomposition with column pivoting from Gnu Scientific Library [13].

We applied the branch-price-and-cut algorithm including basis separation to instances of the following problems: capacitated p-median problem (cpmp) [?], generalized assignment problem (gap) [8,9,18], resource allocation/temporal knapsack problem (rap) [7], and lot sizing problem (lotsizing) [24]. Furthermore, we applied the algorithm to instances of MIPLIB 2003 and MIPLIB 2010 (miplib) that were already successfully tested with a generic branch-price-and-cut code [3]. We only considered instances where separation at the root node could be applied.

6.1   Performance of the Basis Separation Procedure

In Table 1 we compare GCG using the default settings (def), basis separation with face objective (face), basis separation with the combination of face and original objective (face-conv), basis separation with the combination of 8-extended face and original objective (8-ext-conv), and basis separation with original objective (origobj). Additionally, we considered basis separation with k-extended face objective as well as the combination of k-extended face and original objective for $k \in \{4, 8, 12\}$, but preliminary tests have shown that the combination of 8-extended face and original objective outperforms these heuristics.

As we can see, on the majority of the cpmp, lotsizing, and miplib instances cuts are separated no matter which basis heuristic is used. Although the number of applied cuts is in shifted geometrical mean at most 19 over a test set and mostly much smaller, a non negligible part of the integrality gap at the root node is closed in comparison to the default setting. When using basis separation with the combination of 8-extended face and original objective, 8 percent of the

Table 1. Comparison of the percentage of affected instances (aff), i.e., some cuts were separated, the shifted geom. mean with shift value 1 of the int. gap at the root node in percent (gap), the number of applied cuts at the root node of the affected instances (cuts), and the time spent in the basis separation procedure (tm) over the whole testset. The best gap is written bold.

	def	face				face-conv				8-ext-conv				origobj			
	gap	gap	aff	cuts	tm	gap	aff	cuts	tm	gap	aff	cuts	tm	gap	aff	cuts	tm
cpmp_easy	1.21	1.15	78.8	4.3	1.0	1.15	79.8	4.3	1.1	**1.11**	81.7	4.0	1.1	1.18	35.6	1.9	1.7
cpmp_hard	4.18	4.05	87.5	9.3	2.5	4.04	87.5	9.4	2.8	**4.01**	89.1	9.3	3.4	4.13	48.4	1.9	4.3
gap_easy	0.15	0.15	4.2	1.0	0.5	0.15	4.2	1.0	0.6	**0.14**	16.7	2.8	0.5	0.15	8.3	1.0	0.5
gap_hard	0.31	0.31	0.0	0.0	3.3	0.31	0.0	0.0	3.9	0.31	0.0	0.0	3.8	0.31	0.0	0.0	4.5
ls_easy	3.31	2.08	95.2	5.8	0.5	2.11	95.2	5.8	0.6	**2.07**	85.7	6.4	0.6	2.36	85.7	4.0	0.5
ls_hard	13.36	13.04	66.7	14.5	0.5	13.07	66.7	15.5	1.5	**12.83**	66.7	17.5	1.4	12.96	66.7	8.2	0.5
rap_easy	0.04	0.04	0.0	0.0	0.9	0.04	0.0	0.0	1.0	0.04	0.0	0.0	1.0	0.04	6.2	2.0	1.5
rap_hard	0.09	0.09	0.0	0.0	2.4	0.09	0.0	0.0	2.5	0.09	0.0	0.0	2.5	0.09	20.7	1.7	2.9
miplib_easy	1.30	1.13	100.0	2.7	0.5	1.13	100.0	2.7	0.5	**0.87**	100.0	4.1	0.5	1.04	33.3	1.4	0.5
miplib_hard	4.94	4.79	66.7	16.2	1.6	4.77	73.3	14.0	11.9	4.91	73.3	14.4	11.6	**4.56**	73.3	14.0	1.6

gap on easy cpmp, 33 percent of the gap on easy miplib, and 37 of the gap percent on easy lotsizing instances is closed. Using any other basis heuristic closes less of the gap. On the corresponding hard instances up to 8 percent of the gap was closed due to basis separation, where the usage of the original objective or the combination of 8-extended face and original objective perform best.

On the contrary, almost no cuts were separated on gap and rap instances, which is probably due to the already very small integrality gap. Consequently, the usage of basis separation closes hardly anything of the integrality gap.

Note that most often only a few seconds are spent in the basis separation procedure. Only when using basis heuristics that have to compute the number of linear independent inequalities active at the current solution $\bar{x}$ in order to determine the value $\alpha(\bar{x})$, separation can take a bit longer on some hard instances. But in shifted geometrical mean over a test set it does not exceed 12 seconds.

In Fig. 2 the number of nodes and the solution times required by the settings with basis separation are compared to the default settings. On most test sets the solution time and even more significantly the number of nodes is reduced due to

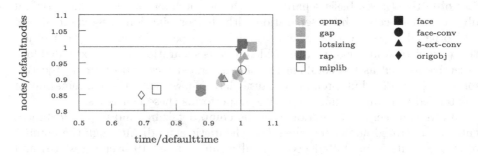

Fig. 2. Ratio between the shifted geom. mean with shift 100 (10) of the number of nodes (solution times) required by the settings with basis separation and the default settings.

separation. Only on gap and rap instances, where almost no cuts were separated, the solution time is increased. But since the basis separation procedure is quite fast, the increase in solution time is relatively small. Note that these results match the previous made observations concerning the integrality gap.

On lotsizing and miplib instances the solution time is decreased by up to 12 and 30 percent, respectively. On cpmp instances the solution time is only marginally decreased. During our computational study, we additionally observed that the number of solved instances of these problems is slightly increased when applying the basis separation procedure.

6.2 Influence of Strengthening the Original LP Relaxation Before Separation

In Table 2 the influence of the valid inequalities presented in section 5 on basis separation with the combination of the 8-extended face and the original objective is investigated. Namely these valid inequalities are the original objective cut (origobjcut), the pricing cuts (ppcuts), and the reduced cost cuts (redcostcuts).

Table 2. Comparison similar to Table 1

	basis-conv-8-ext				+origobjcut				+ppcuts				+redcostcuts			
	gap	aff	cuts	tm	gap	aff	cuts	tm	gap	aff	cuts	tm	gap	aff	cuts	tm
cpmp_easy	1.69	86.2	5.4	1.5	1.72	84.1	5.3	1.5	1.70	81.2	12.0	41.6	1.74	79.0	5.9	33.8
cpmp_hard	3.25	91.3	8.6	3.2	3.27	89.1	8.4	3.3	3.24	78.3	24.6	159.6	3.30	82.6	8.6	261.3
gap_easy	0.18	14.8	2.8	0.6	0.18	14.8	1.9	0.6	0.17	18.5	1.9	0.6	0.17	29.6	2.3	2.7
gap_hard	0.40	0.0	0.0	1.0	0.40	0.0	0.0	0.9	0.40	0.0	0.0	0.5	0.40	0.0	0.0	110.8
ls_easy	2.02	88.2	6.5	0.5	1.72	88.2	5.7	0.5	2.32	94.1	6.5	0.6	1.97	94.1	4.8	0.5
ls_hard	31.18	42.9	30.6	2.1	24.39	42.9	6.3	0.8	29.48	57.1	38.1	3.6	25.34	42.9	6.3	1.2
rap_easy	0.04	0.0	0.0	0.9	0.04	0.0	0.0	0.9	0.04	0.0	0.0	0.9	0.04	0.0	0.0	6.2
rap_hard	0.09	0.0	0.0	1.9	0.09	0.0	0.0	1.9	0.09	0.0	0.0	2.0	0.09	0.0	0.0	86.8
miplib_easy	0.91	100.0	4.3	0.5	0.91	100.0	3.6	0.5	1.36	80.0	5.9	34.6	0.95	100.0	4.1	0.8
miplib_hard	4.91	73.3	14.4	11.6	4.90	73.3	14.0	12.1	5.26	80.0	13.0	15.6	4.77	73.3	13.6	25.3

Notice that the percentage of affected instances is of similar magnitude no matter if the additional valid inequalities were added or not, whereas the number of applied cuts and the integrality gap vary considerably. Surprisingly, every setting that is shown in Table 2 provides on some test set the smallest integrality gap. So the impact of the valid inequalities is not solely positive. The same observation can be made when considering the number of applied cuts. Furthermore, the number of applied cuts and the size of the integrality do not seem to correlate.

On some instances the time spent in the basis separation procedure is noticeably increased due to the valid inequalities that were added before separation.

7   Conclusions and Future Work

We discussed and extended Range's approach [22] to separate cuts in the original problem using a basis in the context of branch-price-and-cut algorithms.

Furthermore, we implemented all ideas in GCG and presented the first computational study on a separation procedure of this kind. The cuts close part of the integrality gap at the root node on instances of various problem types, reducing the number of nodes and the solution time. On instances, where no cuts were found, solution times just slightly increased, because the separation procedure is relatively fast.

Whereas the combination of the 8-extended face and the original objective seems to be the basis heuristic that improves performance the most, computational results concerning the strengthening of the original LP relaxation before separation are not that clear, because they do not solely improve the performance. A task of future research should be finding a selection of valid inequalities that exclusively have a positive influence on the separation procedure.

The presented basis heuristics only compute feasible basic solutions, but Dash and Goycoolea [10] also use basis heuristics that compute infeasible basic solutions in order to heuristically separate rank-1 GMI cuts. Future work should include the implementation of these basis heuristics in our framework.

Since only auxiliary basic solutions and not the solutions we want to separate are used to generate cutting planes, a subject of future research should be the generation of additional valid inequalities as discussed in section 5 such that the solution we want to separate becomes a basic solution in the original LP relaxation whenever this is possible. If we managed to achieve this, we could obtain a corresponding dual solution and apply reduced cost fixing [17,19].

Our experiments suggest that there is a strong relation between the strength of the Dantzig-Wolfe reformulation and the success of separating violated cuts in the original problem. Future research should further examine this relation both computationally and theoretically.

References

1. Achterberg, T.: Constraint Integer Programming. Ph.D. thesis, Technische Universität Berlin (2007)
2. Beasley, J.: OR-Library: Distributing test problems by electronic mail. J. Oper. Res. Soc. **41**(11), 1069–1072 (1990)
3. Bergner, M., Caprara, A., Ceselli, A., Furini, F., Lübbecke, M., Malaguti, E., Traversi, E.: Automatic Dantzig-Wolfe reformulation of mixed integer programs. Math. Prog. **149**(1–2), 391–424 (2015)
4. Bertsimas, D., Tsitsiklis, J.: Introduction to Linear Optimization. Athena Scientific, Belmont (1997)
5. Bixby, R., Rothberg, E.: Progress in computational mixed integer programming - A look back from the other side of the tipping point. Annals of Operations Research **149**(1), 37–41 (2007). http://dx.doi.org/10.1007/s10479-006-0091-y
6. Bode, C., Irnich, S.: Cut-first branch-and-price-second for the capacitated arc-routing problem. Oper. Res. **60**(5), 1167–1182 (2012)
7. Caprara, A., Furini, F., Malaguti, E.: Uncommon Dantzig-Wolfe reformulation for the temporal knapsack problem. INFORMS J. Comput. **25**(3), 560–571 (2013)
8. Cattrysse, D.G., Salomon, M., Wassenhove, L.N.V.: A set partitioning heuristic for the generalized assignment problem. European J. Oper. Res. **72**(1), 167–174 (1994)

9. Chu, P.C., Beasley, J.E.: A genetic algorithm for the generalised assignment problem. Comput. Oper. Res. **24**(1), 17–23 (1997)
10. Dash, S., Goycoolea, M.: A heuristic to generate rank-1 GMI cuts. Math. Program. Comput. **2**(3–4), 231–257 (2010)
11. Desaulniers, G., Desrosiers, J., Spoorendonk, S.: Cutting planes for branch-and-price algorithms. Networks **58**(4), 301–310 (2011)
12. Desrosiers, J., Lübbecke, M.E.: Branch-price-and-cut algorithms. In: Cochran, J.J., Cox, L.A., Keskinocak, P., Kharoufeh, J.P., Smith, J.C. (eds.) Wiley Encyclopedia of Operations Research and Management Science. John Wiley & Sons, Inc. (2010)
13. Galassi, M., et al.: GNU scientific library reference manual. ISBN 0954612078
14. Galati, M.: Decomposition methods for integer linear programming. Ph.D. thesis, Lehigh University (2010)
15. Gamrath, G., Lübbecke, M.E.: Experiments with a generic Dantzig-Wolfe decomposition for integer programs. In: Festa, P. (ed.) SEA 2010. LNCS, vol. 6049, pp. 239–252. Springer, Heidelberg (2010)
16. Goncalves, A.S.: Basic feasible solutions and the Dantzig-Wolfe decomposition algorithm. J. Oper. Res. Soc. **19**(4), 465–469 (1968)
17. Irnich, S., Desaulniers, G., Desrosiers, J., Hadjar, A.: Path-reduced costs for eliminating arcs in routing and scheduling. INFORMS J. Comput. **22**(2), 297–313 (2010)
18. Osman, I.H.: Heuristics for the generalised assignment problem: Simulated annealing and tabu search approaches. OR Spectrum **17**(4), 211–225 (1995)
19. Poggi de Aragão, M., Uchoa, E.: Integer program reformulation for robust branch-and-cut-and-price. In: Mathematical Programming in Rio: A Conference in Honour of Nelson Maculan, pp. 56–61 (2003)
20. Puchinger, J., Stuckey, P., Wallace, M., Brand, S.: Dantzig-Wolfe decomposition and branch-and-price solving in G12. Constraints **16**(1), 77–99 (2011)
21. Ralphs, T., Galati, M.: DIP - Decomposition for integer programming (2009). https://projects.coin-or.org/Dip
22. Range, T.: An integer cutting-plane procedure for the Dantzig-Wolfe decomposition: Theory. Discussion Papers on Business and Economics 10/2006, Dept. Business and Economics. University of Southern Denmark (2006)
23. Rios, J., Ross, K.: Converging upon basic feasible solutions through Dantzig-Wolfe decomposition. Optim. Lett. **8**(1), 171–180 (2014)
24. Tempelmeier, H., Derstroff, M.: A lagrangean-based heuristic for dynamic multilevel multiitem constrained lotsizing with setup times. Management Science **42**(5), 738–757 (1996)
25. Vanderbeck, F.: BaPCod - A generic branch-and-price code (2005). https://wiki.bordeaux.inria.fr/realopt/pmwiki.php/Project/BaPCod
26. Vanderbeck, F., Savelsbergh, M.: A generic view of Dantzig-Wolfe decomposition in mixed integer programming. Oper. Res. Lett. **34**(3), 296–306 (2006)

On a Nonconvex MINLP Formulation of the Euclidean Steiner Tree Problem in n-Space

Claudia D'Ambrosio[1], Marcia Fampa[2], Jon Lee[3](✉), and Stefan Vigerske[4]

[1] LIX CNRS (UMR7161), École Polytechnique, 91128 Palaiseau Cedex, France
dambrosio@lix.polytechnique.fr
[2] Instituto de Matemática and COPPE, Universidade Federal do Rio de Janeiro,
Rio de Janeiro, RJ, Brazil
fampa@cos.ufrj.br
[3] IOE Department, University of Michigan, Ann Arbor, Michigan, USA
jonxlee@umich.edu
[4] ZIB (Konrad-Zuse-Zentrum für Informationstechnik Berlin), Berlin, Germany
vigerske@zib.de

Abstract. The Euclidean Steiner Tree Problem in dimension greater than 2 is notoriously difficult. Successful methods for exact solution are *not* based on mathematical-optimization — rather, they involve very sophisticated enumeration. There are two types of mathematical-optimization formulations in the literature, and it is an understatement to say that neither scales well enough to be useful. We focus on a known nonconvex MINLP formulation. Our goal is to make some first steps in improving the formulation so that large instances may eventually be amenable to solution by a spatial branch-and-bound algorithm. Along the way, we developed a new feature which we incorporated into the global-optimization solver SCIP and made accessible via the modeling language AMPL, for handling piecewise-smooth univariate functions that are globally concave.

1 Introduction

The Euclidean Steiner tree problem (ESTP) in $\mathbb{R}^n$ is: Given a set of finite points in $\mathbb{R}^n$, find a tree of minimal Euclidean length spanning these points, using or not additional points. Original points are *terminals* and additional nodes in the tree are *Steiner points*. The ESTP is NP-Hard [9], and interest in the problem stems from both the mathematical challenge and its potential applications (e.g., communications, infrastructure networks). In biology, [3] gives an application of the ESTP to phylogenetic analysis (i.e., the construction of evolutionary trees).

Basic properties of an optimal solution, called a *Steiner minimal tree (SMT)*, are: (i) A Steiner point in an SMT has degree 3; a Steiner point and its adjacent nodes lie in a plane, and the angles between the edges connecting the point to its adjacent nodes are 120 degrees. (ii) A terminal in an SMT has degree between 1 and 3. (iii) An SMT on p terminals has at most $p-2$ Steiner points (see [5,12]).

The *topology* of a Steiner tree is the tree for which we have fixed the number of Steiner points and the edges between all points, but not the position of the

© Springer International Publishing Switzerland 2015
E. Bampis (Ed.): SEA 2015, LNCS 9125, pp. 122–133, 2015.
DOI: 10.1007/978-3-319-20086-6_10

Steiner points. A topology is a *Steiner topology* if each Steiner point has degree 3 and each terminal has degree 3 or less. A Steiner topology with p terminals is a *full Steiner topology* if there are $p - 2$ Steiner points and each terminal has degree 1. A *full Steiner tree* is a Steiner tree corresponding to a full Steiner topology. A Steiner tree corresponding to some topology, but with certain edges shrunk to zero length, is *degenerate*. Any SMT with a non-full Steiner topology can be associated with a full Steiner topology for which the tree is degenerate.

Many papers have addressed exact solution in $\mathbb{R}^2$, and impressive results were obtained with the GeoSteiner algorithm [18]. But these algorithms cannot be applied when $n \geq 3$, and only a few papers have considered exact solution in this case. [11] proposed solving the problem in $\mathbb{R}^n$ by enumerating all Steiner topologies and computing a min-length tree associated with each topology, which in practice, can only solve very small instances because of the fast growth of the number of topologies as p increases. A branch-and-bound (b&b) algorithm for finding SMTs in $\mathbb{R}^n$ was proposed by Smith [14]. He presented a scheme for implicitly enumerating all full Steiner topologies on a given set of terminals, and he gave computational results sufficient to disprove for all $3 \leq n \leq 9$, an important conjecture of Gilbert and Pollak on the "Steiner ratio". Fampa and Anstreicher [6] used Smiths's enumeration scheme and proposed a conic formulation for the problem of locating the Steiner points for a given topology, to obtain a lower bound on the min tree length and to implement a "strong branching" technique. [15] presented geometric conditions that are satisfied by Steiner trees with a full topology and applied those conditions to eliminate candidate topologies in Smith's scheme. The best computational results for the ESTP for $n \geq 3$ are presented in these two last papers.

None of the works mentioned above have considered a math-programming formulation for the ESTP, which was presented only in [13] and [7]. [13] formulated the ESTP as a non-convex mixed-integer nonlinear programming (MINLP) problem and proposed a b&b algorithm using Lagrangian dual bounds. [7] presented a convex MINLP formulation that could be implemented in a b&b algorithm using bounds computable from conic problems. Both formulations use 0/1 variables to indicate whether the edge connecting two nodes is present in a Steiner topology. The presence of these 0/1 variables leads to a natural branching scheme, however neither [13] nor [7] present computational results.

We did some preliminary experiments with the nonconvex model. We solved some randomly generated instances using SCIP [1,16], and the results are dismal. Two difficulties observed have motivated this research: the weakness of the lower bounds given by the relaxations, and non-differentiability at points where the solution degenerates. In what follows, we investigate strategies to deal with these difficulties: (i) the use of approximate differentiable functions for the Euclidean norm, and (ii) nonconvex cuts based on geometric considerations.

2   A Nonconvex MINLP Formulation

[13] formulates the ESTP as a nonconvex MINLP problem, first defining a special graph $G = (V, E)$. Let $P := \{1, 2, ..., p\}$ be the indices associated with the given

terminals $a^1, a^2, ..., a^p$ and $S := \{p+1, p+2, ..., 2p-2\}$ be the indices associated with the Steiner points $x^{p+1}, x^{p+2}, ..., x^{2p-2}$. Let $V = P \cup S$. Denote by $[i, j]$ an edge of G, with $i, j \in V$ such that $i < j$. Define $E := E_1 \cup E_2$, where $E_1 := \{[i, j] : i \in P, j \in S\}$ and $E_2 := \{[i, j] : i \in S, j \in S\}$. Define a 0/1 y_{ij} for each edge $[i, j] \in E$, where $y_{ij} = 1$ if the edge $[i, j]$ is present in the SMT and 0 otherwise. The ESTP is then formulated as

$$\text{(MMX)} \quad \min \sum_{[i,j]\in E_1} \|a^i - x^j\| y_{ij} + \sum_{[i,j]\in E_2} \|x^i - x^j\| y_{ij}, \tag{1}$$

$$\sum_{j\in S} y_{ij} = 1, \quad \text{for } i \in P, \tag{2}$$

$$\sum_{i\in P} y_{ij} + \sum_{k<j,k\in S} y_{kj} + \sum_{k>j,k\in S} y_{jk} = 3, \quad \text{for } j \in S, \tag{3}$$

$$\sum_{k<j,k\in S} y_{kj} = 1, \quad \text{for } j \in S - \{p+1\}, \tag{4}$$

$$y_{ij} \in \{0,1\}, \ [i,j] \in E, \qquad x^i \in \mathbb{R}^n, \ i \in S, \tag{5}$$

where $\|v\| := \sqrt{\sum_{l=1}^{n} v_l^2}$ is the Euclidean norm of $v \in \mathbb{R}^n$. The constraints model a full Steiner topology for p given terminals in $\mathbb{R}^n$. Constraints (2) enforce that the degree of each terminal node is equal to 1. Constraints (3) enforce that the degree of each Steiner point is equal to 3, and constraints (4) eliminate cycles. Every full Steiner tree corresponds to a feasible solution of the formulation.

We aim to solve MMX using a spatial branch-and-bound (sbb) algorithm, as implemented in SCIP. This is not nearly straightforward, as we have to deal with non-differentiability of the distance function and with poor bounds that arise. In what follows, we propose approaches for handling these difficulties.

3   Dealing with the Non-differentiability

The continuous relaxation of MMX is a nonconvex NLP problem. Convergence of most NLP solvers (e.g. Ipopt [17]) requires that functions be twice continuously differentiable. This is not the case for MMX due to the non-differentiability of the Euclidean norm when the solution degenerates (i.e., when the norm is zero); and it is easy to see examples where the optimal solution does degenerate. In water-network optimization, [2] smooths away non-differentiability at zero of another function (modeling the pressure drop due to friction of the turbulent flow of water in a pipe). There are different ways that we can deal with the non-differentiability that we face. Let $w(x^i - x^j) := \|x^i - x^j\|^2$, so that $\|x^i - x^j\| = \sqrt{w(x^i - x^j)}$. In this way, we can focus on $\sqrt{\cdot}$, the source of the non-differentiability.

3.1   Implicit Square Roots

One possibility is to introduce an auxiliary variable z, and use the additional inequality $-z^2 + w \leq 0$ and the nonnegativity constraint $z \geq 0$. In this way, any optimal solution will have $z = \sqrt{w}$. This looks possibly attractive, but the overhead of so many additional nonnegatively-constrained variables and the difficulty of many additional nonconvex constraints is prohibitive.

On the other hand, while nonconvex, these functions $-z^2 + w$ manifest themselves as $-z_{ij}^2 + \sum_{l=1}^{n} v_l^2$, with $v = x^i - x^j$. The nonconvexity in $-z_{ij}^2 + \sum_{l=1}^{n} v_l^2$ is isolated to $-z_{ij}^2$, so it may be possible to adapt the techniques of [4] (exploiting concave separability), though we have to deal with the multiplication of distance variables z_{ij} by 0/1 variables y_{ij} in the objective function of MMX.

Another view is that $-z^2 + w \leq 0$ is equivalent to $\sqrt{w} \leq z$, which is manifested as the second-order cone constraint $\|x^i - x^j\|^2 \leq z_{ij}$. We can try to exploit methods for handling such constraints, though we have to deal with the multiplication of distance variables z_{ij} by 0/1 variables y_{ij} in the objective function.

3.2 Shifting

A simple fix is to approximate $\sqrt{w}$ by $h(w) := \sqrt{w + \delta} - \sqrt{\delta}$ for some small $\delta > 0$. Then we underestimate all positive distances (via the triangle inequality). Because our objective function is increasing in each distance, we get a *relaxation* of MMX. But a strong downside is that we underestimate *all* positive distances, and the error in each distance calculation rapidly approaches $\sqrt{\delta}$ as w increases.

3.3 Linear Extrapolation

The following approximation, depending on the choice of a $\sigma > 0$, was proposed in [10] to avoid non-differentiability of the Euclidean-distance function for the "traveling-salesman problem with neighborhoods."

$$l(w) := \begin{cases} \sqrt{w}, & \text{if } w \geq \sigma^2; \\ \frac{\sigma}{2} + \frac{1}{2\sigma}w, & \text{if } w \leq \sigma^2. \end{cases}$$

The function l is well-defined at $w = \sigma^2$. It is analytic except when $w = \sigma^2$, and in this case, it is still differentiable once. In fact, l simply uses the tangent at $w = \sigma^2$ of the graph of the strictly concave function $\sqrt{w}$ to overestimate $\sqrt{w}$ on $[0, \sigma^2)$. We already see that because l is not twice continuously differentiable, we should not expect good behavior from most NLP solvers. A strong shortcoming for our context is that l miscalculates zero distances; that is, $l(0) = \sigma/2$, while obviously $\sqrt{0} = 0$. Because $l(w)$ is an upperbound on $\sqrt{w}$, and because our objective function is increasing in distances, using the approximation l, we do *not* get a relaxation of MMX. Moreover, for degenerate Steiner trees, we will systematically overestimate distances that should be zero.

3.4 Smooth Under-Estimation

We propose another piecewise smoothing, using a particular homogeneous cubic depending on the choice of a $\lambda > 0$, that has very nice properties:

$$c(w) := \begin{cases} \sqrt{w}, & \text{if } w \geq \lambda^2; \\ \frac{15}{8\lambda}w - \frac{5}{4\lambda^3}w^2 + \frac{3}{8\lambda^5}w^3, & \text{if } w \leq \lambda^2. \end{cases}$$

We have depicted all of these smoothings in Fig. 1.

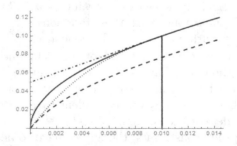

Fig. 1. Behavior of all smoothings ($\lambda^2 = \sigma^2 = 0.01$):

- the solid curve (———) is the true $\sqrt{\cdot}$ function.
- the "smooth underestimation" c, which we advocate, follows the dotted curve (⋯⋯⋯)
  below $w = 0.01$.
- the "linear extrapolation" l follows the dot-dashed line ($\cdot - \cdot - \cdot -$) below $w = 0.01$.
- both piecewise-defined functions c and l follow the true $\sqrt{\cdot}$ function above $w = 0.01$.
- the "shift" h (with $\delta = (4\lambda/15)^2$ chosen so that it has the same derivative as c
  does at 0) follows the consistent underestimate given by the dashed curve (- - - -).

The next result makes a very strong case for the smoothing c.

Thm. 1.

1. $c(w)$ agrees with $\sqrt{w}$ in value at $w = 0$;
2. $c(w)$ agrees with $\sqrt{w}$ in value, derivative and second derivative at $w = \lambda^2$;
   hence c is twice continuously differentiable;
3. c is strictly concave on $[0, +\infty]$ (just like $\sqrt{\cdot}$);
4. c is strictly increasing on $[0, +\infty]$ (just like $\sqrt{\cdot}$); consequently, $c(\|x^i - x^j\|^2)$
   is quasiconvex (just like $\|x^i - x^j\|$);
5. $\sqrt{w} - c(w) \geq 0$ on $[0, +\infty]$;
6. For all $\lambda > 0$, $\max\{(\sqrt{w} - c(w))/\lambda : w \in [0, +\infty]\}$ is the real root γ of
   $-168750 + 1050625x + 996300x^2 + 236196x^3$, which is approximately 0.141106.

Because distances only appear in the objective, and in a very simple manner, the approximation with c is a *relaxation* (due to (5) of Thm. 1). Meaning that the objective value of a global optimum using distance approximation is a lower bound on the true global optimum. And plugging the obtained solution into the true objective function gives an upper bound on the value of a true global optimum. So, in the end we get a solution and a bound on how close to optimal we can certify it to be. To be precise, for any $\lambda > 0$, let MMX(λ) denote MMX with all square roots in the norms replaced by the function c.

Cor. 2. *The optimal value of MMX is between the optimal value of MMX(λ) and the optimal value of MMX(λ) plus $\lambda\gamma(2p - 3)$ ($\gamma \approx 0.141106$).*

We note that the upper bound of Cor. 2 is a very pessimistic worst case — only achievable when the optimal tree has a full Steiner topology and all edges are very short. If such were the case, certainly λ should be decreased. Furthermore, we emphasize that if the length of every edge in the SMT that solves MMX(λ) is either zero (degenerate case) or greater than or equal to λ, then the optimal values of MMX and MMX(λ) are the same.

For $\delta = (4\lambda/15)^2$, we have $c'(0) = h'(0)$. That is, at this value of δ we can expect that c and h will have the same numerical-stability properties near 0.

Prop. 3. *For* $\delta = (4\lambda/15)^2$, *we have* $h(w) < c(w)$ *on* $(0, +\infty)$. *Moreover* $c(w) - h(w)$ *is strictly increasing on* $[0, +\infty)$.

Hence the relaxation provided using c is always at least as strong as the relaxation provided using h. In fact, strictly stronger in any realistic case (specifically, when there is more than 1 terminal). We make a few further observations about c:

Choosing λ. Note that $c'(0) = \frac{15}{8\lambda}$, so we should not be too aggressive in picking λ extremely small. But a large derivative at 0 is the price that we pay for getting everything else; in particular, any concave function f that agrees with $\sqrt{w}$ at $w = 0$ and $w = \lambda^2$ has $f'(0) \geq \frac{1}{\lambda}$. By Cor. 2, choosing λ to be around $2p-1$ would seem to make sense, thus guaranteeing that our optimal solution of MMX(λ) is within a universal additive constant of the optimal value of MMX. If the points are quite close together, either they should be scaled up or λ can be decreased.

Secant Lower Bound. Owing to the strict concavity, in the context of an sbb algorithm, c is always best lower bounded by a secant on *any* subinterval. Of course, it can be an issue whether a sbb solver can recognize and exploit this. In general, such solvers (e.g., `Baron` and `Couenne`) do not yet support general piecewise-smooth concave functions, while conceptually they could *and should*. However, we implemented a feature in `SCIP` (version 3.2) that allowed us to solve the proposed relaxation. In particular, by using the modeling-language `AMPL` interface and its general suffix facility (see [8]), we are able to specify to `SCIP` to treat a piecewise-defined constraint as globally concave. With this new `SCIP` feature we were able to solve the secant relaxation — see Section 5.

4   Tightening Relaxations

4.1   Integer Extreme Points

Here, we examine the continuous relaxation of the feasible region of MMX, just in the space of the y-variables. That is, the set of $y_{ij} \in [0, 1]$ satisfying (2-4). The coefficient matrix of the system (2-4) is *not* totally unimodular (TU). However, for each $j \in S - \{p + 1\}$, we can subtract the equation of (4) from the corresponding equation of (3) to arrive at the system:

$$\sum_{j \in S} y_{ij} \qquad\qquad\qquad\qquad = 1, \quad \text{for } i \in P, \qquad (6)$$

$$\sum_{i \in P} y_{ij} \qquad + \sum_{k>j, k \in S} y_{jk} \;\; = \begin{cases} 3, \text{ for } j = p+1, \\ 2, \text{ for } j \in S - \{p+1\}, \end{cases} \qquad (7)$$

$$\sum_{k<j, k \in S} y_{kj} \qquad\qquad\qquad = 1, \quad \text{for } j \in S - \{p+1\}. \qquad (8)$$

This resulting system is the set of constraints for the ordinary formulation of a *bipartite* 0/1 "*b*-matching" problem. Such a formulation has a TU constraint matrix. So we immediately have the following theorem (also observed in [13, §4.3,pp.217-9] via a much more complicated and less revealing proof) and corollaries:

Thm. 4. *The set of $y_{ij} \in [0,1]$ satisfying (2-4) has integer extreme points.*

Cor. 5. *No valid linear inequality in the y-variables alone can improve the linear relaxation of the equations (2-4) describing full Steiner topologies.*

This is not to say that *optimality*-based (linear) inequalities in the y-variables alone cannot be derived (see §4.2.2).

Cor. 6. *Given a globally-optimal solution of the continuous relaxation of MMX, in polynomial time we can calculate a globally-optimal solution of MMX.*

The relevant *b*-matching problem is depicted in Fig. 2. There is a node for every constraint of (6-8) and an edge for every variable. To the right of each node is its required degree. All possible edges exist between nodes at levels (6) and (7); i.e., those node sets induce a complete bipartite graph. All possible edges extending "down to the right" exist between nodes at levels (7) and (8); i.e., between nodes at level (7) and all nodes of greater number at level (8). A feasible choice of edges (selecting a full Steiner topology) is one that meets the degree requirements. E.g., we can see that we must choose the edge between node $p+1$ of level (7) and node $p+2$ of level (8).

4.2 Geometric Cuts

Based on various geometric considerations concerning optimal solutions, we can derive several families of valid inequalities seeking to improve relaxations of our formulation. Some of these are nonlinear, and it is an ongoing challenge to take advantage of them computationally. Until we address it in §4.2.3, *assume that all norms are taken exactly — not smoothed.*

Let η_i be the distance from terminal a^i to the nearest other terminal; i.e.,

$$\eta_i := \min_{j \in P, \, j \neq i} \{\|a^i - a^j\|\}, \; \forall \, i \in P. \qquad (9)$$

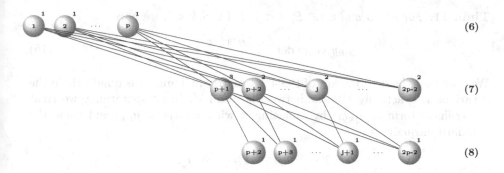

(6)

(7)

(8)

Fig. 2. The bipartite b-matching model for selecting a full Steiner topology

4.2.1 Non-combinatorial Cuts

Thm. 7. *For all $n \geq 2$, we have*

$$y_{ik}\left(\|x^k - a^i\|\right) \leq \eta_i, \ \forall \, i \in P, \ k \in S. \tag{10}$$

Lem. 8. *Among triangles with edge lengths a, b, c and corresponding angles x, y, z, with c and z fixed, the one maximizing $a + b$ is isosceles (that is $a = b$, $x = y$).*

Thm. 9. *For all $n \geq 2$, we have*

$$y_{ik}y_{jk}\left(\|x^k - a^i\| + \|x^k - a^j\|\right) \leq \tfrac{2}{\sqrt{3}}\|a^i - a^j\|, \ \forall \, i, j \in P, \ i < j, \ k \in S. \tag{11}$$

In computations, we treat the bilinear term $y_{ik}y_{jk}$ by replacing it with a variable y_{ijk} and using the standard McCormick inequalities.

Another way to try and use the same principle is as follows:

Thm. 10. *For all $n \geq 2$, we have*

$$y_{ik}y_{jk}\|x^k-a^i\|\|x^k-a^j\|\left(\|x^k - a^i\| - \tfrac{1}{\sqrt{3}}\|a^i - a^j\|\right) = 0, \ \forall \, i, j \in P, \ i < j, \ k \in S. \tag{12}$$

Thms. 9 and 10 easily extend to the case where the Steiner point x^k is adjacent to only 1 terminal in the SMT and also to the case where x^k is not adjacent to any terminal. In the such cases, we have

$$y_{ik}y_{kl}\left(\|x^k - a^i\| + \|x^k - x^l\|\right) \leq \tfrac{2}{\sqrt{3}}\|a^i - x^l\|, \ \forall \, i \in P, \ k, l \in S, \ k < l; \tag{13}$$

$$y_{kl}y_{km}\left(\|x^k - x^l\| + \|x^k - x^m\|\right) \leq \tfrac{2}{\sqrt{3}}\|x^l - x^m\|, \ \forall \, k, l, m \in S, \ k < l < m. \tag{14}$$

Via another geometric principle, we have the following result.

Thm. 11. *For* $n = 3$ *and* $i, j \in P$, $i < j$, $k, l \in S, k < l$, *we have*

$$y_{ik}y_{jk}y_{kl} \cdot \det \begin{bmatrix} a^i & a^j & x^k & x^l \\ 1 & 1 & 1 & 1 \end{bmatrix} = 0. \tag{15}$$

We note that the determinant of the 4×4 matrix in Thm. 11 is quadratic in the x-variables — actually bilinear between x^k and x^l. In computations, we treat the trilinear term $y_{ik}y_{jk}y_{kl}$ by replacing it with a variable y_{ijkl} and using the standard inequalities

$$y_{ijkl} \leq y_{ik}, \quad y_{ijkl} \leq y_{jk}, \quad y_{ijkl} \leq y_{kl},$$
$$y_{ik} + y_{jk} + y_{kl} \leq 2 + y_{ijkl}.$$

Thm. 11 easily extends to dimensions $n > 3$.

Thm. 12. *For* $n \geq 3$, $i, j \in P$, $i < j$, $k, l \in S, k < l$, *and for every* 3×4 *submatrix* $B = [b^i, b^j, \xi^k, \xi^l]$ *of the* $n \times 4$ *matrix* $[a^i, a^j, x^k, x^l]$, *we have*

$$y_{ik}y_{jk}y_{kl} \cdot \det \begin{bmatrix} b^i & b^j & \xi^k & \xi^l \\ 1 & 1 & 1 & 1 \end{bmatrix} = 0. \tag{16}$$

4.2.2 Combinatorial Cuts.

In this section, we introduce some valid linear inequalities satisfied by optimal topologies of the ESTP.

Thm. 13. *For* $n \geq 2$ *and* $i, j \in P$, $i < j$, *we have*

$$\text{If } \|a^i - a^j\| > \eta_i + \eta_j, \text{ then } y_{ik} + y_{jk} \leq 1, \ \forall \, k \in S. \tag{17}$$

Considering the valid inequalities in (17), we note that the inequalities (11) can only be active for $i, j \in P$, $i < j$, $k \in S$, such that $\|a^i - a^j\| \leq \eta_i + \eta_j$. Therefore, only such valid inequalities should be included in MMX. Furthermore, the result in Thm. 13 extends to the more general case addressed in Thm. 14.

Thm. 14. *Let* H *be a graph with vertex set* P, *and such that* i *and* j *are adjacent if* $\|a^i - a^j\| > \eta_i + \eta_j$. *Then for each* $k \in S$, *the set of* $i \in P$ *such that* $y_{ik} = 1$ *in an optimal solution is a stable set of* H. *Therefore, every valid linear inequality* $\sum_{i \in P} \alpha_i y_i \leq \rho$ *for the stable set polytope of* H *yields a valid linear inequality* $\sum_{i \in P} \alpha_i y_{ik} \leq \rho$ *for each* $k \in S$.

Let T be a min-length spanning tree on terminals a^i, $i \in P$. For $i, j \in P$, let

$$\beta_{ij} := \text{length of the longest edge on the path between } a^i \text{ and } a^j \text{ in } T. \tag{18}$$

The proof of the following well-known lemma can be found for example in [12]. We use it to prove the validity of cuts presented in Cor. 16.

Lem. 15. *An SMT contains no edge of length greater than* β_{ij} *on the unique path between* a^i *and* a^j, *for all* $i, j \in P$.

Cor. 16. *For* $n \geq 2$ *and* $i, j \in P$, $i \neq j$, *we have*

$$\text{If } \|a^i - a^j\| > \eta_i + \eta_j + \beta_{ij}, \text{ then } y_{ik} + y_{kl} + y_{jl} \leq 2, \ \forall \, k, l \in S, \ k < l. \tag{19}$$

4.2.3   Smoothing in the Context of Cuts. In the definition and derivation of all cuts, we assumed that norms are taken exactly — not smoothed. Now, we confront the issue that we prefer to work with smoothed norms in the context of mathematical optimization.

First of all, for any norm that just involves data and no variables, we do *not* smooth. This pertains to all occurrences of $\|a^i - a^j\|$, and hence also all occurrences of the parameters defined in (9) and (18). Any valid equation or inequality based *only* on such use of norms (or not based on norms at all) is valid for the original problem. So, these equations and inequalities do not exclude optimal solutions of the original problem, and so these solutions are candidate solutions to the problem where distances are smoothly underestimated (employing h or c), possibly with lower objective value. Therefore, any such equation or inequality is valid for the problem with distances smoothly underestimated. This applies to (15), (16), (17), inequalities based on Thm. 14, and (19).

For (10) and (11), norms involving variables occur only on the low side of the inequalities, so smooth underestimation of these norms, employing h or c, keeps them valid.

Inequalities (13,14) also contain norms involving variables on the high side of those inequalities. For those norms, we should replace them with smooth *overestimates*. For example, we could use an "overestimating shift" $\hat{h}(w) := \sqrt{w + \delta}$, or the linear extrapolation l (possibly with a different breakpoint). In fact, by choosing the breakpoint for l at $\sigma^2 := (4\lambda/15)^2$, where λ^2 is the breakpoint for c, we get $c'(0) = l'(0)$ ($= 15/8\lambda$). That is, if we can numerically tolerate a breakpoint for the underestimate c at $w = \lambda^2$, then we can equally tolerate a breakpoint for the overestimate l at $w = \sigma^2 = (4\lambda/15)^2$. While global sbb solvers do not yet support piecewise functions, at the modeling level we could utilize tangents of the concave $\sqrt{\cdot}$ at a few values of w greater than $(4\lambda/15)^2$. I.e., choose some values $\sigma_r^2 > \cdots > \sigma_1^2 > \sigma_0^2 := (4\lambda/15)^2$, let $\tau_{\sigma_i}(w) := \frac{\sigma_i}{2} + \frac{1}{2\sigma_i}w$, and we can instantiate (13) and (14) $r + 1$ times each, replacing the $\sqrt{\cdot}$ implicit on the high side of these inequalities with τ_{σ_i}, for $i = 0, 1, \ldots, r$.

Finally, we have the *equation* (12). The norms $\|x^k - a^i\|$ and $\|x^k - a^j\|$ can be smoothed in any way that correctly evaluates the norm at 0, and the equation remains valid. So employing h or c leaves the equation valid. The norm involving x^k in the multiplicand $\left(\|x^k - a^i\| - \|a^i - a^j\|/\sqrt{3}\right)$ is thornier. One way to address it is to simply replace it with $\left(\|x^k - a^i\|^2 - \|a^i - a^j\|^2/3\right)$, with these (smooth) squared norms calculated exactly.

5   Experiments

In this section, we demonstrate the impact of our cuts on the solution of the ESTP, with results on 25 instances in $\mathbb{R}^3$, where the terminals are randomly distributed in the cube $[0, 10]^3$. For each value of p from 4 to 8, we created five instances. We show the effect of the cuts by solving the instances a few times with the sbb code SCIP, adding different cuts each time to MMX. In our experiments, we considered cuts (10,11, 13–15, 17). We compare the performance of SCIP on

Table 1. Average % improvements on running time compared to MMX

p	MMX+						(10,14,17)
	(10)	(11)	(13)	(14)	(15)	(17)	
4	51	2	20	0	-21	24	41
5	44	-10	21	0	-82	4	56
6	92	72	91	88	-462	93	93
7	81	-274	48	56	-654	76	84
8	29	0	0	6	0	15	32

eight models, running with a time limit of 2 hours. The first model is MMX, with no cuts, the following six are MMX with the independent addition of cuts, and the last model is MMX with the addition of the three most effective cuts on the previous experiments.

Summarizing our results, we have that adding the cuts have significant effect on improving the lower bounds and decreasing the running time. The three classes of cuts, (10), (14) and (17) together improve the lower bound computed in the time limit, in the only two instances still not solved to optimality in 51% and 52%, and decreased the overall running time in 61% on average. Three extra instances, all with $p = 8$, were solved to optimality in the time limit, after the addition of the cuts. In Table 1, we show for each value of p, the average percentage improvements on the running time of each model tested, when compared to MMX ($100(\text{Time}(\text{MMX}) - \text{Time}(\text{Model}))/\text{Time}(\text{MMX})$). (Negative values indicate worse times with the addition of the cuts)

Although these are still preliminary results, we see that our cuts can potentially improve the quality of the lower bounds computed by SCIP. The two other cuts proposed (16,19) have a positive effect in instances with more terminals or in higher dimensions. Finally, we see that using several families of cuts together can bring further improvements.

6   Conclusions

The ESTP is very difficult to solve for $n > 2$, and the application of sbb solvers to test instances using the MMX formulation points out considerable drawbacks, concerned with the non-differentiability of the Euclidean norm, and also to the extreme weakness of the lower bounds given by relaxations. MMX in its original form, with today's sbb solvers, leads to dismal results.

We presented different approximations for the square-roots — the source of our non-differentiability, which can be judiciously applied in the context of valid inequalities. In particular, we introduced a smooth underestimation function, and we established several very appealing properties. We implemented this smoothing with a new feature of SCIP that we developed. This feature could be specialized to automatically smooth roots and other power functions.

To improve the quality of lower bounds computed in an sbb algorithm, we presented a variety of valid inequalities, based on known geometric properties of a SMT. Preliminary numerical experiments demonstrates the potential of these cuts in improving lower bounds. Many of these cuts are nonconvex, and it is an interesting research direction to apply similar ideas to other nonconvex gloabl-optimization models. We demonstrated that the performance of the MMX model can be significantly improved, though it is still not the best method for ESTP with $n > 2$. So more work is needed to make the MMX approach competitive.

Acknowledgments. J. Lee was partially supported by NSF grant CMMI–1160915 and ONR grant N00014-14-1-0315, and Laboratoire d'Informatique de l'École Polytechnique. M. Fampa was partially supported by CNPq and FAPERJ.

References

1. Achterberg, T.: SCIP: Solving constraint integer programs. Mathematical Programming Computation **1**(1), 1–41 (2009)
2. Bragalli, C., D'Ambrosio, C., Lee, J., Lodi, A., Toth, P.: On the optimal design of water distribution networks. Optim. Eng. **13**(2), 219–246 (2012)
3. Cavalli-Sforza, L., Edwards, A.: Phylogenetic analysis: Models and estimation procedures. Evolution **21**, 550–570 (1967)
4. D'Ambrosio, C., Lee, J., Wächter, A.: A global-optimization algorithm for mixed-integer nonlinear programs having separable non-convexity. In: Fiat, A., Sanders, P. (eds.) ESA 2009. LNCS, vol. 5757, pp. 107–118. Springer, Heidelberg (2009)
5. Du, D., Hu, X.: Steiner tree problems in computer communication networks. World Scientific Publishing Co. Pte. Ltd., Hackensack (2008)
6. Fampa, M., Anstreicher, K.M.: An improved algorithm for computing Steiner minimal trees in Euclidean d-space. Discrete Optimization **5**(2), 530–540 (2008)
7. Fampa, M., Maculan, N.: Using a conic formulation for finding Steiner minimal trees. Numerical Algorithms **35**(2–4), 315–330 (2004)
8. Fourer, R., Gay, D.M., Kernighan, B.W.: AMPL: A Modeling Language for Mathematical Programming. Duxbury Press, November 2002
9. Garey, M., Graham, R., Johnson, D.: The complexity of computing Steiner minimal trees. SIAM J. Applied Mathematics (1977)
10. Gentilini, I., Margot, F., Shimada, K.: The travelling salesman problem with neighbourhoods: MINLP solution. Opt. Meth. and Soft. **28**(2), 364–378 (2013)
11. Gilbert, E., Pollack, H.: Steiner minimal trees. SIAM J. Applied Math. (1968)
12. Hwang, F., Richards, D., Winter, W.: The Steiner tree problem. Ann. of Disc. Math. Elsevier, Amsterdam (1992)
13. Maculan, N., Michelon, P., Xavier, A.: The Euclidean Steiner tree problem in R^n: A mathematical programming formulation. Ann. OR **96**(1–4), 209–220 (2000)
14. Smith, W.D.: How to find Steiner minimal trees in Euclidean d-space. Algorithmica **7**(1–6), 137–177 (1992)
15. Van Laarhoven, J.W., Anstreicher, K.M.: Geometric conditions for Euclidean Steiner trees in R^d. Comput. Geom. **46**(5), 520–531 (2013)
16. Vigerske, S.: Decomposition of multistage stochastic programs and a constraint integer programming approach to MINLP. PhD thesis, Humboldt-U. Berlin (2013)
17. Wächter, A., Biegler, L.: On the implementation of an interior-point filter line-search algorithm for large-scale NLP. Math. Prog. **106**, 25–57 (2006)
18. Warme, D., Winter, P., Zachariasen, M.: Exact algorithms for plane Steiner tree problems: a computational study. Advances in Steiner, 81–116 (1998)

Scheduling and Allocation

Scheduling MapReduce Jobs and Data Shuffle on Unrelated Processors

Dimitris Fotakis[1], Ioannis Milis[2], Orestis Papadigenopoulos[1],
Emmanouil Zampetakis[3], and Georgios Zois[2]([✉])

[1] School of Electrical and Computer Engineering,
National Technical University of Athens, Athens, Greece
`fotakis@cs.ntua.gr, opapadig@corelab.ntua.gr`
[2] Department of Informatics, Athens University of Economics
and Business, Athens, Greece
`{milis,georzois}@aueb.gr`
[3] CSAIL, Massachusetts Institute of Technology, Cambridge, MA, USA
`mzampet@mit.edu`

Abstract. We propose a constant approximation algorithm for generalizations of the Flexible Flow Shop (FFS) problem which form a realistic model for non-preemptive scheduling in MapReduce systems. Our results concern the minimization of the total weighted completion time of a set of MapReduce jobs on unrelated processors and improve substantially on the model proposed by Moseley et al. (SPAA 2011) in two directions: (i) we consider jobs consisting of multiple Map and Reduce tasks, which is the key idea behind MapReduce computations, and (ii) we introduce into our model the crucial cost of the data shuffle phase, i.e., the cost for the transmission of intermediate data from Map to Reduce tasks. Moreover, we experimentally evaluate our algorithm compared with a lower bound on the optimal cost of our problem as well as with a fast algorithm, which combines a simple online assignment of tasks to processors with a standard scheduling policy. As we observe, for random instances that capture data locality issues, our algorithm achieves a better performance.

1 Introduction

The widespread use of MapReduce [6] to implement massive parallelism for data intensive computing motivates the study of new challenging shop scheduling

This research was supported by the projects *"Handling uncertainty in data intensive applications on a distributed computing environment (cloud computing) (DELUGE)"* (D. Fotakis, I. Milis and E. Zampetakis) and *"Energy Efficiency of Road Networks and Vehicles: Measurement, Pricing, Regional and Environmental Effects (EERNV)"* (G.Zois), co-financed by the European Union (European Social Fund - ESF) and Greek national funds, through the Operational Program "Education and Lifelong Learning" of the National Strategic Reference Framework (NSRF) - Research Funding Program: THALES, investing in knowledge society through the European Social Fund. A short extended abstract of this work, including partial results, appeared in EDBT/ICDT 2014 Workshop on Algorithms for MapReduce and Beyond.

© Springer International Publishing Switzerland 2015
E. Bampis (Ed.): SEA 2015, LNCS 9125, pp. 137–150, 2015.
DOI: 10.1007/978-3-319-20086-6_11

problems. Indeed, a MapReduce job consists of a set of Map tasks and a set of Reduce tasks that can be executed simultaneously, provided that no Reduce task of a job can start execution before all the Map tasks of this job are completed. Moreover, a significant part of the processing cost in MapReduce applications is the communication cost due to the transmission of intermediate data from Map tasks to Reduce tasks (a.k.a. data shuffle, see e.g., [1]). To exploit the inherent parallelism, the scheduler, which operates in centralized manner, has to efficiently assign and schedule Map and Reduce tasks to the available processors. In this context, standard shop scheduling problems are revisited to capture key constraints and singularities of MapReduce systems. In fact, a few results have been recently proposed based on simplified abstractions and resulting in known variants of the classical Open Shop and Flow Shop scheduling problems [3,4,12].

In this work, we significantly generalize the Flexible Flow Shop (FFS) model for MapReduce scheduling proposed in [12]. Recall that in the FFS problem, we are given a set of jobs, each consisting of a number of tasks (each task corresponds to a stage), to be scheduled on a set of parallel processors dedicated to each stage. The jobs should be executed in the same fixed order of stages, without overlaps between different stages of the same job. Our generalization extends substantially the model proposed in [12] by taking into account all the important constraints of MapReduce systems: (a) each job has multiple tasks in each stage; (b) the assignment of tasks to processors is flexible; (c) there are dependencies between Map and Reduce tasks; (d) the processors are unrelated to capture data locality; and (e) there is a significant communication cost for the data shuffle. Our goal is to find a non-preemptive schedule minimizing the standard objective of total weighted completion time for a set of MapReduce jobs.

Related Work. Known results for the FFS problem concern the two-stage case on parallel identical processors. For the makepsan objective a PTAS is known [13], while for the objective of total weighted completion time, a simple 2-approximation algorithm was proposed in [8] for the special case where each stage has to be executed on a single processor. For the latter case, [12] recently proposed a QPTAS which becomes a PTAS for a fixed number of task processing times.

In the MapReduce context, most of the previous work concerns the experimental evaluation of scheduling heuristics, from the viewpoint of finding good tradeoffs between different practical criteria (see e.g., [15]). From a theoretical viewpoint, all known results [3,4,12] concern the minimization of total weighted completion time. Chang et al. [3] studied a simple model, equivalent to the well-known *concurrent open shop* problem [11], where there are no dependencies between Map and Reduce tasks and the assignment of tasks to processors is given. Chen et al. [4] generalized the last model by considering dependencies between Map and Reduce tasks and presented an LP-based 8-approximation algorithm. Moreover, they managed to incorporate the data shuffle into their model and to derive a 58-approximation algorithm. Finally, Moseley et al. [12] suggested the connection of MapReduce scheduling to the FFS problem and proposed a 12-approximation algorithm, for the case of identical processors, and a 6-approximation algorithm for the very restricted case of unrelated processors where each job has a single Map and a single Reduce task. For both cases they also proposed constant competitive online algorithms with constant speed augmentation.

Our Results and Contributions. We present constant approximation algorithms which substantially generalize the results of [12] for MapReduce scheduling on unrelated processors in two directions motivated by practical applications of MapReduce systems. In fact, we deal with jobs consisting of multiple Map and Reduce tasks and also incorporate the shuffle phase into our setting. As it has been observed in [12], new ideas and techniques are required for both these directions.

In Section 2, we present a 54-approximation algorithm for the Map-Reduce scheduling problem when jobs consist of multiple Map and Reduce tasks. We first give an interval-indexed LP-relaxation for the problem of minimizing the total weighted completion times separately for Map and Reduce tasks on unrelated processors. Our LP-relaxation is inspired by that proposed by Hall et al. [9] for scheduling a set of single task jobs on unrelated processors under the same objective. However, in our setting, only the task finishing last (instead of all tasks) contributes to the objective value, which complicates the analysis. Recently, Correa et al. [5] proposed a similar LP-relaxation for a more general problem, where, instead of jobs consisting of tasks, they have a set of job orders and the completion time of each order is specified by the completion of the job finishing last. Since scheduling multitask jobs is quite similar to the setting considered in [5], we can apply their approximation result to scheduling the Map and Reduce tasks separately. Next, extending the ideas in [12] for single task jobs, we concatenate the two schedules into a single schedule that respects the task dependencies.

In Section 3, we incorporate the data shuffle phase into our model by introducing an additional set of *Shuffle tasks*, each one associated with a communication cost (expressed as processing time). When the Shuffle tasks are scheduled on the same processors as the corresponding Reduce tasks, we are able to keep the same approximation ratio of 54 for the Map-Shuffle-Reduce scheduling problem. Moreover, we prove an approximation ratio of 81 when the Shuffle tasks can be executed on different processors than their corresponding Reduce tasks. To the best of our knowledge, this is the most general setting of the FFS problem (with a special third stage) for which a constant approximation guarantee is known.

In Section 4, we compare experimentally the performance of our LP-based approximation algorithm with a lower bound on the optimal cost of our problem as well as with a simple and fast algorithm. The latter algorithm combines a simple assignment of the tasks, using an online algorithm for makespan minimization on unrelated processors with logarithmic competitive ratio [2], with the standard Weighted Shortest Processing Time first (WSPT) scheduling policy. As we observe, for instances where the processing times of Map and Reduce tasks are drawn from the same uniform distributions, the simple algorithm performs well enough, while, for instances that capture data locality issues, the more sophisticated LP-based algorithm achieves a better performance also in practise. Moreover, we show that the (empirical) approximation ratio of our algorithms is considerably smaller than the corresponding theoretical upper bound. As far as we know, these are the first experimental results for evaluating the performance guarantee of MapReduce scheduling on unrelated processors.

Problem Statement and Notation. In the sequel we consider a set $\mathcal{J} = \{1, 2, \ldots, n\}$ of n MapReduce jobs to be executed on a set $\mathcal{P} = \{1, 2, \ldots, m\}$ of m unrelated processors. Each job is available at time zero, is associated with a positive weight w_j and consists of a set of Map tasks and a set of Reduce tasks. Let $\mathcal{M}$ and $\mathcal{R}$ be the set of all Map and all Reduce tasks respectively. Each task is denoted by $\mathcal{T}_{k,j} \in \mathcal{M} \cup \mathcal{R}$, where $k \in N$ is the task index of job $j \in \mathcal{J}$ and is associated with a vector of non-negative processing times $\{p_{i,k,j}\}$, one for each processor $i \in \mathcal{P}_b$, where $b \in \{\mathcal{M}, \mathcal{R}\}$. Let $\mathcal{P}_{\mathcal{M}}$ and $\mathcal{P}_{\mathcal{R}}$ be the set of Map and the set of Reduce processors respectively. For convenience, we assume that $\mathcal{P}_{\mathcal{M}} \cap \mathcal{P}_{\mathcal{R}} = \emptyset$, however we are able to extend our results to the case where the two sets of processors are not necessarily disjoint (or are even identical). Each job has at least one Map and one Reduce task and every Reduce task can start its execution after the completion of all Map tasks of the same job.

For a given schedule we denote by C_j and $C_{k,j}$ the completion times of each job $j \in \mathcal{J}$ and each task $\mathcal{T}_{k,j} \in \mathcal{M} \cup \mathcal{R}$ respectively. Note that, due to the precedence constraints between Map and Reduce tasks, $C_j = \max_{\mathcal{T}_{k,j} \in \mathcal{R}} \{C_{k,j}\}$. By $C_{max} = \max_{j \in \mathcal{J}} \{C_j\}$ we denote the makespan of the schedule, i.e., the completion time of the job which finishes last. Our goal is to schedule *non-preemptively* all Map tasks on processors of $\mathcal{P}_{\mathcal{M}}$ and all Reduce tasks on processors of $\mathcal{P}_{\mathcal{R}}$, with respect to their precedence constraints, so as to minimize the total weighted completion time of the schedule, i.e., $\sum_{j \in \mathcal{J}} w_j C_j$. We refer to this problem as Map-Reduce scheduling problem.

Concerning the complexity of Map-Reduce scheduling problem, it generalizes the FFS problem which is known to be strongly $\mathcal{NP}$-hard [7], even when there is a single Map and a single Reduce task that has to be assigned only to one Map and one Reduce processor respectively.

2   The Map-Reduce Scheduling Problem

In this section, we present a 54-approximation algorithm for the Map-Reduce scheduling problem. Our algorithm is executed in the following two steps: (i) it computes a 27/2-approximate schedule for assigning and scheduling all Map tasks (resp. Reduce tasks) on processors of the set $\mathcal{P}_{\mathcal{M}}$ (resp. $\mathcal{P}_{\mathcal{R}}$) and (ii) it merges the two schedules in one, with respect to the precedence constraints between Map and Reduce tasks of each job. Step (ii) is performed by increasing the approximation ratio by a factor of 4.

$$\textbf{LP(b)} : \text{minimize} \sum_{j \in \mathcal{J}} w_j C_{D_j}$$

$$\text{subject to}: \sum_{i \in \mathcal{P}_b, \ell \in \mathcal{L}} y_{i,k,j,\ell} \geq 1, \qquad\qquad \forall T_{k,j} \in b \qquad (1)$$

$$C_{D_j} \geq C_{k,j}, \qquad\qquad \forall j \in \mathcal{J}, T_{k,j} \in b \qquad (2)$$

$$\sum_{i \in \mathcal{P}_b} \sum_{\ell \in \mathcal{L}} (1+\delta)^{\ell-1} y_{i,k,j,\ell} \leq C_{k,j}, \qquad\qquad \forall T_{k,j} \in b \qquad (3)$$

$$\sum_{T_{k,j} \in b} p_{i,k,j} \sum_{t \leq \ell} y_{i,k,j,t} \leq (1+\delta)^{\ell}, \qquad \forall i \in \mathcal{P}_b, \ell \in \mathcal{L} \qquad (4)$$

$$p_{i,k,j} > (1+\delta)^{\ell} \Rightarrow y_{i,k,j,\ell} = 0, \; \forall i \in \mathcal{P}_b, T_{k,j} \in b, \ell \in \mathcal{L} \qquad (5)$$

$$y_{i,k,j,\ell} \geq 0, \qquad\qquad \forall i \in \mathcal{P}_b, T_{k,j} \in b, \ell \in \mathcal{L}$$

Scheduling Map Tasks and Reduce Tasks. To schedule the Map and Reduce tasks separately on the processors $\mathcal{P}_{\mathcal{M}}$ and $\mathcal{P}_{\mathcal{R}}$, respectively, we formulate the interval-indexed LP-relaxation above for minizing the total weighted completion time. For notational convenience, we use an argument $b \in \{\mathcal{M}, \mathcal{R}\}$ to refer either to Map or to Reduce sets of tasks. We define $(0, t_{\max} = \sum_{T_{k,j} \in b} \max_{i \in \mathcal{P}_b} p_{i,k,j}]$ to be the time horizon of potential completion times, where $t_{\max}$ is an upper bound on the makespan of a feasible schedule. We discretize the time horizon into intervals $[1, 1], (1, (1+\delta)], ((1+\delta), (1+\delta)^2], \ldots, ((1+\delta)^{L-1}, (1+\delta)^L]$, where $\delta \in (0, 1)$ is a small constant, and L is the smallest integer such that $(1+\delta)^{L-1} \geq t_{\max}$. Let $I_\ell = ((1+\delta)^{\ell-1}, (1+\delta)^\ell]$, for $1 < \ell < L$, and $\mathcal{L} = \{1, 2, \ldots, L\}$. Note that, interval $[1, 1]$ implies that no job finishes its execution before time 1; in fact, we can assume, w.l.o.g., that all processing times are positive integers. Note also that the number of intervals is polynomial in the size of the instance and in $1/\delta$. For each processor $i \in \mathcal{P}_b$, task $T_{k,j} \in b$ and $\ell \in \mathcal{L}$, we introduce a variable $y_{i,k,j,\ell}$ that indicates if task $T_{k,j}$ is completed on processor i within the time interval I_ℓ. Furthermore, for each task $T_{k,j} \in T$, we introduce a variable $C_{k,j}$ corresponding to its completion time. For every job $j \in \mathcal{J}$, we introduce a dummy task D_j with zero processing time processed after the completion of each task $T_{k,j} \in b$. Note that, the corresponding integer program is a $(1 + \delta)$-relaxation of the original problem.

Our objective is to minimize the sum of weighted completion times of all jobs. Constraints (1) ensure that each task is completed on a processor of the set $\mathcal{P}_b$ in some time interval. Constraints (2) assure that for each job $j \in \mathcal{J}$, the completion of each task $T_{k,j}$ precedes the completion of task D_j. Constraints (3) impose a lower bound on the completion time of each task. For each $\ell \in \mathcal{L}$, constraints (4) and (5) are validity constraints which state that the total processing time of jobs executed up to an interval I_ℓ on a processor $i \in \mathcal{P}_b$ is at most $(1 + \delta)^\ell$, and that if processing a task $T_{k,j}$ on a processor $i \in \mathcal{P}_b$ takes more than $(1 + \delta)^\ell$, $T_{k,j}$ should not be scheduled on i, respectively.

Our algorithm, called Algorithm TASKSCHEDULING(b), starts from an optimal fractional solution $(\bar{y}_{i,k,j,\ell}, \bar{C}_{k,j}, \bar{C}_{D_j})$ to LP(b) and, working similarly to [5, Section 5], rounds it to an integral $27/2$-approximate schedule of the jobs

$\mathcal{J}$ on processors $\mathcal{P}_b$. The idea is to partition the set of tasks $T_{k,j}$ into classes $S(\ell) = \{T_{k,j} \in b \mid (1+\delta)^{\ell-1} \leq a\bar{C}_{k,j} \leq (1+\delta)^{\ell}\}$, where $\ell \in \{1,\ldots,L\}$ and $a > 1$ is a parameter, according to their (fractional) completion time in the optimal solution of LP(b), and to use [14, Theorem 2.1] for scheduling the tasks in each class $S(\ell)$ independently. In fact, Algorithm TASKSCHEDULING(b) can be regarded as a generalization of the approximation algorithm in [9, Section 4], where the objective is to minimize weighted completion time, but each job consists of a single task (see also the discussion in [5, Section 5]).

More specifically, we first observe that by the definition of $S(\ell)$ and due to constraints (1) and (3), for each task $T_{k,j} \in S(\ell)$, $\sum_{i \in \mathcal{P}_b} \sum_{t \leq \ell} y_{i,k,j,t} \geq \frac{a-1}{a}$. Otherwise, it would be $\sum_{i \in \mathcal{P}_b} \sum_{t \geq \ell+1} y_{i,k,j,t} > \frac{1}{a}$, which implies $a\bar{C}_{k,j} > (1+\delta)^{\ell}$. Therefore, if we set $y^*_{i,j,k,t} = 0$, for all $t \geq \ell+1$, and $y^*_{i,j,k,t} = \frac{a}{a-1}\bar{y}_{i,k,j,t}$, for all $t \leq \ell$, we obtain a solution $y^*_{i,k,j,t}$ that satisfies the constraints (1), (4), and (5) of LP(b), if the right-hand side of (4) is multiplied by $a/(a-1)$. Therefore, for each $\ell = 1,\ldots,L$, the tasks in $S(\ell)$ alone can be (fractionally) scheduled on processors $\mathcal{P}_b$ with makespan at most $\frac{a}{a-1}(1+\delta)^{\ell}$. Now, using [14, Theorem 2.1], we obtain an integral schedule for the tasks in $S(\ell)$ alone with makespan at most $(\frac{a}{a-1}+1)(1+\delta)^{\ell}$. By the definition of $S(\ell)$, in this integral schedule, each task $T_{k,j} \in S(\ell)$ has a completion time of at most $a(\frac{a}{a-1}+1)(1+\delta)\bar{C}_{k,j}$. Therefore, if we take the union of these schedules, one after another, in increasing order of $\ell = 1,\ldots,L$, the completion time of each job j is at most $a(\frac{a}{a-1}+1+\frac{1}{\delta})(1+\delta)\bar{C}_{D_j}$. Choosing $a = 3/2$ and $\delta = 1/2$, we obtain that:

Theorem 1. *[5] Algorithm* TASKSCHEDULING*(b) is a 27/2-approximation for scheduling a set of Map tasks (resp. Reduce tasks) on a set of unrelated processors $\mathcal{P}_M$ (resp. $\mathcal{P}_R$), in order to minimize their total weighted completion time.*

Merging Task Schedules. Let σ_M, σ_R be two schedules computed by TASKSCHEDULING(b), for $b = M$ and $b = R$, respectively. Let also $C^{\sigma_M}_j = \max_{T_{k,j} \in M}\{C_{k,j}\}$ and $C^{\sigma_R}_j = \max_{T_{k,j} \in R}\{C_{k,j}\}$ be the completion times of all Map and all Reduce tasks of a job $j \in \mathcal{J}$ within these schedules, respectively. Depending on these completion time values, we assign each job $j \in \mathcal{J}$ a *width* equal to $\omega_j = \max\{C^{\sigma_M}_j, C^{\sigma_R}_j\}$. The following algorithm computes a feasible schedule for Map-Reduce scheduling.

Algorithm MR. Every time a processor $i \in \mathcal{P}_b$ becomes available, schedule: either the Map task, assigned to $i \in \mathcal{P}_M$ in σ_M, with the minimum width, or the available (w.r.t. its "release time" ω_j) Reduce task, assigned to $i \in \mathcal{P}_R$ in σ_R, with the minimum width.

Theorem 2. *Algorithm* MR *is a 54-approximation for Map-Reduce scheduling.*

Proof (Sketch). By the execution of MR, it is immediate to verify the feasibility of the final schedule. So, it suffices to prove that in such a schedule σ, all tasks of a job $j \in \mathcal{J}$ are completed by time $2\max\{C^{\sigma_M}_j, C^{\sigma_R}_j\}$. Let C^{σ}_j, be the completion time of a job $j \in \mathcal{J}$ in σ. Note that, for each of the Map tasks of j, their completion time is upper bounded by ω_j. On the other hand, the completion time of each Reduce task is upper bounded by a quantity equal to $r + \omega_j$, where r is the earliest time when the task is available to be scheduled in σ. However, $r = C_j^{\sigma_M} \leq \omega_j$ and thus $C_j^{\sigma} \leq 2\omega_j = 2\max\{C_j^{\sigma_M}, C_j^{\sigma_R}\}$. Then, the theorem follows from Theorem 1.   □

3   The Map-Shuffle-Reduce Scheduling Problem

In practical MapReduce systems, data shuffle represents a significant cost for the key-value pairs with the same key to be transmitted from their Map tasks to the corresponding Reduce task. Motivated by [4], we model this cost by introducing a number of *Shuffle tasks* for each Map task. However, in contrast to [4], where the assignment of Shuffle tasks to processors is fixed, our model distinguishes between two variants: a) Each Shuffle task is scheduled on the same processor as its corresponding Reduce task and b) the Shuffle tasks are scheduled on a different set of processors. For both variants, we present $O(1)$-approximation algorithms.

The number of different keys is usually significantly larger than the number of Reduce processors. Hence, a Reduce task receives all key-value pairs with key in a set of different keys. Allowing the transmission time of some Shuffle tasks to be 0, we may assume wlog that all Reduce tasks receive key-value pairs from all Map tasks. We also assume that only a single key-value pair can be transferred to a Reduce processor at any time and moreover, the transmission process cannot be interrupted. Thus, since the key-value pairs allocated to the same Reduce task cannot be transmitted in parallel, we can assume that all key-value pairs from a Map task, assigned to the same Reduce task, can be considered as a single Shuffle task. Hence, the number of Shuffle tasks per Map task equals the number of Reduce tasks. The following summarizes the above discussion.

Properties: *i) Each Shuffle task cannot start its execution before the completion of its corresponding Map task.*
(ii) For every Map task of a job j, there are as many Shuffle tasks as j's Reduce tasks. When no key-value pairs are transmitted from a Map task to a Reduce task, the transmission time of the corresponding Shuffle task is equal to 0.
(iii) Each Shuffle task is executed non-preemptively.
(iv) Shuffle tasks transmitting to the same processor do not overlap.

Before presenting our results for the Map-Shuffle-Reduce scheduling problem, we introduce some additional notation. For each Map task $T_{k,j} \in \mathcal{M}$ of a job $j \in \mathcal{J}$, we introduce a set of Shuffle tasks $T_{r,k,j}$, $1 \leq r \leq \tau_j$, with τ_j denoting the number of Reduce tasks of job j. We denote by $\mathcal{H}$ the set of Shuffle tasks; note that for each Map task of a job, there is a bijection between its Shuffle tasks and the job's Reduce tasks. Each Shuffle task $T_{r,k,j} \in \mathcal{H}$ is associated with a *transmission time* $t_{r,k,j}$, which is independent of the processor assignment.

The Shuffle Tasks are Executed on the Reduce Processors. The key step here is the integration of the Shuffle phase into the Reduce phase. In this direction, we consider a Reduce task $T_{r,j}$ of a job j and let $s_j^r = \{T_{r,k,j} \mid T_{k,j} \in \mathcal{M}\}$

be the set of Shuffle tasks that must be completed before task $T_{r,j}$ starts its execution. The tasks in s_j^r are executed in the same processor as Reduce task $T_{r,j}$. Thus, we obtain that:

Lemma 1. *There is an optimal schedule of Shuffle tasks and Reduce tasks on processors of $\mathcal{P_R}$ such that (i) there are no idle periods, and (ii) all the Shuffle tasks in s_j^r are executed together and are completed exactly before the execution of $T_{r,j}$.*

Proof. (i) Let σ be a feasible schedule. There are three cases where an idle time can occur: either between the execution of two Shuffle or two Reduce tasks or between a Shuffle and a Reduce task. Since all Shuffle and Reduce tasks are assumed to be available from time zero and there are no precedence constraints among only Shuffle tasks or only Reduce tasks, skipping the idle times in the first two cases can only decrease the objective value of σ. For the third case, we observe that since the Shuffle tasks precede the corresponding Reduce tasks, skipping the idle intervals can only decrease the completion time of the Reduce tasks. Hence, σ can be transformed into a schedule with no idle periods without increasing the total weighted completion time.

(ii) Let us consider a schedule σ that violates the claim and has the last Reduce task $T_{r,j}$ of a job j completed on some processor $i \in \mathcal{P_R}$. We fix the completion time of $T_{r,j}$ and shift all the Shuffle tasks in s_j^r to execute just before $T_{r,j}$, consecutively and in arbitrary order. Then, the completion time of j remains unchanged, while the completion time of every task preceding $T_{r,j}$ in σ may decrease. After a finite number of shifts, we obtain a schedule that satisfies (ii) and has at most the total weighted completion time of σ.   □

Using Lemma 1, we can incorporate the execution of the Shuffle tasks of each job into the execution of the corresponding Reduce tasks. Namely, for each Reduce task $T_{r,j}$ of a job j, $1 \leq r \leq \tau_j$, we increase its processing time $p_{i,r,j}$, on each processor $i \in \mathcal{P_R}$, by a quantity equal to the total transmission time of the Shuffle tasks in s_j^r, i.e., equal to $p(s_j^r) = \sum_{T_{r,k,j} \in s_j^r} t_{r,k,j}$. Let $p'_{i,r,j} = p_{i,r,j} + p(s_j^r)$ be the increased processing time for each task $T_{r,j} \in \mathcal{R}$ on processor $i \in \mathcal{P_R}$, referred to as Shuffle-Reduce task and let $\mathcal{R_H}$ be the new set of Shuffle-Reduce tasks.

Now, running Algorithm TASKSCHEDULING(b), for $b \in \{\mathcal{M}, \mathcal{R_H}\}$, we obtain two 27/2-approximate schedules, one for the Map tasks and one for the Shuffle-Reduce tasks. Moreover, by considering the same precedence constraints as for the Map and Reduce tasks, we can merge the above schedules by applying Algorithm MR. Despite satisfying Property (i), these dependencies are more general than the precedence constraints among Map and Shuffle tasks of each job, because in order to start the execution of a Shuffle task, we have to wait for all the Map tasks of a job to finish. However, since the completion time of a job j in the optimal schedule is lower bounded by the completion time in optimal schedules of either the Map or the Shuffle-Reduce tasks, regardless of their precedences, we have that:

Theorem 3. *Algorithm* MR *is a* 54-*approximation for Map-Shuffle-Reduce scheduling.*

The Shuffle Tasks are Executed on Different Processors. To deal with this case, we assume that for any processor $i \in \mathcal{P}_{\mathcal{R}}$, there exists an "input" processor which receives data from the Map processors. Therefore, the input processor executes the Shuffle tasks that correspond to the Reduce tasks which have been assigned to processor i. We refer to the set of input processors as $\mathcal{P}_{\mathcal{S}}$.

Lemma 2. *Consider two optimal schedules σ and σ' of Shuffle and Reduce tasks on processors in $\mathcal{P}_{\mathcal{R}} \cup \mathcal{P}_{\mathcal{S}}$ and $\mathcal{P}_{\mathcal{R}}$, respectively. Let also $C_{k,j}^{\sigma}, C_{k,j}^{\sigma'}$ be the completion times of any Reduce task $T_{k,j}$ in σ and σ', respectively. Then, $C_{k,j}^{\sigma'} \leq 2C_{k,j}^{\sigma}$.*

Proof. We start with an optimal schedule σ on the set $\mathcal{P}_{\mathcal{R}} \cup \mathcal{P}_{\mathcal{S}}$ of processors. We fix a Reduce processor i^r, the corresponding input processor i^s and a Reduce task $T_{k,j} \in \mathcal{R}$ of a job j. We build the schedule σ' on i^r by executing the Reduce tasks in the same order as in σ and just before each Reduce task, we execute the corresponding Shuffle task. Let $B(k)$ be the set of Reduce tasks executed on processor i^r, before $T_{k,j}$ and let $Sh(k)$ the set of the shuffle tasks that correspond to the Reduce tasks in $B(k) \cup \{T_{k,j}\}$. Then, we have that

$$C_{k,j}^{\sigma'} = \sum_{T_{l,j} \in B(k)} p_{i^r, l, j} + \sum_{T_{q,l,j} \in Sh(k)} t_{q,l,j},$$

which holds because there are no idle intervals in σ', by Lemma 1. Moreover, since both $B(k)$ and $Sh(k)$ have to be completed before $T_{k,j}$ in σ, we have that

$$C_{k,j}^{\sigma} \geq \max \left\{ \sum_{T_{l,j} \in B(k)} p_{i^r, l, j}, \sum_{T_{q,l,j} \in Sh(k)} t_{q,l,j} \right\},$$

and therefore $C_{k,j}^{\sigma'} \leq 2C_{k,j}^{\sigma}$. □

Combining Lemma 2 and Theorem 1, we obtain a 27-approximation for scheduling the Shuffle-Reduce tasks. Then, running Algorithm MR, we get the following corollary. Here, the Shuffle tasks form a special third stage in the FFS problem.

Corollary 1. *Algorithm* MR *is a* 81-*approximation for Map-Shuffle-Reduce scheduling, in the general case where the Shuffle tasks run on a separate set of processors.*

4   Experimental Evaluation

In this section we experimentally evaluate the performance of Algorithm MR. To deal with data shuffle in our model, in Section 3, we apply Algorithm MR to instances of the Map-Reduce scheduling problem with increased processing times for the Reduce tasks that take the data transmission time of the Shuffle tasks into account. Thus, to simplify the experimental evaluation, we restrict our attention to instances of Map-Reduce scheduling. We compare the solutions of Algorithm MR, for two different families of random instances, with a lower bound on the optimal cost and with the solutions of a simple and fast scheduling algorithm, called Fast-MR, that we propose below.

To compute a lower bound on the optimal solution of Map-Reduce scheduling, we include in the LP-relaxation LP(b) all the Map and Reduce tasks and also the precedence constraints among them. These dependencies can be captured by the following set of constraints in LP(b):

$$C_{k,j} \geq C_{k,j'} + \sum_{i \in \mathcal{P}_\mathcal{R}} \sum_{\ell \in \mathcal{L}} p_{i,k,j} y_{i,k,j,\ell} \qquad \forall j \in \mathcal{J}, T_{k,j} \in \mathcal{R}, T_{k,j'} \in \mathcal{M}.$$

Our Fast-MR algorithm consists of two steps: First it finds a online assignment of tasks to processors and then schedules them using a variant of the well known Weighted Shortest Processing Time first (WSPT) policy.

Fast-MR. *Step A*: Apply the online algorithm ASSIGNU, presented in [2] for makespan minimization on unrelated processors: For an arbitrary order of jobs, arriving one-by-one in an arbitrary order, assign each task $T_{k,j}$ of the current job j to the processor $k = \arg\min_{i \in \mathcal{P}_b} \{\lambda^{L_i + p_{i,k,j}} - \lambda^{L_i}\}$, where L_i is the current load of processor i and $\lambda > 1$ an appropriately chosen constant. The tasks of each job are considered one-by-one in an arbitrary order.
Step B: Order the tasks assigned to each processor, in Step A, by applying the following version of the WSPT policy:

For each pair of jobs $j, j' \in \mathcal{J}$, if $(w_j / \sum_{T_{k,j} \in j} p_{k,j}) > (w_{j'} / \sum_{T_{k,j'} \in j'} p_{k,j'})$, then job j precedes j' in the schedule.

When a processor becomes available, schedule the task that is not yet executed, while respecting the precedences among Map and Reduce tasks.

4.1   Computational Experiments and Results

We performed the experiments on a machine with 4 packages (Intel(R) Xeon(R) E5-4620 @ 2.20GHz) of 8 cores each (16 threads with hyperthreading) and a total memory of 256 GB. The operating system was a Debian GNU/Linux 6.0. We used Python 2.7 for scripting. The solver used for the linear programs was Gurobi Optimizer 6.0.

An instance of the problem consists of a $n \times |\{T_{k,j} \in \mathcal{M} \cup \mathcal{R}\}| \times m$ matrix that describes the processing times of the tasks, a vector of size n that describes the job weights, and a precedence graph for the tasks of the same job. We use two disjoint sets of processors, each consisting of 40 Map and 40 Reduce processors. We consider instances from 5 to 50 jobs; each job has 30 map tasks and 10 reduce tasks. Moreover, we fix $\delta = 0.5$ and $a = 1.5$, for the parameters of Algorithm MR. For each of the n jobs, its weight is uniformly distributed in $[1, n]$.

The quality of the solutions in our experiments depend on whether there is any correlation between jobs and processors. Based on [10], in order to experiment with two representative cases, we generate the task processing times in each processor in two different ways, uncorrelated and processor-job correlated, and test experimentally both Algorithm MR and Fast-MR by running 10 different trials for each possible number of jobs. The instances and the code used in our experiments are available at http://www.corelab.ntua.gr/~opapadig/mrexperiments/.

Uncorrelated Input. The processing times $\{p_{i,k,j}\}_{i\in\mathcal{P}_{\mathcal{M}}}$ of the Map tasks $T_{k,j} \in \mathcal{M}$ of each job j are selected uniformly at random (u.a.r.) from $[1,100]$. Similarly to [3], we set the processing times $\{p_{i,k,j}\}_{i\in\mathcal{P}_{\mathcal{R}}}$ of the Reduce tasks $T_{k,j} \in \mathcal{R}$ to thrice a value selected u.a.r. from $[1,100]$ plus some "noise" selected u.a.r. from $[1,10]$.

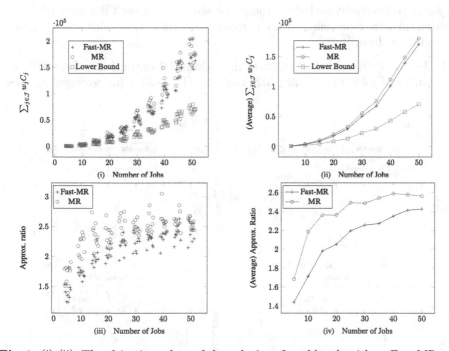

Fig. 1. (i)-(ii): The objective values of the solutions found by algorithms Fast-MR and MR and a lower bound on the optimal. (iii)-(iv): The observed approximation ratios of Fast-MR and MR for instances with uncorrelated processing times.

In Fig. 1.(i)-(ii), we observe that Fast-MR performs better than MR in general when the processing times are uncorrelated. For a small number of jobs, Fast-MR gives up to 21% (on average) better solutions. However, as the number of jobs increases, the gap between Fast-MR and MR decreases, e.g., for $n = 45$ and $n = 50$ Fast-MR gives 6% and 5% (on average) better solutions, respectively.

In fact, since the processing times are selected u.a.r. from identical uniform distributions, the processors tend to behave as essentially identical, rather than unrelated, which gives a significant practical advantage to Fast-MR, especially for small instances. This holds for both the assignment and the scheduling phase of Fast-MR, since WSPT is also known to perform quite well on identical processors. With respect to performance guarantees, as we can see, in Fig. 1.(iii)-(iv), the (empirical) approximation ratio of MR ranges from 1.68 to 2.58 (on average), while the approximation ratio of Fast-MR ranges from 1.43 to 2.42 (on average). These values are far from MR's worst-case approximation guarantee of 54.

Processor-Job Correlated Input. To better capture issues of data locality in our unrelated processors setting, we next focus on instances which use processor and job correlations. In this direction, the processing times $\{p_{i,k,j}\}_{i \in \mathcal{P}_\mathcal{M}}$ of the Map tasks $T_{k,j} \in \mathcal{M}$ of each job j are uniformly distributed in $[\alpha_i \beta_j, \alpha_i \beta_j + 10]$, where α_i, β_j are selected u.a.r. from $[1, 20]$, for each processor $i \in \mathcal{M}$ and each job $j \in \mathcal{J}$ respectively. As before, the processing time of each Reduce task is set to three times a value selected u.a.r. from $[\alpha_i \beta_j, \alpha_i \beta_j + 10]$ plus some "noise" selected u.a.r. from $[1, 10]$.

In Fig. 2.(i)-(ii), we observe that Algorithm MR outperforms Fast-MR for any number of jobs. Specifically, MR leads to $11\% - 34\%$ (on average) smaller

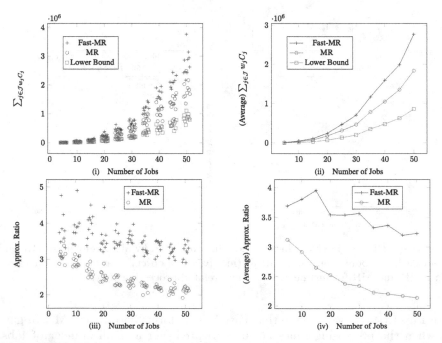

Fig. 2. (i)-(ii): The objective values of the solutions found by algorithms Fast-MR and MR and a lower bound on the optimal. (iii)-(iv): The observed approximation ratios of Fast-MR and MR for instances with correlated processing times.

total weighted completion times than Fast-MR. Due to the processor-job correlation in task processing times, the environment now resembles better that of
MapReduce scheduling with unrelated processors and data locality. Then, the
more sophisticated assignment and the scheduling procedures of Algorithm MR
have a significant advantage over the simple online assignment and WSPT-based
scheduling of Fast-MR. In fact, even for a small number of jobs, $n = 5$, Algorithm MR results in up to 11% (on average) better solutions. The (empirical)
approximation ratio of MR, in Fig. 2.(iii)-(iv), ranges from 2.13 to 3.12 (on
average), while, for Fast-MR, the approximation ratio ranges from 3.19 to 3.95
(on average). Again, both algorithms are far from MR's worst-case approximation guarantee of 54. Furthermore, we observe that the empirical approximation
ratio of MR (and also its advantage over Fast-MR) seem to improve as the number of jobs increases and the assignment and scheduling problem becomes more
demanding.

References

1. Afrati, F.N., Ullman, J.D.: Optimizing multiway joins in a map-reduce environment. IEEE Transactions on Knowledge and Data Engineering **23**(9), 1282–1298
   (2011)
2. Aspnes, J., Azar, Y., Fiat, A., Plotkin, S., Waarts, O.: On-line Routing of Virtual
   Circuits with Applications to Load Balancing and Machine Scheduling. Journal of
   the ACM **44**(3), 486–504 (1997)
3. Chang, H., Kodialam, M.S., Kompella, R.R., Lakshman, T.V., Lee, M., Mukher
   jee, S.: Scheduling in mapreduce-like systems for fast completion time. In: IEEE
   Proceedings of the 30th International Conference on Computer Communications,
   pp. 3074–3082 (2011)
4. Chen, F., Kodialam, M.S., Lakshman, T.V.: Joint scheduling of processing and
   shuffle phases in mapreduce systems. In: IEEE Proceedings of the 31st International Conference on Computer Communications, pp. 1143–1151 (2012)
5. Correa, J.R., Skutella, M., Verschae, J.: The power of preemption on unrelated machines and applications to scheduling orders. Mathematics of Operations
   Research **37**(2), 379–398 (2012)
6. Dean, J., Ghemawat, S.: Mapreduce: simplified data processing on large clusters.
   In: Proceedings of the 6th Symposium on Operating System Design and Implementation, pp. 137–150 (2004)
7. Garey, M.R., Johnson, D.S., Sethi, R.: The complexity of flowshop and jobshop
   scheduling. Mathematics of Operations Research **1**(2), 117–129 (1976)
8. Gonzalez, T., Sahni, S.: Flowshop and jobshop schedules: complexity and approximation. Operations research **26**(1), 36–52 (1978)
9. Hall, L.A., Schulz, A.S., Shmoys, D.B., Wein, J.: Scheduling to minimize average
   completion time: Off-line and on-line approximation algorithms. Mathematics of
   Operations Research **22**, 513–544 (1997)
10. Hariri, A.M., Potts, C.N.: Heuristics for scheduling unrelated parallel machines.
    Computers and Operations Research **18**(3), 323–331 (1991)
11. Mastrolilli, M., Queyranne, M., Schulz, A.S., Svensson, O., Uhan, N.A.: Minimiz-
    ing the sum of weighted completion times in a concurrent open shop. Operations
    Research Letters **38**(5), 390–395 (2010)

12. Moseley, B., Dasgupta, A., Kumar, R., Sarlós, T.: On scheduling in map-reduce and flow-shops. In: Proc. of the 23rd ACM Symposium on Parallel Algorithms and Architectures (SPAA), pp. 289–298 (2011)
13. Schuurman, P., Woeginger, G.J.: A polynomial time approximation scheme for the two-stage multiprocessor flow shop problem. Theoretical Computer Science **237**(1), 105–122 (2000)
14. Shmoys, D.B., Tardos, É.: An approximation algorithm for the generalized assignment problem. Mathematical Programming **62**, 461–474 (1993)
15. Yoo, D.-J., Sim, K.M.: A comparative review of job scheduling for mapreduce. In: IEEE Proc. of the International Symposium on Cloud Computing and Intelligece Systems, pp. 353–358 (2011)

Station Assignment with Reallocation

Miguel A. Mosteiro[1], Yulia Rossikova[1], and Prudence W.H. Wong[2]([⊠])

[1] Department of Computer Science, Kean University, Union, NJ, USA
{mmosteir,rossikoy}@kean.edu
[2] Department of Computer Science, University of Liverpool, Liverpool, UK
pwong@liverpool.ac.uk

Abstract. We study a dynamic allocation problem that arises in various scenarios where mobile clients joining and leaving have to communicate with static stations via radio transmissions. Restrictions are a maximum delay, or laxity, between consecutive client transmissions and a maximum bandwidth that a station can share among its clients. Clients are assigned to stations so that every client transmits under those restrictions. We consider reallocation algorithms, where clients are revealed at its arrival time, the departure time is unknown until they leave, and clients may be reallocated to another station, but at a cost determined by the laxity We present negative results for related previous protocols that motivate the study; we introduce new protocols that expound trade-offs between station usage and reallocation cost; we prove theoretically bounds on our performance metrics; and we show through simulations that, for realistic scenarios, our protocols behave even better than our theoretical guarantees.

1 Introduction

We study a dynamic allocation problem in scenarios where data on mobile devices has to be gathered and uploaded periodically to one of the static access points available[1]. Examples include *wearable health-monitoring systems*, where data gathered via physiological sensors on ambulatory patients must be periodically uploaded, and *participatory sensing*, where mobile device users upload periodically environment information.

Mobile devices, called *clients*, join and leave continuously, and they communicate with the static access points, called *stations*, The clients' ephemeral nature is modeled by the *life interval* of each client (from its arrival to departure), during which the client has to communicate with some station *periodically*. Periodic communication is modeled by the client's *laxity*, which bounds the maximum duration a client is not transmitting to some stations. The intrinsically shared nature of the access to stations is modeled by a maximum shared *station bandwidth*, by a *client bandwidth*, and by the client laxity.

---

[1] We consider an upstream model, but the same results apply to downstream communication.

© Springer International Publishing Switzerland 2015
E. Bampis (Ed.): SEA 2015, LNCS 9125, pp. 151–164, 2015.
DOI: 10.1007/978-3-319-20086-6_12

Based on the above model, we study the problem of assigning clients to stations so that every client transmits to some stations satisfying the laxity and bandwidth constraints. We consider settings where clients are revealed at its arrival time and their departure time is only revealed when they depart (as in online algorithms). Clients may be reassigned from one station to another and we call such reassignment *reallocation*. Intuitively reallocation causes more disturbance to a client with small laxity. Therefore, we assume reallocation incurs a cost inversely proportional to a client's laxity. We aim to reduce the number of active stations and reduce the reallocation cost. However, these two goals are orthogonal, e.g., we can reallocate the clients every time a client arrives/departs so that the number of active stations is minimized while incurring a very high reallocation cost; alternatively we can keep the reallocation cost to zero but we may use many active stations after a sequence of client departures. In this paper, we aim to obtain a balance between the two performance metric. We call this problem *Station Assignment Problem with Reallocation* (SA).

Previous Work. The closest work to the present paper is [12], where reallocation algorithms were presented for Windows Scheduling (WS). The WS problem [6,7,11,12] is a particular case of SA where the bandwidth requirement of each client is the same and each channel (a.k.a. station in our case) can only serve one client at a time. In [12], a unit cost is incurred for each client reallocated and the objective is to minimize an aggregate sum reflecting the amortized reallocation cost and the number of channels used. A protocol called Classified Reallocation is shown to guarantee an amortized constant number of reallocations. This protocol is also evaluated experimentally together with two other protocols Preemptive Reallocation and Lazy Reallocation.

WS [6,7,11] was first studied without reallocation and the objective function was the number of channels. Both static case where clients never depart [6,7] and dynamic case where clients may depart [11] have been studied. For the dynamic case, the comparison is against peak load which may occur at different time in the online algorithm and the optimal offline algorithm. In [12] and this work, we compare against current load. Clients with same laxity were considered in [14]. We also extend the objective function in [12] such that the number of reallocated clients is weighted inversely by their laxity, and we provide trade-off between reallocation cost and number of stations.

SA and Other Assignment Problems. Our problem differs from various scheduling problems. The load balancing problem [4] also assigns tasks of different load to servers, yet does not consider periodic tasks and disallow reallocation. Interval coloring [1] concerns the number of machines used but not periodic tasks. Periodic appearance of tasks in real time scheduling [8] is determined by the input but not by the algorithm to satisfy laxity constraint. The SA problem is also related to b-matching [15], fractional matching [5], and adwords [13]. Among other details, the objective function is different.

There are two typical approaches of handling orthogonal objectives: to minimize the summation of two costs, e.g., energy efficient flow time scheduling minimizes the sum of energy usage and total flow time of the tasks [2]; and to formulate two approximation ratios, e.g., energy efficient throughput scheduling algorithm is t-throughput-competitive and e-energy-competitive [10]. We adopt the later approach.

Reallocation has been considered in scheduling [3,9,16]. In [9], a distinction is made between reassignment within server (reschedule) and between servers (migration). Here, we assume rescheduling within a station is free and we use "reallocation" to refer to reassignment to other stations. It is often that the number/size of jobs reallocated is bounded, e.g., by a function of the number of jobs in the system [9], the size of the arriving job [16] or the number of machines [3]. In our problem, we bound the reallocation by the weight (cumulative inverse laxity) of the clients departed.

2   Our Results

In this paper, we study reallocation algorithms for SA assuming that clients have arbitrary laxities and bandwidth requirements, that clients depart from the system at arbitrary times, and that they may be reallocated, but at some cost proportional to the resources needed. Specifically, our contributions are the following.

- We define a characterization of SA reallocation algorithms, which we call (α, β)- approximation, as a combination of the competitive ratio on station usage (α) and the cost of reallocations contrasted with the resources released by departures (β).
- We show a sequence of negative results proving that worst-case guarantees cannot be provided by previous protocols Classified Reallocation and Preemptive Reallocation [12], even if they are modified to our reallocation cost function.
- We present a novel SA protocol called Classified Preemptive Reallocation (CPR) where clients are *classified* according to laxity and bandwidth requirements, and upon departures the remaining clients are *preemptively* reallocated to minimize station usage, but only within their class. The protocol presented includes a range of classifications that exposes trade-offs between reallocation cost and station usage. In fact, we found first experimentally what is the classification function that better balances these goals, and then we provided theoretical guarantees for all functions.
- In our main theorem, we prove bounds on both of our performance metrics, and we instantiate those bounds into three classifications and for specific scenarios in two corollaries (refer to Section 5 for the specific bounds.)
- Finally, we present the results of our extensive simulations that allowed us to find the function that best balances station usage and reallocation cost. Additionally, our simulations show that, for a variety of realistic scenarios, CPR performs better than expected by the worst-case theoretical analysis, and close to optimal on average.

3   Definitions

Model. We consider a set S of stations and a set C of clients. Each client must transmit packets to some station. Time is slotted so that each time slot is long enough to transmit one packet. A client can be assigned to transmit to only one station in any given time slot. Starting from some initial time slot 1, we refer to the infinite sequence of time slots $1, 2, 3, \ldots$ as **global time**. Each client $c \in C$ is characterized by an **arrival time** a_c and a **departure time** d_c, that define a **life interval** $\tau_c = [a_c, d_c]$ in which c is **active**. That is, client c is active from the beginning of time slot a_c up to the end of time slot d_c. We define $C(t) \subseteq C$ to be the set of clients that are active during time slot t. With respect to resources required, each client c is characterized by a **bandwidth** requirement b_c, and a **laxity** $0 < w_c \leq |\tau_c|$, such that c must transmit to some station in S at least one packet within each w_c consecutive time slots in $\tau_c{}^2$. On the other hand, each station $s \in S$ is characterized by a **station bandwidth** or **capacity** B, which is the maximum aggregated bandwidth of clients that *may* transmit to s in each time slot.

Notation. Let the **schedule** of a client c be an infinite sequence σ_c of values from the alphabet $\{0\} \cup S$. Let $\sigma_c(t)$ be the tth value of σ_c. A **station assignment** is a set σ of schedules that models the transmissions from clients to stations. That is, for each client $c \in C$ and time slot t, it is $\sigma_c(t) = s$ if c is scheduled to transmit to station $s \in S$ in time slot t, and $\sigma_c(t) = 0$ if c does not transmit in time slot t. If a client c is scheduled to transmit to a station s we say that c is **assigned** to station s. We say that a station that has clients assigned is **active**, and **inactive** or **empty** otherwise.

Problem. The **Station Assignment problem (SA)** is defined as follows. For a given set of stations and set of clients, obtain a station assignment such that (i) each client transmits to some station at least once within each period of length its laxity during its life interval, (ii) in each time slot, no station receives from clients whose aggregated bandwidth is more than the station capacity. Notice that, for any finite set of stations, there are sets of clients such that the SA problem is not solvable. We assume in this work that S is infinite and what we want to minimize is the number of *active* stations.

Algorithms. We study **reallocation algorithms** for SA. That is, the parameters w_c and b_c needed to assign the client to some station are revealed at time a_c, but the departure time d_c is unknown to the algorithm until the client actually leaves the system (as in online algorithms). Then, at the beginning of time slot t, an SA reallocation algorithm returns the transmission schedules of all clients that are active in time slot t, possibly reassigning some clients from one station to another. (I.e., the schedules of clients that were already active may be changed from one time slot to another.) We refer to the reassignment of one client as

---

^{2} To maintain low station usage, we will assume that the laxity can be relaxed during reallocation.

a *reallocation*, whereas all the reassignments that happen at the beginning of the same time slot are called a *reallocation event*.

Performance Metric. Previous work [12] has considered the number of clients reallocated as the reallocation cost. In the present work, we consider a different scenario where the cost of reallocating a client is proportional to resources requested by that client. Specifically, we assume a cost for the reallocation of each client c of ρ/w_c, where $\rho > 0$ is a parameter. Then, letting $\mathcal{R}(ALG, t)$ be the cost of the reallocation event incurred by algorithm ALG at time t, and $R(ALG, t)$ be the set of clients being reallocated, the overall cost is the following.

$$\mathcal{R}(ALG, t) = \rho \sum_{c \in R(ALG, t)} \frac{1}{w_c}. \tag{1}$$

We will drop the specification of the algorithm whenever clear from the context.

With respect to performance, we aim for algorithms with low reallocation cost and small number of active stations. Unfortunately, these are contradictory goals. Indeed, the reallocation cost could be zero if no client is reallocated (online algorithm), but the number of active stations could be as big as the number of active clients (e.g. initially multiple clients assigned to each station, and then all but one client from each active station depart). On the other hand, the number of active stations could be minimized applying an offline algorithm on each time slot, but the reallocation cost could be large. Thus, we characterize algorithms with both metrics as follows.

For any SA algorithm ALG, let $S(ALG, t)$ be the number of active stations at time t in the schedule, let $D(ALG, t)$ be the set of clients departed since the last reallocation up to time t. Denoting $\sum_{c \in C'} 1/w_c$ as the *weight* of the clients in $C' \subseteq C$, let $\mathcal{D}(ALG, t)$ be the weight of the clients departed since the last reallocation up to time t, that is, $\mathcal{D}(ALG, t) = \sum_{c \in D(ALG, t)} 1/w_c$. Also, we denote the minimum number of active stations needed at time t as $S(OPT, t)$. Throughout, we will drop the specification of the algorithm whenever it is clear from the context. Then, we say that an SA reallocation algorithm ALG achieves an (α, β)-*approximation* if the following holds for any input.

$$\max_t \frac{S(ALG, t)}{S(OPT, t)} \leq \alpha$$

$$\max_t \frac{\mathcal{R}(ALG, t)}{\mathcal{D}(ALG, t)} \leq \beta.$$

In words, the overhead on the number of stations used by ALG is never more than a multiplicative factor α over the optimal, and the reallocation cost, amortized on the "space" left available by departing clients is never more than β. The latter is well defined since reallocations only occur after departures. Notice that these ratios are strong guarantees, in the sense that they are the maximum of the ratios instead of the ratio of the maxima. (This distinction was called previously in the literature *against current load* versus *against peak load* respectively.) Moreover, the reallocation ratio computed as the maximum *over reallocation events* is also stronger than the ratio of cumulative weights since the system started.

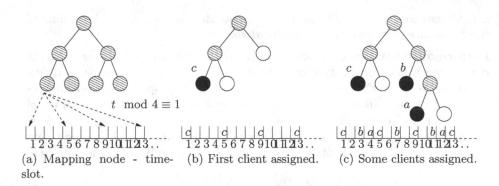

(a) Mapping node - time-slot.

(b) First client assigned.

(c) Some clients assigned.

Fig. 1. Illustration of a binary broadcast tree. (a) A depth-2 tree corresponds to periodic broadcast of period 2^2. (b) Clients are assigned to leaves, e.g., client c with laxity 4 is assigned the black node meaning timeslot $1, 5, 9$, etc. are reserved for it. (c) Open leaf (white node) corresponds to available slot.

4   Algorithms

Broadcast Trees. A common theme in WS algorithms with *periodic* transmission schedules is to represent those schedules with *Broadcast Trees* [6,11,12]. See Figure 1 for illustration. Throughout the paper, we refer to a set of broadcast trees as the ***forest***, and to the distance in edges from a node to the root as the ***depth***. Generalizing, the 2^d nodes at depth d in a complete binary tree represent the time slots $t \bmod 2^d$ (see Figure 1(a)). Then, to indicate that some (periodic) time slot has been reserved for a client c to transmit to a given station s, we say informally that c is assigned to the corresponding node in the broadcast tree of s. Throughout the rest of the paper, we use both indistinctively. Refer to [6,11] for further details on broadcast trees.

WS Algorithms. Chan et al. [11] presented a WS algorithm preserving the following invariant. For each station, the broadcast tree has at most one available leaf at each depth. In order to preserve this invariant, when a client departs, the remaining clients in the same tree are rearranged. If reallocations among trees are possible, the algorithm *Preemptive Reallocation* (PR) [12] extended the same idea to all trees, maintaining the invariant that *throughout all trees* there is at most one available leaf at each depth. For laxities that are powers of 2, PR achieves an optimal station usage. However, we show in Lemma 1 (1) and (2) that simple modification to PR leads to negative results.

A WS algorithm with provable bounded reallocation cost guarantees was shown also in [12]. The protocol, called *Classified Reallocation* (CR), guarantees that all clients assigned to the same station have the same laxity, except for one distinguished station that handles all laxities linear and above. To attain constant amortized reallocation cost, clients are moved to/from the distinguished station only after the number of clients has halved/doubled. However, for the reallocation cost function in Equation 1, CR has an arbitrarily bad reallocation cost ratio, as we show in Lemma 1 (3).

Algorithm 1. Classified Preemptive Reallocation. $\lfloor\lfloor x \rfloor\rfloor$ is the largest power of 2 that is not larger than x. We represent the transmission schedules with broadcast trees. A node with both children available becomes an available leaf. A station with no client assigned becomes non-active. $\langle w_{low}, w_{high}\rangle$ are the boundaries of the class of the input client.

```
 1  Algorithm
 2  │  upon arrival or departure of a client c do
 3  │  │  if arrival then allocate(c, ⟨w_low, w_high⟩)
 4  │  └  else consolidate(c, ⟨w_low, w_high⟩)

 5  Procedure allocate(c, ⟨w_low, w_high⟩)
 6  │  for each depth i = ⌊log w_c⌋ − ⌈log w_low⌉ down to 0 do
 7  │  │  for each active station s of class ⟨w_low, w_high, 1/⌊⌊B/b_c⌋⌋⟩ do
 8  │  │  │  if there is a leaf ℓ available at depth i in the broadcast tree of s then
 9  │  │  │  │  allocate to ℓ a new subtree with client c assigned at depth
    │  │  │  │  ⌊log w_c⌋ − i − ⌈log w_low⌉ of the broadcast subtree
10  │  │  └  │  return

11  │  activate a new station s in class ⟨w_low, w_high, 1/⌊⌊B/b_c⌋⌋⟩
12  │  choose one of the leaves ℓ at depth 0 of the broadcast subtrees of s
13  │  allocate to ℓ a new subtree with client c assigned at depth
    │  ⌊log w_c⌋ − ⌈log w_low⌉ of the broadcast subtree

14  Procedure consolidate(c, ⟨w_low, w_high⟩)
15  │  for each depth i = ⌊log w_c⌋ − ⌈log w_low⌉ down to 1 do
16  │  │  if there are two active stations of class ⟨w_low, w_high, 1/⌊⌊B/b_c⌋⌋⟩ both
    │  │  with a leaf at depth i available then reallocate sibling subtree of
    │  │  smaller weight
17  │  └  else return

    │  // reallocations cleared a whole broadcast subtree
18  │  if there are two active stations of class ⟨w_low, w_high, 1/⌊⌊B/b_c⌋⌋⟩ with
    │  empty broadcast subtrees then reallocate a subtree from the station with at
    │  least one empty subtree to the station with exactly one empty subtree
```

Classified Preemptive Reallocation. The negative results in Lemma 1 apply to WS. Given that WS is a particular case of SA fixing $b_c = B$ for all clients, the same negative results apply to SA. Thus, should the reallocation cost be maintained low, a new approach is needed. We present now an online SA protocol (Algorithm 1), which we call *Classified Preemptive Reallocation* (CPR), that provides guarantees in channel-station usage and reallocation cost. The protocol may be summarized as follows. Clients are classified according to laxity and bandwidth requirements. Upon arrival, a client is allocated to a station within its corresponding class to guarantee a usage excess (with respect to optimal) of at most one station per class plus one station throughout all classes. Upon departure of a client, if necessary to maintain the above-mentioned guarantee, clients are reallocated, but only within the corresponding class. The protocol includes three different classifications providing different trade-offs between reallocation cost and station usage. We recreate the idea of broadcast trees, but now we have multiple trees representing the schedule of each station. On one hand, we use broadcast trees with depth bounded by the class laxities. We call them **broadcast subtrees** to reflect that they are only part of a regular broadcast tree. On the other hand, we have the multiplicity yielded by the shared station capacity B. An example of broadcast subtrees can be seen in Figure 2. Further details follow.

The mechanism to allocate an arriving client can be described as follows. Upon arrival, a client c is classified according to its laxity and bandwidth requirement. Specifically, c is assigned to a class for clients with bandwidth requirement $B/\lfloor\lfloor B/b_c \rfloor\rfloor$ and laxity in $[w_{low}, w_{high})$, for some w_{low} and w_{high} that depend on the classification chosen. Notice that each station has up to $\lfloor\lfloor B/b_c \rfloor\rfloor \cdot \lceil\lceil w_{low} \rceil\rceil$ subtrees. That is, $\lfloor\lfloor B/b_c \rfloor\rfloor$ ways to share its capacity B and $\lceil\lceil w_{low} \rceil\rceil$ ways to share its schedule (see Figure 2). Within its class, we assign c to an available leaf at depth $\lfloor\log w_c\rfloor - \lceil\log w_{low}\rceil$ in any subtree in the forest (see Figure 2(b)). If there is no such leaf available, we look at smaller depths up in the forest one by one. If we find an available leaf at depth $\lceil\log w_{low}\rceil \le i < \lfloor\log w_c\rfloor - \lceil\log w_{low}\rceil$, we allocate to that leaf a new subtree with c assigned at depth $\lfloor\log w_c\rfloor - i$ with respect to the root of the broadcast subtree (see Figures 2(a) and 2(c)). If no such leaf is available at any depth, a new broadcast subtree T is created with c assigned at depth $\lfloor\log w_c\rfloor - \lceil\log w_{low}\rceil$, and T is assigned to a newly activated station. Refer to Algorithm 1 for further details.

The above allocation mechanism maintains the following invariant: (1) there is at most one leaf available at any depth larger than $\lceil\log w_{low}\rceil$ of the forest, and (2) there is at most one station with leaves available at depth $\lceil\log w_{low}\rceil$ (an empty broadcast subtree). When a client departs, this invariant is re-established through reallocations as follows. When a client c departs, if $\lfloor\log w_c\rfloor > \lceil\log w_{low}\rceil$, we check if there was already a leaf ℓ available at depth $\lfloor\log w_c\rfloor - \lceil\log w_{low}\rceil$. If there was one, either the sibling of c or the sibling of ℓ has to be reallocated to re-establish the invariant. We greedily choose to reallocate whichever sibling has smaller weight of the two (see Figure 3(a)). The process does not necessarily stop here because, if $\lfloor\log w_c\rfloor - 1 > \lceil\log w_{low}\rceil$ and there was a leaf already available at depth $\lfloor\log w_c\rfloor - 1 - \lceil\log w_{low}\rceil$, together with the newly available leaf at depth $\lfloor\log w_c\rfloor - 1 - \lceil\log w_{low}\rceil$ due to the reallocation at depth $\lfloor\log w_c\rfloor - \lceil\log w_{low}\rceil$, it yields two leaves available at depth $\lfloor\log w_c\rfloor - 1 - \lceil\log w_{low}\rceil$. Hence, again one of the sibling subtrees has to be reallocated (see Figure 3(b)). This transitive reallocations upwards the forest may continue until a depth where no reallocation is needed or until the depth $\lceil\log w_{low}\rceil + 1$ is reached, when the reallocation leaves a broadcast subtree empty. In the latter case, we reallocate a whole broadcast subtree so that only one station has empty subtrees and the invariant is re-established. Refer to Algorithm 1 for further details.

Notice that when a client is reallocated (even within a station) its laxity may be violated once. Consider for instance the schedule in Figure 1(c). Let $w_a = 4$, that is, a is transmitting at its lowest possible frequency. If at the end of time

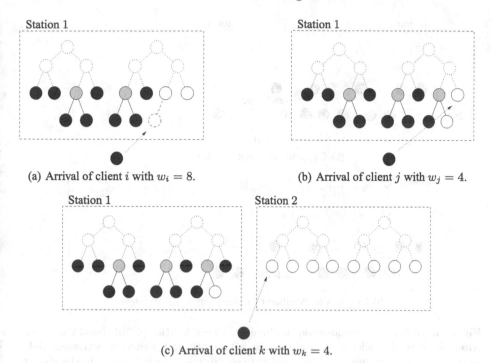

(a) Arrival of client i with $w_i = 8$.   (b) Arrival of client j with $w_j = 4$.

(c) Arrival of client k with $w_k = 4$.

Fig. 2. Illustration of allocation mechanism. Class: laxities $[4, 16)$, bandwidth $1/2$. Subtrees are depicted connected to a broadcast tree to reflect their location in the station schedule.

slot 7 client b departs, at the beginning of time slot 8 client a will be reallocated to the slot of client b, that is, to transmit next in slot 11. This new schedule violates w_a because the previous slot when a transmitted was 5. For WS, in [11] the issue is approached making a client transmit once more within the original schedule. As the authors say, this approach introduces a transition delay. In their model, there is no impact on station usage because their ratio is against peak load. However, for a ratio against current load such as our model, reserving a slot for a client in more than one station implies an overhead on channel usage. Indeed, for any given allocation/reallocation policy, an adversarial input can be shown so that either the laxity is stretched or the channel usage is not optimal. Hence, in our model we assume that when a client is reallocated the laxity may be stretched, folding the cost in the reallocation cost.

5   Analysis

We start with negative results in Lemma 1, which apply to WS, and to SA fixing $b_c = B$ for all clients. The proofs, left to the full paper, are all based on showing an adversarial client set for which the claim holds.

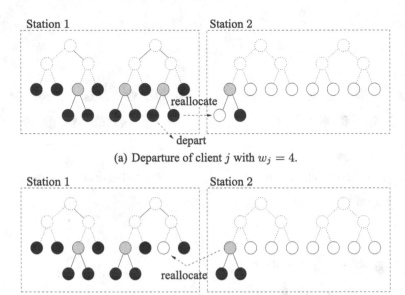

(a) Departure of client j with $w_j = 4$.

(b) Upwards reallocation of sibling with smaller weight.

Fig. 3. Illustration of reallocation mechanism. Class: laxities $[4, 16)$, bandwidth $1/2$. After the second reallocation Station 2 is left empty and, hence, deactivated. Subtrees are depicted connected to a broadcast tree to reflect their location in the station schedule.

Lemma 1.   *1. There exists a client arrival/departure schedule such that, in Preemptive Reallocation [12], the ratio of number of clients reallocated against the number of arrivals plus departures is unbounded.*

   *2. For Preemptive Reallocation [12], modified so that the sibling subtree of smaller **weight** is reallocated to restore the invariant, rather than the subtree with less clients, the following holds. For any $d > 0$, there exists a client arrival/departure schedule such that it is $\max_t \mathcal{R}(t)/\mathcal{D}(t) \geq \rho(2^d - 1)^2/2^d$.*

   3. For any integer $x > 0$ and any $w \geq 2^{x+5}$ arbitrarily big such that w is a power of 2, there exists a client arrival/departure schedule such that, in Classified Reallocation [12], it is $\max_t \mathcal{R}(t)/\mathcal{D}(t) \geq \frac{\rho/4}{7 \cdot 2^x} w$.

The above lemma shows that the application of previous WS reallocation algorithms to SA is not feasible. Theorem 1 gives guarantees on station usage and reallocation cost for CPR. The proof, left to the full paper, shows that the invariant is re-established after each arrival or departure. Then, competitiveness on station usage is derived from the invariant properties. Finally, to bound β, a worst case scenario minimizing the weight of departed clients and maximizing the reallocated weight is shown. To provide intuition and comparison for the simulations, we instantiate Theorem 1 on a setting where all laxities are powers of 2 and all bandwidth requirements are the full capacity of a station.

Theorem 1. *At any time slot t, CPR achieves an (α, β)-approximation as follows.*

$$\alpha = \max_t \frac{4(1 + \Gamma(ALG, t) + S(OPT, t))}{S(OPT, t)}$$
$$\beta = \max_t \rho(2\lfloor\lfloor w_{high_{\max}}(t)\rfloor\rfloor / \lceil\lceil w_{low_{\max}}(t)\rceil\rceil] - 1).$$

Where $\Gamma(ALG, t)$ is the number of classes used by CPR at time t, and $w_{high_{\max}}(t)$ and $w_{low_{\max}}(t)$ are the maximum upper and lower limits of a class at time t.

Corollary 1. *For a set of clients C such that, for all $c \in C$, it is $b_c = B$ and $w_c = 2^i$ for some $i \geq 0$, and for all t it is $w_{\max}(t) > w_{\min}(t) \geq 4$, the following holds. At any time slot t, CPR achieves an (α, β)-approximation as follows.*

1. *If the client classification boundaries are $[w_i, w_{i+1})$, where $w_1 = 1$, and $w_i = 2w_{i-1}$, for any $i > 1$, then*

$$\alpha = 1 + (2 + \log(w_{\max}(t)/w_{\min}(t))) / H(C(t))$$
$$\beta = 3\rho.$$

2. *If the client classification boundaries are $[w_i, w_{i+1})$, where $w_1 = 1, w_2 = 2, w_3 = 4$, and $w_i = w_{i-1} \log w_{i-1}$, for any $i > 3$, then*

$$\alpha = 1 + (2 + \log w_{\max}(t)/\log\log w_{\min}(t)) / H(C(t))$$
$$\beta = \rho(2\log w_{\max}(t) - 1).$$

3. *If the client classification boundaries are $[w_i, w_{i+1})$, where $w_1 = 1, w_2 = 2$, and $w_i = w_{i-1}^2$, for any $i > 2$, then*

$$\alpha = 1 + (2 + \log(\log w_{\max}(t)/\log w_{\min}(t))) / H(C(t))$$
$$\beta = \rho\left(2\sqrt{w_{\max}(t)} - 1\right).$$

Where $H(C(t)) = \lceil\sum_{c \in C(t)} 1/w_c\rceil$, $w_{\max}(t) = \max_{c \in C(t)} w_c$, $w_{\min}(t) = \min_{c \in C(t)} w_c$, $b_{\max}(t) = \max_{c \in C(t)} b_c$, and $b_{\min}(t) = \min_{c \in C(t)} b_c$.

6 Simulations

In this section, we present the main experimental simulations results of the CPR algorithm. We highlight here that the classification factor (logarithmic) that balances station usage and reallocation cost was found through experimentation with various functions. For the specific cases presented (constant, logarithmic, and linear factors) we have focused on a scenario where $\forall c \in C, b_c = B$ and $w_c = 2^i, i \geq 0$ (as in Corollary 1). Simulations for arbitrary bandwidths and laxities are left to the full version of this paper.

To evaluate thoroughly the performance of our protocol, we have produced various sets of clients (recall that each client is characterized by arrival time,

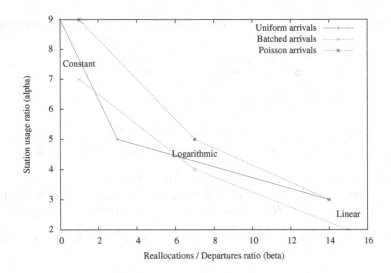

Fig. 4. Worst case α vs. β. $|C| = 4000$, $w_{\max} = 1024$, $w_{\min} = 1$, $\rho = 1$.

departure time, and laxity). The laxity of each client was chosen independently at random from $\{1, 2, 4, \ldots, 1024\}$, with a distribution biased towards large laxities. More precisely, for each client c, $w_c = 1$ with probability $1/1024$, or $w_c = 2^i$ with probability $2^i/2^{11}$, for $1 \le i \le 10$. For $n = 4000$ clients, time was discretized in $2n$ slots. The arrival time of each client was chosen: (a) uniformly at random within $[1, 2n]$; (b) in 3 batches of $n/3$ clients arriving at $t = 1$, $t = n/2$, and $t = n$; and (c) as a Poisson process with rate 0.7. For each client, the departure time was chosen uniformly at random from the interval $[t_a, 2n]$, where t_a is the time of arrival of such client. With respect to the protocol, three different classification factors: constant, logarithmic, and linear, were used.

For each of the nine scenarios arising from the combination of the variants, we evaluated experimentally the (α, β)-approximation of CPR. Our simulations showed that the performance in practical settings is as expected or better than the theoretical bounds. The reallocation vs. departures weight ratio (bounded by β) is around 1 most of the time for all three algorithms. On the other hand, after a period upon initial arrivals and a period before last departures, the station usage ratio against $H(C(t))$, which is only a lower bound of the optimal, (bounded by β) is most of the time below 2.

To evaluate the behavior of our algorithms in adverse conditions, we extended the number of cases considering $|C| = 4000, 8000$, and 16000 clients, and the range of laxities to $\{16, 32, 64, \ldots, w_{\max}\}$, for $w_{\max} = 1024, 4096$, and 16384. The laxities were drawn uniformly at random. These cases, combined with the arrival distributions and the classification factors, yielded 81 scenarios tested. We observed that the trade-offs between α and β according to the algorithm used apply to all these scenarios. Indeed, having more clients and setting higher $w_{\max}$ does not affect the trade-offs, only their magnitude as expected from the functions bounding α and β in Corollary 1. Should the reallocation ratio be minimized, the constant factor classification achieves better performance at a higher station usage. On the other hand, if channel usage must be kept low, the linear factor classification performs better incurring in higher reallocation cost. The logarithmic factor balances both costs. Figure 4 illustrates these trade offs for one of the scenarios. In comparison with the bounds proved in Corollary 1, for the scenarios simulated CPR behaves better than expected.

Acknowledgments. Authors would like to thank Martín Farach-Colton for useful discussions. This work has been supported in part by the National Science Foundation (CCF-1114930); Kean University UFRI grant; U. of Liverpool Departmental Visiting Fellowship; U. of Liverpool Network Sciences & Technologies (NeST).

References

1. Adamy, U., Erlebach, T.: Online coloring of intervals with bandwidth. In: Solis-Oba, R., Jansen, K. (eds.) WAOA 2003. LNCS, vol. 2909, pp. 1–12. Springer, Heidelberg (2004)
2. Albers, S., Fujiwara, H.: Energy-efficient algorithms for flow time minimization. ACM Trans. on Algorithms **3**(4), 49 (2007)
3. Albers, S., Hellwig, M.: On the value of job migration in online makespan minimization. In: Epstein, L., Ferragina, P. (eds.) ESA 2012. LNCS, vol. 7501, pp. 84–95. Springer, Heidelberg (2012)
4. Azar, Y.: On-line load balancing. In: Fiat, A. (ed.) Online Algorithms 1996. LNCS, vol. 1442, pp. 178–195. Springer, Heidelberg (1998)
5. Azar, Y., Litichevskey, A.: Maximizing throughput in multi-queue switches. Algorithmica **45**, 69–90 (2006)
6. Bar-Noy, A., Ladner, R.E.: Windows scheduling problems for broadcast systems. SIAM Journal on Computing **32**(4), 1091–1113 (2003)
7. Bar-Noy, A., Ladner, R.E., Tamir, T.: Windows scheduling as a restricted version of bin packing. ACM Trans. on Algorithms **3**(3), 28 (2007)
8. Baruah, S., Goossens, J.: Scheduling real-time tasks: algorithms and complexity. In: Leung, J. (ed.) Handbook of Scheduling: Algorithms, Models and Performance Analysis, pp. 15-1–15-41. CRC Press (2004)
9. Bender, M.A., Farach-Colton, M., Fekete, S.P., Fineman, J.T., Gilbert, S.: Reallocation problems in scheduling. In: SPAA, pp. 271–279 (2013)
10. Chan, H.-L., Chan, J.W-T., Lam, T.W., Lee, L.-K., Mak, K.-S., Wong, P.W.H.: Optimizing throughput and energy in online deadline scheduling. ACM Trans. on Algorithms **6**(1) (2009)
11. Chan, W.-T., Wong, P.W.H.: On-line windows scheduling of temporary items. In: Fleischer, R., Trippen, G. (eds.) ISAAC 2004. LNCS, vol. 3341, pp. 259–270. Springer, Heidelberg (2004)
12. Farach-Colton, M., Leal, K., Mosteiro, M.A., Thraves, C.: Dynamic windows scheduling with reallocation. In: Gudmundsson, J., Katajainen, J. (eds.) SEA 2014. LNCS, vol. 8504, pp. 99–110. Springer, Heidelberg (2014)
13. Feldman, J., Mehta, A., Mirrokni, V., Muthukrishnan, S.: Online stochastic matching: beating 1–1/e. In: FOCS, pp. 117–126 (2009)

14. Fernández Anta, A., Kowalski, D.R., Mosteiro, M.A., Wong, P.W.H.: Station assignment with applications to sensing. In: Flocchini, P., Gao, J., Kranakis, E., der Heide, F.M. (eds.) ALGOSENSORS 2013. LNCS, vol. 8243, pp. 155–166. Springer, Heidelberg (2014)
15. Kalyanasundaram, B., Pruhs, K.: An optimal deterministic algorithm for online b-matching. Theoretical Computer Science **233**(1–2), 319–325 (2000)
16. Sanders, P., Sivadasan, N., Skutella, M.: Online scheduling with bounded migration. In: Díaz, J., Karhumäki, J., Lepistö, A., Sannella, D. (eds.) ICALP 2004. LNCS, vol. 3142, pp. 1111–1122. Springer, Heidelberg (2004)

Online Knapsack of Unknown Capacity:
Energy Optimization for Smartphone Communications

Daniele Diodati, Alfredo Navarra$^{(\boxtimes)}$, and Cristina M. Pinotti

Dipartimento di Matematica e Informatica,
Università degli Studi di Perugia, Via Vanvitelli 1, 06123 Perugia, Italy
daniele.diodati@dmi.unipg.it,
{alfredo.navarra,cristina.pinotti}@unipg.it

Abstract. We propose a new variant of the more standard online knapsack problem where the only information missing to the provided instances is the capacity B of the knapsack. We refer to this problem as the online *Knapsack of Unknown Capacity* problem. Any algorithm must provide a strategy for ordering the items that are inserted in the knapsack in an online fashion, until the actual capacity of the knapsack is revealed and the last inserted item might not fit in. Apart from the interest in a new version of the fundamental knapsack problem, the motivations that lead to define this new variant come from energy consumption constraints in smartphone communications. We do provide lower and upper bounds to the problem for various cases. In general, we design an optimal algorithm admitting a $\frac{1}{2}$-competitive ratio. When all items admit uniform ratio of profit over size, our algorithm provides a $\frac{49}{86} = .569\ldots$ competitive ratio that leaves some gap with the provided bound of $\frac{1}{\varphi} = .618\ldots$, the inverse of the golden number. We then conduct experimental analysis for the competitive ratio guaranteed algorithms compared to the optimum and to various heuristics.

1 Introduction

In this paper, we consider a new variant of the knapsack problem where we are given with a set of items $\{a_1, a_2, \ldots, a_n\}$, each associated with a profit $p(a_i)$ and a size $s(a_i)$, $1 \leq i \leq n$. The capacity B of the knapsack itself is unknown.

A resolution algorithm must provide the order in which the items have to be inserted in the knapsack. Capacity B is then revealed in an online fashion, so that the capability to substitute items already inserted is not applicable. The solution found by the algorithm is given by the initial sequence of consecutive items that completely fit in the knapsack, while all subsequent items, including the first one that does not fit, are ignored.

We are interested in designing an online algorithm for the proposed variant of the knapsack problem with a 'good' competitive ratio guarantee. We remind that the *competitive ratio* of an online algorithm is defined as the ratio between

Research partially supported by the Research Grant 2010N5K7EB 'PRIN 2010' ARS TechnoMedia (Algoritmica per le Reti Sociali Tecno-mediate) from the Italian Ministry of University and Research.

© Springer International Publishing Switzerland 2015
E. Bampis (Ed.): SEA 2015, LNCS 9125, pp. 165–177, 2015.
DOI: 10.1007/978-3-319-20086-6_13

the gain incurred by such an algorithm and that of an optimal solution provided by an offline algorithm.

It is easy to show that the problem defined so far does not admit any constant competitive ratio. An instance can be provided with two items a_1 and a_2 such that $s(a_1) = 1$, $s(a_2) = 1 + \epsilon$, $p(a_1) = 1$, $p(a_2) = \infty$, and the capacity B chosen by the adversary is either 1 or $1 + \epsilon$. Any deterministic online algorithm will be infinitely worse than the offline algorithm on at least one of the inputs. In fact, if the online algorithm chooses to insert a_1, then $B = 1 + \epsilon$ provides a bad performance, as well as if the algorithm chooses to insert a_2 and $B = 1$ because a_2 does not fit into the knapsack. Further constraints are then necessary in order to be able to guarantee a reasonable competitive ratio. We want to consider the minimal set of constraints that permits to obtain significant results. We simply assume that each item has size not bigger than B. Actually, the constraint on the capacity B arises from practical observations conducted on smartphone communications. In fact, beyond the interest on the new combinatorial problem, it turns out that the issue of saving energy in smartphone communications is well related to our variant of knapsack. It has been observed that while serving some tasks, a smartphone may optimize the available bandwidth by parallelizing other services. Parallel communications (e.g., voice call and data transfer) turn out to require less energy than their stand-alone execution (see, e.g. [2]). This can be modeled by our new knapsack problem, for instance considering a voice call as the knapsack - we do not know the call duration before it ends - and available data to upload or download as items to be inserted in the knapsack. Each item is associated with a profit which represents the energy saved by performing the data transfer in parallel with the voice call, and a size which is the duration of the data transfer adapted to the available bandwidth.

Summarizing, we consider the knapsack problem with unknown capacity B and we assume that each of the items fits within capacity B that will be revealed in an online fashion. So, the only constraint we assume on the capacity of the knapsack is $B \geq s_{max} = \max_{1 \leq i \leq n}\{s(a_i)\}$. We refer to this problem as the online *Knapsack of Unknown Capacity* problem, *KUC* for short.

1.1   Related Work

While there is much literature concerning the variant of the knapsack problem where B is known and the items appear online, or follow some probability distribution, little has been done with respect to the case of unknown B.

The knapsack problem with known B and items appearing online was first studied in [12] where it has been shown that non-trivial competitive algorithms do not exist for the general case. Many special cases of the basic online knapsack version have been studied (see, e.g. [3,17]). These include stochastic online knapsack problem [9,11,16], the removable online knapsack problem [6–8] and the online partially fractional knapsack problems [15]. None of them do imply our assumptions.

To the best of our knowledge, the only paper dealing with the *KUC* problem without any constraint on the capacity of the knapsack is [14]. The authors consider the problem from a robustness point of view, rather than our online acceptation (see [1,10] and references therein for a survey on robustness). Nevertheless, providing a robustness factor for their problem would result in a competitive ratio for our problem. As shown above, without any further constraint the problem defined does not admit any reasonable competitive ratio. The authors of [14] then study the problem with respect to the so called *instance-sensitive performance guarantees*. This is a very interesting way for designing specific algorithms with respect to each input instance. This permits to design a PTAS to achieve a competitive ratio arbitrarily close to the best possible factor for every specific instance but not in general.

In [5], still the capacity B of the knapsack is unknown. In contrast to our model, whenever an item does not fit in the knapsack, it is discarded but the remaining items still can be inserted if they fit in. With this variant, two algorithms are presented: the first provides a general $\frac{1}{2}$-competitive ratio; the second provides a competitive ratio equal to the inverse of the golden number, $\frac{1}{\varphi} = .618\ldots$, holding when all items admit the same profit over size ratio. Both factors are shown to be optimal by providing two lower bounds for the two cases. As we are going to see we borrow from them the lower bounds that hold also for our model, whereas we cannot do the same for their algorithms.

Another paper dealing with a similar variant of our KUC problem is [4]. The authors assume that B is not fixed but it is modeled as a random variable whose distribution is assumed to be known. They provide an optimal formulation in terms of integer linear programming, and then conduct experimental studies.

1.2   Our Results

We do provide lower and upper bounds to the problem for the general case and for instances where the ratio between the profit and the size of each item is uniform. In the former case our algorithm admits an optimal $\frac{1}{2}$-competitive ratio. The lower bound of $\frac{1}{2}$ is borrowed from [5]. In addition, we show that the competitive ratio of our algorithm grows like function $\frac{x-1}{x}$, $x > 2$ as long as B grows like $x \cdot s_{max}$, and that this trend is optimal. The achieved time complexity is $O(n \log n)$, with n being the number of given items.

In the uniform case, our algorithm provides a $\frac{49}{86} = .569\ldots$ competitive ratio while the lower bound is $\frac{1}{\varphi} = .618\ldots$, the inverse of the golden number. The achieved time complexity is linear in the number of given items.

As last theoretical result, we show that the algorithm designed for the uniform case can be sometimes useful also in general. Given an instance I of KUC, let ps_m and ps_M be the minimum and the maximum ratios, respectively, of profit over size provided by the items in I. We can obtain a competitive ratio of $\frac{49}{86}\frac{ps_m}{ps_M}$ that might be better than $\frac{1}{2}$ when the two values ps_m and ps_M are sufficiently close to each other.

We then conduct experimental analysis for the competitive ratio guaranteed algorithms compared to the optimum and to various heuristics. Experiments are conducted on random data and synthetic daily smartphone communications.

2   Solving the Online KUC Problem

In this section, we do provide our theoretical results concerning lower and upper bounds for the online Knapsack Unknown Capacity problem. Actually, our problem admits the same lower bound of $\frac{1}{2}$ provided in [5].

Theorem 1 ([5]). *Any online algorithm cannot achieve a competitive ratio greater than $\frac{1}{2} + \rho$, for any $\rho > 0$.*

The proof proceeds by defining a family of instances. For every item i of an instance of n items, there is a capacity B, such that by inserting item i as first can only lead to a solution worse than the optimum by a factor of at least $\frac{1}{(2-\frac{4}{n})}$. This approaches $\frac{1}{2}$ from above for increasing values of n. The instance is given setting $s(a_i) = F_n + F_i - 1$ and $p(a_i) = 1 + \frac{i}{n}$, with F_i being the i-th Fibonacci number ($F_1 = 1$, $F_2 = 1$, and $F_i = F_{i-1} + F_{i-2}$). If the first item inserted by any online algorithm is a_i, $i \geq 3$, then B is chosen by the adversary equal to $2F_n + F_i - 2$, whereas if $i < 3$ then B is set to $2F_n - 1$. In any case, no further items can be inserted in the knapsack.

Theorem 2. *Any online algorithm cannot achieve a competitive ratio greater than $\frac{x-1}{x} + \rho$, for any $x > 2$ and any $\rho > 0$ (even though items admit uniform ratio of profit over size).*

Proof. Consider the case with items admitting uniform ratio of profit over size. We define a family of instances as follows. There are always two "big" items of size s_{max} and k "small" items of size $\epsilon < \frac{1}{2}$, such that $k \cdot \epsilon = c \cdot s_{max}$, for some $c > 1$. Varying on c defines different instances. We now define the adversary behavior and then we look for the best strategy in terms of competitive ratio guarantee. The adversary fixes B to cut the first big item of size s_{max} inserted by any algorithm $\mathcal{A}$, unless such an item is inserted as first. In this latter case, the adversary fixes B to cut the second big item inserted by $\mathcal{A}$. The cut is such that the item does not fit in the knapsack for just an ϵ space. If SOL is the profit achieved by $\mathcal{A}$, then the optimal solution achieves $SOL + s_{max} - \epsilon$. In order to obtain the best competitive ratio, it is desirable that $\mathcal{A}$ inserts as first element a big one, and then postpones the insertion of the second big item as much as possible, hence obtaining: $\frac{SOL}{OPT} \leq \frac{SOL}{SOL + s_{max} - \epsilon} \leq \frac{s_{max} + k\epsilon}{2s_{max} + (k-1)\epsilon} = \frac{(c+1)s_{max}}{(c+2)s_{max} - \epsilon} = \frac{x-1}{x} + \rho$ for any $x > 2$ and any $\rho > 0$.   □

As absolute value, the above theorem provides a useless bound greater than that achieved by Theorem 1. In fact, function $\frac{x-1}{x}$, tends to one as x grows. However, the importance of the statement resides in observing that when the capacity of the knapsack B grows, the competitive ratio of any algorithm cannot grow better than function $\frac{x-1}{x} + \rho$ as for any $x > 2$ there always exists an instance where any algorithm cannot guarantee a competitive ratio greater than that function.

We now provide a new greedy algorithm, called GREEDY-KUC, that in general guarantees a competitive ratio of $\frac{1}{2}$. The algorithm is optimal with respect to the absolute lower bound. Moreover, we show that it is optimal also whenever the capacity B of the knapsack reveals to be greater than $2s_{max}$. This is particularly important as the $\frac{1}{2}$ ratio can be forced only for small values of B.

The algorithm makes use of the substitution technique usually exploited by greedy algorithms in the offline knapsack problem. This allows to replace an item already inserted in the knapsack with one not yet included in the solution. Since the capacity B of the knapsack is unknown, we cannot apply substitution to the whole final solution but only for that arising by assuming $B = s_{max}$. Indeed, this is the only known limit for B. Algorithm GREEDY-KUC works as follows:

- Items are sorted in non-increasing order according to the profit per size unit, that is, $\frac{p(a_i)}{s(a_i)} \geq \frac{p(a_{i+1})}{s(a_{i+1})}$, for each $1 \leq i < n$;
- Let a_i be the first item not fitting in B when considering $B = s_{max}$;
- If $p(a_i) > \sum_{j=1}^{i-1} p(a_j)$ then move item a_i at the first position of the ordered items;
- Fill the knapsack according to the computed order until the current item does not fit in.

Note that, the last item considered by the above algorithm is always discarded unless B reveals to be at least the sum of the size of all items.

Theorem 3. GREEDY-KUC *is $\frac{1}{2}$-competitive, and runs in time $O(n \log n)$.*

Proof. Let OPT be the optimal solution computed once B is revealed. If $B = s_{max}$ then the competitive ratio is clearly $\frac{1}{2}$. In fact, in this case our algorithm works exactly like the basic greedy algorithm for the off-line knapsack (see, [13]). By increasing B, we can observe the following. Let item a_i be the item detected by GREEDY-KUC at the second step, and let a_j be the first item not fitting in the knapsack once the actual capacity B is revealed. If $j \leq i$, then by the above arguments still the $\frac{1}{2}$ competitive ratio holds, since the items involved are the same. If $j > i$, then by the ordering provided at the first step of the algorithm, all inserted items have the best ratio profit over size. The optimal solution could have fit the same set of items plus filling up the remaining part of the knapsack by means of shorter items with respect to a_j. This implies $OPT < SOL + p(a_j)$. Moreover, by the defined ordering of the items $p(a_j) \leq SOL$ as otherwise item a_j would have been ranked before other items. Hence, $\frac{SOL}{OPT} > \frac{SOL}{SOL + p(a_j)} \geq \frac{1}{2}$.

For the time complexity, it is sufficient to point out that the most expensive step of the algorithm is the sorting at the first step. $\square$

It is worth noting that the competitive ratio of $\frac{1}{2}$ is achieved only for small values of B. To have an idea of the growing of the competitive ratio with respect to the capacity B, the next corollary can be stated.

Corollary 1. *When $B = x \cdot s_{max}$, for any $x > 2$, GREEDY-KUC is $\frac{x-1}{x}$-competitive.*

Proof. It is sufficient to note that when $B = x \cdot s_{max}$, the optimal solution can insert more than GREEDY-KUC at most a set of items such that the sum of their sizes cannot be larger than s_{max}. By the provided ordering, the profit of such a set cannot be greater than the profit that GREEDY-KUC achieves for each preceding portion of size s_{max}. □

This reflects what one expects, that is, the more is the actual capacity B, the less is the possibility for GREEDY-KUC to loose with respect to the optimal solution. Moreover, by Theorem 2 we obtain:

Corollary 2. *When $B \geq x \cdot s_{max}$, for any $x > 2$, GREEDY-KUC is optimal.*

3   Uniform Ratio of Profit Over Size

When all items admit the same ratio of profit over size, better bounds can be provided. For the lower bound, still the one provided in [5] holds.

Theorem 4. *Any online algorithm cannot achieve a competitive ratio greater than $\frac{1}{\varphi} + \rho$, for any $\rho > 0$.*

As competitive ratio guarantee, we now define a new algorithm specific for the case of uniform ratio of profit over size. As we are going to see, we can assure a competitive ratio of $\frac{49}{86}$.

Given $1 < \beta < 2$ and $\alpha > \beta$, Algorithm UNIFORM-KUC works as follows:

- Items are partitioned into three sets: I_1 being items smaller than $\frac{1}{\alpha}s_{max}$, I_2 being items not in I_1 and smaller than $\frac{1}{\beta}s_{max}$, and I_3 being all items greater than or equal to $\frac{1}{\beta}s_{max}$.
- If $I_2 \neq \emptyset$ or $\sum_{a \in I_1} s(a) \geq \frac{1}{\alpha}s_{max}$ then insert in the knapsack first the biggest item of size s_{max}, then
  - If $\sum_{a \in I_1} s(a) \geq \frac{1}{\alpha}s_{max}$ then insert all items from I_1, then all items from I_2
  - *Else* insert all items from I_2, then all items from I_1;
  - finally, insert all items from I_3;
- *Else* insert first all items from I_3 in the non-decreasing order according to their size then insert all items from I_1.

Theorem 5. UNIFORM-KUC *provides a competitive ratio of $\frac{49}{86}$ when items admit uniform ratio of profit over size. It can be implemented to run in $\Theta(n)$ time. Moreover, when $B \geq x \cdot s_{max}$, for any $x > 2$, UNIFORM-KUC is optimal.*

Proof. Consider the *If* branch of the algorithm when $I_2 \neq \emptyset$ or $\sum_{a \in I_1} s(a) \geq \frac{1}{\alpha}s_{max}$. When B is revealed, we distinguish three cases according to the possible item that UNIFORM-KUC looses: the item belongs to a) I_1; b) I_2; or c) I_3. In case a), the algorithm looses an element of size smaller than $\frac{1}{\alpha}s_{max}$ with respect to the optimal solution. Hence the induced competitive ratio is $\frac{SOL}{OPT} \geq \frac{SOL}{SOL + \frac{1}{\alpha}s_{max}} \geq \frac{\alpha}{1+\alpha}$. In fact, since items admit uniform ratio of profit over size, all items inserted by UNIFORM-KUC provide an optimal profit. So the minimum of the above function is obtained for the smallest value of SOL, which is at least s_{max}. Similarly, in case b) the competitive ratio guarantee is $\frac{\beta}{1+\beta}$. Since $\alpha > \beta$, the competitive ratio obtained for case b) dominates the one obtained for case a). In case c), SOL is obtained by inserting in the knapsack at least the element of size s_{max} and one of size $\frac{1}{\alpha}s_{max}$, and the loss item may have been of size s_{max}. Hence, the competitive ratio guarantee is not smaller than $\frac{1+\frac{1}{\alpha}}{2+\frac{1}{\alpha}} = \frac{\alpha+1}{2\alpha+1}$.

In the *Else* branch of the algorithm, when $I_2 = \emptyset$ and $\sum_{a \in I_1} s(a) < \frac{1}{\alpha}s_{max}$, we distinguish among two cases where UNIFORM-KUC looses an element belonging to a) I_1; b) I_3. In case a), the competitive ratio guarantee is again $\frac{\alpha}{1+\alpha}$. In fact, elements from I_3 have been all inserted in the knapsack and I_2 is empty. The smallest value of SOL is then obtained when it is composed of just the element of size s_{max}, while the optimum cannot do better than $SOL + \frac{1}{\alpha}s_{max}$. In case b), since items from I_3 are chosen by UNIFORM-KUC in a non-decreasing order, then the number of items from I_3 in OPT cannot be bigger than the number inserted by our algorithm. It follows that if SOL is composed by considering the first c items from I_3, the optimal solution can be composed of at most c items of size s_{max} plus all items from I_1. Since $\sum_{a \in I_1} s(a) < \frac{1}{\alpha}s_{max}$, this provides a competitive ratio guarantee of $\frac{SOL}{OPT} \geq \frac{c\frac{1}{\beta}s_{max}}{c \cdot s_{max} + \frac{1}{\alpha}s_{max}} = \frac{\frac{c}{\beta}}{c+\frac{1}{\alpha}} \geq \frac{\alpha}{\beta(\alpha+1)}$.

Considering the obtained ratios all together by means of function $f(\alpha, \beta) = \min\{\frac{\beta}{1+\beta}, \frac{\alpha+1}{2\alpha+1}, \frac{\alpha}{\beta(\alpha+1)}\}$, we look for the maximum achievable by $f(\alpha, \beta)$ for $1 < \beta < 2$ and $\alpha > \beta$. By means of numerical evaluations we obtain $\max f(\alpha, \beta) = \frac{49}{86} = .569\ldots$ for $\alpha = \frac{37}{12}$ and $\beta = \frac{49}{37}$.

For the time complexity, it is worth noting that the sorting described at the third step of the algorithm is not fully necessary. In fact, it is sufficient to find out the smallest three items belonging to I_3. If all of them are successfully inserted in the knapsack, then the algorithm starts behaving as described by Corollary 1 regardless the actual order of the items, and by Corollary 2 it is optimal.   □

Note that there is still some gap between the competitive ratio guaranteed by UNIFORM-KUC and the lower bound provided by Theorem 4.

As last result, it is worth nothing that UNIFORM-KUC can be useful also in the non-uniform case. Given an instance I of KUC, we remind that ps_m and ps_M denote the minimum and the maximum ratios, respectively, of profit over size for items in I. UNIFORM-KUC can be applied on I by considering all items of profit over size ratio equal to ps_M. In doing so, the next theorem can be stated.

Theorem 6. UNIFORM-KUC *applied in the non-uniform case provides a competitive ratio of* $\frac{49}{86}\frac{ps_m}{ps_M}$.

Proof. Let SOL' be the profit of the solution provided by UNIFORM-KUC for an input instance I when considering all items of profit over size ratio equal to ps_M. Let SOL be the profit of the same solution when considering the actual profits associated to the items. Clearly, $SOL' \leq \frac{ps_M}{ps_m}SOL$. Moreover, $OPT' \geq OPT$, with OPT' being the optimal profit achievable for the modified instance, and OPT being the optimal profit achievable for the original instance. Then $\frac{SOL}{OPT} \geq \frac{ps_m}{ps_M}\frac{SOL'}{OPT'}$, and by Theorem 5 the claim holds.   □

When $\frac{49}{86}\frac{ps_m}{ps_M} > \frac{1}{2}$, the above result provides a better competitive ratio than that given by Theorem 1. As we are going to see in the next section, this is of practical interest when dealing with smartphones communications.

4   Experimental Analysis

In this section, we present the performance of our algorithms GREEDY-KUC and UNIFORM-KUC compared with the optimal solution and various heuristics:

Randomize (complexity $O(n)$): it chooses items in a totally random way;

Decrease (complexity $O(n \log n)$): it chooses items in a non-increasing order with respect to the size. The rational is to try to insert bigger items at the beginning rather than after having consumed a portion of the knapsack;

Increase (complexity $O(n \log n)$): it chooses items in a non-decreasing order with respect to the size. In this way we increase the probability to let the current item fitting in the knapsack, and to compose a "critical mass" comparable with the optimum;

Balance (complexity $O(n^2)$): Let $B_{max} = \sum_{i=1}^{n} s(a_i)$ and S_t be the size of the items already inserted in the knapsack. Balance inserts the item a that maximizes $(B_{max} - S_t - s(a)) \cdot p(a)$. Differently from the previous two heuristics, Balance depends on both the size and the profit of the items;

Uniform-KUC-H (complexity $O(n)$): let a_j be such that $p(a_j) = \max_{i=1}^{n} p(a_i)$. When items admit different ratios of profit over size, we set a new profit p' of each item a_i as $p'(a_i) = p(a_j)\frac{s(a_i)}{s(a_j)}$. In so doing the new ratios of profit over size are uniform. After executing UNIFORM-KUC, we replace the new profits with the real ones for each item.

In the following, DIFFKNAP denotes experiments where the size of the knapsack vary while keeping the same set of items. DIFFITEMS denotes experiments where the set of the items grows while keeping the same size for the knapsack.

For each configuration, we run the algorithms on 100 different instances, and we plot the average results of the obtained competitive ratios. For each instance, the optimal solution is obtained by the classic Dynamic Programming algorithm [13]. Since the optimal algorithm requires integer values, we run each algorithm on approximated instances multiplying $s(a_i)$ and B by 10^2 and using the ceil of the obtained values. We consider two main scenarios:

Smartphone Scenario. To simulate a real user experience we compare the algorithms on instances extracted from the dataset described in [2]. The dataset is synthetic and contains 100 days of smartphone's usage by a hypothetical user. For each day, the dataset contains all the arisen communications.

A communication service is characterized by a duration τ, a starting time and a type. The services are classified in *real-time* and *delay-tolerant*. The type of a real-time service is characterized by a couple (a, b), with $a \in$ {Voice call, VoIP call, YouTube, Internet Radio} and $b \in$ {3G, 4G, WiFi}. The type j of a delay-tolerant service is in the set {Download, Upload}.

Each knapsack represents a real-time service, say R. Let the duration τ of R be the length B of the knapsack. Each delay-tolerant service represents an item to be inserted in the knapsack. Let a_i be a delay-tolerant service of type j and (stand-alone) duration τ_i. When a_i is processed in parallel with R (that is, item a_i has been inserted in the knapsack R) then the duration of executing a_i may vary depending on its own type j and the type (a, b) of R. We compute $s(a_i) = \tau_i speed_{(a,b)}(j)$, with $speed_{(a,b)}(j)$ being the time increment in percentage necessary for executing a_i paired with R. Finally, we set $p(a_i) = \tau_i \cdot gain_{(a,b)}(j)$, where $gain_{(a,b)}(j)$ represents the gain in percentage of energy saving when a_i is paired with R. A study summarizing the gains obtained by pairing services along with the bandwidth values can be found in [2]. Fixed the type of R, there are only two different ratios of profit over size in the computed instances because the delay-tolerant services admit only two types, namely Download and Upload. This allows to optimize the sorting in GREEDY-KUC so as it works in $O(n)$ time.

In DIFFKNAP experiments, we firstly extract the delay-tolerant services in a 12 hours interval and the first real-time service successive to that interval. The real-time service determines only the type of the knapsack while we vary the size uniformly in the interval $[s_{max}, 10 \cdot s_{max}]$. It is worth remarking that varying on the knapsack type changes the size of the items and hence of s_{max}. The competitive ratios averaged on 100 different instances are shown in the left plot of Figure 1 varying x in $[1, 10]$, with $B = x \cdot s_{max}$.

In DIFFITEMS experiments, we firstly randomly pick a real-time service R as knapsack, and then we extract groups of items that precede R in such a way that the sum of their sizes vary in the interval $[B, 10 \cdot B]$. Precisely, each group will consist of n items such that the *Length Ratio* $LR = \sum_{i=1}^{n} s(a_i)/B$ grows from 1 to 10. The competitive ratios averaged on 100 different instances are shown in the right plot of Figure 1 varying on LR.

For the uniform case, without loss of generality, we choose only items of download type. Some results are shown in Figure 2.

Generic Scenario. In this scenario, we generate items with sizes and profits uniformly distributed in the interval $[1, 50]$, while the knapsack size varies in the interval $[50, 250]$. In the uniform case, profits are set equal to the sizes.

In DIFFKNAP experiments, the set of items is fixed to 10 in each run. The competitive ratio is evaluated by varying B uniformly in the interval $[50, 250]$.

In DIFFITEMS experiments, B is fixed to 125 in each run. The competitive ratio is evaluated using a number of items increasing from 5 to 25.

Figures 3 and 4 show the results for algorithms GREEDY-KUC and UNIFORM-KUC, respectively.

Outcomes from the experiments. In general, our GREEDY-KUC and UNIFORM-KUC algorithms behave really well, providing an approximation ratio most of the times greater than the 90% of the optimum. As shown in the reported results, the standard deviation of GREEDY-KUC and UNIFORM-KUC is also small.

In the DIFFKNAP experiments, when the size of the knapsack is very large, almost all the items can be fit in the knapsack, and thus all algorithms tend to the

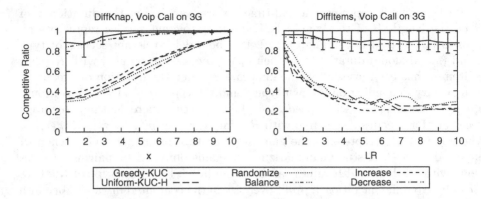

Fig. 1. Experiments from the Smartphone Scenario in the non-uniform case

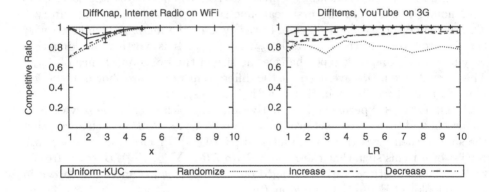

Fig. 2. Experiments from the Smartphone Scenario in the uniform case

optimum value. In contrast, in the DIFFITEMS experiments, when the number of items increases, the selection criterium becomes more important, and the performances of all heuristics, but GREEDY-KUC and Balance, become worse.

In the Smartphone scenario, both GREEDY-KUC and Balance reveal a much better behavior than those of all the other heuristics, see Figure 1. This is due to the fact that GREEDY-KUC and Balance make their item selection on the basis of profit and size, while the other heuristics either select items at random (Randomize) or based only on the size. It follows that the heuristics not taking into account both the criteria poorly perform.

It is worthy to note that when ps_m and ps_M are close, the choice of a 'wrong' item implies a smaller loss than when $ps_M >> ps_m$, and thus all the heuristics improve their performance. In Figure 1, for instance, in Voip Call on $3G$ we have $ps_m = .235$ and $ps_M = 1.9$, and the curves are set apart one from the other. Contrary, in Figure 2 that shows the uniform case (i.e., $ps_m = ps_M$), curves are very close to each other.

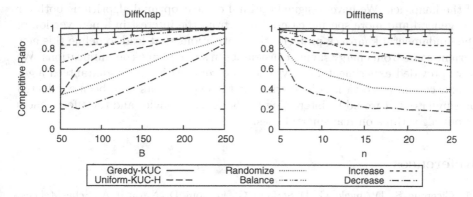

Fig. 3. Experiments from the Generic Scenario in the non-uniform case

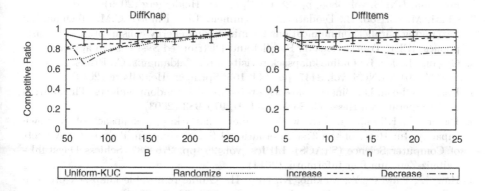

Fig. 4. Experiments from the Generic Scenario in the uniform case

In the Generic scenario, Figure 3 reports the results when the ratios of profit over size assume a larger set of possible values. The different ratios do not seem to influence much GREEDY-KUC, while a deeper impact is faced for the other heuristics. The Decrease heuristic is always the worse since it takes always the highest risk. Clearly, the paid penalty is higher for small values of B. In the uniform case, reported in Figure 4, again the differences become less extreme, and the curves appear flattened towards the up border of the figure. In the DIFFKNAP experiments, since the items are longer it appears that the knapsack is saturated less frequently than in the Smartphone scenario and thus the curves do not easily reach the optimum value.

5   Conclusion

We have proposed a new variant of the more standard online knapsack problem where the only information missing to the provided instances is the capacity B

of the knapsack. We have designed optimal or near optimal algorithms both for the general and more specific cases. Apart from the interest in a new version of the fundamental knapsack problem, the motivations that lead to define this new variant come from energy saving constraints in smartphone communications. We have provided experimental analysis for the competitive ratio guaranteed algorithms compared to various heuristics. As the experiments have been conducted on simulators, it would be interesting to check the behavior and the effectiveness of our algorithms on real smartphones.

References

1. Cicerone, S., D'Angelo, G., Di Stefano, G., Frigioni, D., Navarra, A., Schachtebeck, M., Schöbel, A.: Recoverable robustness in shunting and timetabling. In: Ahuja, R.K., Möhring, R.H., Zaroliagis, C.D. (eds.) Robust and Online Large-Scale Optimization. LNCS, vol. 5868, pp. 28–60. Springer, Heidelberg (2009)
2. Conti, M., Crispo, B., Diodati, D., Nurminen, J.K., Pinotti, C.M., Teemaa, T.: Leveraging parallel communications for minimizing energy consumption on smartphones. IEEE Transactions on Parallel and Distributed Systems (to appear)
3. Cygan, M., Jeż, Ł.: Online knapsack revisited. In: Kaklamanis, C., Pruhs, K. (eds.) WAOA 2013. LNCS, vol. 8447, pp. 144–155. Springer, Heidelberg (2014)
4. Das, S., Ghosh, D.: Binary knapsack problems with random budgets. The Journal of the Operational Research Society 54(9), 970–983 (2003)
5. Disser, Y., Klimm, M., Megow, N., Stiller, S.: Packing a knapsack of unknown capacity. In: Proc. of the 31st International Symposium on Theoretical Aspects of Computer Science (STACS), LIPIcs, vol. 25, pp. 276–287. Schloss Dagstuhl - Leibniz-Zentrum fuer Informatik (2014)
6. Han, X., Kawase, Y., Makino, K., Guo, H.: Online removable knapsack problem under convex function. Theor. Comput. Sci. 540, 62–69 (2014)
7. Iwama, K., Taketomi, S.: Removable online knapsack problems. In: Widmayer, P., Triguero, F., Morales, R., Hennessy, M., Eidenbenz, S., Conejo, R. (eds.) ICALP 2002. LNCS, vol. 2380, pp. 293–305. Springer, Heidelberg (2002)
8. Iwamaa, X., Zhang, G.: Online knapsack with resource augmentation. Information Processing Letters 110(22), 1016–1020 (2010)
9. Kleywegt, A.J., Papastavrou, J.D.: The dynamic and stochastic knapsack problem. Operations Research 46(1), 17–35 (1998)
10. Liebchen, C., Lübbecke, M., Möhring, R., Stiller, S.: The concept of recoverable robustness, linear programming recovery, and railway applications. In: Ahuja, R.K., Möhring, R.H., Zaroliagis, C.D. (eds.) Robust and Online Large-Scale Optimization. LNCS, vol. 5868, pp. 1–27. Springer, Heidelberg (2009)
11. Lueker, G.S.: Average-case analysis of off-line and on-line knapsack problems. J. of Algorithms 29(2), 277–305 (1998)
12. Marchetti-Spaccamela, A., Vercellis, C.: Stochastic on-line knapsack problems. Mathematical Programming 68(1–3), 73–104 (1995)
13. Martello, S., Toth, P.: Knapsack problems. J. Wiley & Sons, Chichester (1990)
14. Megow, N., Mestre, J.: Instance-sensitive robustness guarantees for sequencing with unknown packing and covering constraints. In: Proc. of the 4th International Conference on Innovations in Theoretical Computer Science (ITCS), pp. 495–504. ACM (2013)

15. Noga, J., Sarbua, V.: An online partially fractional knapsack problem. In: Proc. of the 8th Sym. Parallel Architectures, Algorithms and Networks (ISPAN), pp. 108–112. IEEE Computer Society (2005)
16. Papastavrou, J.D., Rajagopalan, S., Kleywegt, A.J.: The dynamic and stochastic knapsack problem with deadlines. Manage. Sci. **42**(12), 1706–1718 (1996)
17. Zhou, Y., Chakrabarty, D., Lukose, R.: Budget constrained bidding in keyword auctions and online knapsack problems. In: Papadimitriou, C., Zhang, S. (eds.) WINE 2008. LNCS, vol. 5385, pp. 566–576. Springer, Heidelberg (2008)

Combinatorial Optimization II

Reoptimization Techniques for MIP Solvers

Gerald Gamrath, Benjamin Hiller, and Jakob Witzig[⊠]

Zuse Institute Berlin, Takustraße 7, 14195 Berlin, Germany
{gamrath,hiller,witzig}@zib.de

Abstract. Recently, there have been many successful applications of optimization algorithms that solve a sequence of quite similar mixed-integer programs (MIPs) as subproblems. Traditionally, each problem in the sequence is solved from scratch. In this paper we consider reoptimization techniques that try to benefit from information obtained by solving previous problems of the sequence. We focus on the case that subsequent MIPs differ only in the objective function or that the feasible region is reduced. We propose extensions of the very complex branch-and-bound algorithms employed by general MIP solvers based on the idea to "warmstart" using the final search frontier of the preceding solver run. We extend the academic MIP solver SCIP by these techniques to obtain a reoptimizing branch-and-bound solver and report computational results which show the effectiveness of the approach.

1   Introduction

In the last decades many powerful decomposition- and reformulation-based techniques for solving hard optimization problems were developed, e.g., column generation and Lagrangian relaxation. These methods decompose a problem into a master problem and several subproblems which are repeatedly solved to update the master problem. Frequently, the subproblems solved in successive iterations differ only in the cost vector, reflecting updated information from the master problem. It is a natural idea to exploit this property in order to improve the running time of the overall algorithm for solving the master problem. Methods to achieve this are known as *reoptimization techniques*. They have been investigated in the context of decomposition methods, e.g., in the context of Lagrangian relaxation [20], column generation [8], and for generic branch-and-bound [17].

In the literature, reoptimization techniques have been investigated for polynomial solvable problems mainly, e.g., the shortest path problem [21] or the min cost flow problem [12]. This is partly due to the fact that traditionally, decomposition methods have been applied such that the resulting subproblems are (pseudo)polynomially solvable. More recently, *Mixed Integer Programs* (MIPs) have been used as subproblems, e.g., for cut generation [10,11] or in generic decomposition schemes [14,22] and corresponding solvers [9,15]. The resulting subproblems are solved by standard MIP solvers, which are very sophisticated branch-and-bound algorithms. Thus there is a need for reoptimization techniques in MIP solvers to benefit from the knowledge obtained in previous iterations.

One of the first investigations on reoptimizing MIPs was done by Güzelsoy and Ralphs [16,23]. They addressed sequences of MIPs that differ only in the

© Springer International Publishing Switzerland 2015
E. Bampis (Ed.): SEA 2015, LNCS 9125, pp. 181–192, 2015.
DOI: 10.1007/978-3-319-20086-6_14

right-hand side. Their approach is mainly based on duality theory, which they employed to develop techniques for "warm starting" using dual information obtained through primal algorithms. Our approach to reoptimizing MIPs is similar to the One-Tree algorithm [6] for generating multiple solutions of a single MIP. Similar techniques have also been used in [18] to benefit from a preliminary restricted branching phase when solving a single MIP instance.

In this paper, we propose a reoptimizing variant of the general LP-based branch-and-bound algorithm used by modern MIP solvers. It is based on the idea [17] to "continue" the search at the last known search frontier of the branch-and-bound tree. As the performance of state-of-the-art MIP solvers is based to a substantial extend on exploiting dual information, we introduce a mechanism to deal with this. This mechanism is in particular applied to cope with strong branching. It is intuitively clear that continuing the solving process is a poor idea if the objective function has changed a lot or the search frontier is rather huge. To deal with the first situation, we use a similarity measure for objective functions to decide whether to reoptimize or to start from scratch. Moreover, we propose heuristics to start with a reduced search frontier that is still based on the previous one. Our ideas have been carefully implemented using the MIP solver SCIP [2,25]. We test our reoptimization techniques on sequences arising from the generic column generation solver GCG [15] and on instances of the k-constrained shortest path problem arising from a ship navigation problem[1]. More details and computational results can be found in the master thesis of the last author[27].

The paper is outlined as follows. Sec. 2 provides a summary of the relevant ingredients of a state-of-the-art MIP solver (i.e., SCIP) together with an in-depth motivation to reoptimization for MIPs. Sec. 3 presents our technical contributions summarized above. Computational results are discussed in Sec. 4. Finally, Sec. 5 concludes the paper.

2   Mixed Integer Programming and Reoptimization

In this paper we consider mixed integer linear programs (MIPs) of the form

$$z_{MIP} = \min\{c^T x : Ax \geq b, \ell \leq x \leq u, x_i \in \mathbb{Z} \text{ for all } i \in \mathcal{I}\} \tag{1}$$

with objective function $c \in \mathbb{R}^n$, constraint matrix $A \in \mathbb{R}^{m \times n}$ and constraint right-hand sides $b \in \mathbb{R}^m$, variable lower and upper bounds $\ell, u \in \bar{\mathbb{R}}^n$ where $\bar{\mathbb{R}} := \mathbb{R} \cup \{\pm\infty\}$, and a subset $\mathcal{I} \subseteq \mathcal{N} = \{1, \ldots, n\}$ of variables which need to be integral in every feasible solution. In the remainder of the paper, we focus on mixed-binary programs, i.e., MIPs with $\ell_i = 0, u_i = 1$ for all $i \in \mathcal{I}$.

When omitting the integrality restrictions, we obtain the *linear program (LP)*

$$z_{LP} = \min\{c^T x : Ax \geq b, \ell \leq x \leq u\}. \tag{2}$$

---

[1] Thanks to Mirai Tanaka from Tokyo Institute of Technology and Kazuhiro Kobayashi from National Maritime Research Institute Japan for providing the instances.

It is a relation of the corresponding MIP and provides a lower bound on its optimum, i.e., $z_{LP} \leq z_{MIP}$. This fact plays an important role in the most widely used method to solve general MIPs, the LP-based branch-and-bound method [5,19]. It is a divide-and-conquer method which starts by solving the LP relaxation of the problem to compute a lower bound and a solution candidate x^*. If x^* fulfills the integrality restrictions, the problem is solved to optimality, if not, it is split into (typically) two disjoint subproblems, thereby removing x^* from both feasible LP regions. Typically, a fractional variable $x_i, i \in \mathcal{I}$ with $x_i^* \notin \mathbb{Z}$ is selected and the restrictions $x_i \geq \lceil x_i^* \rceil$ and $x_i \leq \lfloor x_i^* \rfloor$ are added to the two subproblems, respectively. This step is called branching. While this process is iterated, we store and update the best solution $\bar{x}$ found so far whenever one of the subproblems has an integral LP solution. The key observation is that a subproblem can be disregarded when its lower bound is not smaller than the objective value of $\bar{x}$. This is called bounding. This leads to the following cases that need to be distinguished when regarding a subproblem:

1. the LP relaxation is infeasible: the subproblem can be disregarded
2. x^* is integral: the subproblem is solved to optimality
3. $c^T x^* \geq c^T \bar{x}$: the current subproblem can be disregarded due to bounding
4. else: the current subproblem is split (branching).

The branch-and-bound process is typically illustrated as a tree. The root node represents the original problem and the two subproblems created by the branching step correspond to two child nodes being created for the current node.

This general scheme is extended by various algorithms to enhance the performance (see [1]). We briefly sketch those advanced components that we need to handle specifically in the context of reoptimization. One of these components is *presolving*, which is done before the branch-and-bound process starts and analyzes the problem, removes redundancies and tightens the formulation. A reduced form of this, called *domain propagation*, is also done during the branch-and-bound phase before the LP relaxation of a node is solved. For more details, we refer to [1,24]. Finally, the decision on which variable to branch is of high importance for a fast convergence. It is typically supported by a technique called *strong branching* [3]. Strong branching precomputes lower bounds for potential child nodes of a candidate variable by solving auxiliary LPs with the branching bound change added. Besides providing very accurate lower bound predictions, it can also deduce bound changes or even infeasibility of the current node, if one or both of the regarded children for a candidate variable are infeasible or their lower bound exceeds the upper bound.

All these reductions can be divided into two classes: *primal* and *dual reductions*. The former are based only on feasibility arguments and remove only infeasible parts of the search space. In contrast to that, dual reductions are based on an optimality argument and may exclude feasible solutions as long as they retain at least one optimal solution. Therefore, dual reductions are not necessarily valid anymore if the objective is changed and we need to treat them with care in the reoptimization context.

Reoptimization for General MIP. In recent years, there is a growing interest in reoptimization techniques for MIP solvers. One of the major applications is the repeated solution of pricing problems within generic branch-cut-and-price solvers. Those solvers are based on a Dantzig-Wolfe decomposition [7] of the given original problem. The resulting problem is solved by a column generation approach, i.e., the LP relaxation is solved with just a subset of the variables and improving variables are searched for by solving the pricing problem, which is a MIP whose objective function depends on the current LP solution. For more details on generic branch-cut-and-price, we refer to [13,14,22]. Since the pricing problem is solved repeatedly with updated objective function, this would greatly benefit from an effective reoptimization technique and is thus our main application in the following.

Another application of reoptimization is the computation of the k best solutions of binary programs. This can be accomplished by iteratively solving the problem to optimality and excluding each optimal solution $\bar{x}$ for the next iteration by a *logic-or constraint*.

Definition 1 (Logic-or Constraint). *Let $\mathcal{B} \subseteq \mathcal{I}$ be a set of binary variables and $C^-, C^+ \subseteq \mathcal{B}$ disjoint subsets. A logic-or constraint has the form*

$$\sum_{i \in C^-} x_i + \sum_{j \in C^+} (1 - x_j) \geq 1. \tag{3}$$

As a shortcut, we use the notation $x(C^-, C^+) := \sum_{i \in C^-} x_i + \sum_{j \in C^+} (1 - x_j)$. As can easily be seen, the logic-or constraint with $C^- = \{i \in \mathcal{I} \mid \bar{x}_i = 0\}$ and $C^+ = \{i \in \mathcal{I} \mid \bar{x}_i = 1\}$ forbids the optimal solution $\bar{x}$ and no other solution. Logic-or constraints play an important role in the remainder of this paper. Note that the sets C^- and C^+ do not need to be a partition of the variable set, but can also cover only a part of the variables in order to exclude all solutions with these variables set to the given values.

While there are more efficient ways to compute the k best solutions for a fixed k, the iterated approach proves useful if the limit k is decided during the optimization process. An example for this case is a mixed-integer nonlinear ship navigation problem investigated by Mirai Tanaka and Kazuhiro Kobayashi [26]. They repeatedly solve a MIP relaxation of the problem to compute a lower bound and a solution candidate. For this candidate, exact costs are computed and it is excluded from the MIP relaxation. The procedure continues until the MIP lower bound exceeds the cost of the best solution found. Therefore, the number k of best solutions needed cannot be determined a priori and an iterated approach is preferred. In contrast to the reoptimization for branch-cut-and-price, the difference between two iterations is not in the objective function, but the additional constraint excluding the previously computed optimum.

In the following section, we will discuss how these two applications of reoptimization cases can be handled within a state-of-the-art MIP solver.

3   Extending SCIP to a Reoptimizing Branch-and-Bound Solver

For our approach we follow the ideas of [17]. Consider the sequence of MIPs

$$(P_i) \qquad \min \left\{ c_i^T x \mid Ax \geq b, \ell \leq x \leq u, x_i \in \mathbb{Z} \right\} \qquad \text{for all } i \in \mathcal{I} \qquad (4)$$

for a given sequence of objective vectors $(c_i)_{i \in I}$, for some index set $I = \{1, \ldots, m\}$. Moreover, let $\mathcal{S}$ be the function mapping an optimization problem $\mathcal{P} \subseteq P_i$ to its set of feasible solutions X. Solving one problem of this sequence with a standard LP-based branch-and-bound algorithm up to a given stopping criterion, e.g., optimality, provides search space dividing subsets:

- a set of subproblems with an infeasible LP relaxation,
- a set $\mathcal{P}_{obj}$ of subproblems that have been either pruned due to bounding or are solved to optimality,
- a set Σ of all feasible solutions found so far, and
- the set $\mathcal{P}_o$ of as-yet unprocessed subproblems,

such that $X = \mathcal{S}(\mathcal{P}_o) \cup \mathcal{S}(\mathcal{P}_{obj}) \cup \Sigma$ holds, where $\mathcal{S}(\mathcal{P}_o)$ and $\mathcal{S}(\mathcal{P}_{obj})$ denote the set of solutions of $\mathcal{P}_o$ and $\mathcal{P}_{obj}$, respectively. Usually, the set $\mathcal{P}_o$ is empty at the end of the computation and there are open nodes only if the solving process terminates due to a stopping criterion different to optimality, e.g., time limit.

We call the solving process of (P_i) an iteration. To summarize the basic idea consider two iterations $i, i+1$. The authors of [17] proposed that solving P_{i+1} can be started at the last search frontier of iteration i. The set of leaf nodes generated in iteration i can be divided into the sets of unprocessed nodes ($\mathcal{P}_o$) and nodes pruned because their dual bound exceeds the best known primal bound ($\mathcal{P}_{obj}$). Additionally, all feasible solutions found during the solving process of P_i are collected in Σ. Obviously, by changing the objective function c_i to c_{i+1} from iteration i to $i + 1$, subproblems that are discarded since the LP relaxation is infeasible cannot become feasible. Thus, an optimal solution of P_i has to be in $\mathcal{S}(\mathcal{P}_o)$, $\mathcal{S}(\mathcal{P}_{obj})$ or Σ. Therefore, in the next iteration, we restore all subproblems in $\mathcal{P}_o \cap \mathcal{P}_{obj}$ as children of the root node. Additionally, all solutions from Σ are added to the solution pool. Note that in case additional constraints were added, we can apply this method as well but need to check these solutions for feasibility.

Additionally, we define a set $\mathcal{P}_f \subseteq \mathcal{P}_{obj}$ which contains all subproblems with an integral LP solution. We use this set in Sec. 3.2 where we present two heuristics which operate on the set of feasible nodes $\mathcal{P}_f$ only.

3.1   Handling Dual Information

Most state-of-the-art MIP solvers use, in addition to branch-and-bound, various preprocessing and domain propagation techniques. In connection with reoptimization, we have to be very careful when using these techniques. Dual methods provide bound changes based on the current objective function. Thus, the pruned part, i.e., the subproblem discarded by fixing variables, could contain feasible solutions which might be of interest after changing the objective function. Hence, it is necessary to disable all dual methods or treat them with special care. In the following we focus on binary and mixed binary programs of the form (1), where $\{l_i, u_i\} = \{0, 1\}$ for all $i \in \mathcal{I}$. Let us denote the set of binary variables by $\mathcal{B}$. Moreover, consider a subproblem $\mathcal{P}$ of P and a method D based on dual information. Calling D for $\mathcal{P}$ fixes variables $C = C^- \cup C^+ \subseteq \mathcal{B}$ to 0 and 1, respectively, where $C^- = \{i \in C \mid x_i = 0\}$ and $C^+ = \{i \in C \mid x_i = 1\}$. There are two ways to deal with these fixings. On the one hand, we can remember the subproblem $\mathcal{P}$ as before calling D and forget all decisions in the subtree induced by $\mathcal{P}$. The disadvantage of this approach is that some methods using dual information, e.g., strong branching, provide the most advantage at the root node. Thus, revoking these decisions will lead to the original problem $\mathcal{P}$. Hence, we do not benefit from reoptimizing. On the other hand, we can ensure that the pruned part may be reconstructed in the next iteration. In this paper we will follow the latter approach. The idea is to split problem $\mathcal{P} \subseteq P$ into two nodes at the beginning of the next iteration. The first node corresponds to $\mathcal{P}$ with all fixings $C^- \cup C^+$; the second node corresponds to $\mathcal{P}$ with an additional logic-or constraint $\mathfrak{C}$ depending on C^- and C^+. This logic-or constraint ensures that at least one variable get value different from the fixings $C^- \cup C^+$.

Theorem 2. *Let P be a binary or mixed binary problem, $\mathcal{P} \subseteq P$ a subproblem, and $C^-, C^+ \subseteq \mathcal{B}$ disjoint sets of binary variables. A complete and disjoint representation of the solution space $\mathcal{S}(\mathcal{P})$ of $\mathcal{P}$ is given by $\mathcal{S}(\mathcal{P}_{\mathfrak{F}}) \cup \mathcal{S}(\mathcal{P}_{\mathfrak{C}})$, where*

$$\mathcal{S}(\mathcal{P}_{\mathfrak{F}}) = \mathcal{S}(\mathcal{P}) \cap \{x \in \mathbb{R}^n \mid x_i = 0 \; \forall i \in C^- \text{ and } x_j = 1 \; \forall j \in C^+\} \quad (5)$$
$$\mathcal{S}(\mathcal{P}_{\mathfrak{C}}) = \mathcal{S}(\mathcal{P}) \cap \{x \in \mathbb{R}^n \mid x(C^-, C^+) \geq 1\}, \quad (6)$$

i.e., $\mathcal{S}(\mathcal{P}_{\mathfrak{C}})$ contains all solutions with at least one variable $x_i = 1$ or $x_j = 0$ for $i \in C^-$ and $j \in C^+$.

Proof. We only have to focus on variables x_i with $i \in C^+ \cup C^-$. Assume $\mathcal{S}(\mathcal{P}_{\mathfrak{F}})$ and $\mathcal{S}(\mathcal{P}_{\mathfrak{C}})$ are not disjoint. Let $x \in \mathcal{S}(\mathcal{P}_{\mathfrak{F}}) \cap \mathcal{S}(\mathcal{P}_{\mathfrak{C}})$. By definition of $\mathcal{S}(\mathcal{P}_{\mathfrak{F}})$ the solution x has to fulfill $x_i = 0$ and $x_j = 1$, for all $i \in C^-$ and $j \in C^+$. Thus, the constraint $x(C^-, C^+) \geq 1$ is violated by x. Hence, $x \notin \mathcal{S}(\mathcal{P}_{\mathfrak{C}})$ and $\mathcal{S}(\mathcal{P}_{\mathfrak{F}}) \cap \mathcal{S}(\mathcal{P}_{\mathfrak{C}}) = \emptyset$. Finally, consider $x \in \mathcal{P}$ and assume $x \notin \mathcal{S}(\mathcal{P}_{\mathfrak{F}}) \cup \mathcal{S}(\mathcal{P}_{\mathfrak{C}})$. Since $x \notin \mathcal{S}(\mathcal{P}_{\mathfrak{F}})$ at least one variable $i \in C^-$ and $j \in C^+$ needs to be different from $x_i = 0$ and $x_j = 1$, respectively. Hence, by the definition of $\mathcal{S}(\mathcal{P}_{\mathfrak{C}})$ it follows $x \in \mathcal{S}(\mathcal{P}_{\mathfrak{C}})$, which is a contradiction. $\qquad\square$

3.2   Heuristics

In the following we present a primal heuristic which is fitted to column generation and two heuristics for reducing the size of the search frontier that needs to be reoptimized. The latter heuristics – so-called compression heuristics – are needed because we have to solve the whole stored search frontier to prove optimality. Hence, a small search frontier of "good quality" would be desirable.

A Primal Heuristic: Trivialnegation. Consider a binary or mixed binary optimization problem $\mathcal{P}$ with n variables and two objective vectors $c, \tilde{c} \in \mathbb{R}^n$. Furthermore, let $x^\star \in \mathcal{P}$ be an optimal solution with $c^T x^\star = \min_{x \in \mathcal{P}} c^T x$. The impact of variable $i \in \mathcal{B}$ w.r.t. c and $\tilde{c}$ is defined by

$$\Psi_i = \begin{cases} 1 & \text{if } \operatorname{sgn}(c_i) \cdot \operatorname{sgn}(\tilde{c}_i) \leq 0 \text{ and } c_i \neq 0 \vee \tilde{c}_i \neq 0 \\ 0 & \text{otherwise.} \end{cases}$$

Based on the impact Ψ_i of a variable $i \in \mathcal{B}$ we construct a (not necessarily feasible) solution candidate $\tilde{x}^B$ for each subset $B \subseteq \mathcal{B}$ by setting $\tilde{x}_i^B = 1 - x_i^\star$ if $\Psi_i = 1$ and $\tilde{x}_i^B = x_i^\star$ otherwise.

Compressing the Search Tree. Consider a search tree with nodes V and the set of leaf nodes $V_{leaf} \subseteq V$ resulting from solving a (mixed) binary optimization problem consisting of n variables with a branch-and-bound algorithm. Each node $v \in V$ corresponds to a subproblem

$$(\mathcal{P}^v) \quad \min\{c^T x : A^v x \geq b^v, \ell^v \leq x \leq u^v, x_i \in \mathbb{Z} \text{ for all } i \in \mathcal{I}\}, \qquad (7)$$

where the constraint set $A^v x \geq b^v$ contains all constraints $Ax \geq b$ plus other constraints added during the branch-and-bound procedure, e.g., logic-or constraints. For each subproblem the set of binary variables can be partitioned into a set of unfixed variables, i.e., $\ell_i^v < u_i^v$, and the sets

$$X_v^0 = \{i \in \mathcal{B} \mid u_i^v = 0\} \quad \text{and} \quad X_v^1 = \{i \in \mathcal{B} \mid \ell_i^v = 1\}.$$

The key idea of our compression heuristics is to find a set of subproblems $\mathcal{P}^r, r \in \mathcal{R}$ of the form (7), such that each leaf $v \in V_{leaf}$ is represented by a unique r (for short: $r \succ v$), i.e., $\mathcal{S}(\mathcal{P}^v) \subseteq \mathcal{S}(\mathcal{P}^r)$. Moreover, we classify the quality of a representative by the loss of information w.r.t. the represented leaves

$$loss(V_{leaf}, r) = \sum_{v \in V_{leaf} : \, r \succ v} |(X_v^0 \cup X_v^1) \setminus (X_r^0 \cup X_r^1)| \qquad (8)$$

and the quality of a set $\mathcal{R}$ by the sum of losses of information for its representatives $r \in \mathcal{R}$.

To keep the search frontier small we use two heuristics. The first heuristic is called *largest representative* and reduces the search frontier to two nodes. In this paper we describe the basic idea only and refer to [27] for more details. Consider a subset of leaf nodes $W \subseteq V_{leaf}$. Based on an arbitrary node $v \in W$ we construct a representative r iteratively. First, we set $X_r^0 = X_v^0$ and $X_r^1 = X_v^1$. Afterwards, we add each node $w \in W$ greedily to the set of represented nodes as long as r and w have at least one common fixed variable and we update

$$X_r^0 \leftarrow X_r^0 \cap X_w^0 \quad \text{and} \quad X_r^1 \leftarrow X_r^1 \cap X_w^1.$$

The compressed search frontier is then given by $\mathcal{P}_{\bar{\mathcal{F}}}^r$ and $\mathcal{P}_{\bar{\mathcal{C}}}^r$, where

$$S(\mathcal{P}_{\bar{\mathcal{F}}}^r) = S(\mathcal{P}) \cap \{x \in \mathbb{R}^n \mid x_i = 0 \, \forall i \in X_r^0 \text{ and } x_i = 1 \, \forall i \in X_r^1\}$$
$$S(\mathcal{P}_{\bar{\mathcal{C}}}^r) = S(\mathcal{P}) \cap \{x \in \mathbb{R}^n \mid x(X_r^0, X_r^1) \geq 1\}.$$

Since we are interested in finding good solutions fast, we run the heuristic on the set of feasible nodes $\mathcal{P}_f$ only. Moreover, to ensure that the compressed search frontier is as good as possible we run the procedure for each node $v \in \mathcal{P}_f$ and choose the representation with minimal loss of information.

The second heuristic is called *weak compression* and we have to distinguish between trees where the only difference between nodes are the sets of fixed variables ($A^v = A$) and trees with additionally added constraints, e.g., logic-or constraints. Due to page limitation we give the basic ideas only and refer to [27] for a complete description. Consider a tree without added constraints, its search frontier V_{leaf} and a subset of leaf nodes $W \subseteq V_{leaf}$. For each node $v \in W$ let v be a representative for itself. This leads to $|W|$ disjoint representatives. To ensure that the representation is complete, i.e., no feasible solution gets lost, we need to construct a subproblem covering $S(\mathcal{P}) \setminus \bigcup_{v \in W} S(\mathcal{P}^v)$. Such a representative can be constructed as before using Theorem 2, i.e., by including a logic-or constraint for each $v \in W$. Therefore, a complete representation of the search frontier is given by $\mathcal{R} = \{\mathcal{P}^{v_1}, \dots, \mathcal{P}^{v_k}, \mathcal{P}_{\mathcal{C}_W}\}$, where $W = \{v_1, \dots, v_k\}$ and

$$S(\mathcal{P}_{\mathcal{C}_W}) = S(\mathcal{P}) \cap \bigcap_{v \in W} \{x \in \mathbb{R}^n \mid x(X_v^0, X_v^1) \geq 1\}.$$

If the search tree consists of nodes containing additional constraints, this construction cannot be adapted while guaranteeing completeness and disjointness. Thus, we restrict ourselves to the case $|W| = 1$. Assume $W = \{v\} \subseteq V_{leaf}$ with at least one fixed variable and potentially added logic-or constraints to handle dual reductions in previous iterations (see Sec. 3.1). Each of these constraints corresponds to disjoint variable sets $C_l = C_l^- \cup C_l^+ \subset \mathcal{B}$. For guaranteeing a complete and disjoint representation we have to reconstruct subproblems which were cut off by the added constraints, for an illustration see Figure 1. Therefore, a complete and disjoint representation is given by $\mathcal{R} = \{\mathcal{P}_{\mathcal{C}}^v, \mathcal{P}_{\mathcal{F}}^{v,1}, \mathcal{P}_{\mathcal{F}}^{v,2}, \dots, \mathcal{P}_{\mathcal{F}}^{v,k}, \mathcal{P}_{\mathcal{C}}^{v,k}\}$ (drawn solid in Figure 1), where

$$S(\mathcal{P}_{\mathcal{F}}^v) = S(\mathcal{P}) \cap \{x \in \mathbb{R}^n \mid x_i = 0 \, \forall i \in X_v^0 \text{ and } x_j = 1 \, \forall j \in X_v^1\},$$
$$S(\mathcal{P}_{\mathcal{C}}^v) = S(\mathcal{P}) \cap \{x \in \mathbb{R}^n \mid x(X_v^0, X_v^1) \geq 1\},$$
$$S(\mathcal{P}_{\mathcal{F}}^{v,1}) = S(\mathcal{P}_{\mathcal{F}}^v) \cap \{x \in \mathbb{R}^n \mid x_i = 0 \, \forall i \in C_1^- \text{ and } x_j = 1 \, \forall j \in C_1^+\},$$
$$S(\mathcal{P}_{\mathcal{C}}^{v,1}) = S(\mathcal{P}_{\mathcal{F}}^v) \cap \{x \in \mathbb{R}^n \mid x(C_1^-, C_1^+) \geq 1\},$$

and for all $l = 2, \dots, k$

$$S(\mathcal{P}_{\mathcal{F}}^{v,l}) = S(\mathcal{P}_{\mathcal{C}}^{v,l-1}) \cap \{x \in \mathbb{R}^n \mid x_i = 0 \, \forall i \in C_l^- \text{ and } x_j = 1 \, \forall j \in C_l^+\},$$
$$S(\mathcal{P}_{\mathcal{C}}^{v,l}) = S(\mathcal{P}_{\mathcal{C}}^{v,l-1}) \cap \{x \in \mathbb{R}^n \mid x(C_l^-, C_l^+) \geq 1\}.$$

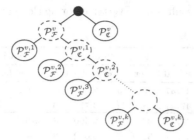

Fig. 1. Construction of the weak compression heuristic

3.3 Similarity of Objective Functions

If two objective functions are quite similar – whatever similar means at this point – we expect the resulting search trees to be similar as well. On the other hand, if two objective functions are quite different the resulting search trees might be very different, too. An immediate consequence is that continuing the solving process at the last search frontier might need much more effort than solving the problem from scratch. The following criterion can be used to estimate the similarity of the resulting search trees a priori.

Definition 3 (Similarity of Objective Functions). *Let $c, \tilde{c} \in \mathbb{R}^n$ be two objective functions. Then the similarity of c and $\tilde{c}$ is given by*

$$\Lambda(c, \tilde{c}) = \frac{\langle c, \tilde{c} \rangle}{\|c\|_2 \|\tilde{c}\|_2}.$$

This corresponds to the cosine between the two objective vectors. We apply reoptimization only if the similarity measure is above a certain threshold (per default 0.8), since reoptimizing the search frontier in spite of very different objective functions seems to be not promising.

4   Computational Results

For evaluating our reoptimization approach we performed computational experiments on vertex coloring instances and on instances of the k-constrained shortest path problem. For the latter problem the sequence consists of k identical objective functions and the consecutive subproblems differ in exactly one logic-or constraint, cf. Sec. 2 and [26]. For generating sequences of objective functions for the vertex coloring problem we use a subset of the `COLOR02/03/04` [4] test set and the generic branch-cut-and-price solver `GCG` [15] which was introduced in [13]. The pricing subproblems that `GCG` needs to solve during the column generation process (see Sec. 2) only differ in the objective function. In order to test `SCIP` and `ReoptSCIP` on these problems, we write out the sequence of pricing problems of a `GCG` run, which avoids side-effects that a different optimal solution computed by any of the two solvers might have.

Table 1. Computational results on the vertex coloring (left) and k-constrained shortest path (right) test sets

algorithm	TN	LR	WC	vertex coloring					k-constrained shortest path				
				solved	faster	slower	nodes	time	solved	faster	slower	nodes	time
SCIP				47/47	0	37	3,896	47	66/66	0	60	1,143	55
	•			47/47	28	6	3,775	43	–	–	–	–	–
ReoptSCIP				47/47	33	4	6,590	38	66/66	48	11	1,310	38
	•			47/47	32	5	6,602	38	–	–	–	–	–
		•		47/47	36	1	4,909	33	66/66	57	1	1,112	35
			•	47/47	36	1	4,825	32	66/66	56	0	1,149	30
		•	•	–	–	–	–	–	66/66	57	3	1,125	36
	•	•	•	47/47	37	0	4,493	28	–	–	–	–	–

We performed all tests on an Intel Xeon CPU E3-1290 V2 @ 3.70GHz with 16GB RAM. Our implementation is based on SCIP 3.1.0.1 using CPLEX 12.6 as LP solver. We limited the solving time of each sequence by 3600 seconds and used the technique described in Sec. 3.1 in order to handle dual reductions obtained by strong branching. The benefits obtained by dual presolving and propagation are less substantial and are outweighted by the enlargement of the search frontier they cause when applying the technique from Sec. 3.1. Therefore, we disabled all dual presolving and propagation techniques. We compute averages of solved nodes and solving time by using the shifted geometric mean [1, Appendix A] with shift $s = 10$ for running times and $s = 100$ for nodes.

In our computational experiments, we skip reoptimizing the search frontier if two consecutive objective functions are not similar enough, i.e., if the similarity is less than 0.8 (see Sec. 3.3). Furthermore, we solve the problem from scratch if the search frontier consists of more than 2000 nodes, because proving optimality needs more effort for a larger search frontier and thus the chance increases that solving from scratch succeeds faster. In order to shrink the search frontier, we run the largest representative (LR) heuristic exclusively on the set of feasible nodes $\mathcal{P}_f$ and we compress the search tree only if the determined representation is better than the last compression of a previous round, i.e., if the loss of information decreases or the number of fixings of the representative increases. We run the weak compression (WC) heuristic for one node only, since in almost every iteration the search tree includes nodes with added constraints. We choose the node with the largest dual bound in the previous iteration and demand that the number of fixed variables is at least one and not less than the number of fixed variables in the last weak compression. Additionally, we do not want to use the same node for weak compression twice in a row.

In Table 1 we compare the results of our reoptimization approach with plain SCIP on both test sets. To this end, we compare different variants, e.g., SCIP in combination with the trivialnegation (TN) heuristic or ReoptSCIP in combination with a single one or all three presented heuristics, as indicated in column "variant". Note that the trivialnegation heuristic is not applied to the k-constrained shortest path instances, since the objective function does not change. We compare the number of instances that are solved within the time limit and those solved significantly (at least 5%) faster or slower than plain SCIP. In a slight abuse of notation, we count an instance to be solved faster by plain SCIP if no other variant solved it at least 5% faster. Finally, we list the average number of branch-and-bound nodes needed to solve a sequence and the average solving time in seconds.

Our computational experiments show that SCIP does not solve any instance faster than our new reoptimizing version ReoptSCIP. The best results on the vertex coloring instances can be achieved by using all three heuristics together. As opposed to that, the best results on the k-constrained shortest path instances can be achieved by using the weak compression heuristic only. Using the trivial-negation heuristic in combination with plain SCIP provides much more benefit than in combination with ReoptSCIP. This is caused by the number of nodes that need to be reoptimized to prove the optimality. Thus, the benefit obtained from constructing an optimal solution is consumed by proving its optimality, i.e., reoptimizing the search frontier. The compression heuristics can speed up the solving process of the vertex coloring instances by a factor of approximately 1.4. Thus, the constructed representatives are of good quality, i.e., they provide good dual and primal bounds. A pleasant observation is that the number of branch-and-bound nodes ReoptSCIP needs to solve the test sets is not much more than that of SCIP, which we did not expect. Summing up, we can state that using reoptimization improves the performance of SCIP significantly on our test sets.

5   Conclusions

We presented a reoptimization approach for MIP solvers which is based on a reconstruction of the branch-and-bound tree. We are able to handle dual reductions and introduced heuristics for compressing the tree. Applied carefully within SCIP, this method is able to reduce the solving time for two applications significantly. Therefore, it will be part of the next SCIP release.

Acknowledgments. The work for this article has been supported by the *Research Campus Modal* funded by the German Federal Ministry of Education and Research (fund number 05M14ZAM). The second author thanks the DFG for their support within project A04 in CRC TRR154. The authors would like to thank the anonymous reviewers for helpful comments on the paper.

References

1. Achterberg, T.: Constraint Integer Programming. Ph.D. thesis, TU Berlin (2007)
2. Achterberg, T.: SCIP: solving constraint integer programs. Mathematical Programming Computation **1**(1), 1–41 (2009)
3. Applegate, D.L., Bixby, R.E., Chvátal, V., Cook, W.J.: On the solution of traveling salesman problems. Documenta Mathematica Journal der Deutschen Mathematiker-Vereinigung Extra Volume ICM III, 645–656 (1998)

4. COLOR02/03/04. http://mat.gsia.cmu.edu/COLOR03/
5. Dakin, R.J.: A tree-search algorithm for mixed integer programming problems. The Computer Journal **8**(3), 250–255 (1965)
6. Danna, E., Fenelon, M., Gu, Z., Wunderling, R.: Generating multiple solutions for mixed integer programming problems. In: Fischetti, M., Williamson, D.P. (eds.) IPCO 2007. LNCS, vol. 4513, pp. 280–294. Springer, Heidelberg (2007)
7. Dantzig, G.B., Wolfe, P.: Decomposition principle for linear programs. Operations Research **8**(1), 101–111 (1960)
8. Desrochers, M., Soumis, F.: A reoptimization algorithm for the shortest path problem with time windows. European J. Oper. Res. **35**(2), 242–254 (1988)
9. DIP - Decomposition for Int. Programming. https://projects.coin-or.org/Dip
10. Fischetti, M., Lodi, A.: Mipping closures: An instant survey. Graphs and Combinatorics **23**, 233–243 (2007)
11. Fischetti, M., Lodi, A.: Optimizing over the first Chvátal closure. Math. Program., Ser. B **110**, 3–20 (2007)
12. Frangioni, A., Manca, A.: A computational study of cost reoptimization for min cost flow problems. INFORMS Journal on Computing **18**(1) (2006)
13. Gamrath, G.: Generic branch-cut-and-price. Master's thesis, TU Berlin (2010)
14. Gamrath, G., Lübbecke, M.E.: Experiments with a generic dantzig-wolfe decomposition for integer programs. In: Festa, P. (ed.) SEA 2010. LNCS, vol. 6049, pp. 239–252. Springer, Heidelberg (2010)
15. GCG - Generic Column Generation. http://www.or.rwth-aachen.de/gcg/
16. Güzelsoy, M.: Dual methods in mixed integer linear programming. Ph.D. thesis, Lehigh University, Bethlehem, Pennsylvania, USA (2009)
17. Hiller, B., Klug, T., Witzig, J.: Reoptimization in branch-and-bound algorithms with an application to elevator control. In: Demetrescu, C., Marchetti-Spaccamela, A., Bonifaci, V. (eds.) SEA 2013. LNCS, vol. 7933, pp. 378–389. Springer, Heidelberg (2013)
18. Karzan, F.K., Nemhauser, G.L., Savelsbergh, M.W.P.: Information-based branching schemes for binary linear mixed integer problems. Mathematical Programming Computation **1**(4), 249–293 (2009)
19. Land, A.H., Doig, A.G.: An automatic method of solving discrete programming problems. Econometrica **28**(3), 497–520 (1960)
20. Létocart, L., Nagih, A., Plateau, G.: Reoptimization in Lagrangian methods for the 0–1 quadratic knapsack problem. Comput. Oper. Res. **39**(1), 12–18 (2012)
21. Miller-Hooks, E., Yang, B.: Updating paths in time-varying networks given arc weight changes. Transportation Science **39**(4), 451–464 (2005)
22. Ralphs, T.K., Galati, M.V.: Decomposition in integer linear programming. In: Karlof, J.K. (ed.) Integer Programming: Theory and Practice. CRC Press (2006)
23. Ralphs, T.K., Güzelsoy, M.: Duality and warm starting in integer programming. In: The Proceedings of the 2006 NSF Design, Service, and Manufacturing Grantees and Research Conference (2006)
24. Savelsbergh, M.W.P.: Preprocessing and probing techniques for mixed integer programming problems. ORSA Journal on Computing **6**, 445–454 (1994)
25. SCIP - Solving Constraint Integer Programs. http://scip.zib.de/
26. Tanaka, M., Kobayashi, K.: MISOCP formulation and route generation algorithm for ship navigation problem. Tech. Rep. 2013–8, Tokyo Inst. of Technology (2013)
27. Witzig, J.: Reoptimization Techniques for MIP Solvers. Master's thesis, TU Berlin (2014)

Submodular Minimization in the Context of Modern LP and MILP Methods and Solvers

Andrew Orso, Jon Lee$^{(\boxtimes)}$, and Siqian Shen

Department of Industrial and Operations Engineering,
University of Michigan, 1205 Beal Ave., Ann Arbor, MI 48109-2117, USA
{orso,jonxlee,siqian}@umich.edu

Abstract. We consider the application of mixed-integer linear programming (MILP) solvers to the minimization of submodular functions. We evaluate common large-scale linear-programming (LP) techniques (e.g., column generation, row generation, dual stabilization) for solving a LP reformulation of the submodular minimization (SM) problem. We present heuristics based on the LP framework and a MILP solver. We evaluated the performance of our methods on a test bed of min-cut and matroid-intersection problems formulated as SM problems.

Keywords: Submodular minimization · Lovász extension · Column generation · Row generation · Dual stabilization

1 Introduction

Let E be a finite (ground) set. A function $f : 2^E \to \mathbb{R}$ is *submodular* if:

$$f(S) + f(T) \geq f(S \cup T) + f(S \cap T) \quad \forall S, T \subseteq E.$$

The goal of submodular minimization (SM) is to choose $S \subseteq E$ such that $f(S)$ is minimized. SM has strong ties to problems in machine learning, graph theory, and matroid theory. For example, in graph theory, a cut function, evaluated as the sum of capacities of arcs that originate from a subset of nodes containing a given source s, but not containing a given sink t, to nodes in the complementary set, is well known to be submodular; thus, the minimum s-t cut problem can be recast as an SM problem (see [11], for example). Additionally, we can recast the max-cardinality matroid-intersection problem, as a SM problem (see [3]).

SM is well known to be solvable in polynomial time using the ellipsoid method (see [7]). But the practicality of such an algorithm is very limited. Iwata, Fleischer, and Fujishige (see [8]) and Schrijver (see [14]) developed the first "combinatorial" polynomial-time algorithms for SM, however, again the practical use of such algorithms is quite limited. An algorithm that has had the most success seeks the minimum-norm point of the "base polyhedron" (see [5]), but even that algorithm has been found to be slow and/or inaccurate (see [9]). So we regard the challenge of developing practically-efficient approaches for SM as open.

© Springer International Publishing Switzerland 2015
E. Bampis (Ed.): SEA 2015, LNCS 9125, pp. 193–204, 2015.
DOI: 10.1007/978-3-319-20086-6_15

The aforementioned algorithms for SM take advantage of the Lovász extension of a submodular function. This function is an extension of a submodular function, viewed as defined on the vertices of the unit hypercube $[0,1]^E$, to the entire hypercube, as a piecewise-linear convex function (see [11]). Using the Lovász extension, one can derive an equivalent linear-programing (LP) problem with a very large number of columns (see [13], for example). Solving this LP problem had been deemed highly impractical (see [1], for example). In what follows, we demonstrate that large-scale LP techniques can be employed with some success, giving evidence that the LP approach should not be abandoned. Finally, we consider the use of modern MILP solvers such as Gurobi for solving SM problems as 0-1 integer programs. We take advantage of heuristics and cutting-plane methods in these solvers to develop heuristics for approximate SM.

The remainder of this paper is organized as follows. In §2, we present an equivalent LP formulation, as well as column- and row-generation procedures for solving it. In §3, we present computational results of these methods to compare row and column generation, as well as their stabilized method variants. In §4, we present experimental results of using MILP solvers for solving the dual of our LP reformulation by utilizing solver cutting-plane methods. Finally, in §5, we make brief conclusions and give future research directions.

2   Large-Scale LP Representation

Without loss of generality, we assume that $f(\emptyset) = 0$. For $c \in \mathbb{R}^E$, we define $c(S) = \sum_{k \in S} c_k$. The *base polyhedron* of f is $B(f) := \{c \in \mathbb{R}^E \mid \forall S \subseteq E, \ c(S) \le f(S), \ c(E) = f(E)\}$. Now, define $c_- \in \mathbb{R}^E$ by $(c_-)_k := \max\{c_k, 0\}$, for $k \in E$. The SM problem $\min\{f(S) : S \subseteq E\}$ can be recast as $\max\{c_-(E) : c \in B(f)\}$ (see [13], for example). This maximization problem can be formulated as an LP problem with a large number of variables as follows. Letting $c^1, \ldots, c^m$ be the extreme points of $B(f)$, and defining the matrix $C := [c^1, \ldots, c^m] \in \mathbb{R}^{E \times m}$, we have the equivalent LP problem

$$\textbf{SMP:} \quad \min \ \mathbf{1}^\top \beta \tag{1}$$
$$\text{s.t.} \ \ Cx - \alpha + \beta = \mathbf{0}, \tag{2}$$
$$\mathbf{1}^\top x = 1, \tag{3}$$
$$x \in \mathbb{R}_+^m, \alpha \in \mathbb{R}_+^E, \beta \in \mathbb{R}_+^E, \tag{4}$$

where $\mathbf{1}$ (resp. $\mathbf{0}$) is a vector of appropriate dimension with all entries equal to 1 (resp., 0). To construct a minimizer of f from a *basic dual optimum* of this LP problem, we consider the dual variables corresponding to constraints (2); these dual variables turn out to be binary and correspond to a minimizer of f (see [13], for example).

2.1   Column Generation

Column generation is a standard technique for handling LP formulations in which we have a manageable number of equations (in a standard-form LP problem) but a very large (but structured) set of variables. In our context, **SMP** has only $|E| + 1$ equations but m variables (which would typically be exponentially large, relative to $|E|$). Define **RSMP** to be a restricted version of **SMP**, in which C is replaced by $\tilde{C}$, having a subset of the columns of the full matrix C. We iteratively solve instances of **RSMP**, further incorporating variables that have negative reduced cost for the current **RSMP**, and possibly dropping variables that are nonbasic, after each iteration. Typically, a basis of **SMP** has $|E| + 1$ variables, and so we maintain at least $|E| + 1$ (but much fewer than m) variables in **RSMP**.

To determine which variable to add to the model at the end of each iteration, we solve a (typically) well-structured problem known as the pricing problem, which determines a variable, not already in the model, that has negative reduced cost. Our pricing problem is $\max\{u^\top c : c \in B(f)\}$, where $u \in \mathbb{R}^E$ is the vector of optimal dual variables corresponding to the constraints $\tilde{C}x - \alpha + \beta = 0$ in **RSMP**. This is simply the maximization of a linear function over the base polyhedron of f, which can be efficiently evaluated via the greedy algorithm (see [10], for example). Furthermore, we can take advantage of the structure of a particular function f to improve the greedy algorithm. For example, for min-cuts, the value of a cut during the greedy algorithm can be used in the subsequent greedy step, after adding one node, by subtracting the capacities of those arcs no longer in the cut and adding the capacities of the new arcs introduced.

The ease of solution of the pricing problem makes column generation a viable method for **SMP**, though a common pitfall of these methods is that they tend to "tail-off", where convergence begins to slow considerably as the solution nears the optimum. To combat such woes, we implemented a dual-stabilization technique, which has been shown to decrease these effects in practice (see [12], for example). In particular, the optimal dual solutions at each iteration tend to fluctuate considerably, rather than following a smooth trajectory. Dual stabilization seeks to localize the dual variables from iteration to iteration, hoping to moderate this behavior (see [12], for example). There exist numerous methods for stabilizing the dual problem (e.g., interior point stabilization (see [4]), bundle methods (see [2]), etc.). One such method is the box-step method: consider a dual optimal solution $(u^i, v^i) \in \mathbb{R}^E \times \mathbb{R}$ at iteration i of the column-generation procedure. We introduce new constraints to the dual, which is equivalent to adding columns to the primal, of the form $u_j^{i+1} \in [u_j^i - \epsilon, u_j^i + \epsilon]$ for all $j \in E$, where $\epsilon > 0$ denotes the user-defined width of the box. In the subsequent iteration, if the optimal dual solution u_j^{i+1} is at either the upper or lower bound of the box, then the box is recentered at that bound and the procedure continues.

2.2 Row Generation

Alternatively, we consider solving the dual of **SMP**

$$\textbf{SMD: } \max \ v \tag{5}$$
$$\text{s.t. } C^\top u + 1v \leq 0, \tag{6}$$
$$0 \leq u \leq 1, \tag{7}$$

with variables $u \in \mathbb{R}^E$ and $v \in \mathbb{R}$ corresponding to equations (3) and (4) respectively. **SMD** typically has a very large number of constraints, as opposed to the large number of variables present in **SMP**. Of course, by the LP strong-duality theorem, the optimal values of **SMP** and **SMD** are the same. Moreover, in some sense **SMD** is more natural to consider since we are really after a dual basic optimum of **SMP**. For solving **SMD**, we consider a row-generation procedure, which is analogous to column generation for **SMP**. We will refer to the constraints generated as *greedy cuts*, as they are the cuts generated using the greedy algorithm at each iteration. Thus, we maintain a relaxed version **RSMD** of **SMD**, replacing C with $\tilde{C}$, as before. We can start with $\tilde{C}$ having no columns, and we employ the pricing problem from before to determine greedy cuts to be added at every iteration. We can apply dual stabilization again, by explicitly bounding the variables u in **RSMD**. The significance of solving **SMD**, as opposed to **SMP**, is not readily apparent, though computational tests on a number of problem instances demonstrates that row generation may perform better than column generation, depending on the structure of the problem instance.

3   Computational Results: LP

We tested our LP-based algorithms on classical min s-t cut and matroid-intersection problems.

For the min-cut problem, we produced RMFGEN networks (see [6]) as b copies of an $a \times a$ grid of vertices in which each vertex within a grid is connected to each of its neighbors, as well as to a random vertex in the adjacent grid. The source is the lower-left vertex of the first grid, and the sink is the upper-right vertex of the last grid.

For the matroid-intersection problem, let $\mathcal{M}_1 = (E, \mathcal{I}_1)$ and $\mathcal{M}_2 = (E, \mathcal{I}_2)$ be matroids with independent sets $\mathcal{I}_1$ and $\mathcal{I}_2$ respectively on the common ground set E. The matroid-intersection problem is the problem of finding the maximum-cardinality independent set that is in both $\mathcal{I}_1$ and $\mathcal{I}_2$. By a result of Edmonds (see [3]), this problem is equivalent to minimizing the submodular function

$$f(S) := r_1(S) + r_2(E \setminus S) \quad \forall S \subseteq E,$$

where r_i is the rank function of $\mathcal{M}_i$, $i = 1, 2$. For our test instances, we chose $\mathcal{M}_1$ to be the graphic matroid of a random connected graph on $p+1$ vertices and $2p$ edges, and $\mathcal{M}_2 = \mathcal{M}_1^*$ to be its dual. This choice of instances has relevance in determining whether a "bar-and-joint framework" in the plane is "minimally generically rigid" (see [10], for example).

We identify our test problems with the following key: **Type_y-z** where **Type** denotes the type of problem [MC = minimum cut, MI = matroid intersection], and **y-z** denotes the size, e.g., for MC, $y = a$, $z = b$, and for MI, $y = p$, $z = 2p$. The problem sizes we consider are given in Table 1.

All computations were performed using Gurobi 5.6.3 with a time limit of 12000 seconds on a Linux cluster with 4GB RAM per core, using four cores for each MC instance, and one core for each MI instance. All results are given as an arithmetic average across 10 instances for each problem type.

Table 1. Test Instances

Problem Sizes	
Minimum Cut	Matroid Intersection
4-32 (512 Nodes, 2032 Arcs)	150-300
8-8 (512 Nodes, 2240 Arcs)	175-350
7-13 (637 Nodes, 2772 Arcs)	190-380
9-9 (729 Nodes, 3200 Arcs)	200-400
5-32 (800 Nodes, 3335 Arcs)	

3.1 Standard LP Formulations

We first tested the column- and row-generation procedures. Results are given in Table 2 and Table 3, for row and column generation, respectively. The $Time$ column reports the total time, in seconds, required to solve the corresponding problem to optimality, the $\#I$ column reports the total number of iterations required to solve to optimality, and the $Time_{j\%}$ and $\#I_{j\%}$ columns, where $j = 1$ or 5, report the amount of time and number of iterations required to reach a $j\%$ optimality gap, respectively.

Table 2. Results of Using Row Generation

Row Generation							
Problem Type	Problem	Time	#I	Time₅%	#I₅%	Time₁%	#I₁%
	MC_8-8	493	826	282	480	360	602
	MC_9-9	2563	1218	1383	687	1769	851
Minimum Cut	MC_4-32	494	803	263	432	364	617
	MC_5-32	3947	1330	1719	695	2688	998
	MC_7-13	1844	1075	938	616	1244	773
	MI_150-300	1044	284	110	28	830	223
Matroid Intersection	MI_175-350	2173	329	13	0	1568	235
	MI_190-380	9172	312	61	0	6937	211
	MI_200-400	3776	397	74	6	2502	259

Concerning the min-cut problem, we observe that while the number of iterations required to prove optimality is similar in the two methods, row generation requires considerably less time. Our tests indicate that the relaxed LP problems solved during each iteration of row generation require less time than the restricted LP problems solved during each iteration of column generation, resulting in an overall decrease in time. Similar results extend to the 5% and 1% optimality gap, where row generation achieves these optimality gaps in fewer iterations and less time. Further, both methods require only about 50% of the total time in order to reach a 5% optimality gap and only about 75% of the

Table 3. Results of Using Column Generation

Problem Type	Problem	Time	#I	Time$_{5\%}$	#I$_{5\%}$	Time$_{1\%}$	#I$_{1\%}$
			Column Generation				
	MC_8-8	1017	819	407	480	621	602
	MC_9-9	4023	1186	1533	683	2303	847
Minimum Cut	MC_4-32	950	816	331	434	575	624
	MC_5-32	9222	1333	2791	691	5238	993
	MC_7-13	2800	1054	1070	623	1739	797
	MI_150-300	1172	334	131	36	841	240
Matroid	MI_175-350	2407	370	13	0	1506	232
Intersection	MI_190-380	3103	377	16	0	1783	217
	MI_200-400	4273	434	62	4	2442	248

total time in order to reach a 1% optimality gap. This quick, early convergence motivates the development of MILP heuristics in the next section.

For the matroid intersection, we see that row generation is only marginally better than column generation for the instances tested. While this result holds across many instances, instance MI_190-380 seems to be difficult for row generation. We will see that dual stabilization, introduced in the next section, helps to reduce this difficulty.

3.2 Dual-Stabilized Formulations

For the dual-stabilized variants of the row and column generation, we ran all problem instances with $\epsilon = 0.25$, where ϵ is the width of the box constraints. The results reported in Table 4 and Table 5 follow the same format as was described for the standard LP formulations.

Both methods derive some benefit from dual stabilization. In fact, both methods see a decrease in the number of iterations required to solve to optimality by approximately 15%. For the min-cut problem, row generation is still a clear front runner, though the improvement derived from stabilizing the dual problem seems to be consistent across both methods, in terms of time and iterations. For the matroid-intersection problem, stabilizing row generation yields a greater improvement in running time. This is in contrast to the standard LP methods, in which neither algorithm proved to be more effective.

4 Computational Results: MILP

In this section, we take advantage of the fact that extreme-point optima of **SMD** are integer and correspond to minima of f. We demonstrate that there exists some number M of greedy cuts at which point **RSMD** can be solved to optimality with a pure integer-programming approach via branch and bound (B&B). We examine the use of this idea in solving **SMD**, as well as for heuristics.

Table 4. Results of Using Dual-Stabilized Row Generation

Dual-Stabilized Row Generation						
Problem Type	Problem	Time	#I	Time$_{5\%}$	#I$_{5\%}$	Time$_{1\%}$	#I$_{1\%}$
Minimum Cut	MC_8-8	432	738	225	466	308	571
	MC_9-9	1659	1124	824	689	1113	836
	MC_4-32	381	769	165	416	252	587
	MC_5-32	3234	1255	1251	686	2126	980
	MC_7-13	982	997	444	605	683	779
Matroid Intersection	MI_150-300	701	194	30	0	30	0
	MI_175-350	1260	209	50	0	50	0
	MI_190-380	1730	208	69	0	69	0
	MI_200-400	2427	250	81	0	81	0

Table 5. Results of Using Dual-Stabilized Column Generation

Dual-Stabilized Column Generation						
Problem Type	Problem	Time	#I	Time$_{5\%}$	#I$_{5\%}$	Time$_{1\%}$	#I$_{1\%}$
Minimum Cut	MC_8-8	1069	745	511	475	713	569
	MC_9-9	6118	1096	3049	702	3917	806
	MC_4-32	1010	752	393	416	677	598
	MC_5-32	8785	1177	3576	678	5861	930
	MC_7-13	3030	1005	1290	614	2027	776
Matroid Intersection	MI_150-300	832	240	29	0	37	2
	MI_175-350	1661	267	51	0	180	22
	MI_190-380	2219	275	66	0	66	0
	MI_200-400	3123	331	77	0	77	0

4.1 Submodular Minimization Using Integer Programming

The goal of this test was to explore the utility of modern integer-programming techniques for SM. We carried this out by a procedure we call "submodular minimization using integer programming" (SMIP). In the first phase of this method, we solve **SMD** to optimality using standard row generation and define K to be the number of greedy cuts necessary to verify optimality. By resolving **RSMD** with the first L greedy cuts generated, for some choice of L, the result is a suboptimal solution of the LP problem **SMD**. But, we can pass this relaxation having L greedy cuts to an MILP solver and allow the MILP solver to run until completion, utilizing B&B, heuristics, cutting planes, etc. At completion, we can then run our greedy-cut-generation procedure again to determine if there exists any violated greedy cuts; if not, then the MILP solution is optimal to **SMD**, otherwise the MILP solver was unable to achieve a true optimum when provided with the L greedy cuts. We carried out such a scheme, continuing to decrement the number of cuts added to **RSMD**, until the MILP solution returned was not optimal to **SMD**, at which point we set M to the previous number of cuts for which the MILP solution was optimal for **SMD**. So, M/K is the proportion of greedy cuts actually needed to solve an instance to optimality in this MILP framework, which is typically very low in practice.

In the second phase, we initialized **RSMD** with the M sufficient constraints from the first phase and formulated **RSMD** as an integer program with variables $u \in \{0, 1\}^E$. We solved the integer variant of **RSMD** as efficiently as possible with varying levels of cut aggressiveness under the MILPFocus Gurobi parameter that controls the focus on proving optimality of the given MILP.

We ran SMIP on MC_8-8, MC_4-32, and MC_7-13 as well as MI_150-300 and MI_175-350. We report the following results for these procedures: the number of iterations K required to solve instances to optimality using row generation, the number of iterations M sufficient to solve instances using a pure integer-programming method, the total amount of time, in seconds, for both phases, $Time$, the amount of time required to get the true optimal solution as an incumbent solution in the B&B search of the second phase, $Time_I$, and the number of Gomory and MIR cuts generated.

The results reported in Tables 6 and 7 demonstrate that a pure integer-programming approach to solving these problems may be beneficial, especially in the case of matroid intersection, where the function evaluations during the greedy algorithm generally require a large number of computationally-expensive matrix-rank calculations. We also note the increase in the amount of time required to solve these problems as the aggressiveness of the cuts increases. The min-cut problem saw little benefit from this method as the average reduction in the number of iterations is not significant enough to warrant the long MILP solve time. In fact, compared to row generation, the times recorded are far worse. Conversely, for matroid intersection, the number of greedy cuts sufficient to solve to optimality in an MILP framework was approximately 60% of the total number required in the row-generation setting, while we required only a fraction of the time that would be required to generate the difference in cuts solving using integer-programming techniques. On the two matroid-intersection problems, the integer-programming approach performed similarly to dual-stabilized row generation.

4.2   Greedy-Cut Integer-Programming Method

In a similar vein, we can enhance SMIP by integrating greedy cuts throughout the B&B search in the second phase. We solved problems MC_4-32, MC_7-13, MC_8-8, MI_150-300, MI_175-350, and MI_190-380 and report the total number of cuts required to solve to optimality using row generation, K, the number of cuts sufficient to solve to optimality using the greedy cut MILP method, M, the average time, in seconds, to solve to optimality, $Time_{IP}$, and the average time to solve the equivalent row-generation problem as a comparison point, $Time_{RG}$.

The results in Table 8 for the min-cut problem demonstrate that this method is competitive with the standard row-generation method proposed previously.

Table 6. Results of Using SMIP on Minimum Cut

Min-Cut SMIP							
Size	Cut Level	K	M	Time	Time$_I$	Gomory	MIR
8-8	0	826	492	1192	15	0	0
	1	826	492	1193	10	6	6
	2	826	492	2220	27	12	2911
	3	826	492	3370	52	12	9780
4-32	0	803	548	1251	20	0	0
	1	803	548	903	23	5	0
	2	803	548	1462	61	13	302
	3	803	548	2060	100	12	1317
7-13	0	1070	761	1831	39	0	0
	1	1070	761	1827	47	4	0
	2	1070	761	2495	87	9	5575
	3	1070	761	2642	135	9	5575

Table 7. Results of Using SMIP on Matroid Intersection

Matroid-Intersection SMIP							
Size	Cut Level	K	M	Time	Time$_I$	Gomory	MIR
150-300	0	284	199	859	10	0	0
	1	284	199	835	4	11	0
	2	284	199	2354	722	29	0
	3	284	199	1710	721	27	0
175-350	0	329	251	1673	6	0	0
	1	329	251	2353	7	11	0
	2	329	251	3089	26	29	0
	3	329	251	3199	23	45	0

Table 8. Collection of Greedy Cut SMIP Data

SMIP Utilizing Greedy Cuts					
Problem Type	Problem	K	M	Time$_{IP}$	Time$_{RG}$
Minimum Cut	MC_4-32	803	507	539	493
	MC_7-13	1072	669	1664	1843
	MC_8-8	826	435	546	492
Matroid Int	MI_150-300	284	185	1077	1043
	MI_175-350	329	218	2076	2173
	MI_190-380	335	209	2426	9171

In fact, for a problem requiring a larger number of rows generated to solve to optimality, e.g., MC_7-13, the IP method performs better than the row-generation
procedure. For matroid intersection, we see little improvement in terms of computation time over the row-generation method, and it is actually worse than
SMIP without greedy cuts.

4.3   Heuristic Methods

We demonstrated not only the practicality of an MILP-based algorithm for solving SM problems, but also the relatively small amount of time in which the true
optimal solution becomes an incumbent of the B&B search. We take advantage
of this fact, in conjunction with local-search heuristics, to develop a fast method
for getting good, approximate solutions to **SMP**, and equivalently **SMD**.

We use a simple local-search method in the heuristics that follow. Given a
vector $u \in \{0,1\}^E$, we repeatedly change at most two components, so as to get
the maximum decrease in f.

The first heuristic we develop can be employed in the context of row or column generation. We initialize **RSMD** (or **RSMP**) and proceed with row (resp.,
column) generation. At some point (say after a fixed number γ of iterations), we
give up and switch to local search, starting that from a rounding of the primal
(resp., dual) solution $u \in [0,1]^E$ to the nearest point in $\{0,1\}^E$. Results for the
row-generation version are given in Table 9 with $\#Cuts$ being the number of
greedy cuts generated before switching to local search, $Time$ being the total
time in seconds, and $Ratio$ being the ratio of the difference between the optimal
and heuristic value with the optimal value.

The second heuristic is identical to the first heuristic until local search is
called. Before switching to local search, we run an MILP solver until some user-
defined state is reached. A good option seems to be to run the MILP solver until
three incumbent solutions of the B&B search have been found, at which point
we start local-search from the best incumbent found. Computational results for
SMIP demonstrate that on average, the true minimum of f is found as an MILP
incumbent very quickly. Results for this method are given in Table 10, with
columns indexed the same as in previous section, except $Time_{IP}$, which is the
time required to find the third incumbent of the B&B search.

From the results in Table 9, we see that the first heuristic competes with
standard row-generation method for the min-cut problem, yielding shorter times
and near optimal solutions in most cases. For the matroid-intersection problem,
optimality is achieved in all cases, though the time benefit does not become
significant until we consider problems of larger size. For the second heuristic,
results in Table 10 indicate poor performance on the min-cut problem, with long
solve times and larger ratios. Conversely, the time required to solve the matroid-
intersection problem is cut down significantly and the ratio is very small.

Table 9. Results from heuristic method without MILP extension

Heuristic w/o MILP Extension				
Problem Type	Problem	# Cuts	Time	Ratio
Minimum Cut	MC_8-8	500	437	7.3
		650	463	0
	MC_9-9	800	1728	0.02
		1000	1964	0
	MC_4-32	600	525	0.37
		700	463	0
	MC_7-13	600	972	0.11
		700	1064	0.11
Matroid Intersection	MI_150-300	100	1243	0
		200	1202	0
	MI_175-350	150	4144	0
		300	2935	0
	MI_190-380	200	4106	0
		300	4482	0

Table 10. Results from heuristic method utilizing MILP extension

Heuristic with MILP Extension					
Problem Type	Problem	# Cuts	Time	Time$_{IP}$	Ratio
Minimum Cut	MC_8-8	500	1515	1110	0.77
		650	1027	453	3.49
	MC_9-9	800	4095	1708	1.55
		1000	3955	667	3.62
	MC_4-32	600	1499	925	1.97
		700	1059	327	1.96
	MC_7-13	600	2112	1127	1.55
		700	3742	2487	2.24
Matroid Intersection	MI_150-300	100	371	4	0.02
		200	667	5	0.05
	MI_175-350	150	935	8	0.02
		300	1812	6	0.04
	MI_190-380	200	1588	8	0.02
		300	2247	11	0.05

5   Conclusion

We explored the applicability of modern LP/MILP methods for SM. For LP, we
saw that row generation typically performs better than column generation, and

we saw that MILP methods can help in an exact or heuristic context. We are currently expanding our tests and exploring how to incorporate side constraints.

Acknowledgments. This research was supported in part by Advanced Research Computing at the University of Michigan, Ann Arbor. J. Lee was partially supported by NSF grant CMMI–1160915 and ONR grant N00014-14-1-0315. S. Shen was partially supported by NSF grants CMMI–1433066 and CCF–1442495.

References

1. Bach, F.: Learning with submodular functions: A convex optimization perspective. Foundation and Trends in Machine Learning **6**, 145–373 (2013)
2. Briant, O., Lemarechal, C., Meurdesoif, P., Michel, S., Perrot, N., Vanderbeck, F.: Comparison of bundle and classical column generation. Math. Prog. **113**, 299–344 (2008)
3. Edmonds, J.: Submodular functions, matroids, and certain polyhedra. In: Combinatorial Structures and Their Applications, pp. 69–87 (1970)
4. Elhedhli, S., Goffin, J.L.: The integration of an interior-point cutting plane method within a branch-and-price algorithm. Math. Prog. **100**(2), 267–294 (2004)
5. Fujishige, S., Isotani, S.: A submodular function minimization algorithm based on the minimum-norm base. Pacific Journal of Optimization **7**(1), 3–17 (2011)
6. Goldfarb, D., Grigoriadis, M.D.: A computational comparison of the dinic and network simplex methods for maximum flow. Ann. of OR **13**(1), 81–123 (1988)
7. Grötschel, M., Lovász, L., Schrijver, A.: The ellipsoid method and its consequences in combinatorial optimization. Combinatorica **1**(2), 169–197 (1981)
8. Iwata, S., Fleischer, L., Fujishige, S.: A combinatorial strongly polynomial algorithm for minimizing submodular functions. JACM **48**, 761–777 (2001)
9. Jegelka, S., Lin, H., Bilmes, J.: On fast approximate submodular minimization. In: Advances in Neural Information Processing Systems (NIPS), pp. 460–468 (2011)
10. Lee, J.: A First Course in Combinatorial Optimization. Cambr. Univ. Press (2004)
11. Lovász, L.: Submodular functions and convexity. In: Mathematical Programming The State of the Art, pp. 235–257. Springer (1983)
12. Lübbecke, M.E., Desrosiers, J.: Selected topics in column generation. Operations Research **53**(6), 1007–1023 (2005)
13. McCormick, S.T.: Submodular function minimization. Handbooks in Operations Research and Management Science **12**, 321–391 (2005)
14. Schrijver, A.: A combinatorial algorithm minimizing submodular functions in strongly polynomial time. JCT, Ser. B **80**(2), 346–355 (2000)

Is *Nearly-linear* the Same in Theory and Practice? A Case Study with a Combinatorial Laplacian Solver

Daniel Hoske[1], Dimitar Lukarski[2], Henning Meyerhenke[1]($^{(\boxtimes)}$), and Michael Wegner[1]

[1] Institute of Theoretical Informatics,
Karlsruhe Institute of Technology (KIT), Karlsruhe, Germany
meyerhenke@kit.edu
http://parco.iti.kit.edu
[2] Paralution Labs UG & Co. KG, Gaggenau, Germany
http://www.paralution.com

Abstract. Linear system solving is one of the main workhorses in applied mathematics. Recently, theoretical computer scientists have contributed sophisticated algorithms for solving linear systems with symmetric diagonally dominant matrices (a class to which Laplacian matrices belong) in provably nearly-linear time. These algorithms are highly interesting from a theoretical perspective, but there are no published results on how they perform in practice.

With this paper we address this gap. We provide the first implementation of the combinatorial solver by [Kelner et al., STOC 2013], which is particularly appealing due to its conceptual simplicity. The algorithm exploits that a Laplacian matrix corresponds to a graph; solving Laplacian linear systems amounts to finding an electrical flow in this graph with the help of cycles induced by a spanning tree with the low-stretch property.

The results of our comprehensive experimental study are ambivalent. They confirm a nearly-linear running time, but for reasonable inputs the constant factors make the solver much slower than methods with higher asymptotic complexity. One other aspect predicted by theory is confirmed by our findings: Spanning trees with lower stretch indeed reduce the solver's running time. Yet, simple spanning tree algorithms perform better in practice than those with a guaranteed low stretch.

1 Introduction

Solving square linear systems $Ax = b$, where $A \in \mathbb{R}^{n \times n}$ and $x, b \in \mathbb{R}^n$, has been one of the most important problems in applied mathematics with wide applications in science and engineering. In practice system matrices are often *sparse*, i.e. they contain $o(n^2)$ nonzeros. Direct solvers with cubic running times do not exploit sparsity. Ideally, the required time for solving sparse systems would grow linearly with the number of nonzeros $2m$. Moreover, approximate

© Springer International Publishing Switzerland 2015
E. Bampis (Ed.): SEA 2015, LNCS 9125, pp. 205–218, 2015.
DOI: 10.1007/978-3-319-20086-6_16

solutions usually suffice due to the imprecision of floating point arithmetic. Spielman and Teng [23], following an approach proposed by Vaidya [26], achieved a major breakthrough in this direction by devising a nearly-linear time algorithm for solving linear systems in symmetric diagonally dominant matrices. *Nearly-linear* means $\mathcal{O}(m \cdot \text{polylog}(n) \cdot \log(1/\epsilon))$ here, where $\text{polylog}(n)$ is the set of real polynomials in $\log(n)$ and ϵ is the relative error $\|x - x_{\text{opt}}\|_A / \|x_{\text{opt}}\|_A$ we want for the solution $x \in \mathbb{R}^n$. Here $\| \cdot \|_A$ is the norm $\|x\|_A := \sqrt{x^T A x}$ given by A, and $x_{\text{opt}} := A^+ b$ is an exact solution. A matrix $A = (a_{ij})_{i,j \in [n]} \in \mathbb{R}^{n \times n}$ is *diagonally dominant* if $|a_{ii}| \geq \sum_{j \neq i} |a_{ij}|$ for all $i \in [n]$. Symmetric matrices that are diagonally dominant (SDD matrices) have many applications: In elliptic PDEs [5], maximum flows [8], and sparsifying graphs [22]. Thus, the problem INV-SDD of solving linear systems $Ax = b$ for x on SDD matrices A is of significant importance. We focus here on Laplacian matrices (which are SDD) due to their rich applications in graph algorithms, e. g. load balancing [10], but this is no limitation [14].

Related Work. Spielman and Teng's seminal paper [23] requires a lot of sophisticated machinery: a multilevel approach [21, 26] using recursive preconditioning, preconditioners based on low-stretch spanning trees [24] and spectral graph sparsifiers [16, 22]. Later papers extended this approach, both by making it simpler and by reducing the exponents of the polylogarithmic time factors.[1] We focus on a simplified algorithm by Kelner et al. [14] that reinterprets the problem of solving an SDD linear system as finding an electrical flow in a graph. It only needs low-stretch spanning trees and achieves $\mathcal{O}(m \log^2 n \log \log n \log(1/\epsilon))$ time.

Another interesting nearly-linear time SDD solver is the recursive sparsification approach by Peng and Spielman [20]. Together with a parallel sparsification algorithm, such as the one given by Koutis [15], it yields a nearly-linear work parallel algorithm.

Spielman and Teng's algorithm crucially uses the low-stretch spanning trees first introduced by Alon et al. [3]. Elkin et al. [11] provide an algorithm for computing spanning trees with polynomial stretch in nearly-linear time. Specifically, they get a spanning tree with $\mathcal{O}(m \log^2 n \log \log n)$ stretch in $O(m \log^2 n)$ time. Abraham et al. [1] as well as Abraham and Neiman [2] later showed how to get rid of some of the logarithmic factors in both stretch and time.

Motivation, Outline and Contribution. Although several extensions and simplifications to Spielman and Teng's nearly-linear time solver [23] have been proposed, none of them has been validated in practice so far. We seek to fill this gap by implementing and evaluating an algorithm proposed by Kelner et al. [14] that is easier to describe and implement than Spielman and Teng's original algorithm. Thus, in this paper we implement the KOSZ solver (the acronym follows from the authors' last names) by Kelner et al. [14] and investigate its practical performance. To this end, we start in Section 2 by settling notation and

---

[1] Spielman provides a comprehensive overview of later work at http://www.cs.yale.edu/homes/spielman/precon/precon.html (accessed on February 10, 2015).

outlining KOSZ. In Section 3 we elaborate on the design choices one can make when implementing KOSZ. In particular, we explain when these choices result in a provably nearly-linear time algorithm. Section 4 contains the heart of this paper, the experimental evaluation of the Laplacian solver KOSZ. We consider the configuration options of the algorithm, its asymptotics, its convergence and its use as a smoother. Our results confirm a nearly-linear running time, but at the price of very high constant factors, in part due to memory accesses. We conclude the paper in Section 5 by summarizing the experimental results and discussing future research directions.

2   Preliminaries

Fundamentals. We consider undirected simple graphs $G = (V, E)$ with n vertices and m edges. A graph is *weighted* if we have an additional function $w : E \rightarrow \mathbb{R}_{>0}$. Where necessary we consider unweighted graphs to be weighted with $w_e = 1 \; \forall e \in E$. We usually write an edge $\{u, v\} \in E$ as uv and its weight as w_{uv}. Moreover, we define the set operations $\cup$, $\cap$ and $\backslash$ on graphs by applying them to the set of vertices and the set of edges separately. For every node $u \in V$ its *neighbourhood* $N_G(u)$ is the set $N_G(u) := \{v \in V : uv \in E\}$ of vertices v with an edge to u and its *degree* d_u is $d_u := \sum_{v \in N_G(u)} w_{uv}$. The *Laplacian matrix* of a graph $G = (V, E)$ is defined as $L_{u,v} := -w_{uv}$ if $uv \in E$, $\sum_{x \in N_G(u)} w_{ux}$ if $u = v$ and 0 otherwise for $u, v \in V$. A Laplacian matrix is always an SDD matrix. Another useful property of the Laplacian is the factorization $L = B^T R^{-1} B$, where $B \in \mathbb{R}^{E \times V}$ is the *incidence matrix* and $R \in \mathbb{R}^{E \times E}$ is the *resistance matrix* defined by $B_{ab,c} = 1$ if $a = c$, $= -1$ if $b = c$ and 0 otherwise. $R_{e_1,e_2} = 1/w_{e_1}$ if $e_1 = e_2$ and 0 otherwise. This holds for all $e_1, e_2 \in E$ and $a, b, c \in V$, where we arbitrarily fix a start and end node for each edge when defining B.

With $x^T L x = (Bx)^T R^{-1} (Bx) = \sum_{e \in E} (Bx)_e^2 \cdot w_e \geq 0$ (every summand is non-negative), one can see that L is positive semidefinite. (A matrix $A \in \mathbb{R}^{n \times n}$ is *positive semidefinite* if $x^T A x \geq 0$ for all $x \in \mathbb{R}^n$.)

Cycles, Spanning Trees and Stretch. A *cycle* in a graph is usually defined as a simple path that returns to its starting point and a graph is called *Eulerian* if there is a cycle that visits every edge exactly once. In this work we will interpret cycles somewhat differently: We say that a cycle in G is a subgraph C of G such that every vertex in G is incident to an even number of edges in C, i.e. a cycle is a union of Eulerian graphs. It is useful to define the addition $C_1 \oplus C_2$ of two cycles C_1, C_2 to be the set of edges that occur in exactly one of the two cycles, i.e. $C_1 \oplus C_2 := (C_1 \backslash C_2) \cup (C_2 \backslash C_1)$. In algebraic terms we can regard a cycle as a vector $C \subseteq \mathbb{F}_2^E$ such that $\sum_{v \in N_C(u)} 1 = 0$ in $\mathbb{F}_2$ for all $u \in V$ and the cycle addition as the usual addition on $\mathbb{F}_2^E$. We call the resulting linear space of cycles $\mathcal{C}(G)$.

In a *spanning tree* (ST) $T = (V, E_T)$ of G there is a unique path $P_T(u, v)$ from every node u to every node v. For any edge $e = uv \in E \backslash E_T$ (an *off-tree-edge with*

Input: Laplacian $L = L(G)$ and vector $b \in \text{im}(L)$.
Output: Solution x to $Lx = b$.
1 $T \leftarrow$ a spanning tree of G
2 $f \leftarrow$ unique flow with demand b that is only nonzero on T
3 **while** *there is a cycle with potential drop* $\neq 0$ *in* f **do**
4 $\quad$ $c \leftarrow$ cycle in $\mathcal{C}(T)$ chosen randomly weighted by stretch
5 $\quad$ $f \leftarrow f - \frac{c^T R f}{c^T R c} c$
6 **return** *vector of potentials in* f *with respect to the root of* T

Algorithm 1. INV-LAPLACIAN-CURRENT solver KOSZ.

respect to T), the subgraph $e \cup P_T(u, v)$ is a cycle, the *basis cycle induced by* e. One can easily show that the basis cycles form a basis of $\mathcal{C}(G)$. Thus, the basis cycles are very useful in algorithms that need to consider all the cycles of a graph. Another notion we need is a measure of how well a spanning tree approximates the original graph. We capture this by the *stretch* $\text{st}(e) := \left(\sum_{e' \in P_T(u,v)} w_{e'} \right) / w_e$ *of an edge* $e = uv \in E$. This stretch is the detour you need in order to get from one endpoint of the edge to the other if you stay in T, compared to the length of the original edge. In the literature the stretch is sometimes defined slightly differently, but we follow the definition in [14] using w_e in the denominator. The *stretch of the whole tree* T is the sum of the individual stretches $\text{st}(T) := \sum_{e \in E} \text{st}(e)$. Finding a spanning tree with low stretch is crucial for proving the fast convergence of the KOSZ solver.

KOSZ (Simple) Solver. We can regard G as an electrical network where each edge uv corresponds to a resistor with conductance w_{uv} and resistance $r_{uv} := 1/w_{uv}$ as well as x as an assignment of potentials to the nodes of G. Then $x_v - x_u$ is the voltage across uv and $(x_v - x_u) \cdot w_{uv}$ is the resulting current along uv. Thus, $(Lx)_u$ is the current flowing out of u that we want to be equal to the right-hand side b_u. Furthermore, one can reduce solving SDD systems to the related problem INV-LAPLACIAN-CURRENT [14]: Given a Laplacian $L = L(G)$ and a vector $b \in \text{im}(L)$, compute a function $f: \widetilde{E} \to \mathbb{R}$ with (i) f being a valid graph flow on G with demand b and (ii) the potential drop along every cycle in G being zero, where a valid graph flow means that the sum of the incoming and outgoing flow at each vertex respects the demand in x and that $f(u, v) = -f(v, u) \; \forall uv \in E$. Also, $\widetilde{E}$ is a bidirected copy of E and the potential drop of cycle C is $\sum_{e \in C} f(e) r_e$. The idea of the algorithm is to start with any valid flow and successively adjust the flow such that every cycle has potential zero. We need to transform the flow back to potentials at the end, but this can be done consistently, as all potential drops along cycles are zero.

Regarding the crucial question of what flow to start with and how to choose the cycle to be repaired in each iteration, Kelner et al. [14] suggest using the cycle basis induced by a spanning tree T of G and prove that the convergence of the resulting solver depends on the stretch of T. More specifically, they suggest starting with a flow that is nonzero only on T and weighting the basis cycles by their stretch when sampling them. The resulting algorithm is shown as Algorithm 1; note that we may stop before all potential drops are zero and we can consistently compute the *potentials induced by f* at the end by only looking at T.

The solver described in Algorithm 1 is actually (our reformulation of) the SimpleSolver by Kelner et al. [14]. They also show how to improve this solver by adapting preconditioning to the setting of electrical flows. In informal experiments we could not determine a strategy that is consistently better than the SimpleSolver, so that we do not pursue this scheme any further here. In the original SimpleSolver description, the error bound ϵ is part of the input and determines the number of loop iterations. Our pseudocode is slightly simplified to emphasize the conceptual ideas, but our implementation is based on the original description. Kelner et al. derive the following running time for KOSZ:

Theorem 1 (comp. [14], Thm. 3.2). *SimpleSolver can be implemented to run in time* $O(m \log^2 n \log \log n \log(\epsilon^{-1}n))$ *while computing an ϵ-approximation of x.*

3   Implementation

While Algorithm 1 provides the basic idea of the KOSZ solver, it leaves open several implementation decisions that we elaborate on in this section.

Spanning Trees. As suggested by the convergence result in Theorem 1, the KOSZ solver depends on low-stretch spanning trees. Elkin et al. [11] presented an algorithm requiring nearly-linear time and yielding nearly-linear average stretch. The basic idea is to recursively form a spanning tree using a star of balls in each recursion step. We use Dijkstra with binary heaps for growing the balls and that we take care not to need more work than necessary to grow the ball. In particular, ball growing is output-sensitive and growing a ball $B(x,r) := \{v \in V : \text{Distance from } x \text{ to } v \text{ is} \le r\}$ should require $\mathcal{O}(d \log n)$ time where d is the sum of the degrees of the nodes in $B(x,r)$. The exponents of the logarithmic factors of the stretch of this algorithm were improved by subsequent papers, but Papp [19] showed experimentally that these improvements do not yield better stretch in practice. In fact, his experiments suggest that the stretch of the provable algorithms is usually not better than just taking a minimum-weight spanning tree. Therefore, we additionally use two simpler spanning trees without stretch guarantees: A minimum-distance spanning tree with Dijkstra's algorithm and binary heaps; as well as a minimum-weight spanning with Kruskal's algorithm using union-find with union-by-size and path compression.

To test how dependent the algorithm is on the stretch of the ST, we also look at a *special ST* for $n_1 \times n_2$ grids. As depicted in Figure 1, we construct this spanning tree by subdividing the $n_1 \times n_2$ grid into four subgrids as evenly as possible, recursively building the STs in the subgrids and connecting the subgrids by a U-shape in the middle.

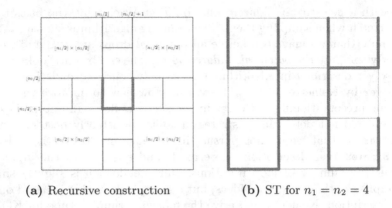

(a) Recursive construction   (b) ST for $n_1 = n_2 = 4$

Fig. 1. Special spanning tree with $\mathcal{O}\big(\frac{(n_1+n_2)^2 \log(n_1+n_2)}{n_1 n_2}\big)$ average stretch for the $n_1 \times n_2$ grid

Proposition 1. *The special ST has* $\mathcal{O}\big(\frac{(n_1+n_2)^2 \log(n_1+n_2)}{n_1 n_2}\big)$ *average stretch on an $n_1 \times n_2$ grid.*

Flows on Trees. Since every basis cycle contains exactly one off-tree-edge, the flows on off-tree-edges can simply be stored in a single vector. To be able to efficiently get the potential drop of every basis cycle and to be able to add a constant amount of flow to it, the core problem is to efficiently store and update flows in T. More formally, we want to support the following two operations for all $u, v \in V$ and $\alpha \in \mathbb{R}$ on the flow f:

- query(u, v): return the potential drop $\sum_{e \in P_T(u,v)} f(e)r_e$
- update(u, v, α): set $f(e) := f(e) + \alpha$ for all $e \in P_T(u, v)$

We can simplify the operations by fixing v to be the root r of T: query(u): return the potential drop $\sum_{e \in P_T(u,r)} f(e)r_e$ and update(u, α): set $f(e) := f(e) + \alpha$ for all $e \in P_T(u, r)$. The itemized two-node operations can then be supported with query$(u, v) :=$ query$(u) -$ query(v) and update$(u, v, \alpha) := \{$update(u, α) and update$(v, -\alpha)\}$ since the changes on the subpath $P_T(r, \text{LCA}(u, v))$ cancel out. Here $\text{LCA}(u, v)$ is the *lowest common ancestor* of the nodes u and v in T, the node farthest from r that is an ancestor of both u and v. We provide two approaches for implementing the operations, first an implementation of the one-node operations that stores the flow directly on the tree and uses the definitions of the operations without modification. Obviously, these operations require $\mathcal{O}(n)$ worst-case time and $\mathcal{O}(n)$ space. With an LCA data structure, one can implement the itemized two-node operations without the subsequent simplification of using one-node operations. This does not improve the worst-case time, but can help in practice. Secondly, we use the improved data structure by Kelner et al. [14] that guarantees $\mathcal{O}(\log n)$ worst-case time but uses $\mathcal{O}(n \log n)$ space. In this case the one-node operations boil down to a dot product (query) and an addition (update) of a dense vector and a sparse vector. We unroll the recursion within the data structure for better performance in practice.

Cycle Selection. The easiest way to select a cycle is to choose an off-tree edge *uniformly at random* in $\mathcal{O}(1)$ time. However, to get provably good results, we need to weight the off-tree-edges by their stretch. We can use the flow data structure described above to get the stretches. More specifically, the data structure initially represents $f = 0$. For every off-tree edge uv we first execute update($u, v, 1$), then query(u, v) to get $\sum_{e \in P_T(u,v)} r_e$ and finally update($u, v, -1$) to return to $f = 0$. This results in $\mathcal{O}(m \log n)$ time to initialize cycle selection. Once we have the weights, we use *roulette wheel selection* in order to select a cycle in $\mathcal{O}(\log m)$ time after an additional $\mathcal{O}(m)$ time initialization.

In the full version of this paper we summarize the implementation choices for Algorithm 1 in a table for the reader's convenience. Note that the convergence theorem requires a low-stretch spanning tree and weighted cycle selection.

4 Evaluation

4.1 Settings

We implemented the KOSZ solver in C++ using NetworKit [25], a toolkit focused on scalable network analysis algorithms. As compiler we use g++ 4.8.3. The benchmark platform is a dual-socket server with two 8-core Intel Xeon E5-2680 at 2.7 GHz each and 256 GB RAM. Only a representative subset of our experiments are shown here. More experiments and their detailed discussion can be found in [13]. We compare our KOSZ implementation to existing linear solvers as implemented by the libraries Eigen 3.2.2 [12] and Paralution 0.7.0 [17]. CPU performance characteristics such as the number of executed FLOPS (floating point operations), etc. are measured with the PAPI library [7].

We mainly use two graph classes for our tests: (i) Rectangular $k \times l$ grids given by $\mathbb{G}_{k,l} := \left([k] \times [l], \left\{ \{(x_1, y_1), (x_2, y_2)\} \subseteq \binom{V}{2} : |x_1 - x_2| = 1 \vee |y_1 - y_2| = 1 \right\} \right)$. Laplacian systems on grids are, for example, crucial for solving boundary value problems on rectangular domains; (ii) Barabási-Albert [4] random graphs with parameter k. These random graphs are parametrized with a so-called *attachment k*. Their construction models that the degree distribution in many natural graphs is not uniform at all. For both classes of graphs, we consider both unweighted and weighted variants (uniform random weights in $[1, 8)$). We also did informal tests on 3D grids and graphs that were not generated synthetically. These graphs did not exhibit significantly different behavior than the two graph classes above.

4.2 Results

Spanning tree. Papp [19] tested various low-stretch spanning tree algorithms and found that in practice the provably good low-stretch algorithms do not yield

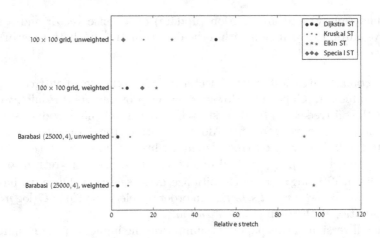

Fig. 2. Average stretch $\mathrm{st}(T)/m$ with different ST algorithms

better stretch than simply using Kruskal. We confirm and extend this observation by comparing our own implementation of Elkin et al.'s [11] low-stretch ST algorithm to Kruskal and Dijkstra in Figure 2. Except for the unweighted 100×100 grid, Elkin has worse stretch than the other algorithms and Kruskal yields a good ST. For Barabási-Albert graphs, Elkin is extremely bad (almost factor 20 worse). Interestingly, Kruskal outperforms the other algorithms even on the unweighted Barabási-Albert graphs, where it degenerates to choosing an arbitrary ST. Figure 2 also shows that our special ST yields significantly lower stretch for the unweighted 2D grid, but it does not help in the weighted case.

Convergence. In Figure 3 we plot the convergence of the residual for different graphs and algorithm settings. We examined a 100×100 grid and a Barabási-Albert graph with 25,000 nodes. While the residuals can increase, they follow a global downward trend. Also note that the spikes of the residuals are smaller if the convergence is better. In all cases the solver converges exponentially, but the convergence speed crucially depends on the solver settings. If we select cycles by their stretch, the order of the convergence speeds is the same as the order of the stretches of the ST (cmp. Figure 2), except for the Dijkstra ST and the Kruskal ST on the weighted grid. In particular, for the Elkin ST on Barabási-Albert graphs, there is a significant gap to the other settings where the solver barely converges at all and the special ST wins. Thus, low-stretch STs are crucial for convergence. In informal experiments we also saw this behavior for 3D grids and non-synthetic graphs.

We could not detect any correlation between the improvement made by a cycle repair and the stretch of the cycle. Therefore, we cannot fully explain the different speeds with uniform cycle selection and stretch cycle selection. For the grid the stretch cycle selection wins, while Barabási-Albert graphs favor uniform cycle selection. Another interesting observation is that most of the convergence

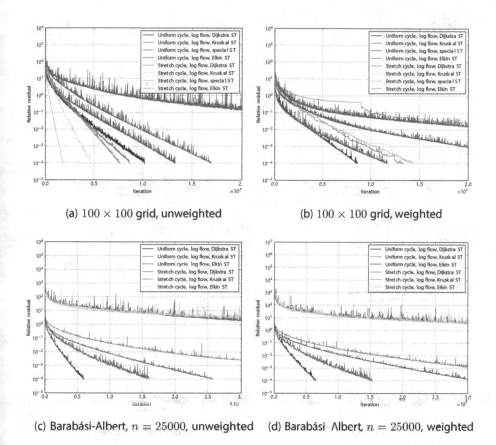

(a) 100×100 grid, unweighted

(b) 100×100 grid, weighted

(c) Barabási-Albert, $n = 25000$, unweighted

(d) Barabási-Albert, $n = 25000$, weighted

Fig. 3. Convergence of the residual. Terminate when residual $\leq 10^{-4}$.

speeds stay constant after an initial fast improvement at the start to about residual 1. That is, there is no significant change of behavior or periodicity. Even though we can hugely improve convergence by choosing the right settings, even the best convergence is still very slow, e.g. we need about 6 million iterations (≈ 3000 sparse matrix-vector multiplications (SpMVs) in time comparison) on a Barabási-Albert graph with 25,000 nodes and 100,000 edges in order to reach residual 10^{-4}. In contrast, conjugate gradient (CG) without preconditioning only needs 204 SpMVs for this graph.

Asymptotics. Now that we know which settings of the algorithm yield the best performance for 2D grids and Barabási-Albert graphs, we proceed by looking at how the performance with these settings behaves asymptotically and how it compares to conjugate gradient (CG) without preconditioning, a simple and popular iterative solver. Since KOSZ turns out to be not competitive, we do not need to compare it to more sophisticated algorithms.

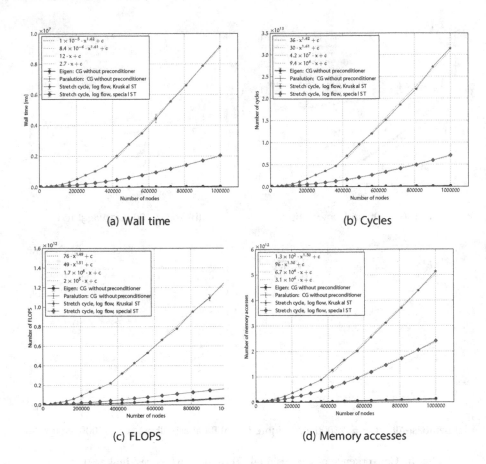

(a) Wall time

(b) Cycles

(c) FLOPS

(d) Memory accesses

Fig. 4. Asymptotic behaviour for 2D grids. Termination when relative residual was $\leq$ 10^{-4}. The error bars give the standard deviation.

In Figure 4 each occurrence of c stands for a new instance of a real constant. We expect the cost of the CG method to scale with $\mathcal{O}(n^{1.5})$ on 2D grids [9], while our algorithm should scale nearly-linearly. This expectation is confirmed in the plot: Using Levenberg-Marquardt [18] to approximate the curves for CG with a function of the form $ax^b + c$, we get $b \approx 1.5$ for FLOPS and memory accesses, while the (more technical) wall time and cycle count yield a slightly higher exponent $b \approx 1.6$. We also see that the curves for our algorithm are almost linear from about 650×650. Unfortunately, the hidden constant factor is so large that our algorithm cannot compete with CG even for a 1000×1000 grid. Note that the difference between the algorithms in FLOPS is significantly smaller than the difference in memory accesses and that the difference in running time is larger still. This suggests that the practical performance of our algorithm is particularly bounded by memory access patterns and not by floating point operations. This is noteworthy when we look at our special spanning tree for the 2D grid. We see that using the special ST always results in performance that is better by a constant factor. In particular, we save a lot of FLOPS (factor 10), while the savings in memory accesses (factor 2) are a lot smaller. Even though the FLOPS when using the special ST are within a factor of 2 of CG, we still have a wide chasm in running time.

The results for the Barabási-Albert graphs are basically the same (and hence not shown in detail): Even though the growth is approximately linear from about 400,000 nodes, there is still a large gap between our algorithm and CG since the constant factor is enormous. Also, the results for the number of FLOPS are again much better than the result for the other performance counters. In conclusion, although we have nearly-linear growth, even for 1,000,000 graph nodes, the KOSZ algorithm is still not competitive with CG because of huge constant factors, in particular a large number of iterations and memory accesses.

Smoothing. One way of combining the good qualities of two different solvers is *smoothing.* Smoothing means to dampen the high-frequency components of the error, which is usually done in combination with another solver that dampens the low-frequency error components. It is known that in CG and most other solvers, the low-frequency components of the error converge very fast, while the high-frequency components converge slowly. Thus, we are interested in finding an algorithm that dampens the high-frequency components, *a good smoother.* This smoother does not necessarily need to reduce the error, it just needs to make its frequency distribution more favorable. Smoothers are particularly often applied at each level of multigrid or multilevel schemes [6] that turn a good smoother into a good solver by applying it at different levels of a matrix hierarchy. To test whether the Laplacian solver is a good smoother, we start with a fixed x with $Lx = b$ and add white uniform noise in $[-1, 1]$ to each of its entries in order to get an initial vector x_0. Then we execute a few iterations of our Laplacian solver and check whether the high-frequency components of the error have been reduced. Unfortunately, we cannot directly start at the vector x_0 in the solver. Our solution is to use *Richardson iteration.* That is, we transform the residual $r = b - Lx_0$ back to the source space by computing $L^{-1}r$ with the Laplacian solver, get the error $e = x - x_0 = L^{-1}r$ and then the output solution $x_1 = x_0 + L^{-1}r$.

Figure 5 shows the error vectors of the solver for a 32×32 grid together with their transformations into the frequency domain for different numbers of iterations of our solver. We see that the solver may indeed be useful as a smoother since the energies for the large frequencies (on the periphery) decrease rapidly, while small frequencies (in the middle) in the error remain.

In the solver we start with a flow that is nonzero only on the ST. Therefore, the flow values on the ST are generally larger at the start than in later iterations, where the flow will be distributed among the other edges. Since we construct the output vector by taking potentials on the tree, after one iteration x_1 will, thus, have large entries compared to the entries of b. In subplot (c) of Figure 5 we see that the start vector of the solver has the same structure as the special ST and that its error is very large. For the 32×32 grid we, therefore, need about 10000 iterations ($\approx$ 150 SpMVs in running time comparison) to get an error of x_1

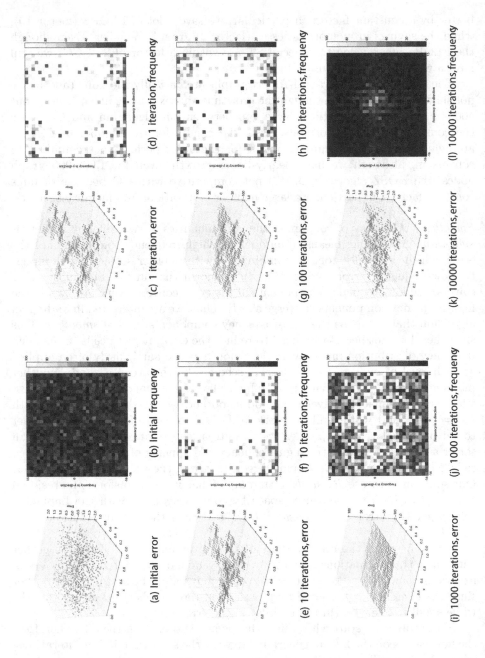

Fig. 5. The Laplacian solver with the special ST as a smoother on a 32 × 32 grid. For each number of iterations of the solver we plot the current error and the absolute values of its transformation into the frequency domain. Note that (a) and (k) have a different scale.

similar to x_0 even though the frequency distribution is favorable. Note that the number of SpMVs the 10000 iterations correspond to depends on the graph size, e.g. for an 100×100 grid the 10000 iterations correspond to 20 SpMVs.

While testing the Laplacian solver in a multigrid scheme could be worthwhile, the bad initial vector creates robustness problems when applying the Richardson iteration multiple times with a fixed number of iterations of our solver. In informal tests multiple Richardson steps lead to ever increasing errors without improved frequency behavior unless our solver already yields an almost perfect vector in a single run.

5   Conclusions

At the time of writing, the presented KOSZ [14] implementation and evaluation provide the first comprehensive experimental study of a Laplacian solver with provably nearly-linear running time. Our study supports the theoretical result that the convergence of KOSZ crucially depends on the stretch of the chosen spanning tree, with low stretch generally resulting in faster convergence. This particularly suggests that it is crucial to build algorithms that yield spanning trees with lower stretch. Since we have confirmed and extended Papp's [19] observation that algorithms with provably low stretch do not yield good stretch in practice, improving the low-stretch ST algorithms is an important future research direction. Even though KOSZ proves to grow nearly linearly as predicted by theory, the constant seems to be too large to make it competitive, even compared to the CG method without preconditioner. Hence, our initial question in the paper title can be answered with "yes" and "no" at the same time: The running time is nearly linear, but the constant factors prevent usefulness in practice. While the negative results may predominate, our effort is the first to provide an answer at all. We hope to deliver insights that lead to further improvements, both in theory and practice. A promising future research direction is to repair cycles other than just the basis cycles in each iteration, but this would necessitate significantly different data structures.

References

1. Abraham, I., Bartal, Y., Neiman, O.: Nearly tight low stretch spanning trees. In: 49th Annual Symposium on Foundations of Computer Science, pp. 781–790 (2008)
2. Abraham, I., Neiman, O.: Using petal-decompositions to build a low stretch spanning tree. In: 44th ACM Symposium on Theory of Computing, pp. 395–406 (2012)
3. Alon, N., Karp, R.M., Peleg, D., West, D.: A graph-theoretic game and its application to the k-server problem. SIAM Journal on Computing **24**, 78–100 (1995)
4. Barabási, A.-L., Albert, R.: Emergence of scaling in random networks. Science **286**(5439), 509–512 (1999)
5. Boman, E., Hendrickson, B., Vavasis, S.: Solving elliptic finite element systems in near-linear time with support preconditioners. SIAM Journal on Numerical Analysis **46**(6), 3264–3284 (2008)
6. Briggs, W.L., Henson, V.E., McCormick, S.F.: A multigrid tutorial. SIAM (2000)

7. Browne, S., Dongarra, J., Garner, N., Ho, G., Mucci, P.: A portable programming interface for performance evaluation on modern processors. Int. J. High Perform. Comput. Appl. **14**(3), 189–204 (2000)
8. Christiano, P., Kelner, J.A., Madry, A., Spielman, D.A., Teng, S.-H.: Electrical flows, laplacian systems, and faster approximation of maximum flow in undirected graphs. In: Proc. 43rd ACM Symp. on Theory of Computing (STOC), pp. 273–282. ACM (2011)
9. Demmel, J.W.: Applied Numerical Linear Algebra. Society for Industrial and Applied Mathematics, Philadelphia (1997)
10. Diekmann, R., Frommer, A., Monien, B.: Efficient schemes for nearest neighbor load balancing. Parallel Computing **25**(7), 789–812 (1999)
11. Elkin, M., Emek, Y., Spielman, D.A., Teng, S.-H.: Lower-stretch spanning trees. In: Proceedings of the Thirty-Seventh Annual ACM Symposium on Theory of Computing, New York, NY, USA, pp. 494–503. ACM (2005)
12. Guennebaud, G., Jacob, B., et al.: Eigen v3 (2010). http://eigen.tuxfamily.org
13. Hoske, D.: An experimental study of a nearly-linear time Laplacian solver. Master's thesis, Karlsruhe Institute of Technology (KIT) (2014)
14. Kelner, J.A., Orecchia, L., Sidford, A., Zhu, Z.A.: A simple, combinatorial algorithm for solving SDD systems in nearly-linear time. In: Proceedings of the Forty-Fifth Annual ACM Symposium on Theory of Computing, New York, NY, USA, pp. 911–920 (2013)
15. Koutis, I.: Simple parallel and distributed algorithms for spectral graph sparsification. In: Proc. 26th ACM Symp. on Parallelism in Algorithms and Architectures (SPAA), pp. 61–66. ACM (2014)
16. Koutis, I., Levin, A., Peng, R.: Improved spectral sparsification and numerical algorithms for SDD matrices. In: Symposium on Theoretical Aspects of Computer Science, vol. 14, pp. 266–277 (2012)
17. Lukarski, D.: Paralution - library for iterative sparse methods (2015). http://www.paralution.com (last accessed February 09, 2015)
18. Marquardt, D.: An algorithm for least-squares estimation of nonlinear parameters. Journal of the Society for Industrial and Applied Mathematics **11**(2), 431–441 (1963)
19. Papp, P.A.: Low-Stretch Spanning Trees. Bachelor thesis, Eötvös Loránd University (2014)
20. Peng, R., Spielman, D.A.: An efficient parallel solver for SDD linear systems. In: Proceedings of the 46th Annual ACM Symposium on Theory of Computing, STOC 2014, New York, NY, USA, pp. 333–342. ACM (2014)
21. Reif, J.H.: Efficient approximate solution of sparse linear systems. Computers & Mathematics with Applications **36**(9), 37–58 (1998)
22. Spielman, D.A., Srivastava, N.: Graph sparsification by effective resistances. In: STOC 2008, New York, NY, USA (2008)
23. Spielman, D.A., Teng, S.-H.: Nearly-linear time algorithms for graph partitioning, graph sparsification, and solving linear systems. In: STOC 2004, New York, NY, USA, pp. 81–90 (2004)
24. Spielman, D.A., Woo, J.: A note on preconditioning by low-stretch spanning trees. CoRR, abs/0903.2816 (2009)
25. Staudt, C.L., Sazonovs, A., Meyerhenke, H.: NetworKit: An interactive tool suite for high-performance network analysis. arXiv:1403.3005 (2014)
26. Vaidya, P.M.: Solving linear equations with symmetric diagonally dominant matrices by constructing good preconditioners. Technical report, University of Illinois at Urbana-Champaign, Urbana, IL (1990)

Efficient and Practical Tree Preconditioning for Solving Laplacian Systems

Luca Castelli Aleardi$^{(\boxtimes)}$, Alexandre Nolin, and Maks Ovsjanikov

LIX - École Polytechnique, Palaiseau, France
{amturing,maks}@lix.polytechnique.fr,
alexandre.nolin@polytechnique.edu

Abstract. We consider the problem of designing efficient iterative methods for solving linear systems. In its full generality, this is one of the oldest problems in numerical analysis with a tremendous number of practical applications. We focus on a particular type of linear systems, associated with Laplacian matrices of undirected graphs, and study a class of iterative methods for which it is possible to speed up the convergence through combinatorial preconditioning. We consider a class of preconditioners, known as tree preconditioners, introduced by Vaidya, that have been shown to lead to asymptotic speed-up in certain cases. Rather than trying to improve the structure of the trees used in preconditioning, we propose a very simple modification to the basic tree preconditioner, which can significantly improve the performance of the iterative linear solvers in practice. We show that our modification leads to better conditioning for some special graphs, and provide extensive experimental evidence for the decrease in the complexity of the preconditioned conjugate gradient method for several graphs, including 3D meshes and complex networks.

1 Introduction

Solving general linear systems of equations is one of the oldest and best studied areas of numerical analysis, with an abundance of both exact and approximate solutions of varying efficiency (see e.g., [12]). In this paper, we focus on iterative methods for solving a particular type of linear systems, associated with Laplacian matrices of undirected graphs. These linear systems arise in a variety of applications, which are related to solving the Poisson equation on discretized domains, including physical (e.g. fluid) simulation, complex system analysis, geometry processing and computer graphics [16,17], among many others. One class of techniques, which is especially useful in solving large systems of equations with Laplacians matrices of certain (sparse) graphs is the conjugate gradient method. This method can be classified as an iterative approach, since it provides progressively better estimates of the final solution, and only requires the ability to compute matrix-vector products. It is also known to terminate in a finite number of steps depending on the quality of the initial guess and the condition number of the matrix in question [25]. The convergence speed of the conjugate gradient method can further be improved significantly using preconditioning, which aims to approximate the given matrix A by another matrix B (a preconditioner),

© Springer International Publishing Switzerland 2015
E. Bampis (Ed.): SEA 2015, LNCS 9125, pp. 219–231, 2015.
DOI: 10.1007/978-3-319-20086-6_17

whose inverse can be readily computed. The quality of the improvement provided by the preconditioner is directly related to the difference between $B^{-1}A$ and identity. Recently a class of preconditioners has been proposed for solving the Poisson equation on undirected graphs, by using the so-called combinatorial (or geometric) preconditioning [5, 24]. The main idea, proposed by Vaidya, is to approximate a given graph by its subgraph, on which the system of equations can be solved easily. The canonical example of this type of preconditioner is a spanning tree of the graph. Since the Poisson equation can be solved in linear time if the graph is a tree, the main idea in Vaidya's approach is to use the Laplacian of the spanning tree as a preconditioner to improve the convergence of iterative methods, such as the conjugate gradient. This basic framework has been extended significantly to both obtain near-optimal trees that can approximate arbitrary graphs, and to use a recursive approach in which a graph can be approximated by a progressively more accurate subgraphs, which can lead to very significant asymptotic speed-up in solving linear systems on general graphs [23]. While the theoretical framework for combinatorial preconditioners has been developed and in some ways settled, with a few notable exceptions, the practical implementations of these ideas are still largely lacking. This can be attributed, in part, to the highly complex nature of the algorithms for obtaining the optimal preconditioners, with potentially very large constants in the asymptotic analysis on the one hand [22, 23], and the relatively little improvement provided by the tree preconditioner on the other hand [9]. As a result, despite the theoretical appeal and the near-optimality in the asymptotic sense of the resulting algorithms [10], the practitioners have not yet fully benefited from the potential practical improvements provided by the combinatorial preconditioners.

Contribution. In this paper, we concentrate on the basic setting of Vaidya's preconditioners where the Laplacian matrix of a single spanning tree is used as a preconditioner for the conjugate gradient method. Indeed, by extending the experiments of Chen et al. [9] to a variety of graphs and large networks, we show empirically that in most cases the improvement given by a single preconditioner is either minor or even non-existent compared to the baseline conjugate gradient approach. In this context, our main contribution is to propose a very simple modification to Vaidya's tree preconditioner, which provides significant practical improvements, with minimal implementation effort. Our modification can be seen as a combination of a basic Jacobi (diagonal) preconditioner with a combinatorial (tree) one. Despite its extreme simplicity, we show that on a set of important special cases, our approach can lead to a decrease in the condition number of the resulting system, compared to the baseline combinatorial preconditioner. Perhaps more importantly, however, we also show via extensive experimentation, that our modification can also lead to practical speedup in the convergence of the conjugate gradient method compared to both the Jacobi and tree preconditioners for a large number of classes of graphs and different target functions. Our approach is not meant to provide a preconditioner structurally very different from the existing ones, or to improve their asymptotic complexity. Rather, by showing that a simple modification can potentially lead to significant practical improvements, we hope to demonstrate the usefulness of such preconditioners and to help eventually bridge theory and practice in this field.

Related works. A tremendous amount of progress has been done in solving linear systems associated to symmetric diagonally dominant matrices in the recent past. Classical iterations, as described in [12], were very sensitive to systems of poor condition number, and until quite recently efficient preconditioning was mostly a matter of heuristics. A major step was done by Spielman and Teng [23], presenting the first nearly linear algorithm (see [20–22]). The feat was made possible by the introduction and refinement of ideas such as spectral sparsification, ultra-sparsifiers and algorithms for constructing low-stretch spanning trees, which they cleverly combined to build a recursively preconditioned iterative solver. The idea of recursive preconditioning is the basis of today's best solvers [10], and its individual parts have been separately improved over the years. For instance, methods for obtaining low-stretch spanning trees, first introduced with no link to preconditioning [2], have been seen a lot of progress over the years. Their use as preconditioners was suggested in [24], in the continuity of the ideas of support theory and combinatorial preconditioning (see [5–7] for early work and formalizations of this theory). Interested readers can read the progression of the stretch in [1], where the currently best algorithm for computing trees of total stretch $O(m \log n \log \log n)$ in time $O(m \log n \log \log n)$ is give. If no better bound has been found since then, recent works introduced a generalization of stretch [11], creating new possibilities of optimization. Spectral sparsification has seen similar improvements, however its progression is less linear than that of spanning trees. Better sparsifiers are described in [3] and fast construction algorithms are given in recent works [14] (see [4] for more details).

2   Preliminaries and Background

Throughout the paper, we consider simple, undirected, unweighted graphs $G = (V, E)$ with $\#V = n$ and $\#E = m$, and $d(i) = \#\{j, (i,j) \in E\}$ the degrees of the vertices. The unweighted (and un-normalized) Laplacian matrix L_G is given via its relation to the diagonal degree matrix D_G and the adjacency matrix A_G:

$$D_G = \begin{cases} d(i) & \text{if } i = j \\ 0 & \text{o/w} \end{cases}, \quad A_G = \begin{cases} 1 & \text{if } (i,j) \in E \\ 0 & \text{o/w} \end{cases}, \quad L_G = D_G - A_G$$

The Laplacian matrix L_G is symmetric, *diagonally dominant*, and only has non-negative eigenvalues. Indeed, it is easy to see that the number of connected components of G equals the dimension of the null space of L_G. Throughout our paper we assume to be working with a connected and unweighted graph G (most of the material can be adapted to the case of positively-weighted edges). The eigenvalues of L_G are given as $0 = \lambda_1 < \lambda_2 \leq \ldots \leq \lambda_n$.

Solving Linear Systems. The canonical problem that we consider is to solve a linear system of equations of the form $Ax = b$, where, in our case $A = L_G$ for some known vector b. Depending on the domain, a problem of this form may also be known as solving the discrete Poisson equation. In general, although the number n of vertices in the graph can be very large, the matrix L_G is typically sparse, which can make direct solvers inefficient or even not applicable, since a full n^2 set of variables can easily exceed the available memory. Instead, iterative solvers have been used to solve this problem, and most notably the *Conjugate Gradient* (CG) method, which is especially useful in cases with limited memory, since it requires only matrix-vector product computations. This method is applicable to symmetric positive (semi)-definite systems, and computes successive approximations of x by taking a step in a direction *conjugate* to previously taken steps, where conjugacy between two vectors x_1, x_2 is defined as $x_1^T A x_2 = 0$ (please see Chap. 11 in [12] for a full discussion of this method, and Figure 1 below for the pseudo-code). It is well-known that in the absence of rounding errors, the conjugate gradient method will converge in at most n iterations in the worst case. A more relevant bound, however, can be given by using the condition number $\kappa(A)$ of the matrix A, given by the ratio of its largest and smallest *non-zero* eigenvalues. After t iterations, the error of the algorithm is bounded by:

$$||x^{(t)} - x||_A \leq 2 \left(1 - \frac{2}{\sqrt{\lambda_n/\lambda_2} + 1}\right)^t ||x||_A \qquad (1)$$

Note that while the Conjugate Gradient method is best suited for positive definite matrices, it can also be easily adapted to positive semi-definite systems, such as the ones including the graph Laplacian. One simply has to make sure that the right hand side of the equation lies in the span of the matrix. For us, this means that the vector b has to sum to zero. Let us also stress that λ_1 in the definition of the condition number is the first *non-zero* eigenvalue. This will become particularly important when we define and analyze the properties of the preconditioned conjugate gradient.

Preconditioning. Since the *condition number* gives a simple bound on the efficiency of iterative solvers, and of the conjugate gradient method in particular, it is natural to try to introduce linear systems equivalent to the original one, but with a lower condition number, and therefore better convergence properties. This process, called *preconditioning*, requires a non-singular matrix M, such that $M^{-1} \approx A^{-1}$. Then, instead of solving $Ax = b$ directly, we solve :

$$C^{-1} A C^{-1} \tilde{x} = C^{-1} b \qquad (2)$$

where $C^2 = M$, and x is found by solving $Cx = \tilde{x}$. Ideally, the preconditioner M should be a positive (semi)-definite matrix, such that the condition number of $M^{-1}A$ is significantly smaller than that of A itself. The design of optimal preconditioners typically involves a trade-off: on the one hand, $M^{-1}A$ should be as close to identity as possible. On the other hand, it should be possible to solve a linear system of the form $Mx = b$ very quickly, since it has to be done at every CG iteration. An example of potentially useful preconditioning is the *Jacobi Preconditioner* for diagonally dominant systems. This consists in

$$\text{Initialization} \begin{cases} r^{(0)} = b - Ax^{(0)} \\ p^{(0)} = r^{(0)} \end{cases}$$

$$\text{Iteration} \begin{cases} \alpha^{(k)} = \frac{||r^{(k)}||^2}{||p^{(k)}||_A^2} \\ x^{(k+1)} = x^{(k)} + \alpha^{(k)}p^{(k)} \\ r^{(k+1)} = r^{(k)} - \alpha^{(k)}Ap^{(k)} \\ \beta^{(k)} = \frac{||r^{(k+1)}||^2}{||r^{(k)}||^2} \\ p^{(k+1)} = r^{(k+1)} + \beta^{(k)}p^{(k)} \end{cases}$$

$$\text{Initialization} \begin{cases} r^{(0)} = b - Ax^{(0)} \\ z^{(0)} = M^{-1}r^{(0)} \\ p^{(0)} = z^{(0)} \end{cases}$$

$$\text{Iteration} \begin{cases} \alpha^{(k)} = \frac{\langle r^{(k)}, z^{(k)} \rangle}{||p^{(k)}||_A^2} \\ x^{(k+1)} = x^{(k)} + \alpha^{(k)}p^{(k)} \\ r^{(k+1)} = r^{(k)} - \alpha^{(k)}Ap^{(k)} \\ z^{(k+1)} = M^{-1}r^{(k+1)} \\ \beta^{(k)} = \frac{\langle z^{(k+1)}, r^{(k+1)} \rangle}{\langle z^{(k)}, r^{(k)} \rangle} \\ p^{(k+1)} = z^{(k+1)} + \beta^{(k)}p^{(k)} \end{cases}$$

Fig. 1. Conjugate Gradient and Preconditioned Conjugate Gradient : this pseudocode shows the general idea of orthogonalization in both algorithms as well as how the preconditioning takes place in PCG

taking the matrix $D = (\delta_{ij}A_{(i,j)})_{(i,j)}$, i.e the diagonal of the original matrix, as preconditioner. This is both very easy to compute, and solving $Dx = b$ takes an optimal $O(n)$ operations. When both A and M are symmetric positive definite, then solving Eq. 2 can be done without explicitly computing the matrix C, by modifying the steps taken during the iterations of the Conjugate Gradient method. This results in the *Preconditioned Conjugate Gradient* for which we provide the pseudo-code in Figure 1. Note that for positive semi-definite systems, one has to use the pseudo-inverse in Eq. 2 above, and make sure that the kernel of M is contained in the kernel of A, for otherwise the system $Mx = b$, may not have a solution. In most cases, when the preconditioning is applied to positive semi-definite systems the kernels of M and A coincide, although the framework can also be applied in a more general case.

Spanning Trees as Preconditioners for Graphs. While both the basic and the preconditioned Conjugate Gradient method can be applied for any positive (semi)-definite linear system, the design of preconditioners can benefit from the knowledge of the structure of the matrix in question. As mentioned above, in this paper, we concentrate on the linear systems arising from the Laplacian matrices of undirected graphs. In this case, a particularly promising idea, first proposed by Vaidya and then extended significantly in the recent years, is to use the Laplacian matrix of a subgraph as a preconditioner to the original system. Note that, if the subgraph is connected over the same set of nodes as the original graph, then the kernels of the Laplacian matrices both have dimension 1, and they contain the constant vector $\mathbf{1}_n$ and all the vectors parallel to it, making the use of preconditioning directly applicable. An appealing candidate for a subgraph to be used as a preconditioner is a spanning tree T of the graph G. This is because if L_T is the Laplacian matrix of the tree T, then the problem of type $L_T x = b$ can be solved very efficiently, in time $O(n)$, with two tree traversals. This makes spanning trees good candidates for preconditioning, because their use keeps the cost per PCG iteration in $O(m)$. It can be shown [25] that for a spanning tree T of G, $\kappa(L_T^\dagger L_G) \leq stretch_T(G)$, where the stretch is defined as the sum of the distances in the tree between any two vertices connected by an edge in G. Together with Eq. 1 this can be used to establish the convergence of the preconditioned Conjugate Gradient for Laplacian matrices.

We note briefly that better bounds can be proved by also looking at the distribution of the eigenvalues. A proof using a lower bound for all eigenvalues and an upper bound on the number of eigenvalues above a certain threshold yields that PCG computes an ϵ−approximation in $O\left((stretch_T(G))^{1/3} log(1/\epsilon)\right)$ iterations (see Lemma 17.2 in [25]). In the past several years, this basic framework for solving linear systems with Laplacian matrices has been extended significantly, with two major research directions: finding trees that can optimize the stretch with respect to arbitrary large graphs [1], and changing this basic framework to use a more sophisticated hierarchical graph approximation scheme in which preconditioners themselves can be solved via iterative (and possibly recursive) schemes [16]. Unfortunately, both of these directions lead to highly complex algorithms and their practical performance has been evaluated only very recently [13]. Rather than trying to improve either of these two directions, our goal is to show that a simple modification to the tree preconditioner can significantly improve the performance of the iterative solver both in theory (for some restricted cases) and in practice (over a large number of graph classes).

3   Contribution: Enhancing Tree-Based Preconditioners

As mentioned in the introduction, our main goal is to show that a simple modification of the combinatorial (tree) preconditioner can have a positive impact on the practical performance of the preconditioned conjugate gradient. Indeed, as has been noted by Chen et al. [9] and we confirm in Section 4, the basic version of Vaidya's approach rarely results in significant practical benefits for PCG. The critical remark behind our work is that it is possible to add positive terms to the diagonal of the preconditioning matrix L_T without changing its combinatorial structure that enables the fast resolution of associated linear systems. Thus, we introduce the matrix $H_T = L_T + D_G - D_T = D_G - A_T$. Note that the matrix H_T has the same diagonal as the Laplacian L_G, but the same sparsity structure as the Laplacian of the subgraph T. Therefore, solving a linear system of equations of the type $H_T x = b$ can still be done in exactly the same time as solving $L_T x = b$. Nevertheless, as we show below theoretically (on some restricted cases) and empirically on a large number of different graphs and linear systems, this simple modification can significantly boost the performance of the PCG method. Before proceeding to the analysis of our modification to Vaidya's preconditioner, we first note that unless $T = G$, the matrix H_T will be full-rank, unlike the L_G which has a kernel consisting of vectors parallel to the constant vector $\mathbf{1}_n$. While in practice, this does not change the method shown in Figure 1, we note that the analysis needs to be adapted slightly. Namely, since we are operating in the space orthogonal to the constant vector $\mathbf{1}_n$, we need to make sure that the condition number of the preconditioned system is calculated correctly. For this, the following Lemma, which is readily verified, is useful:

Lemma 1. *The eigenvalues of the generalized eigenvalue system $L_G x = \lambda H_T x$ are the same as those of the system $L_G x = \lambda P H_T x$, where $P = (I_n - \frac{1}{n} 1_n 1_n^T)$ is the projection onto the space of vectors orthogonal to the constant vector.*

Therefore, computing the condition number $\kappa(L_G, H_T)$ of the preconditioned system can be done by considering the ratio of the largest to smallest non-zero eigenvalues of the matrix $H_T^{-1} L_G$. Equivalently, one can consider the smallest and largest value c such that $x^T (L_G - c H_T) x \geq 0$ for all x, such that $x^T H_T x_c = 0$.

To motivate the use of our preconditioner as well as to provide some intuition on its behavior we proceed in two stages. First, we show some bounds on the condition number for special graphs, and second, we demonstrate empirically that for a very wide range of large scale graphs and linear systems our approach can significantly outperform other baseline preconditioners (Section 4).

3.1   Some Bounds for Special Graphs

Here we provide bounds on the condition number of the preconditioned system for Laplacians and show that we can obtain significant improvement over Vaidya's preconditioners in some important special cases (proofs are given in [8]).

The Complete Graph. Let us first consider $G = K_n$, the complete graph on n vertices and let T be a *star* spanning tree, consisting of one root vertex of degree $n - 1$ which is adjacent to all remaining $n - 1$ vertices.

Lemma 2. *Given the complete graph G and the tree T described above, then for any $n > 2$ we have $\kappa(L_G, H_T) = \frac{n}{n-1} < \kappa(L_G, L_T) = n$.*

Note, in particular that $\kappa(L_G, H_T) \to 1$ whereas $\kappa(L_G, L_T)$ grows with n.

The Ring Graph. Another important example is the cycle (ring) graph with n vertices. Here, the tree T differs from G by a single edge. In this case:

Lemma 3. *If G is a cycle and T is a spanning tree of G, then $\kappa(L_G, H_T) < 2$, while $\kappa(L_G, L_T) = n$ for any n.*

Note that again, the system preconditioned with H_T remains well-conditioned for all n, unlike the system preconditioned by the tree itself, which has an unbounded condition number. Indeed, a strictly more general result holds:

Lemma 4. *Let G be any graph and T be a tree on G, such that the edge-complement T^c of T in G is a star. Then: $\kappa(L_G, H_T) \leq 2$.*

Note that this lemma generalizes the previous one since the complement of the tree in the ring graph is a single edge.

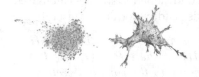

Network/Graph	n	m	$\kappa(L_G)$	$\kappa(L_G, L_T)$ (max tree)	$\kappa(L_G, H_T)$ (max tree)	$\kappa(L_G, L_T)$ (min tree)	$\kappa(L_G, H_T)$ (min tree)
C. Elegans	453	2025	922.53	373.03	20.42	11857	19.67
Email URV	1133	5451	217.44	16476	9.71	24540	10.23
Power grid network	4941	6594	26487	452.89	2366	6445	2331
Random triang.	102	300	115.42	102.65	42.95	286.51	41.59
Random triang.	1002	3000	2009	1206	461	2359	475.11

Fig. 2. Condition numbers for unweighted graphs: we consider a few example of complex networks and random triangulations. Left pictures show the metabolic system of the *C. elegans* worm and a random planar triangulation (picture by Nicolas Curien).

The Wheel Graph. Our final example is the wheel graph, consisting of a cycle with $n - 1$ vertices that are all connected to a central vertex s, which does not belong to the cycle. In this case, let T be the star graph centered around s.

Lemma 5. *Given the graph G and the spanning tree T described above then, for any n odd, $\kappa(L_G, H_T) < \kappa(L_G, L_T) = 5$.*

This example is instructive since the wheel graph can be considered to be a simple case of a triangle mesh, a class of graphs for which we show empirically a significant improvement over Vaidya's preconditioners in the next section.

A Counterexample. We also note that there exist graphs for which the condition number of our system is worse than that of the unmodified approach. The simplest example of such a graph is a path-graph with 6 nodes, with additional edges between nodes $(1, 3)$ and $(4, 6)$, and where the tree is the path. In this case, it can be shown that $\kappa(L_G, L_T) = 3 < \kappa(L_G, H_T)$. Nevertheless our experiments suggest that such cases are rare, and seem to occur when G is very close to the tree T. We leave the characterization of such cases as interesting future work.

4   Experimental Results

We provide experimental evaluations[1] of the performance of our preconditioner against CG (conjugate gradient with no preconditioning), the diagonal preconditioner $JPCG$ (Jacobi preconditioned conjugate gradient) and $TPCG$ (tree-based Vaidya's preconditioned conjugate gradient). As our preconditioner is a combination of the tree-based and diagonal approaches, we denote it by $JTPCG$. We run our experiments on a wide collection of graphs including triangle meshes (obtained from the AIM@SHAPE Shape repository), 2D regular grids, and complex networks (from the Stanford Large Network Dataset Collection). We also consider random planar triangulations, generated by the uniform random sampler by Poulalhon and Schaeffer [19], as well as graphs randomly generated according to the small-world and preferential attachment models.

---

[1] A pure Java implementation of our algorithms is available at www.lix.polytechnique. fr/~amturing/software.html.

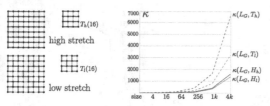

grid size	$\kappa(L_G)$	$\kappa(L_G,T)$ (low stretch)	$\kappa(L_G,H)$ (low stretch)	$\kappa(L_G,T)$ (high stretch)	$\kappa(L_G,H)$ (high stretch)
4	2	4	1.6	4	1.6
16	11.6	9.88	4.52	21.02	4.84
64	50.54	42.64	17.37	95.14	19.63
256	206.17	187.02	77.07	399.89	87.82
1024	828.69	788.56	332.88	1631.64	379.68
4096	3318	3242	1390	6585	1585

Fig. 3. We compute condition numbers for regular (unweighted) grids endowed with different spanning trees: blue and red edges correspond to trees with high and low stretch factors respectively. Our precondition matrix H allows to drastically decrease the condition number in both cases, when compared to standard tree preconditioning (dashed lines).

3D Mesh	n	m	$\kappa(L_G)$	$\kappa(L_G,L_T)$ (max tree)	$\kappa(L_G,H_T)$ (max tree)	$\kappa(L_G,L_T)$ (min tree)	$\kappa(L_G,H_T)$ (min tree)
Sphere	162	480	33.4	723.	25.6	1384	26.06
Helmet	496	1482	245.8	2885	142.4	5341	143.8
Venus	711	2106	411.8	2591	229.6	3950	251.46
Genus 3 mesh	1660	4992	304.9	5862	226.5	13578	227.2
Triceratops	2832	8490	2079	12342	1454	13332	1530
Cow	2904	8706	2964	15184	1853	8868	1982

Fig. 4. Condition numbers for 3D surface meshes: we compare tree preconditioning and Jacobi-tree preconditioning. Meshes are endowed with both minimum (blue) and maximum (red) spanning trees (weights correspond to Euclidean edge length).

4.1 Evaluating the Condition Number

Regular grids. Our first experiments concern the evaluation of the condition numbers for regular grids, for which we know how to construct spanning trees of high and low stretch factors. It is not difficult to see that the total stretch of the blue tree T_h in Fig. 3 is $\Theta(n\sqrt{n})$ (observe that a vertical edge $(u_i, v_i) \in G \setminus T$ belonging to the i-th column contributes with a stretch of $\Theta(i)$, where i ranges from 1 to $\sqrt{n}$). The red edges in Fig. 3 define a spanning tree T_l having a low stretch factor, which can be evaluated to be $O(n \log n)$ using an inductive argument (we refer to [16] for more details). These bounds reflect the numerical evaluation of the condition numbers for both trees T_h and T_l (plotted as dashed lines in Fig. 3). Experimental evaluations show that our Jacobi-tree preconditioner allows to drastically decrease the condition numbers for both trees T_h and T_l. More interestingly, using H_T instead of L_T we obtain new bounds which are extremely close, despite the very different performances of the corresponding spanning trees. Not surprisingly, this behavior does not only concern regular grids, but it is common to a wide class of graph Laplacians, as suggested by experimental evidence provided in next sections.

Mesh graphs and complex networks. We also compute the condition numbers for Laplacians corresponding to several 3D meshes, of different sizes and topology (see Fig. 4). We test and compare our preconditioner against the CG method (without preconditioning) and tree preconditioning, using as test trees both minimum and maximum spanning trees. We consider min spanning trees because

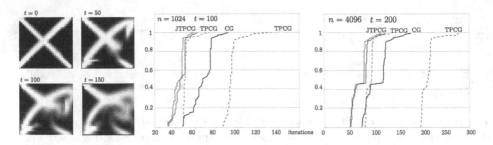

Fig. 5. Fluid simulation: we compare JTPCG against CG and TPCG (the performance of JPCG is very similar to CG). We plot the proportion of the number of iterations required to solve 100 (resp. 200) linear systems with a precision of $1e-5$. For instance, JTPCG (colored curves) takes between 51 and 127 iterations per system on a grid of size 4096, while *CG* (black curve) requires between 75 and 184 iterations.

their performances are in general worse than those of maximum spanning trees: weights are computed according to the Euclidean edge length of the 3D embedding (in the case of unweighted graphs, in order to compute min and max spanning tree, we reweight edges according to a vertex degree driven strategy). We note that our experiments confirm the intuition of Vaidya's seminal work: maximum spanning trees perform in general better as preconditioners than other trees. Once again, our preconditioner is able to get condition numbers which are significantly lower than the ones obtained with the simple preconditioner L_T. We note that this difference in performance is much less prominent when using H_T. These phenomena occur for all tested graphs and result in a significant improvement of the performance of iterative solvers.

4.2   Counting Iterations: Comparison of Iterative Linear Solvers

In this section we provide experimental evidence for the improvement achieved by our JTPCG preconditioner. We test it against other linear solvers (CG, JPCG, and TPCG) on a large set of surface meshes and random graphs. In order to obtain a fair comparison, we measure the convergence rates of linear solvers for $Lx = b$ counting the total number of iterations required to achieve a given error: as metrics we use the standard relative residual error. We use iterative solvers as core linear solvers for simulating fluid diffusion on regular 2D grids of different sizes, while counting the number of iterations required by different solvers at each time step (we use fixed precision $1e-5$). As shown by the plots in Fig. 5, JTPCG is able to drastically decrease the number of iterations, using both the high and low stretch factor spanning trees (red and blue curves). Observe that tree-based preconditioner perform pretty well (even without diagonal modification) when combined with low stretch factors (red curves in Fig. 5).

Solving mesh laplacians. The results reported in Table 1 concern the resolution of linear systems of the form $Ax = b$, where the vector b is a random vector orthogonal to the constant vector. We use the same starting vector as an initial

Table 1. Solving linear systems: we compare the JTPCG against the classical CG method, the JPCG and TPCG preconditioners. We count total number of iterations required to achieve fixed precision $1e - 7$.

Graph	n	m	CG (no prec.)	JPCG	TPCG (max tree)	JTPCG (max tree)	TPCG (min tree)	JTPCG (min tree)
Triceratops	2832	$8K$	225	196	341	188	426	181
Cow	2904	$8K$	214	192	347	170	366	182
Egea	8268	$24K$	305	249	701	219	974	221
Bunny	26002	$78K$	536	432	1632	416	1892	419
Feline	49864	$149K$	962	745	1946	663	2362	682
Eros	476596	$1.4M$	2185	1560	16122	1474	13257	1488
Random triang.	100002	$300K$	2382	1215	1776	1082	1247	1006

Graph	n	m	CG (no prec.)	JPCG	TPCG (max tree)	JTPCG (max tree)	TPCG (min tree)	JTPCG (min tree)
Triceratops	2832	$8K$	5139	5057	6811	4842	7505	4997
Cow	2904	$8K$	5158	5145	6854	4907	6989	4980
Egea	8268	$24K$	7980	7314	12525	6988	15206	7031
Bunny	26002	$78K$	32187	30634	49048	30231	51405	30312
Aphrodite	46096	$138K$	13669	12228	37547	11803	41991	11303
Feline	49864	$149K$	46404	42217	62595	40371	71095	40727
Iphigenia	49922	$149K$	19490	18111	54008	16984	60973	17306

Fig. 6. The picture above shows a mesh together with its 3D spectral embedding. The table reports the total number of iterations performed by the iterative linear solvers during the inverse power iteration (our tests are run with fixed precision $1e - 5$).

Network	n	m	CG	JPCG	TPCG (max tree)	JTPCG (max tree)
C. Elegans	453	2025	8123	7129	7795	7051
Email URV	1133	5451	24395	23540	25684	23435
Facebook social circles	4039	88234	11832	7702	8044	7677
Power grid network	4941	6594	15623	13430	8812	10481
PGP network	10680	24316	64068	54806	55356	53852
Pref. attachment	100000	$500K$	61125	59399	80451	59455
Small world	100000	$500K$	4972	5010	125446	4963
Gowalla	196591	$950K$	202247	146883	176644	147322

Fig. 7. Spectral clustering and complex networks: the picture above shows a partition into five sets of a social network (facebook, 4k nodes) obtained by applying the K-means algorithm to the spectral embedding of the graph.

guess for all linear solvers (tests are repeated several times, in order to take into account the dependency of the convergence speed on the initial guess). As confirmed by the results reported in Table 1, our preconditioner always performs better than other solvers (this has been confirmed for all tested meshes).

Iterative eigensolvers. Spectral methods proved their relevance in various application domains, ranging from graph drawing and data visualization to complex networks analysis (for more details we refer to [15,25]). We also have integrated our preconditioner as the core of an iterative eigensolver (we have implemented a hybrid version of the inverse power iteration). We evaluate its performance by computing the smallest non-trivial eigenvalues of the Laplacian: a fundamental step in problems such as spectral drawing and spectral clustering. Tables in Fig. 6 and 7 report the number of iterations performed by the linear solvers required to compute the first three eigenvalues for 3D meshes and complex networks.

Acknowledgments. This work is supported by the ANR EGOS 12 JS02 002 01, a Google Faculty Research Award, the Marie Curie grant CIG-334283-HRGP, and a CNRS chaire dexcellence, Jean Marjoulet professorial chair.

References

1. Abraham, I., Neiman, O.: Using petal-decompositions to build a low stretch spanning tree. In: Proc. STOC, pp. 395–406. ACM (2012)
2. Alon, N., Karp, R.M., Peleg, D., West, D.B.: A graph-theoretic game and its application to the k-server problem. SIAM J. Comput. **24**(1), 78–100 (1995)
3. Batson, J.D., Spielman, D.A., Srivastava, N.: Twice-ramanujan sparsifiers. SIAM Review **56**(2), 315–334 (2014)
4. Batson, J.D., Spielman, D.A., Srivastava, N., Teng, S.: Spectral sparsification of graphs: theory and algorithms. Commun. ACM **56**(8), 87–94 (2013)
5. Beauwens, R.: Lower eigenvalue bounds for pencils of matrices. Linear Algebra and its Applications **85**, 101–119 (1987)
6. Bern, M.W., Gilbert, J.R., Hendrickson, B., Nguyen, N., Toledo, S.: Support-graph preconditioners. SIAM J. Matrix Analysis Applications **27**(4), 930–951 (2006)
7. Boman, E., Hendrickson, B.: Support theory for preconditioning. SIAM Journal on Matrix Analysis and Applications **25**(3), 694–717 (2003)
8. Castelli Aleardi, L., Nolin, A., Ovsjanikov, M.: Efficient and practical tree preconditioning for solving Laplacian systems (2015). Preprint https://hal.inria.fr/hal-01138603
9. Chen, D., Toledo, S.: Vaidya's preconditioners: implementation and experimental study. Elect. Trans. on Numerical Analysis **16**, 30–49 (2003)
10. Cohen, M.B., Kyng, R., Miller, G.L., Pachocki, J., Peng, R., Rao, A., Xu, S.C.: Solving SDD linear systems in nearly $m \log^{1/2} n$ time. STOC, pp. 343–352 (2014)
11. Cohen, M.B., Miller, G.L., Pachocki, J.W., Peng, R., Xu, S.C.: Stretching stretch (2014). CoRR, abs/1401.2454
12. Golub, G.H., Van Loan, C.F.: Matrix Computations. 4th edn (2013)
13. Hoske, D., Lukarski, D., Meyerhenke, H., Wegner, M.: Is nearly-linear the same in theory and practice? a case study with a combinatorial laplacian solver. In: Proc, SEA (2015)
14. Kolla, A., Makarychev, Y., Saberi, A., Teng, S.-H.: Subgraph sparsification and nearly optimal ultrasparsifiers. In: Proc. STOC, pp. 57–66. ACM (2010)
15. Koren, Y.: Drawing graphs by eigenvectors: Theory and practice. Computers and Mathematics with Applications **49**, 2005 (2005)
16. Koutis, I., Miller, G.L., Peng, R.: A fast solver for a class of linear systems. Commun. ACM **55**(10), 99–107 (2012)
17. Krishnan, D., Fattal, R., Szeliski, R.: Efficient preconditioning of Laplacian matrices for Computer Graphics. ACM Trans. Graph. **32**(4), 142 (2013)
18. Poulalhon, D., Schaeffer, G.: Optimal coding and sampling of triangulations. Algorithmica **46**(3–4), 505–527 (2006)

19. Spielman, D.A., Teng, S.: Spectral sparsification of graphs. SIAM J. Comput. **40**(4), 981–1025 (2011)
20. Spielman, D.A., Teng, S.: A local clustering algorithm for massive graphs and its application to nearly linear time graph partitioning. SIAM J. Comput. **42**(1), 1–26 (2013)
21. Spielman, D.A., Teng, S.: Nearly linear time algorithms for preconditioning and solving symmetric, diagonally dominant linear systems. SIAM J. Matrix Analysis Applications **35**(3), 835–885 (2014)
22. Spielman, D.A., Teng, S.-H.: Nearly-linear time algorithms for graph partitioning, graph sparsification, and solving linear systems. STOC, pp. 81–90 (2004)
23. Vaidya, P.M.: Solving linear equations with symmetric diagonally dominant matrices by constructing good preconditioners. Unpublished manuscript (1991)
24. Vishnoi, N.K.: Lx = b. Foundations and Trends in Theoretical Computer Science **8**(1–2), 1–141 (2013)
25. von Luxburg, U.: A tutorial on spectral clustering. Statistics and Computing **17**(4), 395–416 (2007)

Other Applications I

Efficient Generation of Stable Planar Cages for Chemistry

Dominique Barth, Olivier David, Franck Quessette, Vincent Reinhard,
Yann Strozecki[✉], and Sandrine Vial

Université de Versailles Saint-Quentin, Versailles, France
`yann.strozecki@prism.uvsq.fr`

Abstract. In this paper we describe an algorithm which generates all
colored planar maps with a good minimum sparsity from simple motifs
and rules to connect them. An implementation of this algorithm is available and is used by chemists who want to quickly generate all sound
molecules they can obtain by mixing some basic components.

1 Introduction

Carbon dioxide, as well as methane can be absorbed by large organic cages [1].
These cages are formed by spontaneous assembly of small organic molecules,
called motifs, bearing different reacting centres. The prediction of the overall
shape of the cage that will be obtained by mixing the starting motifs is rather
difficult, especially because a given set of reacting partners can lead to very
different cages. It is hence crucial for chemists to have an operating tool that is
capable of generating the many shapes of cages accessible from predetermined
molecular motifs.

In this paper we present the algorithms we have designed and implemented to
generates molecules that are much larger and less regular that what the chemists
usually design by hand. The molecules are modelled by maps i.e. planar embeddings of planar graphs, as explained in Sec. 2. The use of maps may seem unsuit-
able since they do not represent spatial positions. Though, planar maps are a
good model for spherical topologies and the embedding capture the rigidity of
the motifs. We must also be able to select the most relevant molecules among the
huge number we generate. In Sec. 3.4, we characterize what a "good" molecule is
through graph parameters which are then used to filter the best molecules. The
relevance of our modeling and of our parameters is validated by the results we
obtain: All small molecules (5-10 motifs) we generate and consider to be good
according to our parameters have been studied before by chemists. Some of the
very regular molecules of medium size (10-20 motifs) we generate correspond to
the largest cages chemists have ever produced. We also have produced cages of
shape unknown to chemists that they now try to synthesize.

Authors thank the French Labex CHARMMMAT for the financial support of this
work and David Auger for fruitful discussions about the folding algorithm.

E. Bampis (Ed.): SEA 2015, LNCS 9125, pp. 235–246, 2015.
DOI: 10.1007/978-3-319-20086-6_18

The aim of this paper is the **generation of all colored planar maps up to isomorphism** representing possible molecules obtained from a set of elementary starting motifs (colors). As with all *enumeration problems*, one difficulty is to avoid to produce a solution several times. Moreover the number of solutions may grow exponentially with their size, it is here the case for all bases of motifs but the most contrived. The complexity of such enumeration problems must then take into account the number of produced solutions (see [2] for more details on enumeration).

We say that an algorithm is in *polynomial total time* if its complexity is polynomial in the number of solutions and polynomial in the size of the produced solutions. In our context, where the number of solutions is always exponential in their size, we are interested in *linear total time* algorithms. The best algorithms are in constant amortized time (CAT): the algorithm uses on average a constant time to generate each solution. This kind of efficient algorithms exists for simple enumeration problems such as listing all trees [3]. We may also want to bound the delay that is the time between the production of two consecutive solutions. Good algorithms have a delay polynomial, linear or even constant in the size of the generated solutions.

There exist numerous works on enumeration and generation of planar maps [4], but none of them deals with the generation of planar maps built with a set of starting motifs and color constraints. Moreover, most of the literature deals with non-constructive tools [5] or yields algorithms which are not in polynomial total time. There are a few programs such as plantri [6] and CaGe [7] which generate efficiently some particular class of planar graphs such as cubic graphs or graphs with bounded size of face but they are not general enough for our purposes.

The algorithm we present in Sec. 3 is far from being in polynomial total time since we are not able to bound the number of isomorphic copies of each solution we generate. However, we will present several subroutines used in our algorithm which are either CAT, for instance the generation of paths and almost foldable paths in Sec. 3.1, or in linear delay such as the folding of unsaturated maps of motifs in Sec. 3.2. Moreover, we study several heuristics and improvements which makes the enumeration feasible for maps of medium size. Sec. 4 presents numerical results which supports this assertion and illustrates the relative interest of our heuristics.

2   Modeling of the Problem

In this section, we propose the modeling of our problem by maps. A map is a connected planar graph drawn on the sphere considered up to continuous deformation. Note that by Steinitz's theorem, when a planar graph is 3-connected, there is only one corresponding map, but otherwise there may be several of them. It is relevant to distinguish between two maps with the same underlying graph, since the geometrical informations contained in the maps are useful to the chemist who are interested in their 3D representation. All maps used in this paper are *vertex-colored maps*. The representation of a map is a graph and a cyclic order of the neighbors around each vertex.

We first model the basic chemical elements with maps we call *motifs*. Then the motifs are assembled to form a *map of motifs* and from this map we derive a *molecular map* that is a more faithful model of the molecular cages we try to design.

We use a finite even set of colors $\mathcal{A} = \{a, \bar{a}, b, \bar{b}, c, \bar{c}, \dots\}$ where each positive color a in $\mathcal{A}$ has a unique complementary negative color denoted by $\bar{a}$ and a is the complementary color of $\bar{a}$. Each color represents a different kind of reacting center. Let us give the definition of motifs.

Definition 1. *A map $G = (V_c \sqcup V, E, next)$ is a motif if, (1) V_c contains only one vertex c called the center, (2) each vertex in V is colored with a color in $\mathcal{A}$, (3) $E = \{(c, u),\ u \in V\}$, and (4) next gives an order on the edges of c: $next((c, u)) = (c, v)$ means that the edge (c, v) is "following" the edge (c, u) in a clockwise drawing of G. For all $k < |V|$, $next^k((c, u)) \neq (c, u)$ and $next^{|V|}((c, u)) = (c, u)$.*

Note that a motif is a star graph. We assume as input $\mathcal{M}$ a finite set of motifs all different. Each motif is identified by a distinct color from an alphabet $\mathcal{A}_M$ disjoint from $\mathcal{A}$ induced by the colors existing in $\mathcal{M}$. Fig. 1 gives examples of motifs.

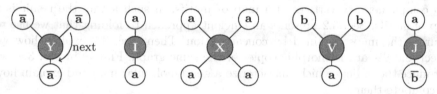

Fig. 1. Example of motifs on $\mathcal{A}_M = \{\mathbf{Y}, \mathbf{I}, \mathbf{X}, \mathbf{V}, \mathbf{J}\}$ and $\mathcal{A} = \{a, \bar{a}, b, \bar{b}\}$

Definition 2. *A connected planar map $G = (V_c \sqcup V, E, next)$ is a map of motifs based on $\mathcal{M}$ if, (1) the closed neighborhood of each vertex in V_c is a motif, (2) each vertex in V is connected to exactly one vertex in V_c and at most one vertex in V. If u and v in V are connected, the colors of u and v must be complementary. The number of vertices in V_c is called the size of G.*

Note that each motif of $\mathcal{M}$ may appear any number of times in a map of motifs, it may also be not present. A motif is a map of motifs of size 1. In a map of motifs, a vertex of degree 1 in V is called a *free vertex*. A map of motifs with no free vertex is called *saturated* otherwise it is called *unsaturated*.

Based on a saturated map of motifs we construct the molecular map that is the graph model of the cages.

Definition 3. *Let $G = (V_c \sqcup V, E_G, next_G)$ be a saturated map of motifs based on $\mathcal{M}$, we define the molecular map M as the map G where all paths of size three between vertices of V_c are replaced by an edge.*

Fig. 2. Example of two maps of motifs based on $\mathcal{M} = \{\mathbf{Y}, \mathbf{I}\}$, the first map is unsaturated while the second map is saturated

3   Description of the Algorithm

The aim of this paper is to solve the following problem: given a base of motifs $\mathcal{M}$ and an integer n, enumerate all molecular maps of size n based on $\mathcal{M}$. The complexity depends only on n since the size of $\mathcal{M}$ and the size of its elements are assumed to be small constants (usually less than 4). In this section, we describe an algorithm which solves this problem and explain in details its two main steps.

The first one, *the concatenation*, consists in adding edges between complementary vertices of two maps of motifs in such a way the result is still a map of motifs. In this paper, we always concatenate a single motif to a map of motifs, see [8] for other concatenations. Sec. 3.1 presents the different strategies of concatenation. The second, *the fold or folding*, consists in adding an edge between two complementary vertices of a map of motifs, in such a way the result is a map of motifs. Sec. 3.2 presents an efficient approach to folding that we use to saturate the maps obtained by concatenation. Then, Sec. 3.3 explain how we detect and discard isomorphic copies of the same graph. Finally in Sec. 3.4, we introduce the indices which characterize a good molecular map and explain how we compute them.

3.1   Backbone Generation

The first step is to generate all *backbones*, that is unsaturated maps of motifs of a given size n which are of a very simple shape. The aim is that, by folding these backbones in a second step, we will recover all saturated maps of motifs. Since every map of motifs have a spanning tree, we can choose trees as backbones and be sure to recover all saturated maps. But for performance reason, we will also use paths and cycles as backbones. This turns out to be good *heuristics*, speeding up considerably our algorithm while only mildly reducing the set of generated maps of motifs. We would also like to restrict the backbones to those which can be folded into some saturated map. We address this problem by enumerating only what we call *the almost foldable backbones*, with a complexity as good as for the generation of regular backbones. This new algorithm greatly improve the computation time.

Spanning tree. In a first version of our algorithm [8], the set of non isomorphic trees of size n was explicitly stored. To produce the set of trees of size $n + 1$, a single motif of every possible color was concatenated to each free vertex of each tree of size n. This generates all trees of size $n + 1$, but the drawback is that some trees are generated several times. The algorithm was thus not in linear total time and we needed to do an isomorphism test on every generated tree. We now generate all trees where the root and its first edge are fixed with a simple CAT algorithm. This method generates a tree as many times as edges in the tree: one for each choice of a vertex as root and for each choice of first edge of this root. Therefore, the implemented algorithm do not need to store the trees which are produced on the fly, and has a linear delay. A way to further improve this would be to use ideas from CAT algorithms which generate unrooted trees [3]. The main idea is to choose as root the centroid of the tree. However we have to deal with a second and harder problem: we generate maps of motifs and their vertices are colored. We can generate all maps of motifs sharing the same underlying tree efficiently but they may turn out to be isomorphic.

Hamiltonian paths. Since generating trees is not easy, we propose to use simpler objects as backbones, here maps of motifs such that all vertices of V_c are on a path. These maps are caterpillar trees, but since the elements of V_c on the central path entirely determine the elements at distance one, we will consider them as paths and call them so. There are two advantages to generating paths instead of trees: they are easier to generate and their number is smaller. The drawback is that not any planar graph has an Hamiltonian path, therefore we could miss some planar maps in our enumeration. However, most small planar graphs have an Hamiltonian path, for instance all planar cubic 3-connected graphs of size less than 38 [9] and, if Barnette's conjecture holds, all fullerene graphs.

The regularity of the graphs (all vertices of the same degree) crucially matters in the existence of an Hamiltonian path. Consider for instance the base of motifs $\mathcal{M} = \{\mathbf{I}, \mathbf{Y}\}$ from Fig. 1. All molecular maps based on $\mathcal{M}$ are bipartite graphs: the $\mathbf{I}$'s in one set of the bipartition and the $\mathbf{Y}$'s in the other. But in saturated maps of motifs, we have twice the number of $\mathbf{Y}$ equal three times the number of $\mathbf{I}$ because all vertices in V must be connected, therefore there are no Hamiltonian path except in graphs with exactly three $\mathbf{I}$ and two $\mathbf{Y}$. This problem can be easily solved by building from $\mathcal{M}$ a new base of motifs which in the end generates the same molecular maps (see [10]).

Let us now explain how we generate all paths based on a set of motifs $\mathcal{M}$. We first build for each letter $a \in \mathcal{A}$ a list L_a of all non isomorphic motifs whose first edge is incident to a vertex of label $\bar{a}$. This data structure allows us to have a complexity independent of the size of $\mathcal{M}$ and of $\mathcal{A}$. Then to build all possible paths of size $n + 1$ from a path of size n, we consider its last vertex $c \in V_c$ and for each of the free vertex v connected to c and of color a, we attach every motif of L_a. Remark that beginning by the empty path, we generate all possible paths of a given size by applying recursively the algorithm. If we consider the paths as rooted at the first vertex produced during the algorithm, every path generated is clearly different. However, we can also consider the last concatenated vertex as the beginning of the path, which means we generate every path but the palindromes twice. To avoid that, we put an ordering on $\mathcal{A}_M$, the colors of the center vertices, and we consider the sequence of colors in a path. If the sequence of colors from the beginning to the end is lexicographically larger than the sequence from the end to beginning we output the path otherwise we do not. This is implemented in our algorithm and adds only in average a constant time.

Proposition 1. *The previous algorithm produces all maps of motifs which are paths without redundancies in constant amortized time, when in the base of motifs no two motifs of degree 2 can be concatenated.*

Proof. The tree of recursive calls of our algorithm can always be seen as of degree at least 3 by merging nodes of degree 2 to nodes of degree larger. Therefore it has at least as many internal nodes as leaves which correspond to output solutions. Since the algorithm needs only a constant time to go from one node to another, the generation of all paths can be done in constant amortized time.   □

In our practical examples, there are never motifs of degree two which can be concatenated. Without this condition, the algorithm has still a linear delay.

Hamiltonian cycles. If we want to further restrict the backbones we generate, a simple idea is to consider cycles instead of paths. Again it is a good choice if all motifs have the same degree or can be made so, since for instance all planar cubic 3-connected graphs of size less than 23 have an Hamiltonian cycle [11]. Moreover, we will only generate 2-connected graphs and not the ones which are only 1-connected. It is a desirable side effect, since those graphs have a bridge they are always the worse for the two main indices we are interested with, i.e. the minimum sparsity and the size of the largest cycle (see Sec. 3.4).

Almost foldable backbones. In each backbone we build, all free vertices will eventually be folded to get a saturated map of motifs. A simple necessary condition on the colors of a saturated map of motifs is that for each color $a \in \mathcal{A}$, there are as many vertices in V labeled by a and $\bar{a}$. A backbone which satisfies this condition is said to be *almost foldable*. Let G be a map of motifs and let $a_1, \ldots, a_k$ be the positive colors of the alphabet $\mathcal{A}$. We denote by C_G the characteristic vector of G, it is of size k and its ith component is the number of elements in V labeled by a_i minus the number of elements labeled by $\bar{a}_i$. Note that a map G is almost foldable if and only if C_G is the zero vector.

We propose here a method to generate in constant amortized time only the almost foldable paths. We introduce a function $F : \mathbb{N} \times \mathcal{A} \times \mathbb{Z}^k \to 2^{\mathcal{A}}$ which has the following semantic: $a' \in F(n, a, (c_1, \ldots, c_k))$ if and only if (1) there is a path P of size n with a free vertex in the first motif labeled by a, (2) $C_P = (c_1, \ldots, c_k)$, (3) a vertex of the last motif is labeled by a'.

Proposition 2. *There is an algorithm which enumerates all almost foldable paths in constant amortized time plus a precomputation in $O(n^{k+1})$, when in the base of motifs no two motifs of degree 2 can be concatenated.*

Proof. First, we explain how to generate all needed values of the function F in time $O(n^{k+1})$ by dynamic programming. Denote by f the maximal number of vertices in a motif labeled by the same color. For a path P of size n, it is clear that the coefficients in C_P are all in the interval $[-nf, nf]$. Therefore, to generate paths of size n, since f and the size of $\mathcal{A}$ are constants, we need to store $O(n^{k+1})$ values of F only.

F is easy to compute for $n = 1$: we consider each motif $M \in \mathcal{M}$ and each v of label a in M, and let $F(1, a, C_M)$ be the set of labels of all vertices of M but v. Assume we have generated the values of F for n, we generate the values for $n + 1$ in the following way. For each a, C and each $a' \in F(n, a, C)$, we consider all motifs $M \in \mathcal{M}$ such that one of their vertex is labeled by $\bar{a}$. We add all the labels of the other vertices to the set $F(n, a, C + C_M)$. This algorithm only does a constant number of operations for each value of F it computes, therefore its complexity is $O(n^{k+1})$.

Now that F is computed, we use it in our path generation algorithm to generate only the almost foldable paths. Assume we have generated a path P of size n', its characteristic vector C_P and we want to add a node at the end by connecting it to a node of label a. Assume we have already computed C_P. The algorithm checks if $F(n - n', \bar{a}, -C_P) \neq \emptyset$. If it is the case the algorithm go on normally otherwise it backtracks since this extension cannot yield a non foldable path. This improvement only adds a single test at each step of the original algorithm, plus an addition of a constant sized vector to maintain the value of C_P. Therefore it is in constant amortized time. $\qquad\square$

The complexity of the precomputation may seem to be large but k must be seen as a small constant (less than 4). It is negligible with respect to the generation of paths, which is exponential in n because of the number of non isomorphic paths. In practice, the precomputation takes only a few milliseconds for size of graphs up to 40 on a regular desktop computer. On the other hand, this optimization makes the time to computes all the backbones much smaller than the time to do the next steps.

The same kind of method has been implemented for trees (see [10]).

3.2   Folding of the Backbones

Let G be a map of motifs, the *fold* operation on the vertices u and v is adding the edge (u, v) to G. The operation is valid if u and v are free, of complementary colors and in the same face of G. Therefore, the graph obtained after the fold is still a map of motifs. In this section we generate from a backbone, by sequences of folds, all possible saturated maps of motifs.

The *outline* of a face is the list in order of traversal of the free vertices. An outline is a circular sequence of vertices $(v_1, \ldots, v_n) \in V^n$. Sequence means that the order is significant and circular means that the starting point is not. For instance, (v_1, v_2, v_3) and (v_3, v_1, v_2) are the same circular sequence but are different from (v_3, v_2, v_1). Remark that a tree or a path has a single outline, a cycle has two and a saturated map has only empty outlines. The color of an outline $(v_1, \ldots, v_n)$ is the word $w_1 \ldots w_n$ with w_i the color of v_i. Folding two vertices v_i and v_j in the same outline of color $W_1 w_i W_2 w_j W_3$ creates two outlines of color $W_3 W_1$ and W_2. The fold operation can then be seen as an operation from words over $\mathcal{A}$ to multiset of words. Remark that this operation is very similar to the reduction of consecutive complementary parentheses which enables to define the classical Dyck language of balanced string parentheses.

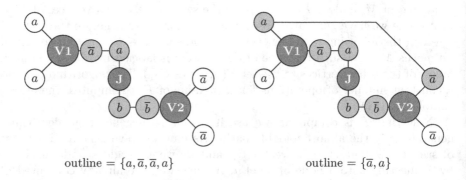

$$\text{outline} = \{a, \bar{a}, \bar{a}, a\} \qquad\qquad \text{outline} = \{\bar{a}, a\}$$

Fig. 3. A map on $\mathcal{A}_M = \{\mathbf{V1}, \mathbf{V2}, \mathbf{J}\}$ and its outline before and after a fold operation

Applying a sequence of fold to a backbone to get a saturated map is the same as applying a sequence of reductions to the colors of an outline so that we obtain only empty words. We work from now on only on the words $w_1 \ldots w_n$ and on sequences of reductions. If in a sequence of reductions, the reduction is applied to w_i and w_j we say that the sequence *pairs i with j*.

Let us call a word (or a multiset of words) which reduces to a multiset of empty words a *foldable word*. As in the case of parentheses languages, we can restrict the reduction to consecutive complementary letters which transforms $W_1 a \bar{a} W_2$ into the word $W_1 W_2$. Indeed, when a word is foldable, it can be reduced to empty words using *the restricted reduction* of consecutive letters only by reordering the sequence of reductions. We call *result* of a sequence of reductions the set of pairs (i, j) such that the sequence has paired i and j. The previous remark shows that it is indeed a set of pairs and not a sequence. Our aim is to generate all different results of sequences of reductions on foldable words without redundancies.

Lemma 1 (Folklore). *The restricted reduction on words is confluent i.e. each sequence of restricted reduction starting from a foldable word can be extended so that we get an empty word.*

As a consequence of this lemma, we get a simple algorithm for testing whether a word is foldable: reduce the word as long as it is possible and if an empty word is obtained, the word is foldable. This algorithm can be implemented in linear time by using a doubly linked list and a good order on the reductions (see [10]). We use this algorithm each time we produce a backbone to test whether it can be folded into a saturated map of motifs. Note that, even if we generate almost foldable backbones only, we may generate some which are not foldable such as those with outline $ba\bar{b}\bar{a}$.

Proposition 3. *There is an algorithm which enumerates all distinct results of sequences of reduction on a foldable word, with a linear delay and a quadratic precomputation.*

Proof. For a given word W we first build the lists L_i which contain the set of indices $j > i$ such that w_i can be folded with w_j and the obtained set of words is still foldable.

The lists L_i are built from a boolean matrix M such that $M_{i,j}$ is true if and only if the word $w_i \ldots w_j$ is foldable. The matrix is computed by dynamic programming: $M_{i,i+1}$ is true if and only if w_i and w_{i+1} are complementary. We compute $M_{i,j}$ once we have computed all $M_{i',j'}$ such that $(j' - i') < (j - i)$ by using the fact that $w_i \ldots w_j$ is foldable if and only if $w_i \ldots w_k$ and $w_{k+1} \ldots w_j$ are foldable for some k in $[i+1, j]$ or w_i and w_j are complementary and $w_{i+1} \ldots w_{j-1}$ is foldable. By this method, the matrix M is computed in time cubic in the size of the word. In fact, by Lemma 1, if there is a k such that $w_i \ldots w_k$ and $w_{k+1} \ldots w_j$ are foldable, then for all l such that $w_i \ldots w_l$ is foldable, then $w_{l+1} \ldots w_j$ is foldable. We store for each i the smallest $k > i$ such that $w_i \ldots w_k$ is foldable. Hence we can decide whether there is a k such that $w_i \ldots w_k$ is foldable in constant time and we compute the matrix M in quadratic time.

Remark that a sequence of reductions applied to a word W yields a set of subwords which are consecutive letters of W. Therefore we can represent the result of several reductions by a set of pairs $\{(l_1, r_1), \ldots, (l_k, r_k)\}$ with (l_i, r_i) representing the word $w_{l_i} \ldots w_{r_i}$ and $l_i < r_i < l_{i+1}$. We build the results of sequences of reductions in a recursive way. Assume we have already built a result R through a sequence of reductions applied to W, which has produced the set $\{(l_1, r_1), \ldots, (l_k, r_k)\}$. We consider l_1, the index of the first letter which has not been reduced and we do the reduction with every possible letter of index $i \in [l_1, r_1] \cap L_{l_1}$ which produces the set $\{(l_1, i), (i + 1, r_1) \ldots, (l_k, r_k)\}$ and the result $R \cup \{(l_1, i)\}$. By using recursively this algorithm starting on W, we obtain all possible results R corresponding to a reduction to a multiset of empty words. It is not possible to generate twice a result since at any point of the algorithm we make recursive calls on $R \cup \{(l_1, i)\}$ for different values of i which makes the results produced by each call disjoint. Between two recursive calls we do only a constant number of operations, therefore the delay is bounded by the depth of the tree of recursive calls, that is the size of the word W.   □

The enumeration algorithm we have described is **exponentially better** than the naive one where each possible letter is folded when it is next to a complementary letter and so on recursively. The complexity of the naive algorithm is proportional to the number of sequences of reductions while our is proportional to the number of results. For instance, on words of the form W^n with $W = a\bar{a}\bar{a}a$, there is only one result but $(2n)!$ sequences of reductions.

3.3   Dealing with Isomorphic Copies

Since the construction process does not guaranty uniqueness of the generated maps, we need to detect during the enumeration the isomorphic copies of already generated maps to discard them. From a theoretical point of view, planar isomorphism is well understood since it has been proved to be solvable in almost linear time [12]. However this algorithm is not practical and hard to implement as observed in [13], especially if we want a signature rather than just an isomorphism test. This is particularly true for our small graphs of size about 20, which is the reason why we rely on a simpler algorithm of quadratic complexity in the spirit of [14] (see [10]).

Moreover, the computed signature allows to detect chiral molecules, a very important notion in chemistry. Two maps are chiral if one is isomorphic to the other when the order of the next predicate is reversed for all neighborhoods.

3.4   Indices Computed on the Molecular Map

A molecular map is a candidate to be a "good" cage for chemistry. The definition of a "good" cage is merely topological: the 3D shape must be close to a sphere, it must be resistant to deformations and cuts and it must have an "entrance". We are able to check if a molecule satisfies or not these requirements only by considering the structure of its molecular map: First the map is *planar* and *connected* by construction. In quadratic time we compute the *equivalence classes of vertices up to automorphism*, using the same technique as to compute a signature, which helps measure the sphericity of the cage. The entrance is given by the *size of its largest face*, which is easily computed in linear time. The resistance of a map is given by its *minimum sparsity*. We now define the sparsity and explain how to compute it, since it is the most relevant index and the hardest to compute.

A cut of a graph $G = (V, E)$ is a bipartition of V. The *size* of a cut $S = (S_1, S_2)$ is the number of edges with one end in S_1 and the other in S_2. The sparsity of a cut is $sparsity(S) = \frac{size(S)}{\min(|S_1|, |S_2|)}$. The Sparsest Cut problem is to find the minimum sparsity over all cuts. We first implemented a brute-force algorithm, using a Gray code which enumerates all possible partitions of the set of vertices in time $O(2^n)$ where n is the number of vertices in our graph.

Although computing the minimum sparsity is NP-complete in general (minimum cut into bounded set in [15]), there is a polynomial time algorithm when the graph is planar [16]. Since the time to compute the minimum sparsity was the limiting factor of our program, we have implemented and adapted to our case this more complicated algorithm (which has never been done as far as we know).

The main idea is that a cut in a graph corresponds exactly to a cycle in the dual graph (see [17] for graph definitions useful in this paragraph). A weight is associated to each cycle of the dual: if the corresponding cut in the primal partitions it into S_1 and S_2, the weight is $\min(|S_1|, |S_2|)$. From a spanning tree of the dual, we build a base of its fundamental cycles. A fundamental cycle is given by any edge not in the spanning tree completed by edges of the spanning tree to form a minimal cycle. From symmetric differences of fundamental cycles, we can generate every cycle and its weight.

For each edge in the dual, we build a graph such that paths from a given vertex correspond to cycles of the dual which use the edge. Moreover, the weight of the cycle can be read in the last vertex of the path, and the size of the corresponding cut is the length of the path. Therefore, computing a single source shortest-path in each of these graphs enables us to compute the value of the sparsest-cut. While in the original article this was done by a modified Dijkstra algorithm, we use a breadth first-search. This is faster and it enables us to use a good heuristic: at any point of one of the breadth first-search, we know the current distance from the source can only increase. We can stop the search, if this distance divided by the maximal weight (equal to the number of vertices) is larger than the current minimum sparsity value. This implementation has **very good practical performances**: on a regular desktop computer the mean time to compute the sparsest cut of a graph of size 30 is 0.2 ms while the brute force algorithm needs 6000 ms.

4   Results

The code and the exhaustive results of our approach can be found at the following address http://kekule.prism.uvsq.fr. For several sets of motifs, one can find the set of generated maps and their indices. We stopped all computations at 300 seconds an put a − in the tables when the algorithm has not finished. All times are given in second, a.f. stands for almost foldable.

In Tab. 1, we give the time to compute *the backbones* and the number of backbones generated (we also count isomorphic copies which are generated). The time to compute cycles is not given since they are computed from paths,

Table 1. Number of backbones and generation time for **J** (a, b), **V1** $(\bar{a}, \bar{a}, b)$, **V2** $(a, \bar{b}, \bar{b})$

Size	Tree		A.f. tree		Path		A.f. path	
	Backbones	Time	Backbones	Time	Backbones	Time	Backbones	Time
9	$5.70\ 10^5$	0.09	$3.85\ 10^5$	0.05	$4.92\ 10^4$	0.01	$9.87\ 10^3$	0.01
12	$1.16\ 10^8$	14.28	$5.55\ 10^7$	7.98	$1.77\ 10^6$	0.28	$2.46\ 10^5$	0.08
15	−	−	−	−	$7.26\ 10^7$	10.88	$6.17\ 10^6$	1.74
18	−	−	−	−	−	−	$1.56\ 10^8$	45.84

Table 2. Number of maps and time to generate them and their indices for **J** (a, b), **V1** $(\bar{a}, \bar{a}, b)$, **V2** $(a, \bar{b}, \bar{b})$

Size	A.f. tree			A.f. path			A.f. cycle		
	A.f. backb.	Maps	Time	A.f. backb.	Maps	Time	A.f. backb.	Maps	Time
9	$3.85\ 10^5$	236	0.32	$9.87\ 10^3$	236	0.03	$8.06\ 10^3$	148	0.01
12	$5.55\ 10^7$	4476	53.99	$2.46\ 10^5$	4463	0.71	$2.03\ 10^5$	1931	0.32
15	−	> 98100	−	$6.17\ 10^6$	97112	28.40	$5.13\ 10^6$	29164	8.81
18	−	−	−	$1.56\ 10^8$	2307686	−	$1.30\ 10^8$	501503	184.48

the difference is seen in the number of folded maps and the time to generate them. In Tab. 2, we give the time to generate *all unique maps* and their indices. Remark that the number of unique maps generated by trees, paths or cycles are different, since only the generation from trees is exhaustive. However, most of the maps with the largest minimum sparsity are generated with paths or cycles as backbones.

References

1. Holst, J., Trewin, A., Cooper, A.: Porous organic molecules. Nature Chem. **2**, 915–920 (2010)
2. Strozecki, Y.: Enumeration complexity and matroid decomposition. Ph.D thesis, Université Paris Diderot - Paris 7 (2010)
3. Li, G., Ruskey, F.: The advantages of forward thinking in generating rooted and free trees. In: ACM-SIAM Symposium on Discrete Algorithms, pp. 939–940 (1999)
4. Liskovets, V.: Enumeration of nonisomorphic planar maps. Selecta Math. Soviet. **4**, 304–323 (1985)
5. Cori, R., Vauquelin, B.: Planar maps are labelled trees. Canadian Journal Math. **33**, 1023–1042 (1981)
6. Brinkmann, G., McKay, B.D.: Fast generation of planar graphs. MATCH Commun. Math. Comput. Chem **58**, 323–357 (2007)
7. Brinkmann, G., Friedrichs, O.D., Lisken, S., Peeters, A., Van Cleemput, N.: Cage– a virtual environment for studying some special classes of plane graphs–an update. MATCH Commun. Math. Comput. Chem **63**, 533–552 (2010)
8. Barth, D., Boudaoud, B., Couty, F., David, O., Quessette, F., Vial, S.: Map generation for CO_2 cages. In: Computer and Information Sciences III, pp. 503–510. Springer (2013)
9. Holton, D.A., McKay, B.D.: The smallest non-hamiltonian 3-connected cubic planar graphs have 38 vertices. Journal of Combinatorial Theory, Series B **45**, 305–319 (1988)
10. Barth, D., David, O., Quessette, F., Reinhard, V., Strozecki, Y., Vial, S.: Efficient generation of stable planar cages for chemistry (2015). arXiv:1503.06610
11. Aldred, R.E., Bau, S., Holton, D.A., McKay, B.D.: Cycles through 23 vertices in 3-connected cubic planar graphs. Graphs and Combinatorics **15**, 373–376 (1999)
12. Hopcroft, J.E., Wong, J.K.: Linear time algorithm for isomorphism of planar graphs (preliminary report). In: ACM Symposium on Theory of Computing, pp. 172–184 (1974)
13. Kukluk, J.P., Holder, L.B., Cook, D.J.: Algorithm and experiments in testing planar graphs for isomorphism. Journal of Graphs Algorithms and Applications **8**, 313–356 (2004)
14. Weinberg, L.: A simple and efficient algorithm for determining isomorphism of planar triply connected graphs. Circuit Theory **13**, 142–148 (1966)
15. Garey, M., Johnson, D.: Computers and intractability: a guide to NP-completeness. WH Freeman and Company, San Francisco (1979)
16. Park, J.K., Phillips, C.A.: Finding minimum-quotient cuts in planar graphs. In: ACM Symposium on Theory of Computing, pp. 766–775 (1993)
17. Diestel, R.: Graph theory. 2005. Grad. Texts in Math (2005)

Accurate and Efficient Methods to Improve Multiple Circular Sequence Alignment

Carl Barton[1], Costas S. Iliopoulos[2], Ritu Kundu[2], Solon P. Pissis[2(✉)], Ahmad Retha[2], and Fatima Vayani[2]

[1] The Blizard Institute, Barts and The London School of Medicine and Dentistry, Queen Mary University of London, London, UK
c.barton@qmul.ac.uk

[2] Department of Informatics, King's College London, London, UK
{c.iliopoulos,ritu.kundu,solon.pissis,ahmad.retha, fatima.vayani}@kcl.ac.uk

Abstract. Multiple sequence alignment is a core computational task in bioinformatics and has been extensively studied over the past decades. This computation requires an implicit assumption on the input data: the left- and right-most position for each sequence is relevant. However, this is not the case for circular structures; for instance, MtDNA. Efforts have been made to address this issue but it is far from being solved. We have very recently introduced a fast algorithm for approximate circular string matching (Barton *et al.*, *Algo Mol Biol*, 2014). Here, we first show how to extend this algorithm for approximate circular dictionary matching; and, then, apply this solution with agglomerative hierarchical clustering to find a sufficiently good rotation for each sequence. Furthermore, we propose an alternative method that is suitable for more divergent sequences. We implemented these methods in BEAR, a programme for improving multiple circular sequence alignment. Experimental results, using real and synthetic data, show the high accuracy and efficiency of these new methods in terms of the inferred likelihood-based phylogenies.

1 Introduction

Sequence comparison is an important step in many important tasks in bioinformatics. It is used in many applications; from phylogenetic reconstruction to genome assembly. Traditional techniques for measuring approximation in sequence comparison are based on the notions of distance or similarity between sequences; and these are computed through sequence alignment.

The computation of optimal alignments is usually realised by dynamic programming techniques. These techniques are efficient for the case of pairwise sequence alignment, however, when it comes to the case of multiple sequence alignment finding the global optimum has been shown to be NP-hard [26]. Due to the difficulty of computing the optimal multiple sequence alignment, many heuristic methods have been proposed, including ClustalW [14], MMAFT [13], MUSCLE [6], and T-Coffee [19]. These heuristics require an implicit assumption

E. Bampis (Ed.): SEA 2015, LNCS 9125, pp. 247–258, 2015.
DOI: 10.1007/978-3-319-20086-6_19

on the input data: the left- and right-most position for each sequence is relevant; so should the sequences have a circular structure, there may be a number of issues with the produced alignment. For instance, taking different rotations of the sequences into account may lead to a better alignment. In other words, in circular sequences the chosen left- and right-most position for each sequence is irrelevant; and not considering rotations could lead to erroneously classifying similar sequences as dissimilar.

Circular DNA can be found in viruses; as plasmids in archaea and bacteria; and in the mitochondria and plastids of eukaryotic cells. Hence, algorithms on circular sequences can be important in the analysis of organisms containing such structures. An application of circular sequences has been in the context of reconstructing phylogenies using Mitochondrial DNA (MtDNA) [9,10,27]. MtDNA is generally conserved from parent to offspring, and so it can be used as an indicator of evolutionary relationships among species. The absence of recombination in these sequences allows it to be used as a simple test of phylogenetic evolution, and the high mutation rate leads to a powerful discriminative feature. However, when sequencing a DNA molecule, the position where a circular genome starts can be totally arbitrary. Due to this arbitrary definition, using conventional tools to align such sequences could yield an incorrectly high genetic distance between closely-related species. For instance, the linearised human (NC_001807) and chimpanzee (NC_001643) MtDNA sequences do not start in the same region. Their pairwise sequence alignment using EMBOSS Needle [20] gives a similarity of 85.1% and consists of 1195 gaps; taking different rotations of these sequences into account yields an alignment with a similarity of 91% and only 77 gaps.

The problem of finding the optimal linear alignment of two circular sequences of length n and $m \leq n$ under the edit distance model can be solved in time $\mathcal{O}(nm \log m)$ [16]. The same problem can trivially be solved in time $\mathcal{O}(nm^2)$ with scoring matrices and affine gap penalty scores [17]. Progressive multiple sequence alignments can be constructed by generalising the pairwise algorithm to profiles, similar to ClustalW [14]. This generalisation is implemented in cyclope [17], a programme for improving multiple circular sequence alignment: this method can be used to first obtain the best-aligned rotations, and then realign the rotations by using conventional alignment tools. The cubic runtime of the pairwise alignment algorithm becomes a bottleneck in practical terms. Other fast heuristic methods were also implemented in cyclope, but they are only based on the first two sequences from the input dataset. Another approach to improve multiple circular sequence alignment was implemented in CSA [7], a tool based on the generalised circular suffix tree construction [25]. The best-aligned rotations are found based on the largest chain of non-repeated blocks that belong to all sequences. Unfortunately, CSA is no longer maintained. It also has the restriction that there can be only up to 32 sequences in the input dataset, and that there must exist a block that occurs in every sequence only once.

Our Contribution. We have very recently introduced a fast average-case linear-time algorithm for approximate circular string matching [2]. A similar result was also presented in [11]; an average-case optimal algorithm was later presented in [3]. In this article, we first show how to extend the algorithm of [2] for approximate circular dictionary matching; and, then, apply this solution with agglomerative hierarchical clustering to find a sufficiently good rotation for each sequence. Furthermore, we propose an alternative method that is suitable for more divergent sequences. It uses state-of-the-art algorithms for solving the fixed-length approximate string matching problem [12]. We have implemented these two methods in BEAR (BEst-Aligned Rotations), a programme for improving multiple circular sequence alignment. Experimental results, using real and synthetic data, establish this improvement in terms of the likelihood-based inferred phylogenies, and show the high accuracy and efficiency of these new methods.

2   Definitions and Notation

To provide an overview of our results, we begin with a few definitions. We think of a *string* x of *length* n as an array $x[0..n-1]$, where every $x[i]$, $0 \leq i < n$, is a *letter* drawn from some fixed *alphabet* Σ of size $\sigma = \mathcal{O}(1)$. The *empty string* of length 0 is denoted by ε. A string x is a *factor* of a string y if there exist two strings u and v, such that $y = uxv$. Consider the strings x, y, u, and v, such that $y = uxv$. If $u = \varepsilon$, then x is a *prefix* of y. If $v = \varepsilon$, then x is a *suffix* of y. Let x be a non-empty string of length n and y be a string. We say that there exists an *occurrence* of x in y, or, more simply, that x *occurs in* y, when x is a factor of y. Every occurrence of x can be characterised by a position in y. Thus we say that x occurs at the *starting position* i in y when $y[i..i+n-1] = x$.

Given a string x of length m and a string y of length $n \geq m$, the *edit distance*, denoted by $\delta_E(x,y)$, is defined as the minimum total cost of operations required to transform one string into the other. For simplicity, we only count the number of edit operations, considering the cost of each to be 1 [15]. The allowed edit operations are as follows: *insertion* of a letter in y, not present in x; *deletion* of a letter in y, present in x; and *substitution* of a letter in y with a letter in x. We write $x \equiv_k^E y$ if the edit distance between x and y is at most k. Equivalently, if $x \equiv_k^E y$, we say that x and y have at most k *differences*. We refer to the *standard dynamic programming matrix* of x and y as the matrix defined by $D[i,0] = i$, $0 \leq i \leq m$, $D[0,j] = j$, $0 \leq j \leq n$

$$D[i,j] = \min \begin{cases} D[i-1,j-1] + (1 \text{ if } x[i-1] \neq y[j-1]) \\ D[i-1,j] + 1 \\ D[i,j-1] + 1 \end{cases} , 1 \leq i \leq m, 1 \leq j \leq n.$$

Similarly, we refer to the *standard dynamic programming algorithm* as the algorithm to compute the edit distance between x and y through the above recurrence in time $\mathcal{O}(mn)$. Given a non-negative integer threshold k for the edit distance, this can be computed in time $\mathcal{O}(mk)$ [24]. We say that there exists an *occurrence* of x in y with at most k differences, or, more simply, that x *occurs in* y with at most k differences, when $u \equiv_k^E x$ and u is a factor of y.

Given a string x of length n, we denote by $x^i = x[i..n-1]x[0..i-1]$, $0 < i < n$, the i-th *rotation* of x and $x^0 = x$. Consider, for instance, the string

$x = x^0 =$ ababababbc; this string has the following rotations: $x^1 =$ babababbca, $x^2 =$ ababbcab, $x^3 =$ babbcaba, $x^4 =$ abbcabab, $x^5 =$ bbcababa, $x^6 =$ bcababab, $x^7 =$ cabababb.

In this article, we consider the following two problems, the solutions for which form the basis of our contribution to multiple circular sequence alignment.

APPROXIMATECIRCULARDICTIONARYMATCHING
Input: a set $\mathcal{D} = \{x_0, x_1, \ldots, x_{d-1}\}$ of patterns of total length M, a text t of length n, such that $n > |x_j|$, $0 \leq j < d$, and an integer threshold $k < |x_j|$
Output: all factors u of t such that $u \equiv_k^E x_j^i$, $0 \leq j < d$, $0 \leq i < |x_j|$

FIXEDLENGTHAPPROXIMATESTRINGMATCHING
Input: a pattern x of length m, a text t of length n, an integer $\ell \leq m$, and an integer threshold $k < \ell$
Output: all factors u of t such that $u \equiv_k^E v$, where v is any factor of length ℓ of x

---

3   Algorithmic Toolbox

In this section, we give a generalisation of our algorithm for approximate circular string matching [2], denoted here by ACSM, and show that it can easily be modified to solve the problem of approximate circular dictionary matching under the edit distance model. We denote this new algorithm by ACDM. Algorithm ACDM follows a similar approach to ACSM but with a few key differences. We will make use of the following important fact and lemmas.

Fact 1 ([2]). *Let x be a string of length m. Any rotation of x is a factor of $x' = x[0 \mathinner{\ldotp\ldotp} m - 1]x[0 \mathinner{\ldotp\ldotp} m - 2]$; and any factor of length m of x' is a rotation of x.*

Lemma 1 ([1]). *Let x and $y = y_0 y_1 \ldots y_k$ be two strings, such that $y_0, y_1, \ldots, y_k$ are $k + 1$ non-empty strings and $x \equiv_k^E y$. Then there exists at least one string y_i, $0 \leq i \leq k$, occurring in x.*

Lemma 2 ([2]). *Let x be a string of length m. If we partition $x' = x[0 \mathinner{\ldotp\ldotp} m - 1]x[0 \mathinner{\ldotp\ldotp} m - 2]$ in $2k + 4$ fragments of length $\lfloor (2m - 1)/(2k + 4) \rfloor$ and $\lceil (2m - 1)/(2k + 4) \rceil$, at least $k + 1$ of the $2k + 4$ fragments are factors of any factor of length m of x'.*

We start by constructing the string $x_j' = x_j[0 \mathinner{\ldotp\ldotp} |x_j| - 1]x_j[0 \mathinner{\ldotp\ldotp} |x_j| - 2]$, for all $0 \leq j < d$. Then string x_j' is partitioned in $2k + 4$ fragments of length $\lfloor (2|x_j| - 1)/(2k + 4) \rfloor$ and $\lceil (2|x_j| - 1)/(2k + 4) \rceil$. By Lemma 2, at least $k + 1$ (application of Lemma 1) of the $2k + 4$ fragments are factors of any rotation of x_j (application of Fact 1). We then match the $2k + 4$ fragments of x_j', for all $0 \leq j < d$, against the text t using an Aho Corasick automaton [5]. Let $\mathcal{L}$ be a list of size Occ of tuples, where $< p_{x'_j}, \ell, j, p_t > \in \mathcal{L}$ is a 4-tuple such that: $p_{x'_j}$ is the position where the fragment occurs in x'_j; ℓ is the length of the corresponding fragment; j identifies the pattern the fragment was extracted from; and $0 \leq p_t < n$ is the position where the fragment occurs in t. These tuples, returned by the automaton, give us the necessary information to perform the verification step if a fragment is indeed matched. For each tuple $< p_{x'_j}, \ell, j, p_t > \in \mathcal{L}$, we try to extend to the right by computing the standard dynamic programming matrix of $x'_j[p_{x'_j} + \ell .. |x'_j|]$ and $t[p_t + \ell .. p_t + |x'_j| - p_{x'_j} - 1 + k]$. Similarly, we try to extend to the left. By merging the results of the left and the right extension, we can report all factors u of t such that $u \equiv^E_k x^i_j$, $0 \leq j < d$, $0 \leq i < |x_j|$.

Theorem 1. *Given a set $\mathcal{D} = \{x_0, x_1, \ldots, x_{d-1}\}$ of patterns of total length M drawn from alphabet Σ, $\sigma = |\Sigma|$, a text t of length $n > |x_j|$, where $0 \leq j < d$, drawn from Σ, and an integer threshold $k < |x_j|$, algorithm ACDM requires average-case time $\mathcal{O}((1 + \frac{dk^2|x_{\max}|}{\sigma^{\frac{2|x_{\min}|-1}{2k+4}}})n + M)$ to solve the APPROXIMATECIRCU-LARDICTIONARYMATCHING problem, where $x_{\min}$ and $x_{\max}$ are the minimum- and maximum-length patterns in $\mathcal{D}$, respectively.*

Proof. Constructing and partitioning the strings $x'_0, x'_1, \ldots, x'_{d-1}$ from $\mathcal{D}$ can be done trivially in time $\mathcal{O}(M)$. Building the Aho Corasick automaton of the $(2k+4)d$ fragments requires time $\mathcal{O}(M)$; and the search time is $\mathcal{O}(n + Occ)$. Computing two standard dynamic programming matrices for each occurrence of fragment $< p_{x'_j}, \ell, j, p_t >$ of x'_j requires time $\mathcal{O}(k|x_j|Occ)$ [24]. For each extended occurrence of some fragment, we may report, in the worst case, $\mathcal{O}(|x_j|)$ valid starting positions. The expected number of occurrences for any fragment of x'_j is $n/\sigma^{(2|x_j|-1)/(2k+4)}$ thus $(2k+4)n/\sigma^{(2|x_j|-1)/(2k+4)}$ for all $2k+4$ fragments. So the total expected number Occ of occurrences is $\sum_{j=0}^{d-1} \frac{(2k+4)n}{\sigma^{\frac{2|x_j|-1}{2k+4}}} = \mathcal{O}(\frac{dkn}{\sigma^{\frac{2|x_{\min}|-1}{2k+4}}})$, where $x_{\min}$ is the minimum-length pattern in $\mathcal{D}$. Since the expected number Occ of occurrences of the fragments in t is $\mathcal{O}(\frac{dkn}{\sigma^{\frac{2|x_{\min}|-1}{2k+4}}})$, algorithm ACDM requires average-case time $\mathcal{O}((1 + \frac{dk^2|x_{\max}|}{\sigma^{\frac{2|x_{\min}|-1}{2k+4}}})n + M)$, where $x_{\max}$ is the maximum-length pattern in $\mathcal{D}$. $\square$

Algorithm ACDM achieves average-case time $\mathcal{O}(n + M)$ iff

$$\frac{d(2k+4)k|x_{\max}|}{\sigma^{\frac{2|x_{\min}|-1}{2k+4}}}n \leq cn$$

for some fixed constant c. So we have

$$\frac{d(2k+4)k|x_{\max}|}{\sigma^{\frac{2|x_{\min}|-1}{2k+4}}} \leq c$$

$$\frac{d(2k+4)k|x_{\max}|}{c} \leq \sigma^{\frac{2|x_{\min}|-1}{2k+4}}$$

$$\log_\sigma(\frac{d|x_{\max}|}{c}) + \log_\sigma((2k+4)k) \leq \frac{2|x_{\min}| - 1}{2k+4}$$

$$\log_\sigma d + \log_\sigma |x_{\max}| - \log_\sigma c + \log_\sigma 2 + \log_\sigma(k+2) + \log_\sigma k \leq \frac{2|x_{\min}| - 1}{2k+4}.$$

Some simple rearrangement and by setting c such that $\log_\sigma c \geq \log_\sigma 2 + 1/4$ gives a sufficient condition for ACDM to achieve average-case time $\mathcal{O}(n + M)$:

$$(k+2)(\log_\sigma d + \log_\sigma |x_{\max}| + \log_\sigma(k+2) + \log_\sigma k) \leq |x_{\min}|.$$

We, therefore, obtain the following result which is relevant to the application of multiple circular sequence alignment.

Corollary 1. *Given a set $\mathcal{D} = \{x_0, x_1, \ldots, x_{d-1}\}$ of patterns of total length $M = dm$, such that $m = |x_0| = |x_1| = \ldots = |x_{d-1}|$, drawn from alphabet Σ, $\sigma = |\Sigma|$, a text t of length $n > |x_j|$, where $0 \leq j < d$, drawn from Σ, and an integer threshold $k = \mathcal{O}(m/\log M)$, algorithm ACDM requires average-case time $\mathcal{O}(n + M)$ to solve the* APPROXIMATECIRCULARDICTIONARYMATCHING *problem.*

Since this approach is online—it avoids the computation of global data structures over t—algorithm ACDM can be implemented in space $\mathcal{O}(M)$.

Corollary 2. *Given a set $\mathcal{D} = \{x_0, x_1, \ldots, x_{d-1}\}$ of patterns of total length M drawn from alphabet Σ, $\sigma = |\Sigma|$, a text t of length $n > |x_j|$, where $0 \leq j < d$, drawn from Σ, and an integer threshold $k < |x_j|$, algorithm ACDM requires space $\mathcal{O}(M)$ to solve the* APPROXIMATECIRCULARDICTIONARYMATCHING *problem.*

4   Implementation

Given a set of strings $x_0, x_1, \ldots, x_{d-1}$ as input, our objective is to compute an array R of size d, such that $R[j]$, for all $0 \leq j < d$, stores a sufficiently good rotation of x_j. Then $x_0^{R[0]}, x_1^{R[1]}, \ldots, x_{d-1}^{R[d-1]}$ could be used as the input dataset for a conventional multiple sequence alignment algorithm to obtain the alignment. In this section, we describe two heuristic methods to compute R: the first one is suitable for less divergent sequences and is based on algorithm ACDM (see Section 3) and standard agglomerative hierarchical clustering; the latter is suitable for more divergent sequences and is based on fixed-length approximate string matching [4,12] and standard agglomerative hierarchical clustering. For clarity of presentation, we assume that $m = |x_0| = |x_1| = \ldots = |x_{d-1}|$.

4.1   Method for Less Divergent Sequences

We start by applying algorithm ACDM with $\mathcal{D} := \{x_0, x_1, \ldots, x_{d-1}\}$, $t := x_0 \# x_1 \# \ldots \# x_{d-1}$, and an integer threshold $k < m$. The efficiency of the proposed method relies on the fact that common fragments (see Section 3) will be matched simultaneously, as strings that share a common prefix also share a corresponding set of ancestor nodes in the automaton. Given the output of algorithm ACDM, we construct a matrix M of size $d \times d$ of pairs, such that $M[i,j]$ stores (e, r), denoting that the edit distance between $x_i^{M[i,j].r}$ and x_j is $M[i,j].e$. This pair can be derived from the output of algorithm ACDM; i.e. $x_i^{M[i,j].r}$ occurs at position $j \times m + j$ of t with $M[i,j].e$ differences.

We then apply agglomerative hierarchical clustering with average linkage [22]. As the input dataset, we use a $d \times d$ matrix N which stores the pairwise edit-distance information stored in matrix M, i.e. $N[i,j] = M[i,j].e$. Consider $S = \{s_1, s_2, \ldots, s_q\}$ and $L = \{l_1, l_2, \ldots, l_p\}$ be clusters that require merging such that $q \leq p$, then we must also maintain the rotations array R as follows:

$$R[s_i] = \begin{cases} (R[s_i] + \rho) \mod m & : \rho > 0 \\ (R[s_i] + \rho + m) \mod m & : \rho \leq 0 \end{cases}$$

for all $1 \leq i \leq q$, such that, for some $1 \leq j \leq p$, $\sum_{i=0}^{q} M[s_i, l_j].e$ is minimal, and $\rho = M[s_i, l_j].r - (R[s_i] - R[l_j])$. We denote this method by LDS.

Example 1. Let the following input dataset.

$$x_0 : \text{GGGTCTA}$$
$$x_1 : \text{TCTAGAG}$$
$$x_2 : \text{CGCGTCT}$$

We start by applying algorithm ACDM with $\mathcal{D} := \{\text{GGGTCTA}, \text{TCTAGAG}, \text{CGCGTCT}\}$, $t := \text{GGGTCTA\#TCTAGAG\#CGCGTCT}$, and $k := 2$. x_0^3 occurs at position 8 with 1 difference. x_0^6 occurs at position 16 with 2 differences. x_1^4 occurs at position 0 with 1 difference. x_1^3 occurs at position 16 with 2 differences. x_2^1 occurs at position 0 with 2 differences. x_2^4 occurs at position 8 with 2 differences. We obtain matrices M and N below. We then apply standard agglomerative hierarchical clustering with average linkage. We first merge $S = \{1\}$ and $L = \{0\}$, set $R[1] := 4$ mod 7 = 4, and update matrix N accordingly. We then merge $S' = \{2\}$ and $L' = \{0, 1\}$ and set $R[2] := 1 \mod 7 = 1$. We therefore obtain the following rotation for each string as output of the proposed approach. This output can now be used as the input dataset for a conventional multiple sequence alignment algorithm.

(a) Matrix M				(b) Matrix N				(c) Output
	x_0	x_1	x_2		x_0	x_1	x_2	$x_0^0 : \text{GGGTCTA}$
x_0	(0,0)	(1,3)	(2,6)	x_0	0	1	2	$x_1^4 : \text{GAGTCTA}$
x_1	(1,4)	(0,0)	(2,3)	x_1	1	0	2	$x_2^1 : \text{GCGTCTC}$
x_2	(2,1)	(2,4)	(0,0)	x_2	2	2	0	

4.2 Method for More Divergent Sequences

In this section, we provide a brief description and analysis of an algorithm to solve the FIXEDLENGTHAPPROXIMATESTRINGMATCHING problem. We then show how approximate circular string matching can be reduced to the fixed-length approximate string matching problem. Finally, we provide an informal structure of our approach.

We consider the FIXEDLENGTHAPPROXIMATESTRINGMATCHING problem under the Hamming distance model. Let $D'[0 .. m, 0 .. n]$ be a DP matrix, where $D'[i, j]$ contains the Hamming distance between factor $x[\max\{0, i - \ell\} .. i - 1]$ of x and factor $t[\max\{0, j - \ell\} .. j - 1]$ of t, for all $1 \leq i \leq n$, $1 \leq j \leq m$. Crochemore, Iliopoulos, and Pissis devised an algorithm [4] that solves the FIXEDLENGTH-APPROXIMATESTRINGMATCHING problem under the Hamming distance model.

Theorem 2 ([4]). *Given a string x of length m, a string t of length n, an integer $\ell \leq m$, and the size of the computer word w, matrix D' can be computed in time $\mathcal{O}(m \lceil \ell/w \rceil n)$ and space $\mathcal{O}(m \lceil \ell/w \rceil)$.*

By using word-level parallelism, we are able to compute matrix D' efficiently. The algorithm requires constant time for computing each cell $D'[i, j]$ by using word-level operations, assuming that $\ell \leq w$. In the general case, it requires $\mathcal{O}(\lceil \ell/w \rceil)$ time. Hence, the algorithm requires time $\mathcal{O}(mn)$, under the assumption that $\ell \leq w$. The space complexity is only $\mathcal{O}(m)$ since each column of D' only depends on the immediately preceding column. The same result can also be obtained under the edit distance model [12, 18].

In order to compute array R, the idea is to apply an algorithm for fixed-length approximate string matching, for every pair of strings (x_i, x_j), with $x_i' = x_i[0 .. m - 1]x_i[0 .. \ell - 1]$ and $x_j' = x_j[0 .. m - 1]$, for some $1 \leq \ell \leq m$. Notice that cell $D'[p, q]$, say $D'[p, q] = e$, denotes that factor $x_i'[p - \ell .. p - 1]$ matches factor $x_j'[q - \ell .. q - 1]$ with e mismatches. Equivalently, suffix $x_i^{p \bmod m}[m - \ell .. m - 1]$ matches suffix $x_j^{q \bmod m}[m - \ell .. m - 1]$ with e mismatches. Hence, setting $\ell := m$ solves *exactly* the approximate circular string matching problem; however, setting ℓ to smaller values can be sufficient in practice for the considered application (see Section 5 in this regard). Given the output of this approach, for all pairs of strings, we construct a matrix M of size $d \times d$ of pairs, such that $M[i, j]$ stores (e, r), denoting that the *minimum* distance between *any* factor of length ℓ of $x_i^{M[i,j].r}$ and some factor of x_j is $M[i, j].e$. This step requires time $\mathcal{O}(d^2 m^2 \lceil \ell/w \rceil)$, which, in practice, is much faster than the $\mathcal{O}(d^2 m^3)$-time algorithm implemented in cyclope [17]. After constructing matrix M, we apply standard agglomerative hierarchical clustering, similar to LDS; we denote this method by MDS.

We implemented the LDS and MDS methods, under the edit and Hamming distance models, as programme BEAR to compute the BEst-Aligned Rotations for a set of circular input sequences. Notice that, the simple *distance* model is sufficient for this purpose as more complex scoring schemes will be applied afterwards, in any case, to obtain the multiple sequence alignment. The programme was implemented in the C programming language and developed under GNU/Linux operating system. It takes, as an input argument, a file in Multi-FASTA format, and any of the two methods, for more or less divergent sequences, can be used based on the expected pairwise diversity—LDS is faster for error ratios below 10%. It then produces a file in MultiFASTA format with the rotated sequences as output. The implementation is distributed under the GNU General Public License (GPL), and it is available at http://github.com/solonas13/bear, which is set up for maintaining the source code and the man-page documentation. A very important feature of the proposed methods, compared to cyclope, is that they both require space linear in the length of the sequences (see Corollary 2 and Theorem 2). Hence, we were also able to implement BEAR using the Open Multi-Processing (OpenMP) PI for shared memory multiprocessing programming to distribute the workload across the available processing threads without a large memory footprint.

Summary of Availability and Requirements

- Project name: BEAR
- Project home page: http://github.com/solonas13/bear
- Operating system: GNU/Linux
- Programming language: C with OpenMP
- Other requirements: compiler gcc version 4.6.3 or higher
- License: GNU GPL

5   Experimental Results

The experiments were conducted on a Desktop PC, using one core of Intel Xeon E5540 CPU at 2.5 GHz and 8GB of main memory under 64-bit GNU/Linux. All programmes were compiled with gcc version 4.8.2. Notice that *all* input datasets, multiple sequence alignments, and phylogenetic trees referred to in this section are publicly maintained at the same web-site.

To test the *reliability* of our programme, we performed the following experiment. First, we simulated a basic dataset of DNA sequences using INDELible [8]. The number of taxa, denoted by α, was set to 12; the length of the sequence generated at the root of the tree, denoted by β, was set to 250bp; and the substitution rate, denoted by γ, was set to 0.05 (LDS was used). We then generated another instance of the basic dataset containing one random rotation of each of the 12 sequences from the basic dataset. We repeated this random generation to obtain 50 such instances of the basic dataset. We then used these 50 datasets as input to BEAR; and the output of BEAR as input to ClustalW [14] to obtain 50 multiple sequence alignments. We then used RAxML [23] to infer the 50 respective phylogenetic trees under the maximum-likelihood criterion. Finally, we computed the pairwise Robinson-Foulds (RF) distance [21] between all pairs of these 50 trees. *All* pairwise RF distances between the 50 provided trees were 0 suggesting that our method is in fact reliable. We repeated the same experiment by setting the substitution rate to 0.35 (MDS was used with $\ell := 15$) to obtain the same results: all pairwise RF distances were 0.

To test the *accuracy* of our programme, we performed the following experiment. First, we simulated a basic dataset of DNA sequences using INDELible for different values of $< \alpha, \beta, \gamma >$. We also used the following parameters: a deletion

Table 1. Accuracy measurements based on relative pairwise RF distance using synthetic data

Dataset $< \alpha, \beta, \gamma, \delta, \epsilon >$	BEAR + ClustalW + RAxML	ClustalW + RAxML	cyclope + ClustalW + RAxML
$< 12, 2500, 0.05, 0.06, 0.04 >$	100%	88.89%	100%
$< 12, 2500, 0.20, 0.06, 0.04 >$	100%	77.77%	100%
$< 12, 2500, 0.35, 0.06, 0.04 >$	100%	55.56%	100%
$< 25, 2500, 0.05, 0.06, 0.04 >$	100%	95.45%	100%
$< 25, 2500, 0.20, 0.06, 0.04 >$	100%	50%	100%
$< 25, 2500, 0.35, 0.06, 0.04 >$	100%	72.72%	100%
$< 50, 2500, 0.05, 0.06, 0.04 >$	100%	93.62%	97.87%
$< 50, 2500, 0.20, 0.06, 0.04 >$	100%	100%	100%
$< 50, 2500, 0.35, 0.06, 0.04 >$	100%	19.15%	100%

rate, denoted by δ, of 0.06 *relative* to substitution rate of 1; and an insertion rate, denoted by ϵ, of 0.04 *relative* to substitution rate of 1. The parameters were chosen based on the genetic diversity standard measures observed for sets of MtDNA sequences from primates and mammals [7]. We generated another instance of the basic dataset, containing one random rotation of each of the α sequences from the basic dataset. We then used this dataset as input to BEAR (LDS was used for $\gamma = 0.05$; otherwise MDS was used with $\ell := 45$); and the output of BEAR as input to ClustalW to obtain a multiple sequence alignment. We then used RAxML to infer the respective phylogenetic tree T_1 under the maximum-likelihood criterion. We also inferred the phylogenetic tree T_2 by following the same pipeline but *without* using BEAR, as well as the phylogenetic tree T_3 by using the basic dataset as input of this pipeline. Hence, notice that T_3 represents the original tree. Finally, we computed the pairwise RF distance between: T_1 and T_3; and T_2 and T_3. Let us define *accuracy* as the difference between 1 and the relative pairwise RF distance. The results in Table 1 suggest the high accuracy of BEAR. Notice that 100% accuracy denotes a (relative) pairwise RF distance of 0. We repeated this procedure by using cyclope instead of BEAR obtaining similar and partially identical results.

To test the *efficiency* of our methods, we compared BEAR (MDS was used, due to divergence, with $\ell := 45$) to cyclope using real data. As input datasets, we used three sets of MtDNA sequences: the first set includes sequences of 16 primates; the second set includes sequences of 12 mammals; and the last one is a set of 19 distantly-related sequences (the 16 primates, plus the *Drosophila melanogaster*, the *Gallus gallus*, and the *Crocodylus niloticus*). The MtDNA genome size for each sequence in the datasets is between 16 and 20 Kbp. To ensure a fair efficiency comparison between the two programmes, we made sure that they both produce a *unique* phylogenetic tree (using the aforementioned pipeline) by computing the pairwise RF distance of the inferred trees. The results in Table 2 using a *single* core show that BEAR can accelerate the computations by more than a factor of 20 compared to cyclope, producing, via ClustalW and RAxML, *identical* trees.

Table 2. Elapsed-time comparison and pairwise RF distance using real data

Dataset	BEAR	cyclope	RF distance
First set (Primates)	2m11s	41m46s	0
Second set (Mammals)	1m15s	26m19s	0
Third set (Primates et al)	3m01s	61m35s	0

6   Final Remarks

In this article, our contribution is threefold:

1. We present an average-case algorithm for approximate circular dictionary matching, which requires linear time (Corollary 1) and linear space (Corollary 2) under realistic conditions, and may be of independent interest.
2. We present and make available BEAR, a programme for improving multiple circular sequence alignment.
3. We present experimental results establishing this improvement in terms of the inferred likelihood-based phylogenies; we show the high accuracy (Table 1) and efficiency (Table 2) of BEAR compared to the state-of-the-art.

Our immediate target is twofold:

1. We plan on extending BEAR by investigating and including different methods for improving multiple circular sequence alignment.
2. We plan on implementing a web service based on BEAR that may be used by researchers for performing analyses on molecular sequences with circular genome structure.

References

1. Baeza-Yates, R.A., Perleberg, C.H.: Fast and practical approximate string matching. Information Processing Letters **59**(1), 21–27 (1996)
2. Barton, C., Iliopoulos, C.S., Pissis, S.P.: Fast algorithms for approximate circular string matching. Algorithms for Molecular Biology **9**(1), 9 (2014)
3. Barton, C., Iliopoulos, C.S., Pissis, S.P.: Average-case optimal approximate circular string matching. In: Dediu, A.-H., Formenti, E., Martín-Vide, C., Truthe, B. (eds.) LATA 2015. LNCS, vol. 8977, pp. 85–96. Springer, Heidelberg (2015)
4. Crochemore, M., Iliopoulos, C.S., Pissis, S.P.: A parallel algorithm for fixed-length approximate string-matching with k-mismatches. In: Elomaa, T., Mannila, H., Orponen, P. (eds.) Ukkonen Festschrift 2010. LNCS, vol. 6060, pp. 92–101. Springer, Heidelberg (2010)
5. Dori, S., Landau, G.M.: Construction of Aho Corasick automaton in linear time for integer alphabets. Information Processing Letters **98**(2), 66–72 (2006)
6. Edgar, R.C.: MUSCLE: a multiple sequence alignment method with reduced time and space complexity. BMC Bioinformatics **5**(1), 113 (2004)
7. Fernandes, F., Pereira, L., Freitas, A.T.: CSA: An efficient algorithm to improve circular DNA multiple alignment. BMC Bioinformatics **10**(1), 1–13 (2009)

8. Fletcher, W., Yang, Z.: INDELible: A flexible simulator of biological sequence evolution. Molecular Biology and Evolution **26**(8), 1879–1888 (2009)

9. Fritzsch, G., Schlegel, M., Stadler, P.F.: Alignments of mitochondrial genome arrangements: Applications to metazoan phylogeny. Journal of Theoretical Biology **240**(4), 511–520 (2006)

10. Goios, A., Pereira, L., Bogue, M., Macaulay, V., Amorim, A.: mtDNA phylogeny and evolution of laboratory mouse strains. Genome Research **17**(3), 293–298 (2007)

11. Hirvola, T., Tarhio, J.: Approximate online matching of circular strings. In: Gudmundsson, J., Katajainen, J. (eds.) SEA 2014. LNCS, vol. 8504, pp. 315–325. Springer, Heidelberg (2014)

12. Iliopoulos, C.S., Mouchard, L., Pinzon, Y.J.: The max-shift algorithm for approximate string matching. In: Brodal, G.S., Frigioni, D., Marchetti-Spaccamela, A. (eds.) WAE 2001. LNCS, vol. 2141, pp. 13–25. Springer, Heidelberg (2001)

13. Katoh, K., Misawa, K., Kuma, K.I., Miyata, T.: MAFFT: a novel method for rapid multiple sequence alignment based on fast Fourier transform. Nucleic Acids Research **30**(14), 3059–3066 (2002)

14. Larkin, M., Blackshields, G., Brown, N., Chenna, R., McGettigan, P., McWilliam, H., Valentin, F., Wallace, I., Wilm, A., Lopez, R., Thompson, J., Gibson, T., Higgins, D.: Clustal W and Clustal X version 2.0 **23**(21), 2947–2948 (2007)

15. Levenshtein, V.I.: Binary codes capable of correcting deletions, insertions, and reversals. Tech. Rep. 8 (1966)

16. Maes, M.: On a cyclic string-to-string correction problem. Information Processing Letters **35**(2), 73–78 (1990)

17. Mosig, A., Hofacker, I.L., Stadler, P.F.: Comparative analysis of cyclic sequences: viroids and other small circular RNAs. In: Huson, D.H., Kohlbacher, O., Lupas, A.N., Nieselt, K., Zell, A. (eds.) German Conference on Bioinformatics. LNI, vol. 83, pp. 93–102. GI (2006)

18. Myers, G.: A fast bit-vector algorithm for approximate string matching based on dynamic programming. Journal of ACM **46**(3), 395–415 (1999)

19. Notredame, C., Higgins, D.G., Heringa, J.: T-Coffee: a novel method for fast and accurate multiple sequence alignment. Journal of Molecular Biology **302**(1), 205–217 (2000)

20. Rice, P., Longden, I., Bleasby, A.: EMBOSS: The European Molecular Biology Open Software Suite. Trends in Genetics **16**(6), 276–277 (2000)

21. Robinson, D., Foulds, L.: Comparison of phylogenetic trees. Mathematical Biosciences **53**(1–2), 131–147 (1981)

22. Sokal, R.R., Michener, C.D.: A statistical method for evaluating systematic relationships. University of Kansas Scientific Bulletin **28**, 1409–1438 (1958)

23. Stamatakis, A.: Raxml version 8: a tool for phylogenetic analysis and post-analysis of large phylogenies. Bioinformatics **30**(9), 1312–1313 (2014)

24. Ukkonen, E.: On approximate string matching. In: Karpinski, M. (ed.) Foundations of Computation Theory. LNCS, vol. 158, pp. 487–495. Springer, Berlin Heidelberg (1983)

25. Ukkonen, E.: On-line construction of suffix trees. Algorithmica **14**(3), 249–260 (1995)

26. Wang, L., Jiang, T.: On the complexity of multiple sequence alignment. Journal of Computational Biology **1**(4), 337–348 (1994)

27. Wang, Z., Wu, M.: Phylogenomic reconstruction indicates mitochondrial ancestor was an energy parasite. PLoS ONE **10**(9), e110685 (2014)

Solving k-means on High-Dimensional Big Data

Jan-Philipp W. Kappmeier[1], Daniel R. Schmidt[2], and Melanie Schmidt[2(✉)]

[1] Technische Universität Berlin, Berlin, Germany
kappmeier@math.tu-berlin.de
[2] Department of Computer Science, Carnegie Mellon University,
Forbes Avenue, Pittsburgh, PA 15213, USA
{schmidtd,melanie.schmidt}@cmu.edu

Abstract. In recent years, there have been major efforts to develop data stream algorithms that process inputs in one pass over the data with little memory requirement. For the k-means problem, this has led to the development of several $(1 + \varepsilon)$-approximations (under the assumption that k is a constant), but also to the design of algorithms that are extremely fast in practice and compute solutions of high accuracy. However, when not only the length of the stream is high but also the dimensionality of the input points, then current methods reach their limits.

We propose two algorithms, piecy and piecy-mr that are based on the recently developed data stream algorithm BICO that can process high dimensional data in one pass and output a solution of high quality. While piecy is suited for high dimensional data with a medium number of points, piecy-mr is meant for high dimensional data that comes in a very long stream. We provide an extensive experimental study to evaluate piecy and piecy-mr that shows the strength of the new algorithms.

Keywords: k-means clustering · Data streams · SVD

1 Introduction

Partitioning points into subsets (*clusters*) with similar properties is an intuitive, old and central question. *Unsupervised* clustering aims at finding structure in data without the aid of class labels or an experts opinion. It has many applications ranging from computer science applications like image segmentation or information retrieval to applications in other sciences like biology or physics where it is used on genome data and CERN experiments. For an overview on the broad subject, see for example the survey by Jain [14]. The *k-means problem* asks to cluster data such that the sum of the squared error is minimized. It has been studied since the fifties [18,22] and optimizing it is likely 'the most commonly used partitional clustering strategy' [15]. It measures the quality of a partitioning of points from $\mathbb{R}^d$ based on the squared Euclidean distance function. Each cluster in the partitioning is represented by a center, and the objective function is the sum of the squared distances of all points to their respective center.

The popularity of the k-means problem is underlined by the fact that the most popular algorithm for it, Lloyd's algorithm, was named one of the ten most influential algorithms in the data mining community by the organizers of

© Springer International Publishing Switzerland 2015
E. Bampis (Ed.): SEA 2015, LNCS 9125, pp. 259–270, 2015.
DOI: 10.1007/978-3-319-20086-6_20

the IEEE International Conference on Data Mining (ICDM) in 2008 [24]. Lloyd's algorithm [18] (independently by Steinhaus [22]) is a local search heuristic that converges to a local optimum in finitely many steps. The quality in terms of the sum of squared errors of the output of Lloyd's algorithm depends on the local optimum that is reached. Arthur and Vassilvitskii [4] propose the k-means++ method as an improved version of Lloyd's algorithm which computes a $\mathcal{O}(\log k)$-approximation in expectation. It chooses the initial solution randomly by iteratively sampling new centers from a probability distribution that makes it likely that most optimal centers have a close center in the start solution.

The k-means++ method therefore provides a great tool for solving the k-means problem in practice, with an (expected) worst-case guarantee, a very good practical performance and the advantage that it is very easy to implement. The theoretically best approximation algorithms for the k-means problem provide a constant factor approximation for the general case [16,23] and a $(1+\varepsilon)$-approximation (even in linear time) if k and ε are assumed to be constants [9].

For big data, running Lloyd's algorithm or k-means++ is less viable. Asymptotically, the running time of both algorithms is $\mathcal{O}(ndk)$ if the number of iterations is bounded to a constant. This looks convincing since a straightforward implementation of finding the closest center for a point takes $\Theta(dk)$ time, so even evaluating a solution then has running time $\Theta(ndk)$. Additionally, the input size is already $\mathcal{O}(nd)$, so the running time is linear for constant values of k. However, both algorithms need random access to the data and iterate over it several times. As soon as the data does not fit into main memory, the algorithms do thus not scale very well. For example, k-means++ needed over seven hours to compute 50 centers for a 54-dimensional data set (*Covertype*) with half a million points [2].

A natural strategy to cope with this problem is to summarize the data before running the respective algorithm. A famous example for this is BIRCH [25], a SIGMOD Test of Time Award winning algorithm that computes a summary by one pass over the input data and then clusters the points in the summary. BIRCH is very fast and thus enables the processing of large data sets. However, the quality in terms of the sum of squared errors can be low [2,11].

A more recent development is the design of fast data stream algorithms based on *coresets*. A coreset S of a point set P is a weighted summary of P that maintains a strong quality guarantee: For any choice C of k centers, the k-means costs of the clustering induced by C on S are within an $(1 + \varepsilon)$-factor of the k-means clustering that C induces on P. Thus, executing any k-means algorithm on the coreset gives a good approximation of what the same algorithm would have produced on P. Coresets are generally designed with a focus on strong theoretic bounds, but can be made viable in practice with slight heuristic changes.

Two examples for this approach are StreamKM++ [2] and BICO [11]. The latter enables the processing of data sets with millions of points in less than an hour. The above mentioned test case needs 27 seconds for BICO and ten minutes for StreamKM++ compared to the seven hours for k-means++, and larger instances show even higher acceleration. The quality of the results is competitive, and the memory usage is low (polylogarithmic in theory and $\mathcal{O}(k)$ in experiments). The C++ source code of both algorithms is available online.

For data sets with up to around 100 dimensions, this is a pleasant state of affair. However, both the analysis of the running time and memory requirement of StreamKM++ and BICO assume that the dimension is a constant. At least for BICO, this is not a theoretically imposed restriction, but does indeed correspond to an unfavorable dependency on the dimension. The reason is that BICO covers the input data by spheres (in order to summarize all points in the same sphere by one point). When the number of spheres is too large, a rebuilding step reduces it by merging some spheres. Covering a set by spheres gets increasingly difficult as the dimension gets higher, which results in several rebuilding steps of BICO, and in a higher running time. On the theoretical level, however, there are several results saying that it is possible to compute a coreset of a point set in one pass and with low memory requirements. For example, Feldman and Langberg [9] propose a one-pass algorithm that computes a coreset with storage size of $\mathcal{O}\left(kd\log^4 n\varepsilon^{-3}\log 1/\varepsilon\right)$. It is thus theoretically possible to compute coresets which scale well with the dimension, but there is no practical algorithm yet that achieves a high quality summary and can cope with very high dimensional, large data sets.

Our Contribution. We develop two new algorithms, *piecy* and *piecy-mr* that can deal with high-dimensional big data. For that, we combine BICO with a dimensionality reduction. This reduction is done by projecting onto the best fit subspace (of a parameterized dimension) which can be computed by the *singular value decomposition* (SVD). This is theoretically supported by recent results [6, 10] that say that projecting onto the best fit subspace of dimension $\lceil k/\varepsilon \rceil$ and then solving the k-means problem gives a $(1 + \varepsilon)$-approximation guarantee. We find that $3k/2$ dimensions are often sufficient to give highly accurate results.

The next challenge is to intertwine the dimensionality reduction with the coreset computation in order to do both in one pass over the data. The first algorithm, *piecy*, reads chunks (pieces) of the data and processes, reduces the dimensionality of each chunk and feeds the resulting points into BICO. The drawback of this approach is that the total dimensionality of the complete point set that is fed into BICO increases with the number of pieces. For large data sets and high input dimension, this approach will eventually run into the same trouble as BICO (but for larger and higher dimensional data sets than BICO). In *piecy-mr*, we resolve this potential limitation by adapting a technique called *Merge-and-Reduce* [13]. It is a method that shows that any coreset computation can be turned into a one-pass algorithm at the cost of additional polylogarithmic factors. We change it to take advantage of the fact that we use a coreset computation (BICO) which already *is* a one-pass algorithm.

As intermediate steps of our work, we evaluate two implementations for the singular value decomposition, [1] and redSVD [20], comparing their speed and quality, and we extend BICO to process weighted inputs (which is necessary for our piecy-mr approach).

Full Version. The full version of this paper can be found at [17].

2   The Algorithms

In the following, we describe BICO and our two new algorithms, *piecy* and *piecy-mr*. For a point set P, we denote the centroid of P by $\mu(P) := \sum_{x \in P} x / |P|$.

BICO. Being a streaming algorithm, BICO cannot store all the input points. Instead the idea of BICO is to cover the input data with spheres and to then replace the points inside of each sphere by sufficient statistics which are stored. The statistics have to be updated on the fly whenever a new point is read. BICO requires three numbers per sphere S: The number of points $|S|$ in S, their sum $\sum_{x \in S} x$ and the sum $\sum_{x \in S} x^t x$ of their squared length. This triple is called a *clustering feature*. This data can be updated in constant time and by the well-known formula $\sum_{x \in P} ||x - c||^2 = |P| \cdot ||\mu(P) - c||^2 + \sum_{x \in P} ||x - \mu(P)||^2$, which holds for any point set P, its centroid $\mu(P)$ and any $c \in \mathbb{R}^d$, it is enough to exactly compute the cost between S and *one* center $c \in \mathbb{R}^d$. In order to keep the overall error of this compression small, the sphere covering must satisfy certain properties that BICO guarantees by managing the sphere in a well-organized cluster tree. Each time a new point is read, the algorithm decides to which clustering feature it should be added. The worst-case running time for this operation is $\Theta(m)$ per point (where m is the core-set size), however, BICO includes several heuristics that reduce the running time of the operation to $O(1)$ in many cases. Still, the quality of the heuristics depends on the dimension of the point set. Whenever the number of spheres exceeds m, BICO performs a rebuilding step on the cluster tree. For high-dimensional data sets, this rebuilding step may occur more often, resulting in a higher running time.

Piecy. Our aim is to compute coresets for large high-dimensional data sets by using BICO and dimensionality reduction techniques, but in *only one pass* over the data. *Piecy* pursues the idea of running only a single instantiation of BICO and subsequently feeding it with chunks of low dimensional points. Thus, piecy reads a piece of p points, reduces its intrinsic dimension and inputs the resulting points into BICO.

   Choice of dimensionality reduction technique and number of dimensions. We use the projection to the best fit subspace of dimension ℓ, where ℓ is a parameter to be optimized. The best fit subspace can be computed by using the *singular value decomposition*. The theoretical background of this approach is that projecting to best fit subspaces and then solving k-means (optimally) yields a good approximation [6,7]. When projecting to k dimensions, a 2-approximation is guaranteed, while projecting to $\lceil k/\varepsilon \rceil$ guarantees a $(1+\varepsilon)$-approximation. Thus, we test values between k and moderate multiples of k to get a reasonable compromise between approximation factor and running time.

   Using SVD to project to the best fit subspace. When we say that we use 'the' SVD, we mean the SVD of the matrix $A \in \mathbb{R}^{n \times d}$ where the input points are stored in the columns. The SVD of A has the form $A = UDV^T$ for matrices $U \in \mathbb{R}^{n \times n}, D \in \mathbb{R}^{n \times d}, V \in \mathbb{R}^{d \times d}$, where U and V are unitary matrices and D is a diagonal matrix. The matrix V contains the right singular vectors of A. The projection of (the points stored in) A to the best fit subspace of dimension ℓ is the matrix $A_\ell = UD_\ell V^T$, where D_ℓ is obtained by replacing all but the first ℓ diagonal elements by zero. Notice that the resulting matrix still contains d-dimensional points, but their *intrinsic* dimension is reduced to ℓ. This still helps since the ℓ-dimensional point set is easier to cover for BICO.

Computation of the SVD. Numerically stable computation of the singular value decomposition is a research field of its own. Basic methods that compute the *full* SVD, e.g. U, V and D, have a running time of $\Omega\left(nd\min\left(n,d\right)\right)$. This full SVD can be used by dropping the appropriate entries of D to obtain a matrix D_ℓ and evaluating the matrix product $UD_\ell V^t$ to obtain the projection onto the best fit subspace of dimension ℓ. However, a variety of more efficient algorithms have been developed for this specific task, which are known as algorithms for the *truncated* SVD that computes a decomposition $A_\ell = U_\ell D_\ell V_\ell^t$ directly without computing the full SVD of A. Additionally, random variations are known that reduce the running time sufficiently at the cost of a small error. Mahoney [19] gives a very nice overview on different methods to compute the singular value decomposition, then continuing with a detailed view on randomized methods and also discussing practical aspects. For this work, we use an implementation that is based on the randomized algorithm presented in [12] that multiplies A with a randomly drawn matrix to reduce the number of its columns before computing the SVD. The implementation is called *redSVD* [20]. In addition to reducing the number of columns, it also reduces the number of rows before computing the SVD. Below, we experimentally compare the performance of redSVD to the performance of the *lapack++* implementation of the full SVD computation.

Parameters. The authors of BICO propose using a coreset size of $200k$ for BICO, which we adopt. That given, there are two parameters to be chosen: The size of the pieces that are the input for one SVD, and the number of dimensions we project to. As we argued above, the latter should be at least k and not more than a reasonable multiple of k.

Memory requirement. At each point in time, we store at most one piece of the input, one SVD object and one BICO object. The memory requirement of BICO is proportional to the output size, i.e., to $200k$.

Obtaining a solution. Running *piecy* computes a summary of the input points. In order to obtain an actuall solution for the k-means problem, we run k-means++[4] on the summary.

Piecy-MR. Notice that each chunk of data that is processed by piecy adds (in the worst case) m dimensions to the intrinsic dimension of the point set that is stored by the BICO instance, as long as the maximum dimension is reached. For large data sets, this is unfavorable.

Helpful coreset properties. Assume that S_1 and S_2 are coresets for points sets P_1 and P_2, i.e., their weighted cost approximates the weighted cost of P_1 or P_2, respectively, for any possible solution, and up to an ε-fraction. Then the weighted cost of their union $S_1 \cup S_2$ approximates the cost of $P_1 \cup P_2$ for any solution up to an ε-fraction as well. Furthermore, if we use a coreset construction to reduce $S_1 \cup S_2$ to a smaller set (since $|S_1 \cup S_2|$ will be larger than the size of one coreset), then we obtain a coreset for $P_1 \cup P_2$. The error gets larger but is bounded by a (3ε)-fraction of the cost of $P_1 \cup P_2$ (which can be compensated by choosing a smaller ε to begin with).

The Merge-and-Reduce technique. Assume for a moment that our aim is solely to compute a coreset with no thoughts about the intrinsic dimension of the points, but given a coreset computation that needs random access to the data. Then an intuitive approach is to read chunks of the data, computing a coreset for each chunk and joining it with previous corsets, until the union becomes too large. Then we could reduce the union by another coreset construction. The problem with this approach is that the first chunk of the data will participate in all following reduce steps, making the error unnecessary high. The Merge-and-Reduce technique [5] (for clustering for example used in [3,13]) organizes the merge and reduce steps in a binary tree such that each point takes part in at most $\mathcal{O}(\log n)$ reduce steps for a stream of n points.

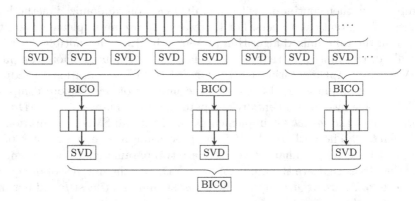

Fig. 1. The Merge-and-Reduce style tree built by piecy-mr with an exemplary *piece size* of 5 and a *number of pieces* of 3. *piece size* many points are fed into an SVD. The result of the SVD contains the same number of points but has a smaller intrinsic dimension. It is then fed into an instantiation of the BICO algorithm. After *number of pieces* many chunks, the BICO algorithm computes a coreset of size *piece size*. Then we continue on the next layer. On each layer, the number of points is reduced by a factor of *number of pieces*. We continue to call the SVD on each layer to keep the intrinsic dimension of the point set small.

Our computation tree. The coreset construction that we use, BICO, does not require random access to the data, so we have a different problem. We wish to keep the dimension of the input data small. Assume we would consider this problem independently from the coreset computation, by just computing the SVD of chunks of the data and keeping the reduced points in memory (maybe performing a second pass over the data to compute the coreset). This is infeasible since the number of points is not reduced and hence we would store the complete data set (with a lower intrinsic dimension). Imagine even that at each point in time, an oracle could provide us with the best fit subspace of dimension ℓ of all points seen so far. We could still not easily use this information since the best fit subspace would change over time. So if we use one instance of BICO, and input each point into it, projected to the best fit subspace of all points seen so far, then we would still get a high intrinsic dimension for the points stored in BICO.

By also embedding BICO into the Merge-and-Reduce tree, we solve these problems. The first way of doing this would be to view the two steps of reducing the dimension and entering the points into BICO as one coreset computation, and just embed this into the Merge-and-Reduce technique. However, this has the drawback that we perform the same number of dimensionality reductions as we use BICO for reducing sets to smaller sets. We do, however, expect that the union of multiple dimensionality reduced sets will not immediately have a high intrinsic dimension. In particular if the data evolves over time, then multiple consecutive pieces of the input data will have approximately the same best fit subspace (but over time, the subspace will change). We add more flexibility to the algorithm by running more than one copy of BICO, while allowing that more than one SVD output is processed by the same BICO instance. The actual computation tree is visualized in Figure 1.

Parameters. The algorithm has three parameters, the dimension that the SVD reduces to, the *piece size* which is the number of points that are read as input for one SVD computation, and the *number of pieces*, which is the number of SVD outputs that are processed by one instance of BICO. When BICO reaches the limit, the computed coreset is given to a SVD instance and then entered into a BICO on a higher level. It is convenient to set the piece size to $200k$, which also means that BICO computes a summary of size $200k$, the summary size suggested in the original BICO publication.

Memory requirement. We store one BICO element for each level of the computation tree. The degree of the tree is equal to the number of pieces b, so we have $\log_b n$ levels. At each point in time, there is at most one SVD object in the memory since there is always at most one SVD computation at the same time. If the piece size is equal to $200k$, then the memory requirement of each BICO element is proportional.

Weighted BICO. In the original implementation, BICO processes unweighted input points. In the *piecy-mr* computation tree, the instances of BICO on higher levels of the computation tree have to process weighted inputs (since the coreset points are weighted). Thus, we extended the source code of BICO to work for weighted inputs. For an input point x with weight w, we have to simulate what BICO would do for w copies of x. The main observation is that in most routines of BICO, multiple copies of the same point can be treated as one. For example, finding the closest reference point that is currently in the data structure can be done once and the result is then valid for all copies of x. Additionally, if we decide to open a new clustering feature with x as the reference point, we can insert all (not yet inserted) copies into this clustering feature at no cost.

What we have to adjust is the insertion process into already existing clustering features, and the initial values for new clustering features. Setting the correct values for a new clustering feature is straightforward: The new clustering feature has reference point x, its sum of points is $w \cdot x$, the sum of squares is $w \cdot x^2$ and the number of points stored in the feature is w. When we add w copies of a point x to an existing clustering feature with centroid μ and s points in it, then the actual increase of the error due to this is $s \cdot \|\mu - \mu_n\|^2 + w\|x - \mu_n\|^2 = s \cdot \left\|\mu - \frac{s\mu + wx}{s+w}\right\|^2 + w\left\|x - \frac{s\mu + wx}{s+w}\right\|^2 = \frac{sw^2}{(s+w)^2}\|x - \mu\|^2 + \frac{ws^2}{(s+w)^2}\|x - \mu\|^2 = \frac{sw}{s+w}\|x - \mu\|^2$ where we denote the new centroid after adding w copies of x by μ_n. We conclude that the total error made in the feature after inserting w points is $c + \frac{sw}{s+w}\|x - \mu\|^2$, where c denotes the original error made in the feature.

The original BICO implementation would have inserted the w copies sequentially into the clustering feature until the features threshold error of T would have been surpassed. It actually uses $\|x - \mu\|^2$ to measure the additional error and thus overestimates it. When adding single points, the effect of this overestimation decreases with each added point such that this works well for BICO. In the weighted version, however, using $w \cdot \|x - \mu\|^2$ is can be off by a large margin.

Instead, we compute how many copies w' of x can be inserted into the feature without surpassing the threshold, which is $w'(s\|x - \mu\|^2 - T + c) \leq sT - sc$. If $s \cdot \|x - \mu\|^2 - T + c \leq 0$, the threshold will not be reached for any $w' \geq 0$. and we insert all w copies. Otherwise, we insert $w' = \min\{w, (sT - sc)/(s\|x - \mu\|^2 - T + c)\}$ copies of x. If the threshold is reached before all w copies of x are inserted, i.e., if $w' < w$, we continue recursively as in the original BICO implementation.

Best Fit Subspace for Weighted Points. The singular value decomposition of a matrix is defined in an unweighted fashion, yet we want to use it for reducing the dimensionality of the weighted coreset points that result from BICO runs. Actually, we want to project the points to the best fit subspace of the point set where each point is replaced by several copies of itself according to its weight. Translated into the matrix notation, this means that we want to compute the projection of A to the best fit subspace of dimension ℓ of a matrix F which contains multiple copies of the points from A according to their (integral) weight[1].

Certainly, we do not want to actually create F. Instead, we construct a matrix A' where each row A_{i*} is replaced by $\sqrt{w_i}A_{i*}$ where w_i is the weight of the ith point. By linear algebra, we can verify that for each pair of left and right singular vectors u and v of F with singular value σ, there exists a vector u' such that u' and v are a pair of left and right singular vectors of A' for the same singular value. The reverse direction also holds. Thus, A' and F have the same best fit subspace and we can compute the SVD of A' in order to obtain it. After obtaining A'_ℓ, we divide each row i by $\sqrt{w_i}$ to get the projection of the points

---

[1] The weights that are computed by BICO are always integral. In fact, they sum up to the number of points BICO has processed.

in A. Their weight does not change. Notice that we cannot replace weighted points by some multiplied version when we input the points into BICO since the clustering behaviour of a weighted point differs from the clustering behaviour of any multiple (imagine a center that lies at the weighted point, so that it has no cost – but any multiplied point would have).

3   Experiments

The experiments were performed in three settings. For class I, all source codes were compiled using gcc 4.9.1, and experiments were performed on 20 identical machines with a 3.2 GHz AMD Phenom II X6 1090T processor and 8GiB RAM. For class II, all source codes were compiled with gcc 4.8.2 and all experiments were performed on 7 identical machines with a 2.8 GHz Intel E7400 processor and 8 GiB RAM. In class III, all source codes were compiled with gcc 4.9.1 and all experiments were performed on one machine with a 2.6 GHz Intel Core i5-4210M CPU processor and 16 GiB RAM. Our testbed consists of the following benchmark instances.

CalTech128. The `Caltech128` instance was created from the Caltech101 image database [8] and consists of 128 SIFT descriptors, resulting in 128 dimensions and about 3.1 million points. The instance was used in [11] for BICO benchmarks and was provided to the authors in private communication.

StructuredWithNoise. The `StructuredWithNoise` instances is to hide $\ell \in \mathbb{N}$ random point sets of $y \in \mathbb{N}$ points in $\mathbb{R}^d$. To build cluster $i \in \{1, \ldots, \ell\}$, select x dimensions $D_i = \{d_1, \ldots, d_x\} \subseteq \{1, \ldots, d\}$ uniformly at random. Then build the y points for cluster i: For point j, choose the coordinates corresponding to D_i uniformly at random from $[-10, 10]$. Select the remaining coordinates, i.e., the noise, uniformly at random from $[-1/2, 1/2]$. We get $\ell \cdot y$ points in d dimensions.

LowerBound. Arthur and Vassilvitskii [4] propose the following class of instances where the `kmeans++` algorithm can achieve no better approximation than $\Omega(\log n)$ in expectation. Define the (affine) (k, Δ)-simplex as the convex combination of the k unit vectors $e_1, \ldots, e_k$ in $\mathbb{R}^k$, scaled by $\Delta > 0$. Now, embed such a (k, Δ)-simplex $\mathfrak{S}$ in the first k dimensions of $\mathbb{R}^{k+n}$. Then use the remaining n dimensions of $\mathbb{R}^{k+n}$ to place a $(n/k, \delta)$-simplex S_i in each vertex i of $\mathfrak{S}$ such that all S_i use disjoint dimensions. We use a generator by Stallmann [21] to generate instance of this type.

We repeated all experiments at least five times.

Redsvd as a Replacement for the Lapack++ SVD. Replacing the exact SVD computation in our algorithm by an approximate one as outlined in Section 2 can only work if the approximation is fast and provides reliable results. We found that the error made by redSVD is indeed very small (at most 7%) while computation times become significantly faster: instances with 30,000 rows in 1000 columns can still be solved by redSVD in about 3s while `lapack++`'s takes 3000s on the same instace. RedSVD was able to compute approximate SVDs of matrices with 500,000 rows and 500 columns in 40s.

3.1   Performance of BICO, Piecy and Piecy-MR

All numerical results discussed in this section are contained in the full version [17]. We use BICO as a base line to compare our results. Notice that we use the current version of the source code from the BICO website. In contrast to the version used in [10], this version has varying running times. This shows both in the BICO experiments itself as in the experiments for piecy and piecy-mr since they both use BICO. Obviously, piecy and piecy-mr will improve when the source code of BICO is updated. For this reason, we will pay most attention to the median of the running times and not the average running time. We denote the number of points by n, the dimension by d and the number of centers by k.

Piecy. For piecy, we test the influence of two parameters, the piece size, abbreviation ps, and the number of dimensions to which we project the points, abbreviation svd. We computed an extensive number of test cases for the data set CalTech128 to study the influence of the parameters. For $k = 5, 10, 50$, piecy is *always* faster than BICO. Larger values of svd increase the running time, which is expected, but it stays below the running time of BICO for these test cases. The accuracy of piecy is high, in particular for larger svd values. At $k = 100$, the situation starts to change as there are three test cases where piecy is slower than BICO. For $k = 250, 1000$ the results by piecy become somewhat unpredictable. Notice that the number of centers is here higher than the input dimension of the points (which is 128). Thus, piecy cannot gain anything from projecting to a number of dimensions $\geq k$, and the SVD processing becomes overhead. It is thus clear that piecy does not perform as well on these test cases.

On the Random instance, piecy performs rather badly. The instance is large (one million points with 1000 dimensions, i.e., a total of 10^9 input numbers). In this case, most of the advantage due to the dimensionality reduction is lost because too many pieces are processed and contribute to the intrinsic dimension of the point set that is given to BICO. A similar behavior can be observed for the three largest StructuredWithNoise data sets. In particular when n reaches a million points, piecys running time goes up.

On the smaller LowerBound test cases though, piecy again outperforms BICO's running time. The LowerBound instances have a huge dimension of 10^5 but the number of points is also bounded by 10^5. Thus, there is less time for piecy to accumulate too many intrinsic dimensions.

Piecy-mr. Piecy-mr also uses ps, the piece size, as a parameter, as well as svd, the number of dimensions to project to. The additional parameter np is the number of pieces that processed into the same BICO instance.

For CalTech128, the overhead of piecy-mr does not pay off and it performs worse than piecy. On the LowerBound test cases, piecy-mr is always slightly faster than BICO and comparable to piecy. On the Random instances, piecy-mr is much faster than BICO, close to a factor of 2 on most test cases. This is in particular a much better running time than for piecy. The fact that Random has both a

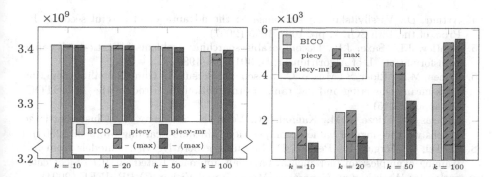

Fig. 2. Results for a `StructuredWithNoise` data set with 10^6 points in 10^3 dimensions. Left side reports quality, right side run times. Variances stem from different parameters.

huge number of points and a high dimension means that the strength of piecy-mr shows and is not dominated by the overhead of the computation tree. The study of the three `StructuredWithNoise` data sets confirms this behaviour. In all three cases, the running time of piecy-mr is much faster or at least comparable to BICO with very few exceptions. This effect is particularly clear for the largest data set with one million points and a dimension of 1000, showing the speed of piecy-mr for large high-dimensional data sets. Figure 2 shows results for this data set. Notice that the large variance for piecy and piecy-mr is due to very different parameter choices. The best parameter choices yield a significant speed-up, particularly for large values of k. See the full version [17] for details.

Conclusion. The experiments show the potential speed-up by using piecy and piecy-mr. When choosing the algorithm, one should take the dimensions of the input matrix into account. For large dimension but a moderate number of points, piecy is ideal since it reduces the dimension effectively with little overhead. For data sets where the dimension is high and the number of points is also high, the additional overhead of piecy-mr pays off.

Acknowledgments. We thank Cameron Musco and Chris Schwiegelshohn for insightful discussions on the topic of this paper, Hendrik Fichtenberger and Lukas Pradel for sharing some pieces of source code and Jan Stallmann and Ren Grzeszick for providing the `LowerBound` and `CalTech` data sets.

References

1. C++ library: Lapack++ v2.5.4. (2010). http://sourceforge.net/projects/lapackpp/ (accessed: February 8, 2015)
2. Ackermann, M.R., Märtens, M., Raupach, C., Swierkot, K., Lammersen, C., Sohler, C.: Streamkm++: A clustering algorithm for data streams. ACM J. of Exp. Algorithmics **17**, 1–30 (2012)
3. Agarwal, P.K., Har-Peled, S., Varadarajan, K.R.: Approximating extent measures of points. J. of the ACM **51**(4), 606–635 (2004)

4. Arthur, D., Vassilvitskii, S.: k-means++: the advantages of careful seeding. In: Proc. of the 18th SODA, pp. 1027–1035 (2007)
5. Bentley, J.L., Saxe, J.B.: Decomposable searching problems i: Static-to-dynamic transformation. J. of Algorithms 1(4), 301–358 (1980)
6. Cohen, M.B., Elder, S., Musco, C., Musco, C., Persu, M.: Dimensionality reduction for k-means clustering and low rank approximation. In: Proc. of the 47th STOC, (to appear 2015)
7. Drineas, P., Frieze, A.M., Kannan, R., Vempala, S., Vinay, V.: Clustering large graphs via the singular value decomposition. Machine Learning 56, 9–33 (2004)
8. Fei-Fei, L., Fergus, R., Perona, P.: Learning generative visual models from few training examples: an incremental bayesian approach tested on 101 object categories. In: Workshop on Generative-Model Based Vision, CVPR. IEEE (2004)
9. Feldman, D., Langberg, M.: A unified framework for approximating and clustering data. In: Proc. of the 43th STOC, pp. 569–578 (2011)
10. Feldman, D., Schmidt, M., Sohler, C.: Turning big data into tiny data: constant-size coresets for k-means, PCA and projective clustering. In: Proc. of the 24th SODA, pp. 1434–1453 (2013)
11. Fichtenberger, H., Gillé, M., Schmidt, M., Schwiegelshohn, C., Sohler, C.: BICO: BIRCH meets coresets for k-means clustering. In: Proc. 21st ESA, pp. 481–492 (2013)
12. Halko, N., Martinsson, P.-G., Tropp, J.A.: Finding structure with randomness: Probabilistic algorithms for constructing approximate matrix decompositions. SIAM Review (SIREV) 53(2), 217–288 (2011)
13. Har-Peled, S., Mazumdar, S.: On coresets for k-means and k-median clustering. In: Proc. of the 36th STOC, pp. 291–300 (2004)
14. Jain, A.K.: Data clustering: 50 years beyond k-means. Pattern Recognition Letters 31(8), 651–666 (2010)
15. Jain, A.K., Dubes, R.C.: Algorithms for Clustering Data. Pr. Hall (1988)
16. Jain, K., Vazirani, V.V.: Approximation algorithms for metric facility location and k-median problems using the primal-dual schema and lagrangian relaxation. J. of the ACM 48(2), 274–296 (2001)
17. Kappmeier, J.-P.W., Schmidt, D.R., Schmidt, M.: Solving k-means on high-dimensional big data (2015). CoRR, abs/1502.04265
18. Lloyd, S.P.: Least squares quantization in PCM. Bell Lab. Tech. Memor (1957)
19. Mahoney, M.W.: Randomized algorithms for matrices and data. Foundations and Trends in Machine Learning 3(2), 123–224 (2011)
20. Okanohara, D.: C++ project: redsvd - RandomizED Singular Value Decomposition (2011). https://code.google.com/p/redsvd/ (accessed: February 2, 2015)
21. Stallmann, J.: Benchmarkinstanzen für das k-means Problem. TU Dortmund University, Bachelorarbeit (2014). In german
22. Steinhaus, H.: Sur la division des corps matériels en parties. Bulletin de l'Académie Polonaise des Sciences IV(12), 801–804 (1956)
23. Kanungo, T., Mount, D.M., Netanyahu, N.S., Piatko, C.D., Silverman, R., Wu, A.Y.: A local search approximation algorithm for k-means clustering. Comp. Geom. 28(2–3), 89–112 (2004)
24. Wu, X., Kumar, V., Quinlan, J.R., Ghosh, J., Yang, Q., Motoda, H., McLachlan, G.J., Ng, A.F.M., Liu, B., Yu, P.S., Zhou, Z.H., Steinbach, M., Hand, D.J., Steinberg, D.: Top 10 algorithms in data mining. Know. and Inf. Sys. 14(1), 1 37 (2008)
25. Zhang, T., Ramakrishnan, R., Livny, M.: BIRCH: A New Data Clustering Algorithm and Its Applications. Data M. and Know. Disc. 1(2), 141–182 (1997)

Transportation Networks

Public Transit Labeling

Daniel Delling[1], Julian Dibbelt[2]([✉]), Thomas Pajor[3],
and Renato F. Werneck[4]

[1] Sunnyvale, CA, USA
daniel.delling@gmail.com
[2] Karlsruhe Institute of Technology, Karlsruhe, Germany
dibbelt@kit.edu
[3] Microsoft Research, New York, NY, USA
tpajor@microsoft.com
[4] San Francisco, CA, USA
rwerneck@acm.org

Abstract. We study the journey planning problem in public transit networks. Developing efficient preprocessing-based speedup techniques for this problem has been challenging: current approaches either require massive preprocessing effort or provide limited speedups. Leveraging recent advances in Hub Labeling, the fastest algorithm for road networks, we revisit the well-known time-expanded model for public transit. Exploiting domain-specific properties, we provide simple and efficient algorithms for the earliest arrival, profile, and multicriteria problems, with queries that are orders of magnitude faster than the state of the art.

1 Introduction

Recent research on route planning in transportation networks [5] has produced several speedup techniques varying in preprocessing time, space, query performance, and simplicity. Overall, queries on road networks are several orders of magnitude faster than on public transit [5]. Our aim is to reduce this gap.

There are many natural query types in public transit. An *earliest arrival* query seeks a journey that arrives at a target stop t as early as possible, given a source stop s and a departure time (e. g., "now"). A *multicriteria* query also considers the number of transfers when traveling from s to t. A *profile* query reports all quickest journeys between two stops within a time range.

These problems can be approached by variants of Dijkstra's algorithm [13] applied to a graph modeling the public transit network, with various techniques to handle time-dependency [18]. In particular, the *time-expanded* (TE) graph encodes time in the vertices, creating a vertex for every *event* (e. g., a train departure or arrival at a stop at a specific time). Newer approaches, like CSA [12] and RAPTOR [11], work directly on the timetable. Speedup techniques [5] such as Transfer Patterns [4,6], Timetable Contraction Hierarchies [14], and ACSA [20] use preprocessing to create auxiliary data that is then used to accelerate queries.

Work done mostly while all authors were at Microsoft Research Silicon Valley.

© Springer International Publishing Switzerland 2015
E. Bampis (Ed.): SEA 2015, LNCS 9125, pp. 273–285, 2015.
DOI: 10.1007/978-3-319-20086-6_21

For aperiodic timetables, the TE model yields a *directed acyclic graph* (DAG), and several public transit query problems translate to reachability problems. Although these can be solved by simple graph searches, this is too slow for our application. Different methodologies exist to enable faster reachability computation [7,15,16,19,21–23]. In particular, the *2-hop labeling* [8] scheme associates with each vertex two labels (forward and backward); reachability (or shortest-path distance) can be determined by intersecting the source's forward label and the target's backward label. On continental road networks, 2-hop labeling distance queries take less than a microsecond [2].

In this work, we adapt 2-hop labeling to public transit networks, improving query performance by orders of magnitude over previous methods, while keeping preprocessing time practical. Starting from the time-expanded graph model (Section 3), we extend the labeling scheme by carefully exploiting properties of public transit networks (Section 4). Besides earliest arrival and profile queries, we address multicriteria and location-to-location queries, as well as reporting the full journey description quickly (Section 5). We validate our Public Transit Labeling (PTL) algorithm by careful experimental evaluation on large metropolitan and national transit networks (Section 6), achieving queries within microseconds.

2   Preliminaries

Let $G = (V, A)$ be a (weighted) *directed graph*, where V is the set of vertices and A the set of arcs. An arc between two vertices $u, v \in V$ is denoted by (u, v). A *path* is a sequence of adjacent vertices. A vertex v is *reachable* from a vertex u if there is a path from u to v. A *DAG* is a graph that is both directed and acyclic.

We consider *aperiodic* timetables, consisting of sets of stops S, events E, trips T, and footpaths F. *Stops* are distinct locations where one can board a transit vehicle (such as bus stops or subway platforms). *Events* are the scheduled departures and arrivals of vehicles. Each event $e \in E$ has an associated stop $\mathtt{stop}(e)$ and time $\mathtt{time}(e)$. Let $E(p) = \{e_0(p), \ldots, e_{k_p}(p)\}$ be the list (ordered by time) of events at a stop p. We set $\mathtt{time}(e_i(p)) = -\infty$ for $i < 0$, and $\mathtt{time}(e_i(p)) = \infty$ for $i > k_p$. For simplicity, we may drop the index of an event (as in $e(p) \in E(p)$) or its stop (as in $e \in E$). A *trip* is a sequence of events served by the same vehicle. A pair of a consecutive departure and arrival events of a trip is a *connection*. *Footpaths* model transfers between nearby stops, each with a predetermined walking duration.

A journey planning algorithm outputs a set of *journeys*. A journey is a sequence of trips (each with a pair of pick-up and drop-off stops) and footpaths in the order of travel. Journeys can be measured according to several criteria, such as arrival time or number of transfers. A journey j_1 *dominates* a journey j_2 if and only if j_1 is no worse in any criterion than j_2. In case j_1 and j_2 are equal in all criteria, we break ties arbitrarily. A set of non-dominated journeys is called a *Pareto set*. Multicriteria Pareto optimization is NP-hard in general, but practical for natural criteria in public transit networks [11,12,17,18]. A journey is *tight* if there is no other journey between the same source and target that dominates it in terms of departure and arrival time, e.g., that departs later and arrives earlier.

Given a timetable, stops s and t, and a departure time τ, the (s, t, τ)-*earliest arrival* (EA) problem asks for an s–t journey that arrives at t as early as possible and departs at s no earlier than τ. The (s, t)-*profile* problem asks for a Pareto set of all tight journeys between s and t over the entire timetable period. Finally, the (s, t, τ)-*multicriteria* (MC) problem asks for a Pareto set of journeys departing at s no earlier than τ and minimizing the criteria arrival time and number of transfers. We focus on computing the *values* of the associated optimization criteria of the journeys (i.e., departure time, arrival times, number of transfers), which is enough for many applications. Section 5 discusses how the full journey description can be obtained with little overhead.

Our algorithms are based on the 2-hop labeling scheme for directed graphs [8]. It associates with every vertex v a *forward label* $L_f(v)$ and a *backward label* $L_b(v)$. In a *reachability labeling*, labels are subsets of V, and vertices $u \in L_f(v) \cup L_b(v)$ are *hubs* of v. Every hub in $L_f(v)$ must be reachable from v, which in turn must be reachable by every hub in $L_b(v)$. In addition, labels must obey the *cover property*: for any pair of vertices u and v, the intersection $L_f(u) \cap L_b(v)$ must contain at least one hub on a u–v path (if it exists). It follows from this definition that $L_f(u) \cap L_b(v) \neq \emptyset$ if and only if v is reachable from u.

In a *shortest path labeling*, each hub $u \in L_f(v)$ also keeps the associated distance $\mathrm{dist}(u, v)$ (or $\mathrm{dist}(v, u)$, for backward labels), and the cover property requires $L_f(u) \cap L_b(v)$ to contain at least one hub on a *shortest* u–v path. If labels are kept sorted by hub ID, a *distance label query* efficiently computes $\mathrm{dist}(u, v)$ by a coordinated linear sweep over $L_f(u)$ and $L_b(v)$, finding the hub $w \in L_f(u) \cap L_b(v)$ that minimizes $\mathrm{dist}(u, w) + \mathrm{dist}(w, v)$. In contrast, a *reachability label query* can stop as soon as any matching hub is found.

In general, smaller labels lead to less space and faster queries. Many algorithms to compute labelings have been proposed [2,3,7,15,21,23], often for restricted graph classes. We leverage (as a black box) the recent RXL algorithm [9], which efficiently computes small shortest path labelings for a variety of graph classes at scale. It is a sampling-based greedy algorithm that builds labels one hub at a time, with priority to vertices that cover as many relevant paths as possible.

Different approaches for transforming a timetable into a graph exist (see [18] for an overview). In this work, we focus on the *time-expanded model*. Since it uses scalar arc costs, it is a natural choice for adapting the labeling approach. In contrast, the *time-dependent model* (another popular approach) associates functions with the arcs, which makes adaption more difficult.

3   Basic Approach

We build the time-expanded graph from the timetable as follows. We group all departure and arrival events by the stop where they occur. We sort all events at a stop by time, merging events that happen at the same stop and time. We then add a vertex for each unique event, a *waiting arc* between two consecutive events of the same stop, and a *connection arc* for each connection (between the corresponding departure and arrival event). The cost of arc (u, v) is $\texttt{time}(v) - \texttt{time}(u)$, i.e., the time difference of the corresponding events. To account for footpaths between two stops a and b, we add, from each vertex at stop a, a *foot arc* to the first reachable vertex at b (based on walking time), and vice versa. As events and vertices are tightly coupled in this model, we use the terms interchangeably.

Any label generation scheme (we use RXL [9]) on the time-expanded graph creates two (forward and backward) *event labels* for every vertex (event), enabling *event-to-event* queries. For our application *reachability* labels [21], which only store hubs (without distances), suffice. First, since all arcs point to the future, time-expanded graphs are DAGs. Second, if an event e is reachable from another event e' (i.e., $L_f(e') \cap L_b(e) \neq \emptyset$), we can compute the time to get from e' to e as $\texttt{time}(e) - \texttt{time}(e')$. In fact, *all* paths between two events have equal cost.

In practice, however, event-to-event queries are of limited use, as they require users to specify both departure *and* arrival times, one of which is usually unknown. Therefore, we discuss earliest arrival and profile queries, which *optimize* arrival time and are thus more meaningful. See Section 5 for multicriteria queries.

Earliest Arrival Queries. Given event labels, we answer an (s, t, τ)-EA query as follows. We first find the earliest event $e_i(s) \in E(s)$ at the source stop s that suits the departure time, i.e., with $\texttt{time}(e_i(s)) \geq \tau$ and $\texttt{time}(e_{i-1}(s)) < \tau$. Next, we search at the target stop t for the earliest event $e_j(t) \in E(t)$ that is reachable from $e_i(s)$ by testing whether $L_f(e_i(s)) \cap L_b(e_j(t)) \neq \emptyset$ and $L_f(e_i(s)) \cap L_b(e_{j-1}(t)) = \emptyset$. Then, $\texttt{time}(e_j(t))$ is the earliest arrival time. One could find $e_j(t)$ using linear search (which is simple and cache-friendly), but binary search is faster in theory and in practice. To accelerate queries, we *prune* (skip) all events $e(t)$ with $\texttt{time}(e(t)) < \tau$, since $L_f(e_i(s)) \cap L_b(e(t)) = \emptyset$ always holds in such cases. Moreover, to avoid evaluating $L_f(e_i(s))$ multiple times, we use *hash-based queries* [9]: we first build a hash set of the hubs in $L_f(e_i(s))$, then check the reachability for an event $e(t)$ by probing the hash with hubs $h \in L_b(e(t))$.

Profile Queries. To answer an (s, t)-profile query, we perform a coordinated sweep over the events at s and t. For the current event $e_i(s) \in E(s)$ at the source stop (initialized to the earliest event $e_0(s) \in E(s)$), we find the first event $e_j(t) \in E(t)$ at the target stop that is reachable, i.e., such that $L_f(e_i(s)) \cap L_b(e_j(t)) \neq \emptyset$ and $L_f(e_i(s)) \cap L_b(e_{j-1}(t)) = \emptyset$. This gives us the earliest arrival time $\texttt{time}(e_j(t))$. To identify the latest departure time from s for that earliest arrival event (and thus have a tight journey), we increase i until $L_f(e_i(s)) \cap L_b(e_j(t)) = \emptyset$, then add $(\texttt{time}(e_{i-1}(s)), \texttt{time}(e_j(t)))$ to the profile. We repeat the process starting from the events $e_i(s)$ and $e_{j+1}(t)$. Since we increase either i or j after each intersection test, the worst-case time to find all tight journeys is linear in the number of events (at s and t) multiplied by the size of their largest label.

4   Leveraging Public Transit

Our approach can be refined to exploit features specific to public transit networks. As described so far, our labeling scheme maintains reachability information for *all pairs* of events (by covering all paths of the time-expanded graph, breaking ties arbitrarily). However, in public transit networks we actually are only interested in *certain paths*. In particular, the labeling does *not* need to cover any path ending at a departure event (or beginning at an arrival event). We can thus discard forward labels from arrival events and backward labels from departure events.

Trimmed Event Labels. Moreover, we can disregard paths representing dominated journeys that depart earlier and arrive later than others (i. e., journeys that are not tight, cf. Section 2). Consider all departure events of a stop. If a certain hub is reachable from event $e_i(s)$, then it is also reachable from $e_0(s), \ldots, e_{i-1}(s)$, and is thus potentially added to the forward labels of all these earlier events. In fact, experiments show that on average the same hub is added to 1.8–5.0 events per stop (depending on the network). We therefore compute *trimmed event labels* by discarding all but the latest occurrence of each hub from the forward labels. Similarly, we only keep the earliest occurrence of each hub in the backward labels. (Preliminary experiments have shown that we obtain very similar label sizes with a much slower algorithm that greedily covers tight journeys explicitly [2,9].)

Unfortunately, we can no longer just apply the query algorithms from Section 3 with trimmed event labels: if the selected departure event at s does not correspond to a tight journey toward t, the algorithm will not find a solution (though one might exist). One could circumvent this issue by also running the algorithm from subsequent departure events at s, which however may lead to quadratic query complexity in the worst case (for both EA and profile queries).

Stop Labels. We solve this problem by working with *stop labels*: For each stop p, we merge all forward event labels $L_f(e_0(p)), \ldots, L_f(e_k(p))$ into a forward stop label $SL_f(p)$, and all backward event labels into a backward stop label $SL_b(p)$. Similar to distance labels, each stop label $SL(p)$ is a list of pairs $(h, \text{time}_p(h))$, each containing a hub and a time, sorted by hub. For a forward label, $\text{time}_p(h)$ encodes the latest departure time from p to reach hub h. More precisely, let h be a hub in an event label $L_f(e_i(p))$: we add the pair $(h, \text{time}(e_i(p)))$ to the stop label $SL_f(p)$ only if $h \notin L_f(e_j(p)), j > i$, i. e., only if h does not appear in the label of another event with a later departure time at the stop. Analogously, for backward stop labels, $\text{time}_p(h)$ encodes the earliest arrival time at p from h.

By restricting ourselves to these entries, we effectively discard dominated (non-tight) journeys to these hubs. It is easy to see that these stop labels obey a *tight journey cover property*: for each pair of stops s and t, $SL_f(s) \cap SL_b(t)$ contains at least one hub on each tight journey between them (or any equivalent journey that departs and arrives at the same time; recall from Section 2 that we allow arbitrary tie-breaking). This property does *not*, however, imply that the label intersection *only* contains tight journeys: for example, $SL_f(s)$ and $SL_b(t)$ could share a hub that is important for long distance travel, but not to get from s to t. The remainder of this section discusses how we handle this fact during queries.

Stop Label Profile Queries. To run an (s,t)-profile query on stop labels, we perform a coordinated sweep over both labels $SL_f(s)$ and $SL_b(t)$. For every matching hub h, i.e., $(h, \text{time}_s(h)) \in SL_f(s)$ and $(h, \text{time}_t(h)) \in SL_b(t)$, we consider the journey induced by $(\text{time}_s(h), \text{time}_t(h))$ for output. However, since we are only interested in reporting tight journeys, we maintain (during the algorithm) a tentative set of tight journeys, removing dominated journeys from it on-the-fly. (We found this to be faster than adding all journeys during the sweep and only discarding dominated journeys at the end.) We can further improve the efficiency of this approach in practice by (globally) reassigning hub IDs by the time of day. Note that every hub h of a stop label is still also an event and carries an event time $\text{time}(h)$. (Not to be confused with $\text{time}_s(h)$ and $\text{time}_t(h)$.) We assign sequential IDs to all hubs h in order of increasing $\text{time}(h)$, thus ensuring that hubs in the label intersection are enumerated chronologically. Note that this does not imply that journeys are enumerated in order of departure or arrival time, since each hub h may appear anywhere along its associated journey. However, preliminary experiments have shown that this approach leads to fewer insertions into the tentative set of tight journeys, reducing query time. Moreover, as in shortest path labels [9], we improve cache efficiency by storing the values for hubs and times separately in a stop label, accessing times only for matching hubs.

Overall, stop and event labels have different trade-offs: maintaining the profile requires less effort with event labels (any discovered journey is already tight), but fewer hubs are scanned with stop labels (there are no duplicate hubs).

Stop Label Earliest Arrival Queries. Reassigned hub IDs also enable fast (s, t, τ)-EA queries. We use binary search in $SL_f(s)$ and $SL_b(t)$ to find the earliest relevant hub h, i.e., with $\text{time}(h) \geq \tau$. From there, we perform a linear coordinated sweep as in the profile query, finding $(h, \text{time}_s(h)) \in SL_f(s)$ and $(h, \text{time}_t(h)) \in SL_b(t)$. However, instead of maintaining tentative profile entries $(\text{time}_s(h), \text{time}_t(h))$, we ignore solutions that depart too early (i.e., $\text{time}_s(h) < \tau$), while picking the hub h^* that minimizes the tentative best arrival time $\text{time}_t(h^*)$. (Note that $\text{time}(h) \geq \tau$ does not imply $\text{time}_s(h) \geq \tau$.) Once we scan a hub h with $\text{time}(h) \geq \text{time}_t(h^*)$, the tentative best arrival time cannot be improved anymore, and we stop the query. For practical performance, *pruning* the scan, so that we only sweep hubs h between $\tau \leq \text{time}(h) \leq \text{time}_t(h^*)$, is very important.

5   Practical Extensions

So far, we presented stop-to-stop queries, which report the departure and arrival times of the quickest journey(s). In this section, we address multicriteria queries, general location-to-location requests, and obtaining detailed journey descriptions.

Multicriteria Optimization and Minimum Transfer Time. Besides optimizing arrival time, many users also prefer journeys with fewer transfers. To solve the underlying multicriteria optimization problem, we adapt our labeling approach by (1) encoding transfers as arc costs in the graph, (2) computing shortest path labels based on these costs (instead of reachability labels on an unweighted graph), and (3) adjusting the query algorithm to find the Pareto set of solutions.

Reconsider the earliest arrival graph from Section 3. As before, we add a vertex for each unique event, linking consecutive events at the same stop with waiting arcs of cost 0. However, each connection arc (u, w) in the graph is subdivided by an intermediate *connection vertex* v, setting the cost of arc (u, v) to 0 and the cost of arc (v, w) to 1. By interpreting costs of 1 as leaving a vehicle, we can count the number of trips taken along any path. To model staying in the vehicle, consecutive connection vertices of the same trip are linked by zero-cost arcs.

A shortest path labeling on this graph now encodes the number of transfers as the shortest path distance between two events, while the duration of the journey can still be deduced from the time difference of the events. Consider a fixed source event $e(s)$ and the arrival events of a target stop $e_0(t), e_1(t), \ldots$ in order of increasing time. The minimum number of transfers required to reach the target stop t never increases with arrival times. (Hence, the whole Pareto set P of multicriteria solutions can be computed with a single Dijkstra run [18].)

We exploit this property to compute (s, t, τ)-EA multicriteria (MC) queries from the labels as follows. We initialize P as the empty set. We then perform an (s, t, τ)-EA query (with all optimizations described in Section 3) to compute the *fastest* journey in the solution, i.e., the one with most transfers. We add this journey to P. We then check (by performing distance label queries) for each subsequent event at t whether there is a journey with fewer transfers (than the most recently added entry of P), in which case we add the journey to P and repeat. The MC query ends once the last event at the target stop has been processed. We can stop earlier with the following optimization: we first run a distance label query on the *last* event at t to obtain the *smallest* possible number of transfers to travel from s to t. We may then already stop the MC query once we add a journey to P with this many transfers. Note that, since we do not need to check for domination in P explicitly, our algorithm maintains P in constant time per added journey.

Minimum Transfer Times. Transit agencies often model an entire station with multiple platforms as a single stop and account for the time required to change trips inside the station by associating a *minimum transfer time* $mtt(p)$ with each stop p. To incorporate them into the EA graph, we first locally replace each affected stop p by a *set* of new stops p^*, distributing *conflicting* trips (between which transferring is impossible due to mtt(p)) to different stops of p^*. We then add footpaths between all pairs of stops in p^* with length mtt(p). A small set p^* can be computed by solving an appropriate coloring problem [10]. For the MC graph, we need not change the input. Instead, it is sufficient to *shift* each arrival event $e \in E(p)$ by adding mtt(p) to time(e) before creating the vertices.

Location-to-Location Queries. A query between arbitrary locations s^* and t^*, which may employ walking or driving as the first and last legs of the journey, can be handled by a two-stage approach. It first computes sets S and T of relevant stops near the origin s^* and destination t^* that can be reached by car or on foot. With that information, a *forward superlabel* [1] is built from all forward stop labels associated with S. For each entry $(h, \text{time}_p(h)) \in SL_f(p)$ in the label of stop $p \in S$, we adjust the departure time $\text{time}_s^*(h) = \text{time}_p(h) - \text{dist}(s^*, p)$ so that the journey starts at s^* and add $(h, \text{time}_s^*(h))$ to the superlabel. For duplicate hubs that occur in multiple stop labels, we keep only the latest departure time from s^*. This can be achieved with a coordinated sweep, always adding the next hub of minimum ID. A *backward superlabel* (for T) is built analogously. For location-to-location queries, we then simply run our stop-label-based EA and profile query algorithms using the superlabels. In practice, we need not build superlabels explicitly but can simulate the building sweep during the query (which in itself is a coordinated sweep over two labels). A similar approach is possible for event labels. Moreover, point-of-interest queries (such as finding the closest restaurants to a given location) can be computed by applying known techniques [1] to these superlabels.

Journey Descriptions. While for many applications it suffices to report departure and arrival times (and possibly the number of transfers) per journey, sometimes a more detailed description is needed. We could apply known path unpacking techniques [1] to retrieve the full sequence of connections (and transfers), but in public transit it is usually enough to report the list of trips with associated transfer stops. We can accomplish that by storing with each hub the sequences of trips (and transfer stops) for travel between the hub and its label vertex.

6   Experiments

Setup. We implemented all algorithms in C++ using Visual Studio 2013 with full optimization. All experiments were conducted on a machine with two 8-core Intel Xeon E5-2690 CPUs and 384 GiB of DDR3-1066 RAM, running Windows 2008R2 Server. All runs are *sequential*. We use at most 32 bits for distances.

We consider four realistic inputs: the metropolitan networks of London (data.london.gov.uk) and Madrid (emtmadrid.es), and the national networks of Sweden (trafiklab.se) and Switzerland (gtfs.geops.ch). London includes all modes of transport, Madrid contains only buses, and the national networks contain both

Table 1. Size of timetables and the earliest arrival (EA) and multicriteria (MC) graphs

						EA Graph		MC Graph	
Instance	Stops	Conns	Trips	Footp.	Dy.	$|V|$	$|A|$	$|V|$	$|A|$
London	20.8 k	5,133 k	133 k	45.7 k	1	4,719 k	51,043 k	9,852 k	72,162 k
Madrid	4.7 k	4,527 k	165 k	1.3 k	1	3,003 k	13,730 k	7,530 k	34,505 k
Sweden	51.1 k	12,657 k	548 k	1.1 k	2	8,151 k	34,806 k	20,808 k	93,194 k
Switzerland	27.1 k	23,706 k	2,198 k	29.8 k	2	7,979 k	49,656 k	31,685 k	170,503 k

Table 2. Preprocessing figures. Label sizes are averages of forward and backward labels.

	Earliest Arrival						Multicriteria			
		Event Labels			Stop Labels			Event Labels		
Instance	RXL [h:m]	Hubs p. lbl	Hubs p. stop	Space [MiB]	Hubs p. stop	Space [MiB]	RXL [h:m]	Hubs p. lbl	Hubs p. stop	Space [MiB]
London	0:54	70	15,480	1,334	7,075	1,257	49:19	734	162,565	26,871
Madrid	0:25	77	49,247	963	9,830	403	10:55	404	258,008	10,155
Sweden	0:32	37	5,630	1,226	1,536	700	36:14	190	29,046	12,637
Switzerland	0:42	42	11,189	1,282	2,970	708	61:36	216	58,022	12,983

long-distance and local transit. We consider 24-hour timetables for the metropolitan networks, and two days for national ones (to enable overnight journeys). Footpaths were generated using a known heuristic [10] for Madrid; they are part of the input for the other networks. See Table 1 for size figures of the timetables and resulting graphs. The average number of unique events per stop ranges from 160 for Sweden to 644 for Madrid. (Recall from Section 3 that we merge all coincident events at a stop.) Note that no two instances dominate each other (w. r. t. number of stops, connections, trips, events per stop, and footpaths).

Preprocessing. Table 2 reports preprocessing figures for the unweighted earliest arrival graph (which also enables profile queries) and the multicriteria graph. For earliest arrival (EA), preprocessing takes well below an hour and generates about one gigabyte, which is quite practical. Although there are only 37–70 hubs per label, the total number of hubs per stop (i. e., the combined size of all labels) is quite large (5,630–49,247). By eliminating redundancy (cf. Section 4), stop labels have only a fifth as many hubs (for Madrid). Even though they need to store an additional distance value per hub, total space usage is still smaller. In general, *average* labels sizes (though not total space) are higher for metropolitan instances. This correlates with the higher number of daily journeys in these networks.

Preprocessing the multicriteria (MC) graph is much more expensive: times increase by a factor of 26.2–54.8 for the metropolitan and 67.9–88 for the national networks. On Madrid, Sweden, and Switzerland labels are five times larger compared to EA, and on London the factor is even more than ten. This is immediately reflected in the space consumption, which is up to 26 GiB (London).

Table 3. Evaluating earliest arrival queries. Bullets (•) indicate different features: profile query (Prof.), stop labels (St. lbs.), pruning (Prn.), hashing (Hash), and binary search (Bin.). The column "=" indicates the average number of matched hubs.

Prof. St.lbs. Prn. Hash Bin.	London				Sweden				Switzerland			
	Lbls.	Hubs	=	[µs]	Lbls.	Hubs	=	[µs]	Lbls.	Hubs	=	[µs]
o o o o o	108.4	6,936	1	14.7	68.0	2,415	1	6.9	89.0	3,485	1	8.7
o o • o o	16.1	1,360	1	5.9	34.4	1,581	1	5.4	33.5	1,676	1	5.8
o o • • o	16.1	1,047	1	4.2	34.4	1,083	1	3.6	33.5	1,151	1	3.8
o o • • •	7.0	332	4	2.8	6.5	179	3	2.1	7.6	204	4	2.1
o • o o o	2.0	13,037	1,126	54.8	2.0	2,855	81	10.0	2.0	5,707	218	20.4
o • • o o	2.0	861	62	6.2	2.0	711	16	3.6	2.0	699	19	3.8
• o o o o	658.5	40,892	211	141.7	423.7	13,590	118	39.4	786.6	29,381	240	81.4
• • o o o	2.0	13,037	1,126	74.3	2.0	2,855	81	12.1	2.0	5,707	218	24.5

Queries. We now evaluate query performance. For each algorithm, we ran 100,000 queries between random source and target stops, at random departure times between 0:00 and 23:59 (of the first day). Table 3 reports detailed figures, organized in three blocks: event label EA queries, stop label EA queries, and profile queries (with both event and stop labels). We discuss MC queries later.

We observe that event labels result in extremely fast EA queries (6.9–14.7 µs), even without optimizations. As expected, pruning and hashing reduce the number of accesses to labels and hubs (see columns "Lbls." and "Hubs"). Although binary search cannot stop as soon as a matching hub is found (see the "=" column), it accesses fewer labels and hubs, achieving query times below 3 µs on all instances.

Using stop labels (cf. Section 4) in their basic form is significantly slower than using event labels. With pruning enabled, however, query times (3.6–6.2 µs) are within a factor of two of the event labels, while saving a factor of 1.1–2.4 in space. For profile queries, stop labels are clearly the best approach. It scans up to a factor of 5.1 fewer hubs and is up to 3.3 times faster, computing the profile of the full timetable period in under 80 µs on all instances. The difference in factors is due to the overhead of maintaining the Pareto set during the stop label query.

Comparison. Table 4 compares our new algorithm (indicated as *PTL*, for Public Transit Labeling) to the state of the art and also evaluates multicriteria queries. In this experiment, PTL uses event labels with pruning, hashing and binary search for earliest arrival (and multicriteria) queries, and stop labels for profile queries. We compare PTL to CSA [12] and RAPTOR [11] (currently the fastest algorithms without preprocessing), as well as Accelerated CSA (ACSA) [20], Timetable Contraction Hierarchies (CH) [14], and Transfer Patterns (TP) [4,6] (which make use of preprocessing). Since RAPTOR always optimizes transfers (by design), we only include it for the MC problem. Note

Table 4. Comparison with the state of the art. Presentation largely based on [5], with some additional results taken from [6]. The first block of techniques considers the EA problem, the second the MC problem and the third the profile problem.

Algorithm	Name	Stops [·10³]	Conns [·10⁶]	Dy.	Arr.	Tran.	Prof.	Prep. [h]	Jn.	Query [ms]
CSA [12]	London	20.8	4.9	1	●	○	○	—	n/a	1.8
ACSA [20]	Germany	252.4	46.2	2	●	○	○	0.2	n/a	8.7
CH [14]	Europe (LD)	30.5	1.7	p	●	○	○	<0.1	n/a	0.3
TP [5]	Madrid	4.6	4.8	1	●	○	○	19	n/a	0.7
TP [6]	Germany	248.4	13.9	1	●	○	○	249	0.9	0.2
PTL	London	20.8	5.1	1	●	○	○	0.9	0.9	0.0028
PTL	Madrid	4.7	4.5	1	●	○	○	0.4	0.9	0.0030
PTL	Sweden	51.1	12.7	2	●	○	○	0.5	1.0	0.0021
PTL	Switzerland	27.1	23.7	2	●	○	○	0.7	1.0	0.0021
RAPTOR [11]	London	20.8	5.1	1	●	●	○	—	1.8	5.4
TP [5]	Madrid	4.6	4.8	1	●	●	○	185	n/a	3.1
TP [6]	Germany	248.4	13.9	1	●	●	○	372	1.9	0.3
PTL	London	20.8	5.1	1	●	●	○	49.3	1.8	0.0266
PTL	Madrid	4.7	4.5	1	●	●	○	10.9	1.9	0.0643
PTL	Sweden	51.1	12.7	2	●	●	○	36.2	1.7	0.0276
PTL	Switzerland	27.1	23.7	2	●	●	○	61.6	1.7	0.0217
CSA [12]	London	20.8	4.9	1	●	○	●	—	98.2	161.0
ACSA [20]	Germany	252.4	46.2	2	●	○	●	0.2	n/a	171.0
CH [14]	Europe (LD)	30.5	1.7	p	●	○	●	<0.1	n/a	3.7
TP [6]	Germany	248.4	13.9	1	●	○	●	249	16.4	3.3
PTL	London	20.8	5.1	1	●	○	●	0.9	81.0	0.0743
PTL	Madrid	4.7	4.5	1	●	○	●	0.4	110.7	0.1119
PTL	Sweden	51.1	12.7	2	●	○	●	0.5	12.7	0.0121
PTL	Switzerland	27.1	23.7	2	●	○	●	0.7	31.5	0.0245

that the following evaluation should be taken with a grain of salt, as no standardized benchmark instances exist, and many data sets used in the literature are proprietary. Although precise numbers are not available for several competing methods, it is safe to say they use less space than PTL, particularly for the MC problem.

Table 4 shows that PTL queries are very efficient. Remarkably, they are faster on the national networks than on the metropolitan ones: the latter are smaller in most aspects, but have more frequent journeys (that must be covered). Compared to other methods, PTL is 2–3 orders of magnitude faster on London than CSA and RAPTOR for EA (factor 643), profile (factor 2,167), and MC (factor 203) queries. We note, however, that PTL is a point-to-point algorithm (as are ACSA, TP, and CH); for one-to-all queries, CSA and RAPTOR would be faster.

PTL has 1–2 orders of magnitude faster preprocessing and queries than TP for the EA and profile problems. On Madrid, EA queries are 233 times faster while preprocessing is faster by a factor of 48. Note that Sweden (PTL) and Germany (TP) have a similar number of connections, but PTL queries are 95 times faster. (Germany does have more stops, but recall that PTL query performance depends more on the frequency of trips.) For the MC problem, the difference is smaller, but both preprocessing and queries of PTL are still an order of magnitude faster than TP (up to 48 times for MC queries on Madrid).

Compared to ACSA and CH (for which figures are only available for the EA and profile problems), PTL has slower preprocessing but significantly faster queries (even when accounting for different network sizes).

7   Conclusion

We introduced PTL, a new preprocessing-based algorithm for journey planning in public transit networks, by revisiting the time-expanded model and adapting the Hub Labeling approach to it. By further exploiting structural properties specific to timetables, we obtained simple and efficient algorithms that outperform the current state of the art on large metropolitan and country-sized networks by orders of magnitude for various realistic query types. Future work includes developing tailored algorithms for hub computation (instead of using RXL as a black box), compressing the labels (e. g., using techniques from [6] and [9]), exploring other hub representations (e. g., using trips instead of events, as in 3-hop labeling [21]), using multicore- and instruction-based parallelism for preprocessing and queries, and handling dynamic scenarios (e. g., temporary station closures and train delays or cancellations [5]).

References

1. Abraham, I., Delling, D., Fiat, A., Goldberg, A.V., Werneck, R.F.: HLDB: location-based services in databases. In: SIGSPATIAL, pp. 339–348. ACM (2012)
2. Abraham, I., Delling, D., Goldberg, A.V., Werneck, R.F.: Hierarchical hub labelings for shortest paths. In: Epstein, L., Ferragina, P. (eds.) ESA 2012. LNCS, vol. 7501, pp. 24–35. Springer, Heidelberg (2012)
3. Akiba, T., Iwata, Y., Yoshida, Y.: Fast exact shortest-path distance queries on large networks by pruned landmark labeling. In: SIGMOD, pp. 349–360 (2013)
4. Bast, H., Carlsson, E., Eigenwillig, A., Geisberger, R., Harrelson, C., Raychev, V., Viger, F.: Fast routing in very large public transportation networks using transfer patterns. In: de Berg, M., Meyer, U. (eds.) ESA 2010, Part I. LNCS, vol. 6346, pp. 290–301. Springer, Heidelberg (2010)
5. Bast, H., Delling, D., Goldberg, A.V., Müller-Hannemann, M., Pajor, T., Sanders, P., Wagner, D., Werneck, R.F.: Route planning in transportation networks. Technical Report MSR-TR-2014-4, Microsoft Research (2014)
6. Bast, H., Storandt, S.: Frequency-based search for public transit. In: SIGSPATIAL, pp. 13–22. ACM (2014)
7. Cheng, J., Huang, S., Wu, H., Fu, A.W.-C.: TF-Label: a topological-folding labeling scheme for reachability querying in a large graph. In: SIGMOD, pp. 193–204 (2013)
8. Cohen, E., Halperin, E., Kaplan, H., Zwick, U.: Reachability and distance queries via 2-hop labels. SIAM Journal on Computing 32(5), 1338–1355 (2003)

9. Delling, D., Goldberg, A.V., Pajor, T., Werneck, R.F.: Robust distance queries on massive networks. In: Schulz, A.S., Wagner, D. (eds.) ESA 2014. LNCS, vol. 8737, pp. 321–333. Springer, Heidelberg (2014)

10. Delling, D., Katz, B., Pajor, T.: Parallel computation of best connections in public transportation networks. ACM JEA **17**(4), 4.1–4.26 (2012)

11. Delling, D., Pajor, T., Werneck, R.F.: Round-based public transit routing. Transportation Science (2014). Accepted for publication

12. Dibbelt, J., Pajor, T., Strasser, B., Wagner, D.: Intriguingly simple and fast transit routing. In: Demetrescu, C., Marchetti-Spaccamela, A., Bonifaci, V. (eds.) SEA 2013. LNCS, vol. 7933, pp. 43–54. Springer, Heidelberg (2013)

13. Dijkstra, E.W.: A note on two problems in connexion with graphs. Numerische Mathematik **1**, 269–271 (1959)

14. Geisberger, R.: Contraction of timetable networks with realistic transfers. In: Festa, P. (ed.) SEA 2010. LNCS, vol. 6049, pp. 71–82. Springer, Heidelberg (2010)

15. Jin, R., Wang, G.: Simple, fast, and scalable reachability oracle. VLDB **6**(14), 1978–1989 (2013)

16. Merz, F., Sanders, P.: PReaCH: a fast lightweight reachability index using pruning and contraction hierarchies. In: Schulz, A.S., Wagner, D. (eds.) ESA 2014. LNCS, vol. 8737, pp. 701–712. Springer, Heidelberg (2014)

17. Müller-Hannemann, M., Weihe, K.: On the cardinality of the Pareto set in bicriteria shortest path problems. Ann. Oper. Res. **147**(1), 269–286 (2006)

18. Pyrga, E., Schulz, F., Wagner, D., Zaroliagis, C.: Efficient models for timetable information in public transportation systems. ACM JEA **12**(2.4), 1–39 (2008)

19. Seufert, S., Anand, A., Bedathur, S., Weikum, G.: Ferrari: flexible and efficient reachability range assignment for graph indexing. In: ICDE, pp. 1009–1020 (2013)

20. Strasser, B., Wagner, D.: Connection scan accelerated. In: ALENEX, pp. 125–137. SIAM (2014)

21. Yano, Y., Akiba, T., Iwata, Y., Yoshida, Y.: Fast and scalable reachability queries on graphs by pruned labeling with landmarks and paths. In: CIKM, pp. 1601–1606. ACM (2013)

22. Yildirim, H., Chaoji, V., Zaki, M.J.: GRAIL: Scalable reachability index for large graphs. VLDB **3**(1), 276–284 (2010)

23. Zhu, A.D., Lin, W., Wang, S., Xiao, X.: Reachability queries on large dynamic graphs: a total order approach. In: SIGMOD, pp. 1323–1334. ACM (2014)

On Balanced Separators in Road Networks

Aaron Schild[1] and Christian Sommer[2]([⊠])

[1] UC Berkeley, Berkeley, USA
aschild@berkeley.edu
[2] Apple Inc., Cupertino, USA
csommer@apple.com

Abstract. The following algorithm partitions road networks surprisingly well: *(i)* sort the vertices by longitude (or latitude, or some linear combination) and *(ii)* compute the maximum flow from the first k nodes (forming the source) to the last k nodes (forming the sink). Return the corresponding minimum cut as an edge separator (or recurse until the resulting subgraphs are sufficiently small).

1 Introduction

Graph Partitioning is the well-studied problem of cutting a graph into disjoint regions of approximately equal size while minimizing the number of edges between regions. An example partition of a road network is shown as Fig. 1.

Fig. 1. Recursive bisection using our separator algorithm *Inertial Flow* (with balance 1/4) for the road network of the United States (24M nodes, 29M edges) into 27 regions by cutting a total of 1,413 edges (0.005%). The largest region contains less than 6% of the original graph.

Delling, Goldberg, Razenshteyn, and Werneck [DGRW11] discovered that road networks have remarkably small separators. Prior to our work, their patented method called *PUNCH* appeared to be the only one capable of efficiently computing these separators (*Buffoon* [SS12], another high-quality partitioner for road networks, uses *PUNCH* as a subroutine).

© Springer International Publishing Switzerland 2015
E. Bampis (Ed.): SEA 2015, LNCS 9125, pp. 286–297, 2015.
DOI: 10.1007/978-3-319-20086-6_22

1.1 Problem Statement

Given a graph $G = (V, E)$, a *Graph Partition* is a partition of the vertex set V into disjoint subsets $V_0, V_1, \ldots V_{k-1}$ such that the regions (V_i, E_i) (subgraph induced by V_i) are of *roughly equal size*, and for all V_i, V_j ($i \neq j$) the set of edges between V_i and V_j (denoted by $E(V_i, V_j)$) is *as small as possible*. A main challenge of graph partitioning is the combined objective of minimizing the cut size while keeping good balance. Various objective functions combine the two quantities. Problem variants include *balanced k–partitioning*, where partitions must satisfy $\forall i : |V_i| \leq (1 + \epsilon) |V| / k$ for some imbalance parameter $\epsilon > 0$, or the relaxed (and significantly easier) variant, where only $\forall i : |V_i| \leq r$ for some region size constraint r (as considered in this paper).

One way of obtaining such a partition is by cutting G into two pieces V_0, V_1 and then recursing on each subgraph V_0, V_1. The recursion ends when the resulting subgraphs are sufficiently small. For these bisections, there are also various objective functions.

Definition 1 (Cuts and Balanced Cuts). *Given a graph $G = (V, E)$, a cut is a partition of V into two disjoint subsets V_0, V_1. A b–balanced cut (for any $0 < b \leq 1/2$) is a cut such that $|V_i| \geq \lfloor b \cdot |V| \rfloor$ for both $i \in \{0, 1\}$.*

At each level of the recursion, the objective is to find a b–balanced cut (for some b, say $b = 1/4$) that minimizes the number of cut edges, i.e. $\min |E(V_0, V_1)|$, where $E(V_0, V_1) := \{(u, v) \in E : u \in V_0, v \in V_1\}$.

Other well-known objective functions for cuts include the *minimum st cut* and the *sparsest cut*. A *minimum st cut* is a cut minimizing $|E(V_0, V_1)|$ with the condition that s and t are separated, i.e., $s \subset V_0$ and $t \in V_1$ (no balance requirements). It can be found efficiently using *maximum flow* algorithms [GR98, BK04, GHK+11, Mad13]. A *sparsest cut* is a cut minimizing $|E(V_0, V_1)| / (|V_0| \cdot |V_1|)$. Sparsest cuts are hard even to approximate [KV05, CKK+06].

1.2 Related Work

Theory. Various approximation algorithms for sparsest cut use maximum flow computations [KRV09, AK07, OSVV08, She09]. Roughly speaking, these algorithms iteratively refine an embedding of V by choosing source s and sink t at extremal points (of the embedding), computing st flow, followed by re-arranging V. A simplified statement of these results is that a poly-logarithmic number of carefully chosen maximum-flow computations provides a logarithmic approximation for sparsest cut (details in the corresponding papers). Previously, Lang and Rao [LR04] and Andersen and Lang [AL08] also showed how to improve cuts using maximum flow. Bui, Chaudhuri, Leighton, and Sipser [BCLS87] used maximum flow to compute bisections of regular graphs.

Some graphs are guaranteed to have small balanced cuts. For example, any planar, bounded-genus, or minor-free graph on n nodes has a balanced separator of size $O(\sqrt{n})$ [Ung51, LT79, Dji85, GHT84, And86, AST90], and recursive application yields partitions [LT79, Fre87, HKRS97, vWZA13,

KMS13]. Partitions obtained by recursive bisection may be far from optimal though [ST97].

Practice. The literature on graph partitioning is vast, see e.g. [BMSW13, BMS$^+$13] and references therein. In this brief review, we focus on recent work on partitioning road networks. Delling, Goldberg, Razenshteyn, and Werneck [DGRW11] introduce *PUNCH*, which first computes candidate cuts using maximum flows between sources and sinks chosen as follows: for a node $v \in V$, all nodes within distance $< r$ form the source, and all nodes at distance $> R$ form the sink (for two parameters $r < R$; distance can be measured in terms of BFS, shortest-path, or rank distance). These candidate cuts are then aggregated in various ways to form the final partition. Sanders and Schulz [SS11, SS12, SS13] contribute *KaFFPa[E]* and *KaHIP* (following earlier partitioners such as *Ka{SPar,PPA}*), all general-purpose partitioners, with a variant called *Buffoon* optimized for road networks. Their methods are based on the multi-level graph partitioning framework, where the input graph is first contracted, followed by a partitioning step on the smaller graph, and a refinement step to obtain a partition of the original graph. In *KaFFPa*, one of the refinement steps is called *adaptive flow iterations*, which enforces a balance constraint and computes maximum flow with source and sink chosen as BFS balls in two adjacent regions. Similar refinements using maximum flows had also been used by Boykov, Veksler, and Zabih [BVZ01].

Applications. Road network partitions can be used for applications such as shortest-path queries [Som14] or data distribution [KLSV10]. In particular, the performance of separator-based shortest-path algorithms [Fre87, Dji96, HKRS97, FR06, HSW08, DHM$^+$09, KKS11, MS12, DGPW13] depends on the size of the separator. Most prominently, Delling, Goldberg, Pajor, and Werneck [DGPW13] recently demonstrated that separator-based methods built upon a quality partition (such as those described in their joint work with Razenshteyn [DGRW11]) are highly practical. Their method recursively partitions the graph into a multi-level partition and then, for each region and level, precomputes matrices representing shortest-path costs between boundary nodes. For each region, memory requirements are therefore proportional to the square of the number of boundary nodes, which makes the quality of the partition particularly important. Partitioning is the most time-consuming step in their preprocessing algorithm (approximately 10 minutes to compute a multi-level partition for the US road network). Dibbelt, Strasser, and Wagner [DSW14] compute metric-independent *Contraction Hierarchies* based on nested dissection, which in turn is based on recursive bisection (corresponding theory in [BCRW13]). Finding good bisections is the most time-consuming step in their preprocessing algorithm.

1.3 Contribution

Our main contribution is a simple and efficient method to find sparse balanced cuts in embedded graphs such as road networks. The method, which we call

Inertial Flow, uses the embedding, initially sorts nodes geometrically (like the well-known *Inertial Partitioning*), and then computes a maximum flow. The corresponding minimum cut is used as the separator. *Inertial Flow* is straightforward to implement, yet its partitions are reasonably good. Our experiments using such a straightforward implementation demonstrate that it is competitive with the state-of-the-art partitioner *PUNCH* [DGRW11]. If the *Natural Cut Heuristic* is interpreted as the heart of *PUNCH* then the objective of this paper is to describe a new heart, and not the effects of its transplantation. We speculate that, in combination with the assembly phase of *PUNCH* or *Buffoon* [SS12], partitions might improve further (particularly in terms of balance).

In addition to simplicity, another advantage of recursive bisection is that, after computing the separator tree once, it contains the information for an entire multi-level partition (see e.g. [KMS13]).

As discussed in the section on related work, various partitioners employ a maximum-flow algorithm as an important subroutine. Their main differentiator is the choice of source and sink. On one hand, when terminals consist of too few nodes, minimum cuts may be highly unbalanced. On the other hand, when terminals consist of too many nodes, the best cuts may be violated by the initial source/sink assignment. Many methods use BFS balls to assign terminals, where the choice of radii is particularly delicate: obviously, balls must not intersect, but they should also be reasonably far apart. Such kind of tuning is fairly straightforward for our method, as there is just the balance parameter b to be configured. State-of-the-art theoretical algorithms for sparsest cut first embed the graph and then refine using maximum flow. The main observation leading to our method is that a road network's embedding (which is typically provided as part of the input) may be sufficiently good to serve as the initial embedding in an analogous algorithm.

2   Inertial Flow

We present an efficient heuristic to find b–balanced cuts in road networks. For the sake of exposition, let us consider a simplified road network, defined as an undirected graph $G = (V, E)$ with an embedding $f : V \hookrightarrow \mathbb{R}^2$. We may assume that G is connected, as typically partitioning algorithms are applied to each connected component independently. Our method is rather simple as it merely applies two standard primitives: sorting and maximum flow. (The well-known *Inertial Partitioning* uses sorting, followed by sweeping, hence the name of our method.)

1. Pick a line $\ell \in \mathbb{R}^2$ and orthogonally project V onto ℓ
   (more precisely, for each vertex v, project its point in the embedding $f(v)$ onto ℓ).
2. Sort V by order of appearance on ℓ (ties broken arbitrarily but consistently).
3. Let the first $\lfloor b \cdot |V| \rfloor$ vertices (in projection order) be the *source* s, and let the last $\lfloor b \cdot |V| \rfloor$ vertices be the *sink* t.
4. Compute a maximum flow between source s and sink t.
5. Return a corresponding minimum st cut.

Key Properties

- By choice of s and t, all minimum st cuts are b–balanced.
- The running time is bounded by the time required to sort V plus the time required to compute one maximum flow in G. Computing the entire separator tree (recursive bisection) requires time proportional to sort plus $\log_{1/(1-b)} |V|$ times flow.
- A basic implementation using standard libraries is straightforward.

Choice of ℓ

The quality of the cut depends on the line $\ell \in \mathbb{R}^2$ chosen in the first step of the algorithm. Obvious choices include random lines as well as simple fixed directions such as horizontal, vertical, or diagonal. A natural heuristic is then to try multiple lines and increasing balance values and return the best cut (for some objective function that may involve balance and cut size).

Let us demonstrate the effect of ℓ on the cut using the road network of New York[1] as an example. The cut sizes range from 5 (best) to 44 edges (see Fig. 2). The choice of source and sink forces the cut to be in a corridor that, for $b = 1/4$, contains half the graph. If the source/sink assignment violates a sparse cut and the corridor is relatively dense, then *Inertial Flow* finds a suboptimal cut.

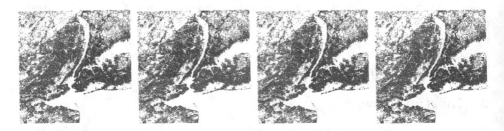

Fig. 2. The road network of New York (264K nodes) cut with balance 1/4 and four different line values. From left to right: horizontal (5 edges cut), vertical (44 edges cut), and diagonal (35 and 25 edges cut, respectively). *Inertial Flow* using horizontal sorting provides the best cut, both visually (along the Hudson) as well as in terms of the number of cut edges. The other sort orders yield comparatively large cuts as the minimum balance criterion forces unfortunate source/sink assignments violating the *Hudson cut*. Note that, compared to a typical worst-case guarantee on the order of $\sqrt{n} \approx 514$, all cuts are smaller by at least an order of magnitude.

---

[1] The NY network contains 264K nodes and 734K arcs (interpreted as 367K undirected edges). All US road networks used for experiments in this paper can be downloaded from http://www.dis.uniroma1.it/challenge9/download.shtml

3 Experiments

3.1 Setup

The main datasets we consider are the road networks of the United States and Europe, respectively. The USA graph (as used for the 9th DIMACS Implementation Challenge on Shortest Paths [DGJ08]) has 24M nodes and 58M directed arcs, which are typically interpreted as 29M undirected edges. The EUR graph (as made available by PTV AG, and also used in [DGJ08]) has 18M nodes and 21M edges (42M arcs).

The method used for comparison is *PUNCH* [DGRW11]. Note that *PUNCH* does not read the embedding, so *Inertial Flow* is given an unfair advantage. A main convenience of *Inertial Flow* as compared to *PUNCH* (and *Buffoon* [SS12]) is that it is straightforward to implement.

Our experiments are meant as a proof of concept, and we use a vanilla implementation (in C++) without any additional heuristics. For this paper, our focus is not on running times, and we also refrain from tuning parameters to experimental data. Unless indicated otherwise, balance is set to $1/4$, the lines are chosen to be horizontal, vertical, and diagonal ($\ell \in \{(1,0),(0,1),(1,1),(-1,1)\}$), and the objective function is simply to minimize the number of cut edges. The main subroutines employed are std::sort and maximum flow using Dinic's algorithm (augmenting paths, in the unit-capacity case computed by breadth-first search) [Din70]. Our implementation is parallel in the most obvious ways: separators for each line ℓ are computed by separate threads (with cross-notification of minimum cut upper bounds), and recursive calls are handled by a thread pool. For recursive bisections, we run 16 threads on two 2.20GHz Intel Xeon CPUs with 8 cores each. We encourage interested readers to combine *Inertial Flow* with other heuristics and/or to write more efficient implementations.

3.2 Results

Graph size vs. separator size and boundary size. Worst-case bounds for planar graphs on n nodes (and more general graph classes) guarantee the existence of a $1/3$–balanced cut/separator of size $O(\sqrt{n})$ [LT79]. Recursive separation yields a partition into $O(n/r)$ regions of size $\leqslant r$ with total boundary size $O(n/\sqrt{r})$. With some more work one can obtain an r–*division* [Fre87], where each region has *worst-case* boundary $O(\sqrt{r})$. Road networks appear to have significantly smaller separators: Delling, Goldberg, Razenshteyn, and Werneck [DGRW11] compare the average boundary size to $\sqrt[3]{r}$ instead (confirmed later by Dibbelt, Strasser, and Wagner [DSW14]). We provide plots for region size vs. total boundary size in Fig. 3. For specific numbers on region size vs. total boundary size, see Table 1 and Table 2.

Running Time. As mentioned above, our main focus is not on running time. Our implementation computes multi-level partitions for USA and EUR in minutes. Specific numbers are provided in Tables 1, 2, and 3. Note that, as expected, the initial cuts on the largest graphs are the most expensive ones. Subsequent cuts operate on smaller graphs and, by maintaining nodes in sorted order(s), do not require sorting the nodes again. For example, cutting USA into 2 regions requires 81 seconds (Table 3, $b = 1/4$). Recursive bisection into 6K regions takes only roughly twice as long (165.8 seconds, Table 2). Using this recursive bisection tree, reading off an entire multi-level partition is straightforward (see e.g. [KMS13]). By contrast, the *Natural Cut Heuristic* of *PUNCH* [DGRW11] depends on the target region size and is run separately for each level.

3.3   Comparison

Comparing partitions is not straightforward [BMS+14]. We compare against various partitions reported for *PUNCH* in Table 1 and observe that *PUNCH* partitions are significantly more balanced. For example, when partitioning USA into 27 regions as in Fig. 1, *Inertial Flow* cuts 1,413 edges with maximum region size 1.4M, while *PUNCH* cuts only 1,404 edges and obtains maximum region size 1M (2^{20}). While recursive bisection with *Inertial Flow* typically uses around 50% more regions than a perfectly balanced partition, *PUNCH* reportedly needs only about 15% more regions. For most partition granularities, the average numbers of cut edges per region are comparable.

We also compare our bisections against the optimal ones, obtained by an efficient algorithm of Delling, Fleischman, Goldberg, Razenshteyn, and Werneck [DFG+14]. Their algorithm guarantees *optimal* bisections for fairly large graphs, so comparing our method without any guarantees on optimality (only balance and running time have worst-case bounds) against their algorithm is not fair. However, we believe that the value of an optimal bisection adds an interesting perspective on cut quality (see Table 3).

Let us restate that the main advantage of *Inertial Flow* over *PUNCH* is simplicity. Another advantage is that multi-level partitions can be computed faster. As cut sizes are comparable, these advantages come at the cost of worse balance. Depending on the application, if better balance is required, a post-processing step (as in *PUNCH* or *Buffoon*) may further improve partitions.

Acknowledgments. Thanks to Ramana Idury for interesting discussions as well as contributions to the experimental framework. Thanks also to Daniel Delling and the anonymous reviewers for their feedback on earlier versions of this paper.

Table 1. An attempt at comparing partitions obtained by *PUNCH* and recursive bisection using *Inertial Flow*. Values for *PUNCH* were extracted from [DGRW11, Table 1(averagevalues)]. Each *PUNCH* average is compared to two *Inertial Flow* partitions: a partition with the same region-size constraint r, and a partition with the same number of regions. Center: when computing a partition with the same upper bounds for the maximum region size r, *PUNCH* requires fewer regions; the average number of cut edges per region is comparable. Right: when computing a partition with the same number of regions, the two partitioners cut a similar number of edges (with some *PUNCH* boundaries slightly smaller, particularly for Europe, and partitions more balanced). The running times for *PUNCH* are fairly uniform; for recursive bisection, the smaller r, the longer the computation. Note that a recursive bisection tree with regions of size at most r also contains a partition for any $r' \geqslant r$ (enabling plots like Fig. 3 with thousands of r values), hence it also contains multi-level partitions. Using the USA values in this table as an example, *Inertial Flow* simultaneously computes all 14 partitions ($r = 4{,}338{,}122$ through 2^{10}) in 4.1 minutes.

Graph	r	PUNCH			Inertial Flow, fixed r			Inertial Flow, target regions		
		regions	boundary	time	regions	boundary	time	r	boundary	time
Europe	1,024	20,129	168,767	79.7	27,129	208,280	209.5	1,378	171,064	216.0
	4,096	5,000	69,304	62.5	6,808	84,291	211.0	5,536	69,016	204.5
	16,384	1,248	28,448	61.6	1,708	34,839	214.0	22,367	28,236	194.7
	65,536	314	11,403	80.5	431	14,054	218.4	88,856	11,317	199.3
	262,144	81	4,194	106.1	106	5,275	210.5	349,449	4,246	209.2
	1,048,576	22	1,464	147.9	28	2,036	213.9	1,299,633	1,694	202.8
	4,194,304	6	371	196.6	7	573	176.3	4,861,623	461	171.8
USA	1,024	26,725	222,636	104.6	36,267	274,756	246.9	1,389	223,531	186.6
	4,096	6,643	87,762	79.9	9,000	107,170	173.2	5,570	87,193	181.8
	16,384	1,661	34,345	75.0	2,233	41,782	157.7	22,310	34,138	172.0
	65,536	418	12,767	89.9	563	15,862	166.0	87,960	12,971	168.5
	262,144	109	4,556	103.3	140	5,578	163.0	336,843	4,557	166.6
	1,048,576	27	1,504	117.6	33	1,716	148.5	1,407,053	1,413	148.3
	4,194,304	7	383	138.7	8	478	128.4	4,338,122	388	128.7

Fig. 3. Average boundary sizes for partitions with various maximum region sizes r (number of edges, logarithmic scale) for the BAY, CAL, and USA road networks, respectively. Worst-case results (such as those for planar graphs) guarantee average boundary sizes proportional to $r^{1/2}$. Delling, Goldberg, Razenshteyn, and Werneck [DGRW11] compare the average boundary size to $r^{1/3}$.

Table 2. Recursive bisection using *Inertial Flow* (balance $1/4$) on the road networks of California and Nevada (CAL), the United States (USA), and Europe (EUR), respectively, for various values of granularity (maximum region size r). Total region boundaries (cut sizes) reported correspond to the number of edges. Note that these partitions typically have around 50% more regions than necessary due to imperfect balance. Time (in seconds) corresponds to the time of recursive bisection (in particular, reading the graph and its embedding from disk is not included) as required by 16 threads (one bisection occupies 4 threads, one per slope). The variance in running times is rather substantial: even though we report the median among 11 consecutive runs, that median running time, e.g., for USA with $\lceil |V|/r \rceil = 64$ is slower than that for 1,024 even though only a relatively small subset of cuts is computed. The initial cuts of comparably large (sub-)graphs are the most expensive ones.

	CAL $|V| = 1.89$M			USA $|V| = 23.9$M			EUR $|V| = 18.0$M		
$\lceil |V|/r \rceil$	regions	boundary	time	regions	boundary	time	regions	boundary	time
2	3	53	3.3	3	140	121.9	3	276	106.2
4	6	103	4.6	6	324	114.7	7	573	135.5
8	12	215	5.8	12	648	125.9	13	1,058	157.8
16	25	441	5.6	24	1,223	120.1	26	1,867	164.3
64	99	1,437	6.0	96	4,234	165.7	98	4,990	169.2
256	387	4,418	6.3	397	12,482	141.5	399	13,280	172.2
1,024	1,561	12,957	6.1	1,593	33,197	145.5	1,592	33,228	170.5
4,096	6,326	36,129	6.9	6,307	84,274	165.8	6,321	80,332	174.7
16,384	25,543	98,232	11.1	25,401	216,078	188.6	25,208	198,433	175.3

Table 3. Bisection of various road networks: perfectly balanced bisections were obtained by Delling, Fleischman, Goldberg, Razenshteyn, and Werneck [DFG$^+$14, Table 4]. The balance of bisections found by *Inertial Flow* depends on the slope and the parameter b and there is no guarantee on optimality. In this table, for each $b \in \{1/5, 1/4, 1/3, 2/5\}$ we provide the minimum number of cut edges among 4 slopes. Balance is reported as the number of nodes in the smaller subgraph divided by the total number of nodes. As in Table 2, times reported are for the bisection (in seconds). When b is close to $1/2$, good balance is guaranteed, but cut sizes may be significantly higher, see e.g. NY at 40 edges for $b = 2/5$, which is more than double the size of an optimal bisection. When accepting worse balance, cuts may be substantially smaller.

Graph	$|V|$	Perfect		$b = 2/5$			$b = 1/3$			$b = 1/4$			$b = 1/5$		
		Cut	Time	Cut	Bal.	Time	Cut	Bal.	Time	Cut	Bal.	Time	Cut	Bal.	Time
NY	264K	18	381	40	0.48	0.1	5	0.43	0.1	5	0.43	0.1	5	0.43	0.1
BAY	321K	18	248	28	0.48	0.2	15	0.46	0.1	12	0.46	0.2	12	0.46	0.2
COL	436K	29	2,164	27	0.43	0.2	20	0.36	0.2	14	0.32	0.3	12	0.29	0.3
FLA	1.1M	25	1,640	28	0.42	0.6	22	0.40	0.7	17	0.29	0.9	15	0.27	1.0
NW	1.2M	18	463	24	0.49	0.7	17	0.50	0.6	17	0.50	0.7	17	0.50	0.9
NE	1.5M	24	751	20	0.49	1.3	20	0.49	1.4	20	0.49	1.7	20	0.49	2.3
CAL	1.9M	32	2,658	29	0.49	2.0	29	0.47	2.4	27	0.30	2.2	26	0.30	2.5
EUR	18M	NA	NA	229	0.46	69.3	201	0.45	95.3	188	0.45	124.9	95	0.30	81.4
USA	24M	NA	NA	61	0.48	58.3	61	0.48	63.9	61	0.48	81.2	61	0.48	84.3

References

[AK07]   Arora, S., Kale, S.: A combinatorial, primal-dual approach to semidefinite programs. In: 39th ACM Symposium on Theory of Computing (STOC), pp. 227–236 (2007)

[AL08]   Andersen, R., Lang, K.J.: An algorithm for improving graph partitions. In: 19th ACM-SIAM Symposium on Discrete Algorithms (SODA), pp. 651–660 (2008)

[And86] Andreae, T.: On a pursuit game played on graphs for which a minor is excluded. Journal of Combinatorial Theory, Series B 41(1), 37–47 (1986)

[AST90] Alon, N., Seymour, P.D., Thomas, R.: A separator theorem for nonplanar graphs. Journal of the American Mathematical Society 3(4), 801–808 (1990). Announced at STOC 1990

[BCLS87] Bui, T.N., Chaudhuri, S., Leighton, F.T., Sipser, M.: Graph bisection algorithms with good average case behavior. Combinatorica 7(2), 171–191 (1987). Announced at FOCS 1984

[BCRW13] Bauer, R., Columbus, T., Rutter, I., Wagner, D.: Search-space size in contraction hierarchies. In: Fomin, F.V., Freivalds, R., Kwiatkowska, M., Peleg, D. (eds.) ICALP 2013, Part I. LNCS, vol. 7965, pp. 93–104. Springer, Heidelberg (2013)

[BK04]   Boykov, Y., Kolmogorov, V.: An experimental comparison of min-cut/max-flow algorithms for energy minimization in vision. In: Figueiredo, M., Zerubia, J., Jain, A.K. (eds.) EMMCVPR 2001. LNCS, vol. 2134, pp. 359–374. Springer, Heidelberg (2001)

[BMS+13] Buluç, A., Meyerhenke, H., Safro, I., Sanders, P., Schulz, C.: Recent advances in graph partitioning (2013) arXiv, abs/1311.3144

[BMS+14] Bader, D.A., Meyerhenke, H., Sanders, P., Schulz, C., Kappes, A., Wagner, D.: Benchmarking for graph clustering and partitioning. In: Encyclopedia of Social Network Analysis and Mining, pp. 73–82 (2014)

[BMSW13] Bader, D.A., Meyerhenke, H., Sanders, P., Wagner, D. (eds.): Graph Partitioning and Graph Clustering. In: 10th DIMACS Implementation Challenge Workshop of Contemporary Mathematics, vol. 588 (2013)

[BVZ01] Boykov, Y., Veksler, O., Zabih, R.: Fast approximate energy minimization via graph cuts. IEEE Transactions on Pattern Analysis and Machine Intelligence 23(11), 1222–1239 (2001). Announced at ICCV 1999

[CKK+06] Chawla, S., Krauthgamer, R., Kumar, R., Rabani, Y., Sivakumar, D.: On the hardness of approximating multicut and sparsest-cut. Computational Complexity 15(2), 94–114 (2006). Announced at CCC 2005

[DFG+14] Delling, D., Fleischman, D., Goldberg, A.V., Razenshteyn, I., Werneck, R.F.: An exact combinatorial algorithm for minimum graph bisection. Mathematical Programming Series A (2014)

[DGJ08] Demetrescu, C., Goldberg, A.V., Johnson, D.S.: Implementation challenge for shortest paths. In: Encyclopedia of Algorithms (2008)

[DGPW13] Delling, D., Goldberg, A.V., Pajor, T., Werneck, R.F.: Customizable route planning. In: Pardalos, P.M., Rebennack, S. (eds.) SEA 2011. LNCS, vol. 6630, pp. 376–387. Springer, Heidelberg (2011)

[DGRW11] Delling, D., Goldberg, A.V., Razenshteyn, I., Werneck, R.F.F.: Graph partitioning with natural cuts. In: 25th IEEE International Symposium on Parallel and Distributed Processing (IPDPS), pp. 1135–1146 (2011)

[DHM⁺09] Delling, D., Holzer, M., Müller, K., Schulz, F., Wagner, D.: High-performance multi-level routing. In: The Shortest Path Problem: 9th DIMACS Implementation Challenge, vol. 74, pp. 73–92 (2009)

[Din70] Dinic, E.A.: Algorithm for solution of a problem of maximum flow in a network with power estimation. Doklady Akademii Nauk SSSR; Translation in Soviet Mathematics Doklady **11**(5), 1277–1280 (1970)

[Dji85] Djidjev, H.N.: A linear algorithm for partitioning graphs of fixed genus. Serdica. Bulgariacae mathematicae publicationes **11**(4), 369–387 (1985)

[Dji96] Djidjev, H.N.: Efficient algorithms for shortest path queries in planar digraphs. In: D'Amore, F., Marchetti-Spaccamela, A., Franciosa, P.G. (eds.) WG 1996. LNCS, vol. 1197, pp. 151–165. Springer, Heidelberg (1997)

[DSW14] Dibbelt, J., Strasser, B., Wagner, D.: Customizable contraction hierarchies. In: Gudmundsson, J., Katajainen, J. (eds.) SEA 2014. LNCS, vol. 8504, pp. 271–282. Springer, Heidelberg (2014)

[FR06] Fakcharoenphol, J., Rao, S.: Planar graphs, negative weight edges, shortest paths, and near linear time. Journal of Computer and System Sciences **72**(5), 868–889 (2006). Announced at FOCS 2001

[Fre87] Frederickson, G.N.: Fast algorithms for shortest paths in planar graphs, with applications. SIAM Journal on Computing **16**(6), 1004–1022 (1987)

[GHK⁺11] Goldberg, A.V., Hed, S., Kaplan, H., Tarjan, R.E., Werneck, R.F.: Maximum flows by incremental breadth-first search. In: Demetrescu, C., Halldórsson, M.M. (eds.) ESA 2011. LNCS, vol. 6942, pp. 457–468. Springer, Heidelberg (2011)

[GHT84] Gilbert, J.R., Hutchinson, J.P., Tarjan, R.E.: A separator theorem for graphs of bounded genus. Journal of Algorithms **5**(3), 391–407 (1984). Announced as TR82-506 in 1982

[GR98] Goldberg, A.V., Rao, S.: Beyond the flow decomposition barrier. Journal of the ACM **45**(5), 783–797 (1998). Announced at FOCS 1997

[HKRS97] Henzinger, M.R., Klein, P.N., Rao, S., Subramanian, S.: Faster shortest-path algorithms for planar graphs. Journal of Computer and System Sciences **55**(1), 3–23 (1997). Announced at STOC 1994

[HSW08] Holzer, M., Schulz, F., Wagner, D.: Engineering multilevel overlay graphs for shortest-path queries. ACM Journal of Experimental Algorithmics **13**(2008). Announced at ALENEX 2006

[KKS11] Kawarabayashi, K., Klein, P.N., Sommer, C.: Linear-space approximate distance oracles for planar, bounded-genus and minor-free graphs. In: Aceto, L., Henzinger, M., Sgall, J. (eds.) ICALP 2011, Part I. LNCS, vol. 6755, pp. 135–146. Springer, Heidelberg (2011)

[KLSV10] Kieritz, T., Luxen, D., Sanders, P., Vetter, C.: Distributed time-dependent contraction hierarchies. In: Festa, P. (ed.) SEA 2010. LNCS, vol. 6049, pp. 83–93. Springer, Heidelberg (2010)

[KMS13] Klein, P.N., Mozes, S., Sommer, C.: Structured recursive separator decompositions for planar graphs in linear time. In: 45th ACM Symposium on Theory of Computing (STOC), pp. 505–514 (2013)

[KRV09] Khandekar, R., Rao, S., Vazirani, U.V.: Graph partitioning using single commodity flows. Journal of the ACM **56**(4) (2009). Announced at STOC 2006

[KV05] Khot, S., Vishnoi, N.K.: The unique games conjecture, integrality gap for cut problems and embeddability of negative type metrics into ℓ_1. In: 46th IEEE Symposium on Foundations of Computer Science (FOCS), pp. 53–62 (2005)

[LR04] Lang, K., Rao, S.: A flow-based method for improving the expansion or conductance of graph cuts. In: Bienstock, D., Nemhauser, G. (eds.) IPCO 2004. LNCS, vol. 3064, pp. 325–337. Springer, Heidelberg (2004)

[LT79] Lipton, R.J., Tarjan, R.E.: A separator theorem for planar graphs. SIAM Journal on Applied Mathematics 36(2), 177–189 (1979)

[Mad13] Madry, A.: Navigating central path with electrical flows: from flows to matchings, and back. In: 54th IEEE Symposium on Foundations of Computer Science (FOCS), pp. 253–262 (2013)

[MS12] Mozes, S., Sommer, C.: Exact distance oracles for planar graphs. In: 23rd ACM-SIAM Symposium on Discrete Algorithms (SODA), pp. 209–222 (2012)

[OSVV08] Orecchia, L., Schulman, L.J., Vazirani, U.V., Vishnoi, N.K.: On partitioning graphs via single commodity flows. In: 40th ACM Symposium on Theory of Computing (STOC), pp. 461–470 (2008)

[She09] Sherman, J.: Breaking the multicommodity flow barrier for $O(\sqrt{\log n})$-approximations to sparsest cut. In: 50th IEEE Symposium on Foundations of Computer Science (FOCS), pp. 363–372 (2009)

[Som14] Sommer, C.: Shortest-path queries in static networks. ACM Computing Surveys 46, 45:1–45:31 (2014)

[SS11] Sanders, P., Schulz, C.: Engineering multilevel graph partitioning algorithms. In: Demetrescu, C., Halldórsson, M.M. (eds.) ESA 2011. LNCS, vol. 6942, pp. 469–480. Springer, Heidelberg (2011)

[SS12] Sanders, P., Schulz, C.: Distributed evolutionary graph partitioning. In: 14th Meeting on Algorithm Engineering & Experiments, (ALENEX), pp. 16–29 (2012)

[SS13] Sanders, P., Schulz, C.: Think locally, act globally: highly balanced graph partitioning. In: Demetrescu, C., Marchetti-Spaccamela, A., Bonifaci, V. (eds.) SEA 2013. LNCS, vol. 7933, pp. 164–175. Springer, Heidelberg (2013)

[ST97] Simon, H.D., Teng, S.-H.: How good is recursive bisection? SIAM Journal on Scientific Computing 18, 1436–1445 (1997)

[Ung51] Ungar, P.: A theorem on planar graphs. Journal of the London Mathematical Society s1–26(4), 256–262 (1951)

[vWZA13] van Walderveen, F., Zeh, N., Arge, L.: Multiway simple cycle separators and I/O-efficient algorithms for planar graphs. In: 24th ACM-SIAM Symposium on Discrete Algorithms (SODA), pp. 901–918 (2013)

SALT. A Unified Framework for All Shortest-Path Query Variants on Road Networks

Alexandros Efentakis[1][✉], Dieter Pfoser[2], and Yannis Vassiliou[3]

[1] Research Center "Athena", Marousi, Greece
efentakis@imis.athena-innovation.gr
[2] Department of Geography and GeoInformation Science,
George Mason University, Faifax, US
dpfoser@gmu.edu
[3] National Technical University of Athens, Zografou, Greece
yv@cs.ntua.gr

Abstract. Although recent scientific literature focuses on multiple shortest-path (SP) problem definitions for road networks, none of the existing solutions can efficiently answer all the different SP query variations. This work proposes SALT, a novel framework that not only efficiently answers most SP queries but also k-nearest neighbor queries not tackled by previous methods. Our solution offers excellent query performance and very short preprocessing times, thus making it also a viable option for dynamic, live-traffic road networks and all types of practical use-cases. The proposed SALT framework is a deployable software solution capturing a range of graph-related query problems under one "algorithmic hood".

Keywords: Shortest-paths · k-nearest neighbors · kNN · Salt framework

1 Introduction

During the last decades, recent scientific literature has produced efficient methods for shortest-path (SP) queries on road networks (cf. [1] for the latest overview). Unfortunately, most aforementioned algorithms are tuned to solving a specific problem efficiently, but are rather inefficient when used in a different context. Contrarily, engineering a framework that efficiently solves multiple shortest-path problems, would be the first step towards the direction of a *grand unified SP toolkit*. To this end, the GRASP algorithms [11], solve most variants of the single-source shortest-path problems on road networks, including *one-to-all* (finding SP distances from a source vertex s to all other vertices), *one-to-many* (computing the SP distances between the source vertex s and a set of target vertices T) and *range queries* (find all vertices reachable from s within a given timespan). GRASP requires minimal preprocessing and provides excellent query performance needed in the context of practical and commercial applications.

E. Bampis (Ed.): SEA 2015, LNCS 9125, pp. 298–311, 2015.
DOI: 10.1007/978-3-319-20086-6_23

Another fundamental problem frequently encountered in location-based services is the kNN query, i.e., given a query location and a set of objects on the road network, the kNN search finds the k-nearest objects to the query location. Unfortunately, even the latest work of [21] is not scalable with the network size, since it requires *several hours* for preprocessing continental road networks. In addition, for a large number of randomly distributed objects, an efficient Dijkstra implementation could answer kNN queries by settling a few hundreds nodes and requiring $< 1ms$. Moreover, most previous methods require a *target-selection phase*, i.e., they need to mark the objects location within the underlying index. This phase requires a few seconds, hence having limited appeal for applications involving moving objects (e.g., vehicles). Therefore, it only makes sense to use a complex (non-Dijkstra) kNN processing framework in cases of either rather "small" numbers of objects or objects following skewed distributions (e.g., POIs located near the city center), i.e., for cases in which Dijkstra does not perform well.

The contribution of this work is to provide a unified algorithmic solution that may be used in a *dynamic road network* context, while covering a *wide range of shortest-path problems*, such as (i) single-pair, (ii) one-to-all, (iii) one-to-many, (iv) range and (v) kNN queries. Specifically, we aim at combining the fragmented approaches related to the various shortest-path problem definitions and instead propose a unified framework that tackles all of them. Our proposed **SALT** (graph Separators + ALT) framework requires seconds for preprocessing continental road networks and provides excellent query performance for a wide range of problems. We will show that SALT is (i) $3 - 4\times$ faster for point-to-point queries when compared to existing methods of similar preprocessing times, (ii) it answers one-to-all, one-to-many and range queries with comparable performance to state-of-the-art approaches, and most importantly, (iii) it may also answer kNN queries in $< 1ms$, for both, static or moving objects. As such, our SALT framework could be a *swiss-army-knife for tackling all shortest-path problem variants*, making it a serious contender for use in commercial applications.

The outline of this work is as follows. Section 2 describes previous related work. Section 3 describes our novel SALT framework and algorithms. Experiments establishing SALT's benefits are provided in Section 4 and Section 5 concludes the paper.

2   Related Work

Throughout this work, we use directed weighted graphs $G(V, E, w)$, where V is the set of vertices, $E \subseteq VxV$ is the set the arcs and w is a positive weight function $E \rightarrow R^+$. The reverse graph $\overline{G} = (V, E)$ is the graph obtained from G by substituting each arc $(u, v) \in E$ by (v, u). A partition of V is a family of sets $C = \{c_0, c_1, \ldots c_M\}$, such that each node $u \in V$ is contained in exactly one set c_i. An element of a partition is called a *cell*. A multilevel partition of V is a family of partitions $\{C^0, C^1, \ldots C^L\}$ where ℓ denotes the level of a partition C^ℓ. Similar to [4], level 0 refers to the original graph, L is the highest partition level and in this work we use *nested multilevel partitions*, i.e., for each $\ell < L$ and each cell c_i^ℓ there exists a unique cell $c_j^{\ell+1}$ (called the supercell of c_i^ℓ) with $c_i^\ell \subseteq c_j^{\ell+1}$. Accordingly, c_i^ℓ is a subcell of $c_j^{\ell+1}$. In this notation, $c^\ell(v)$ is the cell containing the vertex v on level ℓ. Likewise, the number of cells of the partition C^ℓ is denoted as $|C^\ell|$. For a boundary arc on level ℓ, the tail and head vertices are located in different level-ℓ cells; a boundary vertex on level ℓ is connected with at least one vertex in another level-ℓ cell. Note that for nested multilevel partitions, a boundary vertex/arc at level ℓ is also a boundary vertex/arc for all levels below.

In kNN queries, given a query location s and a set of objects O, the kNN search problem finds k-nearest objects to the query location. Throughout this work, similar to [21], we assume that the query location and the objects are both located at vertices.

The ALT algorithm. In the ALT algorithm [13], a small set of vertices called landmarks is chosen. Then, during preprocessing, we precompute distances to and from every landmark for each vertex. Given a set $S \subseteq V$ of landmarks and distances $d(L_i, v)$, $d(v, L_i)$ for all vertices $v \in V$ and landmarks $L_i \in S$, the following triangle inequalities hold: $d(u, v) + d(v, L_i) \geq d(u, L_i)$ and $d(L_i, u) + d(u, v) \geq d(L_i, v)$. Hence, the function $\pi_f = max_{L_i} max\{d(u, L_i) - d(v, L_i), d(L_i, v) - d(L_i, u)\}$ provides a lower-bound for the graph distance $d(u, v)$. Later works [18], showed that landmarks may also provide upper-bounds on the graph distance between any two vertices. Overall, landmarks may be used to approximate graph distances, according to Eq. 1 and 2. ALT then combines the classic A* algorithm with the aforementioned lower-bounds. For bidirectional search, ALT uses the average potential function, defined as $p_f(v) = (\pi_f(v) - \pi_r(v))/2$ for the forward and $p_r(v) = (\pi_r(v) - \pi_f(v))/2 = -p_f(v)$ for the backward search.

$$d(u,v) \geq max_{L_i} max\{d(u,L_i) - d(v,L_i), d(L_i,v) - d(L_i,u)\} \tag{1}$$

$$d(u,v) \leq min_{L_i}(d(u, L_i) + d(L_i,v)) \tag{2}$$

Graph separators. In Graph Separator (GS) methods, such as CRP [4,6], a partition C of the graph is computed. Then, the preprocessing phase builds an *overlay graph* H containing all boundary vertices and arcs of G. It also contains a clique for each cell c: for every pair (u, v) of boundary vertices in c, a clique arc (u, v) is created whose cost is the same as the shortest path (restricted to the inner arcs of c) between u and v. For a SP query between s and t, the Dijkstra algorithm must be run on the graph consisting of the union of H, $c^0(s)$ and $c^0(t)$. To further accelerate queries, we may use multiple levels of overlay graphs. Currently, CRP is the most efficient SPSP algorithm in terms of preprocessing time (since the recent Customizable Contraction Hierarchies [9] is only tested on undirected networks) and is thus suitable for dynamic road networks.

SSSP queries. Recently, Efentakis et al. [11] expanded graph separators and proposed GRASP, a novel set of algorithms for handling all variants of single-source shortest-path (SSSP) queries, including one-to-all, one-to-many and range queries. All three algorithms, namely GRASP (one-to-all), isoGRASP (range) and reGRASP (one-to-many) use the exact same data structures and share all the advantages of graph-separator methods, such as very short preprocessing times and excellent parallel query performance. Unfortunately, parallel reGRASP requires a few ms for one-to-many que-ries on continental road networks and hence is not fast enough for handling kNN queries.

kNN queries There are many works on kNN queries for static objects on road networks. Unfortunately, even the most recent G-tree [21] cannot *scale for continental road networks, requiring 16 hours of preprocessing for the full USA network*. Moreover, all index-based approaches require a *target selection phase* to index which tree-nodes contain objects (requiring few seconds) and thus, they cannot be used for moving objects. There is also previous work around kNN queries for moving objects on road networks. However, they are either disk-based [20], have not been tested on continental road networks [14,17,20] and cannot address dynamic road networks. Recently, CRP was also expanded [7] to handle kNN queries. Unfortunately, (i) CRP also requires a target selection phase and hence, cannot be applied to moving objects and (ii) it may only perform well for objects near the query location (otherwise the entire upper level of the overlay graph must be traversed). Hence, this solution is also not optimal.

3   The SALT Framework

The main contribution of this work is to propose SALT, a unified framework for answering single-pair, single-source (one-to-all, one-to-many and range) and especially kNN queries which are not handled efficiently by existing approaches. The main advantage of SALT is, that the exact same data structures may service all the different type of SP queries on road networks and thus, SALT may be easily integrated into commercial, real-world applications. What follows is a detailed discussion of the SALT framework.

3.1   Preprocessing

SALT's preprocessing consists of two distinct phases, (i) the *graph-separator (GS) phase* and (ii) the *landmarks preprocessing phase*.

The *GS phase* of SALT mimics the preprocessing of GRASP [11] (see Fig. 1). During this phase, we use the Kafpaa/Buffoon [19] partitioning tool to create nested multilevel partitions of the road network graph in a top-down fashion. This initial partitioning phase is *metric independent* and needs to be executed only once, i.e., even in the case of arc-weights changes or for different metrics. Following partitioning, the *customization stage* builds the overlay graph H containing all boundary vertices and arcs of G. The graph H also contains a clique

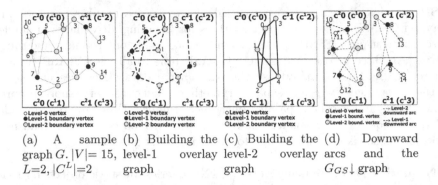

(a)   A   sample (b) Building the (c) Building the (d)   Downward
graph G. $|V|$= 15, level-1   overlay level-2   overlay arcs   and   the
$L=2, |C^L|=2$ graph   graph $G_{GS}\downarrow$ graph

Fig. 1. SALT's GS customization phase. Building the overlay H and the $G_{GS}\downarrow$ graphs

for each cell c: for every pair (u, v) of boundary vertices in c, we create a shortcut
arc (u, v) whose cost is the same as the shortest-path (restricted to inner edges
of c) between u and v (see Fig. 1(b), 1(c)). Similar to [11], we also calculate the
SP distances between all border vertices of level ℓ and all vertices of level $\ell-1$
within each cell c^ℓ (see Fig. 1(d)). To differentiate between the two kinds of arcs
computed, we will denote as (i) *clique arcs* the added overlay arcs that connect
border vertices of the same level ℓ and (ii) *downward arcs* of level ℓ the vertices
connecting different levels, i.e., ℓ and $\ell-1$. For added efficiency, *downward arcs*
are stored as a separate graph, referred to, as $G_{GS}\downarrow$. Both types of arcs are
computed bottom-up and starting at level one. To process a cell, the GS customization stage for SALT executes a Dijkstra algorithm from each boundary
vertex of the cell. We also apply the arc-reduction optimization of [12], which
reports only distances of boundary vertices that are direct descendants of the
root of each executed Dijkstra algorithm.

Although SALT's GS preprocessing phase is similar to GRASP, there are
two major differences. (i) In SALT, H and $G_{GS}\downarrow$ have the same number of
levels ($L = 6$ in our experiments) with $|C^L| = 16$ (cf. the original GRASP
paper with $|C^L| = 128$ and $L = 16$). Using a smaller number of cells at the
upper level slightly lowers one-to-all query parallel performance, but *accelerates
point-to-point queries and reduces preprocessing time.* Hence, it is a very logical
compromise, since our focus is on increased versatility. (ii) Moreover, we have
to repeat SALT's GS customization stage twice, one for the forward and one for
the reverse graph. This is necessary for the landmarks phase of SALT, but it also
allows to answer, both, forward and reverse single-source queries. Thus, at the
end of SALT's GS preprocessing we have built two versions of the overlay graphs,
H and $G_{GS}\downarrow$, one for forward and one for reverse graph queries, respectively.

The *landmarks preprocessing phase* for SALT extends the preprocessing proposed by [10], which optimized and tailored the ALT algorithm for dynamic road
networks. Landmarks are selected by the *partition - corners* landmarks selection
strategy, in which we use the cells created by Kafpaa and from each cell we
select the four corner-most vertices as landmarks. For SALT, we accelerate the

computation of distances of all graph vertices from and to landmarks by executing two sequential GRASP algorithms (forward and reverse) instead of using plain Dijkstra (as in all previous approaches). Moreover, we may perform those $2 \times |S|$ GRASP algorithms in parallel. By using these optimizations, the landmarks preprocessing phase of SALT never takes more than $4s$ *for 24 landmarks* and is therefore *at least* $6 \times$ *faster than any existing work.*

Thus, at the end of the preprocessing stage of SALT, we have built the overlay graphs H and $G_{GS} \downarrow$ for both forward and reverse searches and calculated distances for all vertices from and to the selected landmarks. For *dynamic road networks*, we only need to repeat the GS customization stage and the computation of distances of all vertices from and to the landmarks. Both phases *require less than 19s* for the benchmark road networks we used. This makes SALT suitable for dynamic scenarios, as well.

3.2 Single-Pair Shortest-Path Queries

Using SALT's preprocessing data, we can accelerate single-pair SP queries by our SALT-p2p algorithm, that combines CRP (with arc-reduction) with the ALT's adaptation with SIMD instructions of [10]. In CRP, to perform a SP query between s and t, Dijkstra's algorithm must be run on the graph consisting of the union of H, $c^0(s)$ and $c^0(t)$. The difference in SALT-p2p is that, instead of Dijkstra, we use the ALT-SIMD algorithm on the aforementioned graph. Note that both ALT and CRP may also be used in a unidirectional or a bidirectional setting. A similar combination of CALT [2] and CRP was unofficially introduced in [4], which uses the landmark lower-bounds strictly on the upper-level of the GS overlay graph. Thus, *local searches* could not be accelerated. Local search is crucial for kNN queries, since the kNN results for small values of k are usually located close to the query location. In contrast, our *SALT-p2p algorithm*, combining the ALT-SIMD algorithm of [10] and CRP (with arc-reduction), will be much more efficient than stand-alone ALT or CRP. Moreover, since both methods are extremely robust to the metric used [2,4], their combination will provide excellent performance for both travel times and travel distances.

Theorem 1. *The SALT-p2p algorithm is correct. (Proof omitted for space restrictions)*

3.3 kNN Queries

SALT's preprocessing data may also be used to answer kNN queries. Instead of initiating a kNN search from a query location s to objects O, we start a search *from* all the objects at the same time *to* the query location in the reverse graph. Hence, we take advantage of, both, GS and ALT acceleration for guiding the search towards the query location. The SALT-kNN algorithm's query phase is divided in two independent stages. The *Pruning phase* excludes objects that cannot possibly belong to the kNN set by using the upper and lower-bounds provided by the landmarks preprocessing data. The *Main phase* executes a unidirectional

A. Efentakis et al.

SALT-p2p algorithm in the reverse graph from all remaining objects at the same time to the query location until the query location is settled. Now we have found the first nearest-neighbor. This process has to be repeated another $k - 1$ times until all kNN are discovered. The algorithm is detailed in the following.

Pruning phase. To prune objects that cannot belong to the k-nearest neighbors set, we must (i) calculate the k-th lowest upper-bound of graph distances between the query location and the objects (cf. Equation 2) and (ii) exclude objects whose distance lower-bounds between them and the query location (cf. Equation 1) exceed the k-th lowest upper-bound. To the best of our knowledge, this is the *first work to utilize upper and lower landmark bounds in the context of kNN queries.*

Theorem 2. *SALT-kNN's pruning phase is correct. (Proof omitted for space restrictions)*

For computing the k-th lowest upper-bound between the query location and the objects we use a bounded max-heap Q of size k and procedure GETK-THLOWUPBOUND:

GETKTHLOWUPBOUND($s, O, landDist$)

```
 1   Q = emptyMaxHeap
 2   m = 0
 3   for each o in O
 4       if m < k
 5           Q.push(upperBound(s, o))
 6           m = m + 1
 7       elseif (upperBound(s, o) < Q.top())
 8           Extract − max(Q)
 9           Q.push(upperBound(s, o))
10   return Extract − max(Q)
```

PRUNEPHASE($s, O, landDist$)

```
 1   O_small = {}
 2   kBound = getKthLowBound(s, O, landDist)
 3   for each o in O
 4       if lowerBound(s, o) ≤ kBound
 5           O_small.add(o)
 6   return O_small
```

Since the bounded max-heap Q only stores k-upper-bound distances, we only need to compare the next objects's upper-bound with the top of the heap. If we have found a lower upper-bound, we remove the top of the heap and add the new upper-bound to Q. At the end of the procedure, the top of the max-heap is the k-th lowest upper bound of distances between the query location and the objects.

At the end of the pruning phase (see procedure PRUNINGPHASE), instead of using the objects in O, we only need to check for the k-nearest neighbors within the objects in O_{small}. Our experimentation has shown that the pruning phase is very effective, since it efficiently prunes more than 60% of the total number of objects in O.

Main phase. Following the pruning phase, to find the first nearest-neighbor we start by performing a search simultaneously from all objects in O_{small} to the query location s in the reverse graph. To do so, we use the idea of [16]. We add a new vertex T' connected to all objects in O_{small} using zero-weight edges and then perform a unidirectional SALT-p2p algorithm from T' to s in the reverse graph. At the end of this process, we have found the first NN of query location s. Then we eliminate this vertex from O_{small} and repeat for another $k-1$ iterations to retrieve the full kNN set (see procedure MAINPHASE).

Theorem 3. *SALT-kNN's main phase is correct. (Proof omitted for space restrictions)*

MAINPHASE$(s, O_{small}, k, \overline{G})$
1   **for** $i = 0$ **to** $k-1$
2       $T' =$ newVertex
3   **for** each $o \in O_{small}$
4   Conn. T' to o with 0-weight edges
5           $(iNN, iNNdist) = SALT\text{-}p2p(T', s, \overline{G})$
6           $O_{small} = O_{small} - iNN$

To retrieve not only the SP distance between the query location s and the objects in O_{small} but also the actual kNN vertices, we need to maintain for each labeled vertex a reference that points to the originating vertex in the objects' set O_{small}. Thus, when we extract s from the priority queue and terminate the SALT-p2p algorithm at the i-th iteration, we know not only the i-th SP distance but the i-th NN as well. Moreover, for each object o in O_{small}, we need to store the cell ID $c^1(o)$ of the cell this object belongs at the lowest level of the GS hierarchy, to traverse the overlay graph H during each iteration of the SALT-p2p algorithm. Note it is sufficient to store only the $c^1(o)$, since cell IDs for higher levels may be calculated from that.

Although SALT-kNN will be very fast for retrieving the first NN object, it will become progressively slower when retrieving the additional $k-1$ NN, since at each iteration, the SALT-p2p algorithm will start from scratch. To remedy this, at the beginning of the i-th iteration, we reload the corresponding priority queue with all vertices labeled during the $i-1$ iteration except those originating from the previous NN vertex found, since most of those labeled vertices were already assigned correct SP distances. Since we use a min-heap priority queue (as all Dijkstra variants), this optimization significantly improves query times and still ensures correctness of the SALT-kNN algorithm.

3.4   Summary and Expectations

Although SALT is very efficient for most SP queries, the main phase of SALT-kNN could be performed with any valid unidirectional SP algorithm. However, using SALT-p2p has multiple benefits: (i) Its constituent algorithms, ALT and CRP have very fast preprocessing times suitable for *dynamic road networks*. (ii) Unidirectional SALT-p2p provides better performance than bidirectional SALT-p2p, contrary to existing hierarchical methods that may only be used in a bidirectional setting. (iii) SALT-p2p and hence SALT-kNN are very *robust to the metric used*. This is an important property for kNN queries identifying Points-Of-Interest (POIs) based on walking distance. (iv) SALT-kNN's pruning phase is very crucial for a fast implementation. *Only the landmarks preprocessing data could provide this type of functionality.* (v) Lastly, the main phase of the SALT-kNN algorithm initially expands vertices closer to the query location s. As such, "unattractive" objects furthest from s (as estimated by the lower-bounds) that cannot be excluded during the pruning phase, do not slow down SALT-kNN queries. In fact, experiments showed that finding the first NN is as fast as a plain SALT-p2p query. Hence, it is hard to provide a much better theoretical solution, using standard SP techniques, with fast enough preprocessing times suitable for dynamic road networks.

4   Experiments

The experimentation that follows, assesses the performance of the SALT-p2p and SALT-kNN algorithms. For completeness, we also report the performance of sequential and parallel GRASP [11] algorithm within the SALT framework for single-source (one-to-all) queries. Experiments were performed on a workstation with a 4-core i7-4771 processor clocked at 3.5GHz with 32Gb of RAM, running Ubuntu 14.04 64bit. Our code was written in C++ and GCC 4.8 (with OpenMP). Query times are executed on one core and augmented with SSE instructions. We used the European road network (18M vertices / 42M arcs) and the full USA road network (24M vertices / 58M arcs) [8] and experimented with both travel times and travel distances.

For partitioning the graph into nested-multilevel partitions, similarly to [11], we used Buffoon / KaFFPa [19] in a top-down approach. We use a partitioning setup similar to the best recorded CRP results of [3] with total number of overlay levels set to $L=6$ and $|C^1|=1048576$, $|C^2|=65536$, $|C^3|=8192$, $|C^4|=1024$, $|C^5|=128$ and $|C^6|=16$. We also used 24 landmarks, since adding more landmarks did not offer significant performance benefits for either SALT-p2p or SALT-kNN algorithms.

4.1   Preprocessing

Table 1. SALT, GRASP and G-tree preprocessing

	Preprocessing time (s)			
	Travel Times (TT)		Travel Distances (TD)	
	EUR	USA	EUR	USA
SALT (GS customiz.)	11.1 (5.5)	14.82 (7.4)	11.3 (5.7)	15.4 (7.7)
SALT (Landmarks)	2.6 (1.3)	3.6 (1.8)	2.7 (1.4)	3.6 (1.8)
SALT (Total)	13.7 (6.9)	18.4 (9.2)	14.0 (7.0)	18.9 (9.5)
GRASP (Orig)	8 (8)	12 (12)	10 (10)	13 (13)
G-tree	(198,479)	(5,736)	(25,918)	(5,001)

In this section we report the preprocessing times for SALT, in comparison to the original GRASP version [11]) and G-tree [21] (G-tree source code was provided by its authors). Note, that contrary to the SALT framework that may simultaneously answer single-pair, single-source (one-to-all, one-to-many, range) and kNN queries, GRASP only focuses on single-source queries and G-tree may only be used for undirected networks and kNN queries. SALT and GRASP preprocessing times refer to parallel execution and G-tree preprocessing time is sequential. For GRASP and SALT and its graph-separator subphase we only report preprocessing times for the customization stage, similar to [4] and [11], since this is the preprocessing that must be repeated when arc-weights change, for live-traffic road networks. For a fair comparison, for G-tree we do not report the partitioning time required for the building of the G-tree index (which uses METIS [15]) and we only report the preprocessing time for calculating the SP distances inside the respective index structure. Results are presented on Table 1. Numbers inside parentheses represent preprocessing times for undirected versions of the benchmark road networks.

Results show that: (i) G-tree preprocessing times are very disappointing, especially for Europe and travel times, when more than 24h are required for preprocessing, contrary to SALT's preprocessing time *that never exceeds 19s* for all networks and metrics. (ii) In comparison to GRASP, SALT may calculate both forward and reverse graph SSSP queries. If GRASP was to be extended for reverse graph SSSP queries, its preprocessing time would double and hence *it would be 16−43% slower than SALT.* (iii) SALT's preprocessing time *is very robust to the metric used* and preprocessing time is similar for both metrics. (iv) For undirected versions of the road networks (for comparing results to G-tree), SALT's preprocessing time drops in half, both for the GS customization and landmarks phase. Note that although SALT's total preprocessing time is better than any other previous ALT based approach including [10], the GS customization phase could be potentially further accelerated by using the optimizations of [6], namely SIMD instructions or contraction. Furthermore, SALT's memory requirements still remain quite modest, since it requires less than $8.5Gb$ (including the original graph G) for both benchmark road networks and metrics.

4.2   Single-Pair/Single-Source Shortest-Path Queries

Table 2 compares unidirectional and bidirectional SALT-p2p query performance for single-pair shortest-path (SPSP) queries, compared to its algorithmic components, namely the ALT-SIMD algorithm of [10] and CRP [4] with the arc-reduction of [12], within the SALT framework, for a total of 10,000 queries with the pair of vertices selected uniformly at random. Regarding SSSP queries, we report sequential and parallel performance of GRASP for one-to-all queries within the SALT framework and compare it with the original version of GRASP [11]. For both GRASP versions, the number in parentheses represent sequential times. Results are presented in Table 2.

Results show that: (i) Unidirectional SALT-p2p is always faster than bidirectional SALT-p2p. Thus, to the best of our knowledge, *uniSALT-p2p is the faster unidirectional algorithm for road networks, with preprocessing times of few seconds.* (ii) uniSALT-p2p is $100 − 266\times$ faster than ALT and $3 − 4\times$ faster than CRP. Note that our CRP's query performance is almost identical to the best CRP implementation of [3]. UniSALT-p2p path unpacking (i.e., providing full paths) would also be faster than CRP, since it uses bidirectional ALT instead of bidirectional Dijkstra used by CRP [3]. Moreover, uniSALT-p2p provides comparable performance to recent Customizable Contraction Hierarchies [9] which was only tested on undirected networks. (iii) SALT-p2p is very robust to the metric used. In fact, *uniSALT-p2p is slightly faster for travel distances.*

Table 2. SALT-p2p and GRASP query performance

	SPSP Query times (ms)			
	Travel Times (TT)		Travel Distances (TD)	
	EUR	USA	EUR	USA
biALT	103	60	133	89
CRP (+AR)	1.6	1.8	2	2
uniSALT-p2p	0.6	0.6	0.5	0.5
biSALT-p2p	0.9	0.9	0.9	0.9
	SSSP Query times (ms)			
GRASP (Orig)	43 (150)	58 (207)	46 (156)	66 (218)
GRASP (SALT)	50 (169)	65 (224)	53 (175)	68 (228)

For SSSP queries, the GRASP implementation within the SALT framework is 5−12% slower for sequential and 3−16% slower for parallel execution than the original GRASP implementation. Still, it is fast enough for most practical cases and the SALT framework may also execute forward and reverse SSSP queries, which is a considerable advantage. Note, that the slightly less efficient implementation of GRASP within SALT is attributed to the fact that now $|C_L| = 16$ (in comparison to $|C_L| = 128$ in [11]). However, setting $|C_L| = 16$ is the optimal setting for SPSP and kNN queries and thus, we kept the setting that benefits the most frequent type of queries.

4.3 kNN Queries

Next, we compare SALT-kNN, Dijkstra and G-tree [21] performance for kNN queries. For each experiment, we generate 100 sets of random objects of varying size $|O|$ and for each such set we generate 100 random query locations, for a total of 10,000 kNN queries per $|O|$. Figure 2 reports average query times for $k = 1$ and $k = 4$. Note, that G-tree requires a *target selection phase*, for each set of objects $|O|$ (requiring 1.9−2.4s). Thus, contrary to Dijkstra and SALT-kNN, G-tree cannot be used for moving objects.

Results show that SALT-kNN provides stable performance and query times significantly below 1ms for $k=1$. Contrarily, G-tree is almost *two - three orders of magnitude slower* and cannot compete with either SALT-kNN or Dijkstra. Dijkstra starts very slow for small values of $|O|$ but manages to surpass SALT-kNN performance for $|O| > 8192$. These results are similar to [7], where Dijkstra also outperforms online CRP for $k = 10$ and $|O| = 0.01 \times |V|$. Still, since for static points of interest we are usually interested in a specific type of objects (e.g., gas stations) and in the case of moving objects we rarely have such large vehicle fleets (i.e., taxis, trucks) to monitor and we usually aim for kNN queries among the *available* vehicles (a much smaller subset of total vehicles), then the SALT-kNN algorithm is surely to perform better for most practical applications.

After establishing the superiority of SALT-kNN over G-tree, we next evaluate the impact of objects distribution to SALT-kNN and Dijkstra's performance. To this end, similar to [5], we pick a vertex at random and run Dijkstra's algorithm from it until reaching a predetermined number of vertices $|B|$. If B is the set of vertices visited during this search, we pick our objects O as a random subset of B. We keep the number of objects $|O|$ steady at 2^{14} and we experiment with different values of $|B|$ ranging from $2^{14} \ldots 2^{24}$, to simulate cases of either: (i) POIs mainly located near the city-center or (ii) vehicle fleets which may service an entire continent but operate mainly on a particular country. Results are presented in Figure 3.

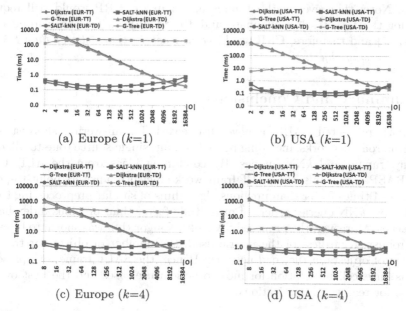

Fig. 2. SALT-kNN Dijkstra and G-tree comparison for varying values of $|O|$

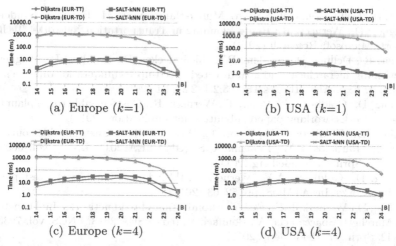

Fig. 3. SALT-kNN and Dijkstra comparison for $|O| = 2^{14}$ and varying values of $|B|$

Results show, that SALT-kNN is one - two orders of magnitude faster than Dijkstra when objects are not uniformly located in the road network (as is the typical case, either for static or moving objects). Thus, SALT-kNN guarantees excellent and stable performance, regardless of: (i) the number of objects and (ii) the objects distribution. Moreover, it does not need a target selection phase, such as G-tree or CRP and therefore, it may be used for either static or moving objects. Note, than even without building an index, CRP would still require 10ms for the target selection phase and 16,384 objects for the Europe road network [7] and therefore, CRP would *be at least 10 times slower than SALT-kNN for moving objects.*

5   Summary and Conclusions

This work presented SALT, a novel framework for answering shortest-path queries on road networks, including point-to-point, single-source (one-to-all, one-to-many, range) and kNN queries. By combining ideas from the ALT, CRP and GRASP algorithms, the SALT framework efficiently answers point-to-point queries 3−4 times faster than previous algorithms of similar preprocessing times and answers kNN queries orders of magnitude faster than previous index-based approaches. Moreover, the proposed SALT-kNN algorithm was shown to be especially robust, regardless of the metric used, the number of objects or the distribution of objects in the road network. Hence, the SALT framework presents itself as an excellent solution for most practical use-cases and the best overall solution for real-world applications.

References

1. Bast, H., Delling, D., Goldberg, A., Müller-Hannemann, M., Pajor, T., Sanders, P., Wagner, D., Werneck, R.: Route Planning in Transportation Networks. Technical report, Microsoft Research (2014)
2. Bauer, R., Delling, D., Sanders, P., Schieferdecker, D., Schultes, D., Wagner, D.: Combining hierarchical and goal-directed speed-up techniques for dijkstra's algorithm. J. Exp. Algorithmics 15, 2.3:2.1–2.3:2.31 (2010)
3. Delling, D., Goldberg, A., Pajor, T., Werneck, R.: Customizable route planning in road networks. working paper, submitted for publication (2013)
4. Delling, D., Goldberg, A.V., Pajor, T., Werneck, R.F.: Customizable route planning. In: Pardalos, P.M., Rebennack, S. (eds.) SEA 2011. LNCS, vol. 6630, pp. 376–387. Springer, Heidelberg (2011)
5. Delling, D., Goldberg, A.V., Werneck, R.F.F.: Faster batched shortest paths in road networks. In: ATMOS, pp. 52–63 (2011)
6. Delling, D., Werneck, R.F.: Faster customization of road networks. In: Demetrescu, C., Marchetti-Spaccamela, A., Bonifaci, V. (eds.) SEA 2013. LNCS, vol. 7933, pp. 30–42. Springer, Heidelberg (2013)
7. Delling, D., Werneck, R.F.: Customizable point-of-interest queries in road networks. IEEE Transactions on Knowledge and Data Engineering (2014) (to appear)
8. Demetrescu, C., Goldberg, A.V., Johnson, D.: The shortest path problem. Ninth DIMACS implementation challenge, Piscataway, NJ, USA, November 13–14, 2006. Proceedings. DIMACS Book 74. AMS (2009)
9. Dibbelt, J., Strasser, B., Wagner, D.: Customizable contraction hierarchies. In: Gudmundsson, J., Katajainen, J. (eds.) SEA 2014. LNCS, vol. 8504, pp. 271–282. Springer, Heidelberg (2014)

10. Efentakis, A., Pfoser, D.: Optimizing landmark-based routing and preprocessing. In: Proceedings of the 6th ACM SIGSPATIAL International Workshop on Computational Transportation Science. IWCTS 2013, New York, NY, USA, pp. 25–30 (2013)

11. Efentakis, A., Pfoser, D.: GRASP. Extending graph separators for the single-source shortest-path problem. In: Schulz, A.S., Wagner, D. (eds.) ESA 2014. LNCS, vol. 8737, pp. 358–370. Springer, Heidelberg (2014)

12. Efentakis, A., Theodorakis, D., Pfoser, D.: Crowdsourcing computing resources for shortest-path computation. In: Proceedings of the 20th International Conference on Advances in Geographic Information Systems, SIGSPATIAL 2012. ACM (2012)

13. Goldberg, A.V., Harrelson, C.: Computing the shortest path: a* search meets graph theory. In: 16th ACM-SIAM Symposium on Discrete Algorithms, pp. 156–165 (2004)

14. Jensen, C.S., Kolářvr, J., Pedersen, T.B., Timko, I.: Nearest neighbor queries in road networks. In: Proceedings of the 11th ACM International Symposium on Advances in Geographic Information Systems, GIS 2003, pp. 1–8. ACM, New York (2003)

15. Karypis, G., Kumar, V.: A fast and high quality multilevel scheme for partitioning irregular graphs. SIAM J. Sci. Comput. 20, 359–392 (1998)

16. Maue, J., Sanders, P., Matijevic, D.: Goal directed shortest path queries using precomputed cluster distances. In: Àlvarez, C., Serna, M. (eds.) WEA 2006. LNCS, vol. 4007, pp. 316–328. Springer, Heidelberg (2006)

17. Mouratidis, K., Yiu, M.L., Papadias, D., Mamoulis, N.: Continuous nearest neighbor monitoring in road networks. In: Proceedings of the 32Nd International Conference on Very Large Data Bases, VLDB 2006, pp. 43–54. VLDB Endowment (2006)

18. Potamias, M., Bonchi, F., Castillo, C., Gionis, A.: Fast shortest path distance estimation in large networks. In: Proceedings of the 18th ACM Conference on Information and Knowledge Management, CIKM 2009, pp. 867–876. ACM, New York (2009)

19. Sanders, P., Schulz, C.: Engineering multilevel graph partitioning algorithms. In: Demetrescu, C., Halldórsson, M.M. (eds.) ESA 2011. LNCS, vol. 6942, pp. 469–480. Springer, Heidelberg (2011)

20. Wang, H., Zimmermann, R.: Processing of continuous location-based range queries on moving objects in road networks. IEEE Transactions on Knowledge and Data Engineering 23(7), 1065–1078 (2011)

21. Zhong, R., Li, G., Tan, K.-L., Zhou, L.: G-tree: an efficient index for knn search on road networks. In: Proceedings of the 22nd ACM International Conference on Information Knowledge Management, CIKM 2013, pp. 39–48. ACM, New York (2013)

Other Applications II

Critical Applications 11

Huffman Codes versus Augmented Non-Prefix-Free Codes

Boran Adaş[1], Ersin Bayraktar[1], and M. Oğuzhan Külekci[2](✉)

[1] Department of Computer Enginering, İstanbul Technical University,
Istanbul, Turkey
{adas,bayrakterer}@itu.edu.tr
[2] ERLAB Software Co., ITU ARI Teknokent, Istanbul, Turkey
oguzhan.kulekci@erlab.com.tr

Abstract. Non–prefix–free (NPF) codes are not uniquely decodable, and thus, have received very few attention due to the lack of that most essential feature required in any coding scheme. Augmenting NPF codes with compressed data structures has been proposed in ISIT'2013 [8] to overcome this limitation. It had been shown there that such an augmentation not only brings the unique decodability to NPF codes, but also provides efficient random access. In this study, we extend this approach and compare augmented NPF codes with the 0th–order Huffman codes in terms of compression ratios and random access times. Basically, we benchmark four coding schemes as NPF codes augmented with wavelet trees (NPF–WT), with R/S dictionaries (NPF–RS), Huffman codes, and sampled Huffman codes. Since Huffman coding originally does not provide random access feature, sampling is a common way in practice to speed up access to arbitrary symbols in the encoded stream. We achieve sampling by simply managing an additional array that marks the beginnings of the codewords in steps of the sampling ratio, and keeping that sparse bit array compressed via R/S dictionary data structure. The experiments revealed that augmented NPF codes achieve compression very close to the Huffman with the additional advantage of random access. When compared to sampled Huffman coding both the compression ratios and random access performances of the NPF schemes are superior.

1 Introduction

Representing more frequent symbols with less number of bits is the central idea in variable–length coding for compression. Since its invention in 1952, the Huffman codes [7] along with its numerous variants have been used in many areas. In Huffman coding, none of the codewords is a prefix of another. This prefix–free structure avoids ambiguities in the decoding phase, and once the encoded stream is received on the decoder side, the symbols are easily decoded one-by-one, from left–to–right, following the order of their appearance in the original data. It had been shown in the original paper [7] that Huffman codes are asymptotically optimal, which means they represent the source sequence within its entropy.

© Springer International Publishing Switzerland 2015
E. Bampis (Ed.): SEA 2015, LNCS 9125, pp. 315–326, 2015.
DOI: 10.1007/978-3-319-20086-6_24

The main limitation in Huffman coding has been the lack of efficient random access. In the Huffman–encoded bit sequences, the boundary positions of the codewords are unknown, and thus, to extract the kth codeword, one needs to decode all preceding $k - 1$ items after which decoding of the kth symbol can be achieved. Particularly in today's big data era, this limitation may cause severe problems while retrieving a random portion of the data from an archive.

One quick solution to overcome that problem is to maintain an additional array, which tells us the beginning positions of the each bth codeword on the encoded stream. With such an information, instead of decoding the whole sequence from its very beginning, accessing the randomly selected kth symbol requires at most b decoding operations by first landing to the $\lfloor \frac{k}{b} \rfloor$th block directly, and then extracting the remaining symbols sequentially until reaching the desired item. The sampling ratio b plays an important role here as smaller b provides faster access with increased overhead, and larger b results in reverse.

The question in such a scenario is how to keep that additional array that includes the starting bit positions of the sampled items. One way is to keep it as an array of $\frac{n}{b}$ integers, where that integer array may also be represented via compact integer encoding schemes [9]. Previous results reported in that direction [8] seem not very promising in terms of space usage.

A second approach might be keeping another bit sequence of length equal to the encoded stream, where the beginning of the each bth symbol is marked with 1 and rest with 0 [1]. Depending on the b value this bit stream is expected to be sparse composed of many 0s and $\frac{n}{b}$ 1s, which reminds us to integrate rank/select (R/S) dictionaries that are especially helpful in compressing such redundant bit arrays with efficient rank and select operations support. In this study we investigate this option to support random access in Huffman encoded sequences. We augment Huffman codes with R/S dictionaries and refer to that Huffman–RS(S), which represents the sampled Huffman coding with sampling ratio S.

Prefix–free condition, which ensures self delimiting of codewords during the decoding phase, is the indispensable obligation in any coding scheme for unique decodability. In that sense, non–prefix–free codes are not of interest since they are not uniquely decodable, and thus, normally received very few attention till today [2,8]. In [8], it is proposed to use wavelet trees to provide unique decodability and random access in non-prefix free codes. In this study we extend this approach and besides the wavelet trees we consider incorporating R/S dictionaries also. We refer to non–prefix-free codes augmented with wavelet trees as NPF–WT, and similarly NPF–RS corresponds to the other augmentation option with R/S dictionaries.

We analyse the compression ratios and random access times of the investigated schemes. We show in practice that non–prefix–free codes may catch the compression ratio of regular Huffman with better random access times than the sampled Huffman codes, particularly on large alphabets.

2   Preliminaries and Notation

Let a given sequence of n symbols be shown by $T = t_1 t_2 \ldots t_n$, where each t_i, $1 \le i \le n$, is drawn from the alphabet $\Sigma = \{\epsilon_1, \epsilon_2, \ldots, \epsilon_\sigma\}$ of size σ. The array $F = \{f_1, f_2, \ldots, f_\sigma\}$ represents the number of occurrences of each symbol in T. Without loss of generality we assume the symbols of the alphabet are listed in decreasing order of their frequencies in T such that ϵ_1 is the most frequent symbol, and ϵ_σ is the least frequent one. As an example $T = \text{NONPREFIXFREE}$, $\Sigma = \{\text{E,R,F,N,I,O,P,X}\}$, and $F = \{3, 2, 2, 2, 1, 1, 1, 1\}$.

The coding scheme C assigns distinct codewords $W = \{w_1, w_2, \ldots, w_\sigma\}$ for each $\epsilon_i \in \Sigma$ such that w_i is the codeword corresponding to symbol ϵ_i. Each codeword w_i is composed of varying number of bits. Thus, the encoding of T with C creates the sequence $C(T) = c_1 c_2 \ldots c_n$, where $c_i \in W$ is the codeword of $t_i \in \Sigma$ according to coding scheme $C : \Sigma \to W$.

The `rank(i)` and `select(i)` operations on a bitmap $B = b_1 b_2 \ldots b_n$, $b_i \in \{0, 1\}$, return the number of bits set to 1 (or 0) in $b_1 b_2 \ldots b_i$, and the index of the ith bit set to $1(0)$, respectively. Several studies have appeared on efficient calculation of rank and select queries [10,11]. The general purpose in any R/S dictionary data structures is to represent the underlying bit sequence entropy compressed while supporting constant time `rank` and `select` operations.

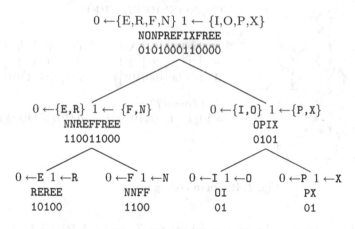

Fig. 1. Wavelet tree example

The wavelet tree data structure introduced by Grossi $et.\,al$ [6] makes the rank and select operations available on non–binary alphabets in logarithmic time. The main idea is to first create a bitmap from T by splitting the alphabet Σ into two equal parts and representing $t_i \in \{\epsilon_1, \epsilon_2, \ldots, \epsilon_{\lfloor \sigma/2 \rfloor}\}$ with 0 and others by 1 for all $1 \le i \le n$. We proceed by creating left and right children of the root node by collecting the symbols represented by 0 on the left and others on the right. We repeat the same procedure by again splitting the corresponding alphabets into two for each child, and continue this until the alphabet for a node becomes less than or equal to two. An example of such a wavelet tree created for $T =$ NONPREFIXFREE is depicted in Figure 1. The height of the balanced WT is $\log \sigma$ and the rank/select queries over the bitmaps of the wavelet tree can be achieved in constant time via the R/S dictionary data structures. Thus, by traversing the tree in $O(\log \sigma)$ time one can achieve rank, select, and access operations on non-binary alphabets.

3   Huffman Coding

The Huffman codeword generation algorithm can be found in any data compression textbook, e.g., [3]. Mainly, the symbols are listed in decreasing order of their frequencies, and initially the codewords are all empty. Bits 0 and 1 are concatenated to the codewords of the bottom two symbols, and those symbols are packed into a meta-symbol by summing their frequencies. This meta-symbol is inserted into the correct position in the list of all symbols. Thus, at each step, the number of symbols decreases by one, and the procedure is repeated σ times. Notice that when we add a bit to a meta-symbol, we add the same bit to the all codewords included in it.

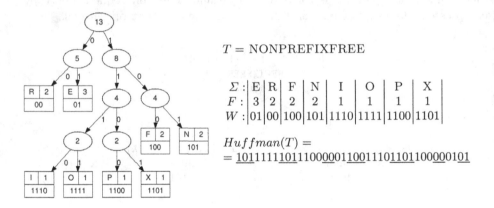

$T = \text{NONPREFIXFREE}$

$$\Sigma : \begin{array}{|c|c|c|c|c|c|c|c|} E & R & F & N & I & O & P & X \end{array}$$
$$F : \begin{array}{|c|c|c|c|c|c|c|c|} 3 & 2 & 2 & 2 & 1 & 1 & 1 & 1 \end{array}$$
$$W : \begin{array}{|c|c|c|c|c|c|c|c|} 01 & 00 & 100 & 101 & 1110 & 1111 & 1100 & 1101 \end{array}$$

$Huffman(T) =$
$= \underline{10}\underline{1111}\underline{11}\underline{10}\underline{111}\underline{000}\underline{00}\underline{1}\underline{100}\underline{1110}\underline{1101}\underline{101}\underline{00}\underline{000}\underline{101}$

Fig. 2. Huffman coding example

The Huffman codeword tree generated[1] for $T = NONPREFIXFREE$ is depicted in Figure 2. Due to the prefix–free property of Huffman, the bit stream $Huffman(T)$ is uniquely decodable from left-to-right without any need to a delimiter between the variable length codewords.

3.1   Huffman–RS(S): Sampled Huffman Codes with R/S Dictionaries

The main limitation in Huffman coding is the lack of an efficient method to support random access. We can simply overcome this by managing an additional

---

[1] The figure is generated by the website http://huffman.ooz.ie/.

bit array of size equal to the $|C(T)|$, where the beginnings of the each code word in steps of a predetermined sampling frequency S are marked with 1 and the rest is set to 0. This additional bit array can be stored compressed with the appropriate R/S dictionary data structures that allow constant time `select` queries. To access the ith keyword, we simply query the additional bit array with `select`($\lfloor \frac{i}{S} \rfloor$), and begin decoding the symbols until we extract the ith one. Since each block contains S codewords, in the worst case one needs to decode at most S items for any random access query.

$$Huffman(T) \ \underline{101}\underline{111}\underline{1101}110\underline{0000}11\underline{1001}110 \ \underline{1101}1000\underline{00101}$$

$$A(Huffman(T)) \ \ 1000000100000010001000000100000100010$$

Fig. 3. The example case given Figure 2 is augmented with sampling to support random access. The beginning bit positions of the $1, 3, 5, 7, 9, 11, 13$th codewords are marked with 1. The $A(Huffman(T))$ array is to be represented by a R/S dictionary data structure to answer `select` queries in constant time.

In the additional bit array there are $\lceil \frac{n}{S} \rceil$ 1 bits and the entropy of that bit array is then $\binom{\lceil \frac{n}{S} \rceil}{n}$. Larger sampling frequencies produce more sparse arrays, which require less space in R/S dictionary representations, with an obvious performance degrade in random access.

4   Non–Prefix–Free (NPF) Codes

In NPF codes we let the codewords to be a prefix of others. The only restriction we need to obey is that the codewords should be distinct.

Definition 1. *Non–Prefix–Free Codes: Assuming that the symbols in the alphabet $\Sigma = \{\epsilon_1, \epsilon_2, \ldots, \epsilon_\sigma\}$ are listed in decreasing order of their occurrences in a given text T, the non-prefix-free codeword assigned to symbol ϵ_i, $1 \leq i \leq \sigma$, is the bit sequence obtained by removing the leftmost 1 bit from the binary representation of integer $(i + 1)$.*

Given Σ and F arrays, we assign codeword 0 to the most frequent symbol, and codeword 1 to the next frequent one. We then continue assigning the 2–bits long codewords 00, 01,10, and 11 to the next most frequent four symbols. In a mathematical notation, what we do is to assign number $1 + i$ to the ith, $1 \leq i \leq \sigma$, most frequent character, and then remove the leftmost significant 1 bit from its minimum–length binary representation. For example, the most frequent one is assigned number $2_{10} = (10)_2$, and when we drop the leftmost 1 bit, the corresponding codeword is simply the bit 0. Notice that the shorter length codewords may be prefixes of longer codewords.

Lemma 1. *The length of the longest codeword in the described NPF coding scheme is at most* $\lfloor \log(\sigma + 1) \rfloor$ *bits.*

Proof. According to the definition 1, the least frequent symbol in Σ will be assigned the integer $\sigma + 1$, whose bit length is $\lceil \log(\sigma + 1) \rceil$. The number of bits that remains after removing the leftmost 1 bit from the binary representation of $(\sigma + 1)$ is $\lceil \log(\sigma + 1) \rceil - 1 = \lfloor \log(\sigma + 1) \rfloor$.

Lemma 2. *Given the number of occurrences of symbols in Σ as $F = \{f_1, f_2, \ldots, f_\sigma\}$, where $f_1 \geq f_2 \geq \ldots \geq f_\sigma$, total length of the NPF encoded bit stream is*

$$(f_{2^{\lfloor \log(\sigma+1) \rfloor} - 1} + \ldots + f_\sigma) \cdot \lfloor \log(\sigma + 1) \rfloor + \sum_{k=1}^{\lfloor \log(\sigma+1) \rfloor - 1} (f_{2^k - 1} + \ldots + f_{2^{k+1} - 2}) \cdot k$$

Proof. The first two most frequent ones will have codewords of length 1 bit long. The next four will be assigned codewords of length 2 bits, as there can be at most $2^2 = 4$ distinct two bits long sequences. Continuing in the same way the last chunk will begin from the $2^{\lfloor \log(\sigma+1) \rfloor} - 1$ th symbol up to the last σth symbol, where all will be assigned codewords of length $\lfloor \log(\sigma + 1) \rfloor$.

Lemma 3. *The NPF codes described in definition 1 are not uniquely decodable.*

Proof. The codewords are not prefix–free as the shorter length codewords are possibly the prefixes of the longer codewords. Thus, NPF is itself ambiguous and cannot be correctly decoded without using additional data structures.

4.1   NPF–RS: NPF Codes Augmented with R/S Dictionaries

We can maintain an additional bit array over the NPF encoded sequence such that the starting positions of each codeword are marked with 1 and rest set to 0. We store this additional array compressed in a R/S dictionary structure supporting constant time **select** operation.

$$\mathbf{T} = NONPREFIXFREE, \mathbf{\Sigma} = \{E, R, F, N, I, O, P, X\}, \mathbf{F} = \{3, 2, 2, 2, 1, 1, 1, 1\}$$
$$\mathbf{\Sigma} \rightarrow \mathbf{W} : E \rightarrow 0, R \rightarrow 1, F \rightarrow 00, N \rightarrow 01, I \rightarrow 10, O \rightarrow 11, P \rightarrow 000, X \rightarrow 001,$$

$$\text{NPF–RS(T)} = \underline{0}11\underline{1}\underline{0}1\underline{0}00\underline{1}0\underline{0}0\underline{1}\underline{0}00\underline{1}00\underline{1}00$$
$$\text{A(NPF–RS(T))} = 10101010011101010010111$$

Fig. 4. NPF–RS coding example

We can extract the ith codeword w_i by running two select queries on the additional bit array to retrieve the beginning of the ith and $i + 1$th codeword. Assuming a word-RAM model, where the word size is larger than or equal to the maximum codeword length, the bits in between these two indices can be retrieved in constant time and be decoded by a table ($W \rightarrow \Sigma$) lookup.

4.2   NPF–WT: NPF Codes Augmented with Wavelet Trees

Another method to bring unique decodability and random access capability
to the NPF codes is to use the wavelet tree structure [8]. After mapping the
symbols to the codewords via the $NPF : \Sigma \to W$ scheme described in definition 1, we create an array $L = \ell_1\ell_2\ldots\ell_n$ such that ℓ_i denotes the number of bits in the codeword corresponding to character t_i. As an example, for
T=NONPREFIXFREE, the $NPF : \Sigma \to W$ scheme maps E, R, F, N, I, O, P, X to
codewords $0, 1, 00, 01, 10, 11, 000, 001$, respectively. Thus, the L array becomes
L=2223112232111, and the alphabet of L is $\{1, 2, 3\}$.

We create a *Huffman–shaped* wavelet tree [4,6] over L. In Huffman–shaped
wavelet tree, the topology of the tree is created in a way that each symbol is
represented in levels equal to its corresponding Huffman code length. For our
sample sequence L, the Huffman code lengths for the symbols 1, 2, and 3 can
be computed as 2, 1, and 2 respectively[2]. Thus, in the Huffman–shaped wavelet
tree the items with label 2 are represented with one level and rest with two levels
as shown in Figure 5. The leaf nodes are reserved for the actual NPF codewords
corresponding that level's code length, e.g., the left child of the root node in
Figure 5 collects the codewords that are two bits long in $NPF(T)$ preserving
their order of appearance.

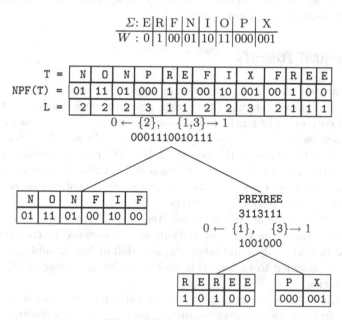

Fig. 5. NPF–WT coding example

---

[2] The numbers 1, 2, and 3 occurs 5, 6, and 2 times in L. Running the standard Huffman
code generation assigns codewords 11 to 1, 0 to 2, and 01 to 3.

In NPF–WT coding scheme the ith symbol can be reached by traversing the tree. Assume we want to access the 13th element in the NPF–WT encoded stream. We check the root bitmap and observe that 13th bit is the 5th 1 bit. Therefore, we follow the right child and check the 5th bit in that node which is the 3rd zero. We move to left child since it is 0, and arrive a leaf node with the knowledge that this leaf node includes the 1 bit long codewords. We go to 3rd bit and extract codeword 1, which represents letter R, and actually the 13th letter in the original sequence. Notice that all rank/select operations in the inner nodes can be achieved in constant time by R/S dictionary data structures.

The overhead we have by the wavelet tree is equal to the 0–order entropy of the L sequence. In other words the number of bits we used in the tree (excluding the leaf nodes that represent the NPF codes of T) is close to the size of the 0–order Huffman compressed L. The $T \to L$ transformation is a down sampling that Σ is mapped onto set $\{1, 2, \ldots, \lfloor \log(\sigma + 1) \rfloor\}$, and thus, the entropy of L is less than the entropy of T. Especially on large alphabets the difference becomes more significant.

The random access time in NPF–WT is not $O(1)$ as it was in previously described NPF–RS. There can be at most $\log(\sigma + 1)$ levels in the Huffman–shaped tree since the Huffman code tree created over L with $\log(\sigma + 1)$ symbols may in worst be of depth equal to its alphabet size. Therefore, in the worst case, NPF–WT guarantees random access in $O(\log \sigma)$ time, where we expect faster response time while retrieving more frequent items.

5   Experimental Results

We have implemented the coding schemes by using the sdsl [5] library, and performed tests on *English* language text, *dna*, and *protein* sequences obtained from Pizza-Chili corpus[3]. The sdsl library provides several alternatives for compressed bitmap representation, where we preferred to use the standard rrr structure that is based on the study of Raman *et. al.* [11]. We considered files of sizes 1MB, 10MB, 50MB, 100MB for each. On each file we also benchmarked the behaviour of the coding schemes on enlarged alphabets by combining consecutive q, $1 \le q \le 8$, symbols into a meta–symbol.

For Huffman–RS(S) scheme, the sampling frequency S was taken to be 1, 10, 50, and 100. Notice that $S = 1$ means marking each codeword in the Huffman encoded text, which actually eliminates the self–delimiting advantage of the Huffman itself. We decided to include this option to give the sense of difference in compression ratios.

We benchmarked the schemes in compression ratio measured in *bits/symbol* and the random access performance measured in μsec/access on a computer running LinuxMint14 operating system with Intel i5 processor and 32GB of memory. We performed the same 1 million random access operations for randomly selected positions on each encoded file and reported their mean value.

---

[3] http://pizzachili.dcc.uchile.cl

We observed very similar behaviour on each file size and decided to include the results on the largest file size, 100MB, where the overhead in the encoded stream effects the coding performance least.

	q σ	1 215	2 8380	3 65861	4 302992	5 845006	6 1710207	7 2801684	8 3965112
NPF-WT	bits/char.	4.770	4.205	3.801	3.504	3.335	3.295	3.362	3.480
	μsec./acc.	2.016	2.999	3.678	4.327	4.934	5.913	5.891	5.726
NPF-RS	bits/char.	5.089	4.617	4.202	3.895	3.710	3.662	3.727	3.856
	μsec./acc.	1.088	1.307	1.616	2.108	2.736	3.843	3.429	3.487
Huffman	bits/char.	4.593	4.112	3.738	3.466	3.314	3.290	3.367	3.509
	μsec./acc.			*Efficient random access not available*					
Huffman-RS(1)	bits/char.	8.438	6.659	5.701	5.060	4.680	4.488	4.432	4.468
	μsec./acc.	1.094	1.376	1.662	2.215	3.872	5.302	7.048	7.877
Huffman-RS(10)	bits/char.	5.720	4.903	4.380	4.017	3.804	3.730	3.770	3.878
	μsec./acc.	2.823	12.194	25.241	39.927	54.941	69.940	82.229	90.402
Huffman-RS(50)	bits/char.	5.258	4.655	4.213	3.891	3.700	3.645	3.695	3.813
	μsec./acc.	10.696	52.087	111.744	175.808	276.680	303.509	354.375	390.062
Huffman-RS(100)	bits/char.	5.193	4.620	4.190	3.873	3.684	3.632	3.683	3.803
	μsec./acc.	20.368	102.648	221.471	346.821	524.471	600.441	687.150	801.020

Fig. 6. Compression and random access performances on 100MB English text

Figure 6 gives the results on the English text. The compression performance of the NPF–WT is very close to the regular Huffman coding, and becomes better especially on larger q values, in which we have larger alphabets. Notice that NPF–WT provides random access which is not supported in Huffman. The random access times in Huffman–RS(50) and Huffman–RS(100) are not competitive and best sampling ration seems to be 10. When compared to NPF–RS and NPF–WT, the Huffman–RS(10) is weaker in both the compression and random access performance even in small alphabet sizes.

	q σ	1 4	2 16	3 64	4 256	5 1024	6 4096	7 16384	8 65491
NPF-WT	bits/char.	2.428	2.093	2.013	1.994	1.981	1.970	1.960	1.955
	μsec./acc.	0.933	1.567	2.037	2.256	2.444	2.709	2.966	3.452
NPF-RS	bits/char.	2.734	2.438	2.414	2.410	2.391	2.384	2.362	2.338
	μsec./acc.	0.949	0.959	1.021	1.112	1.180	1.319	1.492	1.937
Huffman	bits/char.	2.001	1.978	1.956	1.945	1.939	1.934	1.930	1.935
	μsec./acc.			*Efficient random access not available*					
Huffman-RS(1)	bits/char.	4.152	3.749	3.411	3.181	3.051	2.927	2.837	2.784
	μsec./acc.	0.917	0.962	1.005	1.071	1.167	1.308	1.518	1.916
Huffman-RS(10)	bits/char.	2.769	2.496	2.387	2.326	2.288	2.262	2.244	2.236
	μsec./acc.	0.895	1.574	3.053	6.174	10.779	17.091	25.435	38.039
Huffman-RS(50)	bits/char.	2.357	2.270	2.224	2.201	2.188	2.177	2.170	2.171
	μsec./acc.	2.026	5.194	11.849	25.693	46.463	74.878	112.954	170.510
Huffman-RS(100)	bits/char.	2.295	2.238	2.202	2.184	2.174	2.165	2.159	2.162
	μsec./acc.	3.450	9.783	22.668	50.095	90.602	148.513	226.115	335.800

Fig. 7. Compression and random access performances on 100MB DNA data

The results obtained on DNA sequence are reported in Figure 7. The interesting point here is the small alphabet size which makes original Huffman coding to achieve best compression ratios. However, the NPF codecs are very competitive to that with the additional advantage of providing random access. When compared to Huffman–RS(10, the compression of NPF–WT is always superior, and NPF–RS is slighty better for $q = 1$ and $q = 2$. On $q = 1$, the random access timing of the Huffman–RS(10) is better then both NPF codecs, where on all remaining ones the NPF codes make a better job and the NPF–RS is the leader.

The number of symbols in the protein sequence data is in between the English text and DNA sequences, and thus, the results provided in Figure 8 give clues about middle scale alphabets. We observe here that both the compression and random access achieved by the NPF codecs are significantly better in all cases.

	q	1	2	3	4	5	6	7	8
	σ	20	400	8000	159932	2588365	10115106	11707427	10696098
NPF-WT	bits/char.	4.505	4.308	4.230	4.208	4.533	5.673	5.802	5.647
	μsec./acc.	1.699	2.402	2.843	4.247	5.287	5.836	5.683	5.443
NPF-RS	bits/char.	5.108	5.090	5.090	5.020	5.322	6.435	6.516	6.280
	μsec./acc.	0.980	1.166	1.451	2.393	3.603	4.504	4.365	4.196
Huffman	bits/char.	4.196	4.180	4.170	4.185	4.613	5.751	5.921	5.777
	μsec./acc.				*Efficient random access not available*				
Huffman-RS(1)	bits/char.	7.861	6.754	6.217	5.950	6.162	7.098	7.082	6.793
	μsec./acc.	1.033	1.143	1.480	2.729	7.565	15.736	16.975	15.106
Huffman-RS(10)	bits/char.	5.268	4.979	4.866	4.826	5.213	6.294	6.395	6.189
	μsec./acc.	1.702	7.549	20.571	51.948	89.293	122.323	129.657	122.618
Huffman-RS(50)	bits/char.	4.811	4.729	4.696	4.696	5.108	6.205	6.318	6.122
	μsec./acc.	5.685	31.599	91.758	218.822	379.034	510.066	539.238	509.973
Huffman-RS(100)	bits/char.	4.747	4.695	4.672	4.676	5.093	6.191	6.306	6.112
	μsec./acc.	10.746	61.587	183.345	447.087	761.633	1004.790	1009.100	998.729

Fig. 8. Compression and random access performances on 100MB protein sequence

6   Conclusions

In this study we described the NPF codes augmented with R/S dictionaries and wavelet trees, which gave us the opportunity for efficient random access. We compared them against the original Huffman coding, in which random access is not available. We considered a similar augmentation of Huffman codes with the R/S dictionaries to add this support.

Interestingly, the augmented NPF codecs compression capabilities are very close to the original Huffman with the additional random access ability that is missing in Huffman. NPF–WT is the choice for best compression with reasonable random access performance, and NPF–RS behaves the reverse as providing best random access timing with a little bit penalty paid in compression ratio compared to NPF–WT. When compared to Huffman–RS(S) schemes, it seems NPF codes are more advantageous in both metrics.

As a summary, the graphs in Figure 9 depicts the results obtained on 100MB files for $q = 1$ (smallest alphabet size) and also for the q value that provides best compression ratios for the NPF codecs.

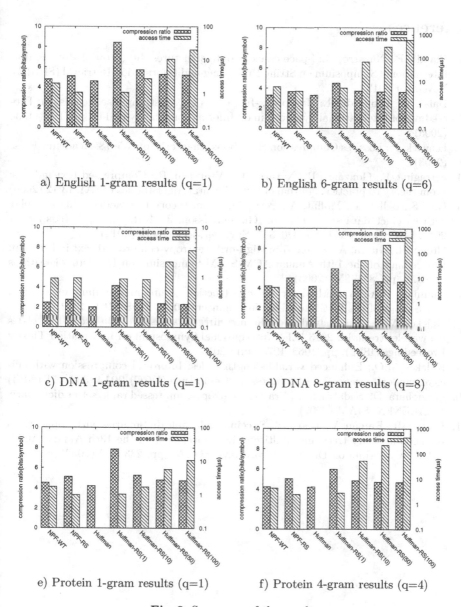

a) English 1-gram results (q=1)   b) English 6-gram results (q=6)

c) DNA 1-gram results (q=1)   d) DNA 8-gram results (q=8)

e) Protein 1-gram results (q=1)   f) Protein 4-gram results (q=4)

Fig. 9. Summary of the results

The NPF codes have received very few attention upto date. It is shown in this study that augmenting them with the compressed data structures they may achieve quite well. The experiments conducted in this study are all static 0–order compression studies. New NPF schemes for higher order compression might be an avenue of future research.

References

1. Claude, F., Navarro, G.: Space efficient data structures. In: Tutorial Presented at International Symposium on String Processing and Information Retrieval (SPIRE), November 2012
2. Dalai, M., Leonardi, R.: Non prefix-free codes for constrained sequences. In: Proceedings of International Symposium on Information Theory (ISIT), pp. 1534–1538 (2005)
3. Fenwick, P.: Lossless Compression Handbook, chapter 3, pp. 55–78. Academic Press (2003)
4. Ferragina, P., González, R., Navarro, G., Venturini, R.: Compressed text indexes: From theory to practice. Journal of Experimental Algorithmics (JEA), 13:12 (2009)
5. Gog, S., Beller, T., Moffat, A., Petri, M.: From theory to practice: plug and play with succinct data structures. In: Gudmundsson, J., Katajainen, J. (eds.) SEA 2014. LNCS, vol. 8504, pp. 326–337. Springer, Heidelberg (2014)
6. Grossi, R., Gupta, A., Vitter, J.S: High-order entropy-compressed text indexes. In: Proceedings of the 14th Annual ACM-SIAM Symposium on Discrete Algorithms (SODA), pp. 841–850 (2003)
7. Huffman, D.A.: A method for the construction of minimum-redundancy codes. Proceedings of the Institute of Radio Engineers 40(9), 1098–1101 (1952)
8. Kulekci, M.O.: Uniquely decodable and directly accessible non-prefix-free codes via wavelet trees. In: 2013 IEEE International Symposium on Information Theory Proceedings (ISIT), pp. 1969–1973, July 2013
9. Kulekci, M.O.: Enhanced variable-length codes: Improved compression with efficient random access. Data Compression Conference (DCC) **2014**, 362–371 (2014)
10. Okanohara, D., Sadakane, K.: Practical entropy-compressed rank/select dictionary. In: ALENEX. SIAM (2007)
11. Raman, R., Raman, V., Rao, S.S.: Succinct indexable dictionaries with applications to encoding k-ary trees and multisets. In: Proceedings of the 13th Annual ACM-SIAM Symposium on Discrete Algorithms (SODA), pp. 233–242 (2002)

Experimental Analysis of an Online Dictionary Matching Algorithm for Regular Expressions with Gaps

Riku Saikkonen, Seppo Sippu, and Eljas Soisalon-Soininen(✉)

Department of Computer Science, School of Science, Aalto University,
P.O. Box 15400, FI-00076 Aalto, Finland
{riku.saikkonen,eljas.soisalon-soininen}@aalto.fi,
seppo.sippu@alumni.aalto.fi

Abstract. Dictionary matching for regular expressions has gained recent interest because of a multitude of applications, including DNA sequence analysis, XML filtering, and network traffic analysis. In some applications, allowing wildcard and character class gaps in strings is enough, but usually the full expressive power of regular expressions is needed. In this paper we present and analyze a new algorithm for online dictionary matching for regular expressions. The unique feature of our algorithm is that it builds upon an algorithm for dictionary matching of string patterns with wildcard gaps, but is also capable of treating more complex regular expressions. In our experiments we used real data from expressions used for filtering spam e-mail. The size of the dictionary, that is, the number of different regular expressions to be matched varied from one to 3080. To find out how our algorithm scales to much larger numbers of patterns, we made small random changes to these patterns to produce up to 100000 patterns that are similar in style. We found out that the scalability of our algorithm is very good, being at its best for 10000–20000 patterns. Our algorithm outperforms the tested competitors for large dictionaries, GNU grep already for tens of patterns and Google's RE2 for hundreds of patterns.

Keywords: Regular-expression matching · Aho–Corasick · Wildcard gaps · Character classes · Spam filtering

1 Introduction

We consider dictionary matching for regular expressions with a large set of operations, not only including wildcards (.), union (|), concatenation and the Kleene star (*), but also many operations that appear in practical applications. These include character classes (e.g., [A-Z] matches any capital letter), bounded and unbounded iteration ($\{l,h\}$ iterates at least l and at most h times and $\{l,\}$ at

---

This research was partially supported by the Academy of Finland.

E. Bampis (Ed.): SEA 2015, LNCS 9125, pp. 327–338, 2015.
DOI: 10.1007/978-3-319-20086-6_25

least l times), and subexpressions enclosed in parentheses. (Our examples follow the standard syntax of POSIX extended regular expressions [11].)

For example, in a set of regular expressions used by an e-mail spam filter the following two patterns appear:

```
urgent.{0,16}(assistance|business|buy|confidential|notice|
              proposal|reply|request|response)
free.{0,12}((instant|express|
            online|no.?obligation).{0,4})+.{0,32}quote
```

Here ? means that the previous character is possibly missing (same as {0,1}), and + that the argument is repeated at least once ({1,}).

In addition to e-mail spam filters, applications include network traffic analysis in general, such as load balancing and intrusion detection and prevention. Other application areas are XML filtering (see e.g. [7]) and computational biology, one example being patterns with variable-length gaps in protein searching [5,6,9].

We require that the matching algorithm works online, that is, the text is scanned only once, and the matches (at most one match for a pattern at a text position) are reported at the point of occurrence. Online solutions are needed for constantly arriving possibly long inputs in data streaming applications, such as XML document filtering.

Our new algorithm is based on a previous solution of the string matching problem, in which the dictionary contains patterns with gaps of the forms $.\{l,\}$ and $.\{l,h\}$ [8]. The algorithm of Haapasalo et al. [8] is based on locating "keywords" of the patterns in the input text, that is, maximal substrings in the patterns that contain only input characters. For computing the keyword matches the Aho–Corasick multiple string matching algorithm [1] is used [8]. For a single pattern, the same approach was previously used by Pinter [12] for fixed-length gaps, and by Morgante et al. [9], Rahman et al. [13], and Bille et al. [3] for variable-length gaps, but not unbounded gaps $.\{l,\}$. A recent work by Amir et al. [2] on dictionary matching for patterns with one gap is based on keyword matching without using Aho–Corasick automata, but its generalization to multiple gaps is open. Bille and Thorup [4] have presented several results on matching regular expressions with gaps, but their algorithm is not directly applicable to multiple patterns.

The idea in the matching of patterns with gaps [8] was that once a prefix of a pattern ending with a keyword has been found, the keyword with preceding gap allowed to arrive next will be searched for. How can we then extend this idea to expressions containing union and iteration operations? We simply number the keywords in the order they appear in the expression, and match prefixes of expression instances that end with a keyword. In this process we need to compute as a preprocessing task for all keywords i the sets $follow(i)$ containing the numbers of all keywords that can follow keyword i. That is, once we have found an instance prefix ending with keyword i we start searching for all keywords in $follow(i)$, observing the preceding gaps. If the found next keyword is a possible last keyword of an instance, a match of the regular pattern is reported.

Another recent multiple regular-expression-matching algorithm [14] has similar properties, but it cannot handle unbounded gaps as elementary operations.

For our experiments we used 3080 patterns extracted from the popular Spam-Assassin software (which uses large numbers of complex regular expressions to detect e-mail advertising) and actual e-mail messages. We compared the performance of our algorithm with GNU grep and Google's RE2. We could not find any research article reporting an implemented multi-pattern regular expression matching algorithm that would have been close to the expressive power of ours. In the online case, that is, the input is scanned only once within the matching process, grep became intolerably slow already for more than about 20 patterns, and RE2 after about 200 patterns, but our algorithm scaled up very well up to the workload of the 3080 real-world patterns. (See Fig. 2 in the Experiments section.) Both grep and RE2 improve if they are allowed to work "offline", that is, they read the input separately for each pattern. For 1000–3080 patterns the best offline competitor took at least 4.5 times as much time as our algorithm.

2   Decomposing Regular Expressions into Keywords

Assume that we are given a string T (called the *text*) over a character alphabet Σ, whose size is assumed to be bounded, and a finite set D (called a *dictionary*) of regular expressions (called *patterns*) P_i.

We decompose each pattern into keywords and gaps: the *keywords* are maximal substrings in Σ^+ of patterns, and the *gaps* maximal substrings of wildcards, character classes and iteration. We assume that each keyword is always preceded by a gap, which thus may be ϵ, the empty string. Complex patterns can be supported by introducing empty keywords; for instance, if a pattern ends at a gap, then an empty keyword can be inserted to follow the gap. (Actually, introducing empty keywords makes matching inefficient, and we present a more efficient solution for these complex cases in Section 5.)

Our task is to determine all occurrences of all patterns $P_i \in D$ in text T. We report a pattern occurrence by a pair of a pattern number and the character position in T of the last character of the occurrence. A pattern may have many occurrences that end at the same character position; all these occurrences are reported once by the same pair of pattern number and character position.

For each pattern P_i, we number the occurrences of its gaps and keywords, so that $gap(i,j)$ denotes the jth gap and $keyword(i,j)$ denotes the jth keyword, that is, the keyword following $gap(i,j)$, $j = 1, 2, \ldots, m_i$, where m_i denotes the number of keywords in pattern P_i. For pattern P_i, we denote by $mingap(i,j)$ and $maxgap(i,j)$, respectively, the minimum and maximum lengths of strings that can be matched by $gap(i,j)$. The length of the jth keyword of pattern P_i is denoted by $length(i,j)$.

To keep track of how the keywords relate to each other, we define, for each pattern P_i, the set $begin(i)$ to contain all j such that $gap(i,j)keyword(i,j)$ appears as a prefix of some instance of P_i, and the set $end(i)$ to contain all j such that $gap(i,j)keyword(i,j)$ appears as a suffix of some instance of P_i. Furthermore, for each $j = 1, \ldots, m_i$, we define the set $follow(i, j)$ to contain all k such that $gap(i, k)keyword(i, k)$ immediately follows $gap(i, j)keyword(i, j)$ in some instance of P_i.

For example, for pattern $P_i = \mathtt{aa(bb|.\{1,3\}c)*d}$, $begin(i) = \{1\}$, $end(i) = \{4\}$, $follow(i, 1) = follow(i, 2) = follow(i, 3) = \{2, 3, 4\}$, $follow(i, 4) = \emptyset$.

The sets $begin(i)$, $end(i)$, and $follow(i, j)$ can be computed using a depth-first traversal of the abstract syntax tree of the pattern P_i (during traversal, keep track of the keywords that precede the current position, and when reaching a keyword, add its number to the $follow$ sets of the preceding keywords). If P_i contains union or iteration, the combined size of the sets may be quadratic in the number of keyword occurrences in P_i, as is the case with the pattern $\mathtt{a?b?c?\cdots z?}$, for example. In the special case in which P_i contains neither union nor iteration, the sets $begin(i)$, $end(i)$, and $follow(i, j)$ are all singletons and their computation takes only linear time.

3   Aho–Corasick with Dynamic Output

We construct an Aho–Corasick pattern-matching automaton from the set of all keywords in the patterns. While the output function of the standard Aho–Corasick PMA is a fixed mapping from states to keywords, we instead use a dynamically changing output function, denoted $current\text{-}output(q)$ for state q. This allows us to keep only relevant keywords enabled in the PMA: for instance, at the beginning the PMA only recognizes the first keywords from each pattern.

The dynamically changing output function is represented by sets $current\text{-}output(q)$ containing $output\ tuples$ of the form (i, j, b, e), where $q = state(keyword (i, j))$, the state corresponding to the jth keyword of the ith pattern (the state reached from the initial state upon reading the keyword), and b and e are the earliest and latest character positions in text T at which some partial match of pattern P_i up to and including an instance of the jth keyword can possibly be found. The latest possible character position e may be ∞, meaning the end of the text. Tuples (i, j, b, e) are inserted into $current\text{-}output(q)$ only at the point when $char\text{-}count$ (the current text position) has reached the value b, so that tuples (i, j, b, e) are stored and often denoted as triples (i, j, e).

The output tuples are inserted to the sets $current\text{-}output(q)$ through a set called $pending\text{-}output$, using an array of $maxdist = \max\{mingap(i, j) + length(i, j) \mid i \geq 1, j \geq 1\}$ elements such that for any character position b in the input text, the element $pending\text{-}output(b \bmod maxdist)$ contains an unordered set of tuples (i, j, e), called $pending\ output\ tuples$. Before reading the first character of the text, the initial pending output tuples (i, j, e) for all patterns P_i, with $j \in begin(i)$ and $e = maxgap(i, j) + length(i, j)$, are inserted into the sets $pending\text{-}output(b \bmod maxdist)$, where $b = mingap(i, j) + length(i, j)$. Before reading each new character of the text, all tuples (i, j, e) from the set $pending\text{-}output(char\text{-}count \bmod maxdist)$ are distributed into the sets $current\text{-}output(q)$, $q = state(keyword(i, j))$.

Detailed algorithms for matching string patterns with gaps are given in our earlier article [8]. The main difference here is that we may have many possibilities

traverse-output-path(*state*):
   $q \leftarrow state$
   traversed ← false
   while not *traversed* **do**
      for all $(i,j,e) \in current\text{-}output(q)$ **do**
         if $e < char\text{-}count$ **then**
            Delete (i,j,e) from $current\text{-}output(q)$.
         else if gap and boundaries match (see Section 5) **then**
            if $j \in end(i)$ **then**
               Report a match of pattern P_i at position *char-count*.
            end if
            for all $k \in follow(i,j)$ **do**
               $q' \leftarrow state(keyword(i,k))$
               $b' \leftarrow char\text{-}count + mingap(i,k) + length(i,k)$
               $e' \leftarrow char\text{-}count + maxgap(i,k) + length(i,k)$
               Insert (i,k,e') into $pending\text{-}output(b' \bmod maxdist)$.
               if $e' = \infty$ and $follow(i,j) = \{k\}$ and $j \notin end(i)$ **then**
                  Delete (i,j,e) from $current\text{-}output(q)$.
               end if
            end for
         end if
      end for
      if $q = initial\text{-}state$ **then**
         traversed ← true
      else
         $q \leftarrow output\text{-}fail(q)$
      end if
   end while

Fig. 1. The *traverse-output-path* algorithm

of continuation given by $follow(i,j)$, not only the following keyword preceded by a gap.

When visiting a state q (Alg. 1), in addition to $current\text{-}output(q)$ we need to process output tuples from states in the fail path from q for which the current output is nonempty, denoted $output\text{-}fail$ in the algorithm.

If $current\text{-}output(q)$ contains a tuple (i,j,e), where $char\text{-}count \leq e$, then a match of the jth keyword of pattern P_i is obtained. This keyword match must then be validated by examining character classes in $gap(i,j)$ and some other restrictions, as described later in Section 5. For now, we assume that there are no character classes nor other restrictions. If $j \in end(i)$, then a match of the entire pattern P_i is obtained. Otherwise, output tuples (i,k,e') with $k \in follow(i,j)$ are inserted into the sets $pending\text{-}output(b' \bmod maxdist)$, where

$$b' = char\text{-}count + mingap(i,k) + length(i,k), \text{ and}$$
$$e' = char\text{-}count + maxgap(i,k) + length(i,k).$$

Here $e' = \infty$ if $maxgap(i,k) = \infty$. If $e' = \infty$ and $follow(i,j) = \{k\}$ and $j \notin end(i)$, we can delete (i,j,e) from $current\text{-}output(q)$. This is because (i,j,e) can no longer lead to any complete match that we could not obtain from (i,k,e').

In general, it is sufficient to store for each (i,j) only one output tuple (i,j,e), namely the one with the greatest e determined thus far. (For details see [8].)

4   Correctness and Complexity

Each instance of pattern P_i is of the form

$$gap(i, j_1)keyword(i, j_1) \ldots gap(i, j_m)keyword(i, j_m),$$

where $j_1 \in begin(i)$, $j_m \in end(i)$, and $j_{k+1} \in follow(i, j_k)$, for $k = 1, \ldots, m - 1$. Let us denote by P_{i,j_k} the prefix of such an instance that ends at $keyword(i, j_k)$:

$$P_{i,j_k} = gap(i, j_1)keyword(i, j_1) \ldots gap(i, j_k)keyword(i, j_k).$$

Whenever the prefix P_{i,j_k} has been recognized at some character position (indicated by the global variable *char-count* in the algorithm), a pending output tuple (i, j_{k+1}, b', e') will be created by inserting (i, j_{k+1}, e') into the set *pending-output*$(b' \bmod maxdist)$ in Alg. 1. This tuple will be further moved to *current-output*$(state(keyword(i, j_{k+1})))$ when *char-count* has reached the value b'. It is easy to see that Alg. 1 finds exactly all prefixes of the pattern instances that match the text, and reports as matches all those that are complete instances of some pattern in the dictionary.

Within each iteration of the **while** loop in Alg. 1 the outer **for** loop is performed for all tuples (i, j, e) that belong to *current-output*(q). Because for any pair (i, j) the set *current-output*(q) for $q = state(keyword(i, j))$ can contain at most one output tuple (i, j, e), we can conclude that the number of iterations performed in the outer **for** loop for state q is bounded by the number of different occurrences of keywords equal to $keyword(i, j) = string(q)$ found at state q. Moreover, as output tuples for the same pair (i, j) cannot occur in different states, we conclude that the total number of iterations of the performed **for** loops at character position *char-count* is bounded by the number of keyword occurrences at *char-count*.

The inner **for** loop in Alg. 1 will be traversed as many times as there are elements in $follow(i, j)$. Altogether we have:

Theorem 1. *Let D be a dictionary of regular patterns P_i with gaps as defined above, and T a text to be matched using Alg. 1. After preprocessing in time $O(|D| + K^2)$, the time complexity of the algorithm is*

$$O(|T| + K \sum_{c=1}^{|T|} K_c + occ(pattern\text{-}prefixes)),$$

where K denotes the maximal number of keywords in one pattern, K_c is the number of different keyword strings in D that match at text position $c = $ char-count, and occ(pattern-prefixes) denotes the number of occurrences of pattern-instance prefixes $P_{i,j_k} = gap(i, j_1)keyword(i, j_1) \ldots gap(i, j_k)keyword(i, j_k)$ in text T.

For an individual text position $c = $ *char-count* the factor K is usually overly pessimistic, because the size of the set $follow(i, j)$ is often very small. Notice that there may be many different keyword occurrences in D that are the same as a string, called a keyword string.

5   Character Classes

A character class specifies a range of characters that can match at a particular point. For instance, the pattern [a-f]{3} matches abc and fff but not a01. Since character classes commonly occur together with iteration, we store information on character classes in the gap associated with each keyword.

For supporting character classes (and for line-based matching, Section 6), we need to store b in the output tuples, i.e., (i, j, b, e) instead of just (i, j, e). Assume that the PMA finds an output tuple (i, j, b, e) in *current-output*, denoting a match of keyword j of pattern P_i. Before accepting the match, we check that any character-class constrains in $gap(i, j)$ are satisfied by the text preceding the keyword – if not, we can simply ignore the keyword match. The gap needs to match text positions starting from what *char-count* was at the point where the output tuple was added (this is $b - mingap(i, j) - length(keyword(i, j))$ due to how b is calculated) up to the current $char\text{-}count - length(keyword(i, j))$.

To match the gap efficiently, we split the pattern that forms the gap into three parts: fixed-length parts at the beginning and end of the gap, and a variable-length part in the middle. For each gap, we store (position, character class) pairs from the fixed length parts, and the character class of the variable length part. For instance, for . [a-z] .. [0-9]{3}[a-z]*[0-9]{5} we store:

> first part: length 7, [a-z] in position 2, [0-9] in positions 5 to 7
> middle part: [a-z]
> last part: length 5, [0-9] in all positions
> $$mingap = 7 + 5 = 12, \quad maxgap = \infty.$$

Length information for the variable-length middle part can be derived from *maxgap*. Instances of the match-anything character class . need not be stored explicitly, and a middle part with . lets us skip the corresponding text. The required storage per gap is proportional to *mingap*.

We store a character class simply as a bit array of $|\Sigma|$ bits specifying which characters are acceptable (other representations are also possible). We can reuse the bit array for all character classes (in any pattern) that have the same content. The required storage is thus only $c \cdot |\Sigma|$ bits, where c is the number of distinct character classes occurring in all the patterns.

The above representation is sufficient for many real-world patterns, but it is too simple to match complex gaps such as [a-k]{5,10}[f-z]{3,7} or ([0-9]{3}|[0-9a-f]{2,4})*. A simple solution would be to introduce empty keywords that split such gaps into simpler parts. However, empty keywords are inefficient for the PMA: they match at every character as long as they are active.

Instead of handling these complex cases ourselves, we decided to employ a *fallback matcher*[1]: we form a conservative approximation of the complex gap and

---

[1] We got the idea of using a fallback matcher from GNU grep, which has three levels of matchers: a very fast Boyer–Moore implementation that looks for just one fixed keyword, an automaton that does not support backreferences, and finally the full-featured regular expression matcher from the C library, which is called only if the other automaton matches and the pattern contains backreferences.

set a flag specifying that a fallback matcher needs to be run if the full pattern matches. This can be implemented using a bit array with one bit per pattern, initialized to 0. In preprocessing, construct a set

$$F = \{(i,j) \mid gap(i,j) \text{ requires the fallback matcher}\};$$

while matching, set the pattern-specific bit when a keyword $(i,j) \in F$ is matched.

The approximation needs to be conservative in the sense that it matches all texts that the actual gap would match. For instance, the above examples could be approximated (using our three-part gaps) as `[a-k]{5}[a-z]{0,9}[f-z]{3}` and `[0-9a-f]{2,}`. Union operations can produce slightly more complex cases; for instance, in `(foo[a-z]|bar[a-z0-9]{2,5})[h-z]baz` the gap before `baz` can be conservatively approximated by `[a-z0-9]{1,5}[h-z]` or more easily just `.{1,5}[h-z]` or even `.{2,6}`.

In line-based matching mode (see below), the fallback matcher can be given the full line, or the line up to where the pattern matched. Otherwise, we need to keep track of the first keyword match of each pattern. If keyword (i,j) matches at *char-count* and this is the first keyword match for pattern i, the first character given to the fallback matcher is at position $char\text{-}count - mingap(i,j) - length(i,j)$.

6   Line-Based Matching

Many real-world tools for regular expression searching (e.g., grep) operate in a line-based fashion: matches cannot span lines, and often the default output of the tool simply lists the matching lines. Supporting this requires the full state of the PMA to be reinitialized to the initial state whenever a line break character is read. This is enough to support lines as well as other kinds of records with a single-character record separator.

However, this trivial implementation is too slow to be practical for our purposes: in the SpamAssassin public corpus we used in our experiments, line breaks occur about every 45 characters. As described in Section 3, reinitializing the state of the PMA requires creating pending output tuples from the *begin* set of every pattern; moreover, they will all be distributed into *current-output* while reading the first *maxdist* characters of each line. Most of these states will normally not be visited before the next line break, so updating *current-output* is unnecessary.

Instead, our solution is to keep track of line numbers and have a separate array *state-age* that stores the line number when each state was last visited. Also, *initialize-output* should distribute the initial output tuples directly into *current-output* (leaving *pending-output* empty), and make a copy of the initial state of *current-output* in a new array *initial-current-output*. Before visiting a state q, if the line number in *state-age*(q) is not current, *current-output*(q) is reinitialized from *initial-current-output*(q). The only thing remaining to be done at a line break is to empty *pending-output*. Since initial output tuples now do not use *pending-output*, Alg. 1 must now ignore output tuples with $b < char\text{-}count$. This does not have an adverse effect on running time, since these only occur in states visited when reading the first *maxdist* characters of each line. In practice, these modifications make line-based matching work at about the same speed as when line breaks are ignored.

Our implementation also supports other real-world features from POSIX extended regular expressions (the same feature set as GNU grep), either directly or through the fallback matcher. For instance, almost all patterns in our experiments specify word or line boundaries (aka anchors, ^ $ \< \> \b). They usually occur next to keywords, so we accept a keyword match only after checking for such a boundary if one is specified.

7 Experiments

For our experiments, we used real-world regular expression patterns from the SpamAssassin software[2], which detects e-mail advertisements mainly by using a large number of complex regular expressions. A reason for selecting this application is that the regular expressions are more varied than typical patterns from protein searching or XML filtering applications [6,7,9]: for instance they contain complex character classes, short and long keywords, and several levels of union and iteration in various contexts.

We compared our algorithm with GNU grep[3] and Google's RE2 library[4]. We used these real-world matchers, as we could not find implementations of "academic" algorithms that would support even a reasonable subset of the features required by the real-world regular expressions; for instance, we could not find a nrgrep [10] implementation that would have supported $\{l, h\}$ or anchors.

Our own implementation was written in C++, compiled using the GNU C++ compiler version 4.7.2, and run on a 3.2 GHz Intel i5-3470 CPU. As the fallback matcher, we used the regular expression matcher from the GNU C library (version 2.13; grep uses the same as its final fallback matcher).

All experiments were repeated 10 times, each time with a new selection of patterns; we show averages. All running times include preprocessing.

We optimized the Aho–Corasick PMA in our algorithm as follows. Our state transition function is a simple array of 256 entries (this was faster than binary search from an array of active entries), and we copied transitions from fail states into each state, so that the fail chain does not need to be traversed during matching. For this input data, it was somewhat faster not to use *pending-output* (instead distributing all output tuples directly to *current-output*), though this increases the worst-case complexity. We also converted one-character keywords to character classes, so that they become parts of gaps instead of keywords to be matched by the PMA (this last optimization gave a moderate increase in speed for large numbers of patterns).

---

[2] SpamAssassin standard ruleset version 3.4.0.r1565117 plus rules used by Debian mailing lists, svn://svn.debian.org/svn/pkg-listmaster revision 450.

[3] Version 2.12 from Debian GNU/Linux 7. We used the "C" locale, avoiding Unicode characters which make matching much slower in some cases. We used grep options -Ec: support extended regular expressions and count the number of matched lines.

[4] RE2 version 20140304 as packaged in Debian.

SpamAssassin uses Perl regular expressions, which have a slightly different syntax; we converted them to POSIX syntax. We had to leave out about 440 patterns that used various non-regular features, mostly look-ahead matching, that are only supported by Perl (though we could support them by making our algorithm as well as grep and RE2 use the Perl matcher as their final fallback matcher). In addition, we left out 17 patterns that use backreferences (RE2 does not support them), 8 patterns which contain no keywords at all (these are not supported by our keyword-based matcher) and 92 patterns that have only single-character keywords (not supported after the last optimization described above); we use the remaining 3080 patterns. SpamAssassin actually preprocesses its input and matches patterns one by one, some against short portions of the e-mails (e.g., only the Subject header); since we use multi-pattern matching, we instead match all patterns together against the full raw e-mails. As input data,

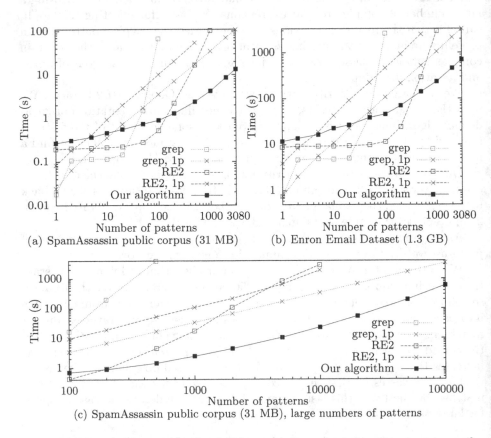

(a) SpamAssassin public corpus (31 MB)

(b) Enron Email Dataset (1.3 GB)

(c) SpamAssassin public corpus (31 MB), large numbers of patterns

Fig. 2. Results from experiments: (a–b) actual SpamAssassin patterns on e-mails, (c) SpamAssassin-derived synthetic patterns. Some values for grep and RE2 with large numbers of patterns are missing, because they take too long to run. The "1p" variants avoid multi-pattern matching by running the matcher one pattern at a time, re-reading the input file (for RE2, each line) for each of the single-pattern matchers.

we used the SpamAssassin public corpus[5] (31 megabytes of e-mails collected for testing spam filtering systems) and the 1.4-gigabyte Enron Email Dataset[6].

Figures 2(a–b) show results from our experiment. As seen in the figure, grep (resp. RE2) becomes very slow when given more than about 20 (resp. 200) patterns. To get useful results for a larger number of patterns, we also ran a separate instance of grep for each pattern (thus avoiding its multi-pattern mode; this is "1p" in the figures). For RE2, we could control the matcher more closely, so for its "1p" mode we precompiled all n patterns and gave each input line separately to each of the n matchers.

Our SpamAssassin input data gave us only 3080 patterns. To find out how our algorithm scales to much larger numbers of patterns, we made small random changes to these patterns to produce more patterns that are similar in style. Specifically, for each original pattern, we changed one randomly selected character in a keyword of at least 5 characters to a random character. We did this for two randomly selected keywords in each pattern, since the original patterns contain about two new keywords per pattern. This was repeated until we had a total of more than 100000 distinct patterns. Results from this experiment are shown in Figure 2(c).

It is clear from Figure 2 that our algorithm scales up much better than the others. The best competitors are normal RE2 (for 2–100 patterns, takes 25–68% of the time of our algorithm, but gets very slow for hundreds or thousands of patterns) and one-pattern-at-a-time (1p) grep. The best competitor takes 12–16 times as much time for 2000–20000 patterns in Figure 2(c) (in Figure 2(a–b), at least 4.5 times for 1000–3080 patterns). Our algorithm is not quite as fast proportionally with more than 20000 patterns, but even for 100000 patterns the best competitor takes 5.7 times as much time.

8   Conclusions

We have presented a new online algorithm for dictionary matching when patterns are regular expressions over keywords and gaps. In addition to the basic operations—catenation, union, and iteration—we support all POSIX extended regular expression features, such as character classes and word boundaries.

Our experiment with SpamAssassin e-mail filters shows that our algorithm scales up well with the size of the dictionary, and outperforms GNU grep for dictionaries of size more than 50, and Google's RE2 for dictionaries of size more than 400. After these dictionary sizes the time needed by grep and RE2 seems to grow very fast, and we could not measure their running times for large numbers of patterns and/or for long inputs (cf. Fig. 2).

In order to measure the scalability of our algorithm for larger numbers of patterns than the 3080 we got from SpamAssassin, we created up to 100000

---

[5] http://spamassassin.apache.org/publiccorpus/, the file labeled "20050311".

[6] The version dated August 21 2009 from https://www.cs.cmu.edu/~./enron/.

distinct patterns by slightly varying the ones from SpamAssassin. The results show that the scalability is very good up to 10000–20000 patterns, but degrades slightly thereafter.

References

1. Aho, A.V., Corasick, M.J.: Efficient String Matching: an Aid to Bibliographic Search. Commun. of the ACM **18**, 333–340 (1975)
2. Amir, A., Levy, A., Porat, E., Shalom, B.R.: Dictionary matching with one gap. In: Kulikov, A.S., Kuznetsov, S.O., Pevzner, P. (eds.) CPM 2014. LNCS, vol. 8486, pp. 11–20. Springer, Heidelberg (2014)
3. Bille, P., Gørtz, I.L., Vildhøj, H.W., Wind, D.K.: String Matching with Variable Length Gaps. Theoretical Computer Science **443**, 25–34 (2012)
4. Bille, P., Thorup, M.: Regular expression matching with multi-strings and intervals. In: Proc. of the 21st Annual ACM-SIAM Symposium on Discrete Algorithms, pp. 1297–1308 (2010)
5. Bucher, P., Bairoch, A.: A generalized profile syntax for biomolecular sequence motifs and its function in automatic sequence interpretation. In: Proc. of Intelligent Systems for Molecular Biology, ISMB, pp. 53–61 (1994)
6. De Castro, E., Sigrist, C.J.A., Gattiker, A., Bulliard, V., Langendijk-Genevaux, P.S., Gasteiger, E., Bairoch, E.A., Hulo, N.: ScanProsite: Detection of PROSITE Signature Matches and ProRule-Associated Functional and Structural Residues in Proteins. Nucleic Acids Res. **34**, 362–365 (2006)
7. Diao, Y., Rizvi, S., Franklin, M.J.: Towards an internet-scale XML dissemination service. In: Proc. of Very Large Data Bases, VLDB, pp. 612–623 (2004)
8. Haapasalo, T., Silvasti, P., Sippu, S., Soisalon-Soininen, E.: Online dictionary matching with variable-length gaps. In: Pardalos, P.M., Rebennack, S. (eds.) SEA 2011. LNCS, vol. 6630, pp. 76–87. Springer, Heidelberg (2011)
9. Morgante, M., Policriti, A., Vitacolonna, N., Zuccolo, A.: Structured Motifs Search. J. Comput. Biol. **12**, 1065–1082 (2005)
10. Navarro, G.: NR-Grep: A Fast and Flexible Pattern-Matching Tool. Software - Practice and Experience - SPE **31**, 1265–1312 (2001)
11. The Open Group, Regular Expressions, Chapter 9 of The Open Group Base Specifications Issue 6, Base Definitions volume, IEEE Std 1003.1, The IEEE and the Open Group (2004)
12. Pinter, R.Y.: Efficient string matching. In: Combinatorial Algorithms on Words, NATO Advanced Science Institute Series F: Computer and System Sciences, vol. 12, pp. 11–29 (1985)
13. Rahman, M.S., Iliopoulos, C.S., Lee, I., Mohamed, M., Smyth, W.F.: Finding patterns with variable length gaps or don't cares. In: Chen, D.Z., Lee, D.T. (eds.) COCOON 2006. LNCS, vol. 4112, pp. 146–155. Springer, Heidelberg (2006)
14. Sippu, S., Soisalon-Soininen, E.: Online matching of multiple regular patterns with gaps and character classes. In: Dediu, A.-H., Martín-Vide, C., Truthe, B. (eds.) LATA 2013. LNCS, vol. 7810, pp. 523–534. Springer, Heidelberg (2013)

An Empirical Study of Finding Approximate Equilibria in Bimatrix Games

John Fearnley[(⊠)], Tobenna Peter Igwe, and Rahul Savani

Department of Computer Science, University of Liverpool, Liverpool, UK
john.fearnley@liverpool.ac.uk

Abstract. While there have been a number of studies about the efficacy of methods to find exact Nash equilibria in bimatrix games, there has been little empirical work on finding approximate Nash equilibria. Here we provide such a study that compares a number of approximation methods and exact methods. In particular, we explore the trade-off between the quality of approximate equilibrium and the required running time to find one. We found that the existing library GAMUT, which has been the de facto standard that has been used to test exact methods, is insufficient as a test bed for approximation methods since many of its games have pure equilibria or other easy-to-find good approximate equilibria. We extend the breadth and depth of our study by including new interesting families of bimatrix games, and studying bimatrix games upto size 2000×2000. Finally, we provide new close-to-worst-case examples for the best-performing algorithms for finding approximate Nash equilibria.

1 Introduction

The computation of Nash equilibria in bimatrix games is one of the central topics in game theory, which has received much attention from a theoretical point of view. It has been shown that the problem of finding a Nash equilibrium is PPAD-complete [3,5], which implies that we are unlikely to find a polynomial-time algorithm for this problem. Naturally, this has led to a line of work studying the complexity of finding *approximate* Nash equilibria [2,6,7,11,19].

Most of the work on approximation algorithms has focussed on additive approximations, where two different notions are used. An ϵ-*Nash equilibrium* requires that both players achieve an expected payoff that is within ϵ of a best response, while the stronger notion of ϵ-*well supported Nash equilibrium* (ϵ-WSNE) requires that both players only place probability on strategies that are within ϵ of a best response. The current state of the art for ϵ-Nash equilibria is the algorithm of Tsaknakis and Spirakis [19], which finds a 0.3393-Nash equilibrium in polynomial time, and the current state of the art of ϵ-WSNE is the algorithm of Fearnley et al. [7], which finds a $(\frac{2}{3} - 0.00591)$-WSNE in polynomial time.

So far, most of the work on approximate equilibria has been theoretical in nature. The goal of this paper is to answer the following question: *Are approximate equilibria relevant to the problem of solving bimatrix games in practice?* To answer this, we must study several related questions.

© Springer International Publishing Switzerland 2015
E. Bampis (Ed.): SEA 2015, LNCS 9125, pp. 339–351, 2015.
DOI: 10.1007/978-3-319-20086-6_26

- Firstly, how good are the algorithms for finding exact Nash equilibria in practice? If they are good enough, then there is no need for approximation. Otherwise, how much faster are the approximation algorithms?
- Secondly, what quality of approximation do the approximation algorithms provide in practice? If the best theoretical guarantee of a 0.3393-Nash equilibrium is not beaten in practice, it is unlikely to be useful.
- Finally, is there a trade off between running time and approximation? We have a wide variety of approximation algorithms, from those that solve a single linear program, to those that perform complicated gradient descent procedures. Do fast algorithms generally produce worse approximate equilibria? Should our desired quality of equilibrium impact our choice of algorithm?

Our contribution. While there have been several empirical studies on computing exact equilibria [4,8,15,16], the empirical work for approximate equilibria has so far been limited to a paper [20] that evaluates the algorithm of Tsaknakis and Spirakis (TS) [19], and one that looks exclusively at symmetric games [12]. We address this by performing a comprehensive study of approximation algorithms. We compare the performance of five different algorithms for finding approximate equilibria on 15 different types of game. Moreover, we include two algorithms for finding exact equilibria: the Lemke-Howson algorithm and support enumeration.

This allows us to answer the questions posed earlier. Firstly, we find that approximation algorithms can tackle instances that exact algorithms cannot. With a timeout of 15 minutes, we found that exact algorithms were mostly unable to solve instances of size 1000×1000, whereas approximation algorithms could easily tackle instances of size 2000×2000. Secondly, we find that approximation algorithms often perform much better than their theoretical worst case guarantees: in agreement with the experimental study of Tsaknakis et al. [20], we find that the TS algorithm often finds 0.01-Nash equilibria or better. In answer to the third question: while our data shows that the TS algorithm clearly wins in terms of quality of approximation, it is usually the slowest, and if we only require weaker approximate equilibria, then other algorithms can find one faster.

To obtain our results, we tested the algorithms on a wide variety of games. Previous work on exact equilibria has typically used the GAMUT library [14]. However, almost all of the games provided by GAMUT have exact pure Nash equilibria, so using only these games could skew our results. For example, the work of Porter et al. [15] concluded that support enumeration typically outperforms the Lemke-Howson algorithm, based on the fact that support enumeration can quickly find the pure equilibria in the games provided by GAMUT.

There are many practical applications of game theory where all equilibria use mixed strategy profiles. Our second contribution is to define several natural classes of games that do not have pure Nash equilibria. All of our game generators and algorithm implementations are open source and freely available[1]. Our results show that algorithms perform very differently on these games, so they should be included in any future study of the practical aspects of computing equilibria.

---

[1] http://bimatrix-games.github.io/

The excellent performance of the TS algorithm raises the question of whether the upper bound of 0.3393 on the quality of approximation of this algorithm is tight. We used a genetic algorithm to search for worst-case examples, and we found a 5×5 bimatrix game in which the TS algorithm gives a 0.3385-Nash equilibrium, which shows that the performance guarantee is essentially tight. We applied this technique to the algorithm of Fearnley et al. [7] for finding ϵ-WSNE, which likewise had no good theoretical lower bound. However, we were only able to find an example for which the algorithm gives a 0.4799-WSNE, which is far from the upper bound of ($\frac{2}{3} - 0.00591375$). Finally, to test the limits of approximation techniques, we ran the same procedure against the combination of our three best approximation algorithms (the TS algorithm, the algorithm of Bosse et al. [2], and the best pure strategy pair). Here the fitness function was the minimum of the approximations provided by the algorithms. We found a game for which all of the algorithms gave no better than a 0.3189-Nash equilibrium, which is relatively close to the theoretical upper bound of 0.3393, and indicates that new techniques will be needed to advance the theoretical state of the art.

Related Work. For exact equilibria, Porter et al. [15] use GAMUT to compare support enumeration (SE) with the Lemke-Howson (LH) algorithm. They showed that SE performs well when compared to LH, because many of the games in the library have small support equilibria. They also highlight random games and covariant games as the most challenging GAMUT games for SE and LH, which our results also confirm. Most of their experiments considered games of size 600×600, but they did consider random games at game sizes up to 1000×1000.

Sandholm at al. [16] describe four ways of solving games via mixed integer programming (MIP). Experiments were carried out games of size 150×150 provided by GAMUT. It was found that MIP performed better than LH but was outclassed by SE.

Codenotti et al. [4] studied of LH algorithm, on random and covariant games, where it was found that LH has a running time of $O(n^7)$ for $n \times n$ covariant games. They presented a heuristic which involves running different LH paths in parallel, which showed an improvement over LH for random games, but not covariant games. They looked at games of size up to 1000×1000.

Tsaknakis et al. [20] performed an experimental analysis of their algorithm (TS) for finding a 0.3393-Nash equilibrium in polynomial time. They studied games of size 100×100, and they also constructed games with no small support equilibria, to prevent easy solutions. It was found that TS always finds a 0.015-Nash equilibrium or better. We confirm the result that in general TS performs well, however among some of our games TS only finds a 0.14-Nash equilibrium.

Gatti et al. [8] evaluated the performance of LH, MIP and SE. They introduced a number of heuristics, and compared their performance on the games from GAMUT. They found that none of the methods was superior for all games. They did look at the quality of approximation achieved by their heuristics and algorithms, but the largest games they looked at were of size 150×150.

2 Experimental Setup

Algorithms. Now, we describe the algorithms that are studied in this paper. For the computation of exact equilibria, we study the following two algorithms.

- **LH.** The Lemke-Howson algorithm is a widely-used pivoting algorithm [13]. It has exponential worst case behaviour [17], and it is PSPACE-hard to compute the equilibrium that it finds [10]. We use the implementation from [4], modified so that degeneracy is resolved by the lexicographic minimum ratio test.
- **SE.** Support enumeration is brute force algorithm that, for every possible pair of supports, solves a system of linear equations to check for an equilibrium. Our implementation goes through supports from small to large cardinality.

There is a wide variety of algorithms for finding ϵ-Nash equilibria, and we have implemented the following (polynomial-time) algorithms:

- **Pure.** This algorithm checks all pure strategy profiles and returns one that gives the best ϵ-Nash equilibrium.
- **DMP.** This algorithm was given by Daskalakis et al. [6]. It finds a 0.5-Nash equilibrium using a very simple approach that starts with an arbitrary pure strategy, and then makes two best response queries.
- **BBM1.** This is the first of two algorithms given by Bosse et al. [2]. It finds a 0.3819-Nash equilibrium. Given a bimatrix game (R, C), BBM1 solves the zero-sum game $(R - C, C - R)$ using linear programming. Then it proceeds in a similar manner to DMP, but uses the LP solution as the initial strategy.
- **BBM2.** This is the second of the two algorithms given by Bosse et al. [2]. It finds a 0.3639-Nash equilibrium. It is an adaptation of BBM1 that contains some extra steps to deal with cases where the first algorithm performs poorly.
- **TS.** This algorithm was given by Tsaknakis and Spirakis [19]. It finds a $(0.3393 + \delta)$-Nash equilibrium, where δ is an arbitrary positive constant. The algorithm uses gradient descent over the space of mixed strategy profiles. The objective function is the quality of approximate Nash equilibrium. The algorithm finds a stationary point. If the stationary point is not a $(0.3393 + \delta)$-Nash equilibrium, then it can be used to find a second point that is a $(0.3393+\delta)$-Nash equilibrium. To investigate the dependence of the algorithm on δ, we use two versions with $\delta = 0.2$ and $\delta = 0.001$. We refer to these as **TS2** and **TS001**, respectively. At the end of Section 3, we analyze the effect of the choice of δ.

There has been comparatively less study of algorithms to find approximate well-supported Nash equilibria. We implemented the following two algorithms:

- **KS.** This algorithm was given by Kontogiannis and Spirakis, and it finds a $\frac{2}{3}$-WSNE [11]. The algorithm first checks all pure strategy profiles in order to determine if there is a pure $\frac{2}{3}$-WSNE. If not, then the algorithm solves the same zero-sum game as BBM1/BBM2, and the equilibrium of this game is a $\frac{2}{3}$-WSNE.

- **KS+.** This algorithm gives an improved approximation guarantee compared to KS [7] of ($\frac{2}{3}$ − 0.00591). It combines the KS algorithm with two extra procedures: one that finds the best WSNE with 2 × 2 support, and one that finds the best WSNE on the supports from an equilibrium to the KS zero-sum game.

Game Classes. We now describe the classes of games used in our study. In order to have a consistent meaning of approximation guarantees, all games are scaled to have payoffs in $[0, 1]$. Firstly, we used games provided by the GAMUT library. We used every class of games in GAMUT that could be scaled indefinitely. We eliminated the classes of games that have fixed size, and we were also forced to eliminate some classes of games because their generators either crashed or produced invalid games when asked to produce games with more than 1000 strategies per player, which included BidirectionalLEG and RandomLEG. We were left with the games shown below.

BertrandOligopoly	CournotDuopoly	CovariantGame
GrabTheDollar	GuessTwoThirdAve	MinimumEffortGame
TravelersDilemma	RandomGame	WarOfAttrition

In covariant games each pure strategy profile is drawn from a multivariate normal distribution with covariance ρ. When $\rho = 1$ we have a coordination game and when $\rho = -1$ we have a zero-sum game. Previous work [4,15] indicates that these games are easy to solve when $\rho > 0$, with the hardest games in the range $[-0.9, -0.5]$. We study 5 classes of games CovariantGame-p for $p = 1, 3, 5, 7, 9$ where $\rho = -0.1, -0.3, -0.5, -0.7, -0.9$ respectively.

With the exception of covariant and random games, the other bimatrix game classes provided by GAMUT have pure equilibria, and are therefore easily solved by support enumeration. In order to broaden our study, we chose to implement generators for the following games, which generally do not have pure equilibria.

- **Non-Zero Sum Colonel Blotto Games.** The players have an equal number of soldiers T that must be assigned simultaneously to n hills. Each player has a value for each hill that he receives if he assigns strictly more soldiers to the hill than his opponent (ties are broken uniformly at random.) Each player's payoff is the sum of the value of the hills won by that player. To avoid pure equilibria, the hill values are drawn from a multivariate normal distribution with covariance of $\rho > 0$. In our experiments, we study families of games with $n = 3, 4$ and $\rho = 0.5, 0.7, 0.9$, which we denote by Blotto-n-p for $p = 5, 7, 9$, respectively. The number of soldiers T was varied in order to generate a scalable family of games.
- **Ranking Games [9].** Each player chooses an effort level with an associated cost and score. A prize is given to the player with the higher score, or is split in the case of a tie. The payoff of a player is the value of the prize minus the cost of the effort. We generated scores and costs as increasing step functions of effort with random step sizes. We denote these games by Ranking.
- **SGC games.** Sandholm et al. [16] also noted that most GAMUT games have small support equilibria. They introduced a family of games where, in

all equilibria, both players use half of their actions. We denote their games as SGC. In these game the only equilibrium in a $(2k - 1) \times (k - 1)$ game has support sizes k for both players, which makes these games hard for support enumeration.

- **Tournament Games** [1]. Starting with a random tournament, an asymmetric bipartite graph is constructed where one side corresponds to the nodes of the tournament, and the other corresponds to subsets of nodes. The bipartite graph is transformed into a win-lose game where the actions of each player are the nodes on their side of the graph. We denote these games as Tournament.

- **Unit Vector Games (UVG)** [18]. The payoffs for the column player are chosen randomly from the range $[0, 1]$, but for the row player each column j contains exactly one 1 payoff with the rest being 0. In order to avoid pure equilibria, we generated these games by placing the 1s uniformly at random in the rows that do not generate a pure equilibrium. We denote these games as Unit.

Some of these games are not square. So, in our results we use instances that have roughly the same name number of payoffs as the corresponding square games, e.g., we compare 100×100 games to non-square games with roughly 10000 payoff entries.

Unlike other studies [8,16], we do not include the exponential-time examples for LH devised by Savani and von Stengel [17]. They are not suitable for an experimental study on approximate equilibria because the games have a number of large payoffs, so when they are normalised to the range $[0, 1]$, almost all payoffs in the game are close to 0. Hence these games are very easy to approximate. For example, the 30×30 instance has a pure 1.63×10^{-10}-WSNE. Furthermore, for instances larger than 16×16, these games exhaust the precision of floating point.

Implementation details. All implementations are written in C. We used CPLEX to solve the linear programs used in some of the algorithms. For our runtime results, we only measured the amount of time spent by the solver, and discarded the time taken to read the game from its input file. Our experiments were carried out on a cluster of 8 identical machines running Scientific Linux 6.6, which each have an Intel Core i5-2500k processor clocked at 3.30GHz with 16GB of RAM.

To verify our results, we implemented three programs that compute the quality of exact, approximate, and well-supported Nash equilibria, respectively. All of these programs carry out their calculations in *exact arithmetic*. Our exact equilibrium checker takes a pair of supports, and checks whether there exists a Nash equilibrium on these supports. Both of the approximate checkers take a mixed strategy profile, and output the value of ϵ that this profile achieves.

3   Experimental Results

Exact Algorithms. We tested LH and SE against our library of games with a timeout of 15 minutes. Table 1 part A shows the percentage of games that were solved by these two algorithms for various game sizes. We have divided the games into three classes. Firstly we have GAMUT games that always have pure equilibria. As expected, SE performs well on these games, while LH is also able to tackle the majority of instances. Secondly, we have the GAMUT games that do not always have pure equilibria. Both algorithms performed very poorly for covariant games, which is in agreement with previous studies. Finally, we have the games that we proposed. Both algorithms struggle with the games in this class, which supports the idea that GAMUT's existing library does not give a comprehensive picture of possible games. Ranking games provide an interesting case that differentiates between LH and SE: these games only had equilibria with medium sized support so SE was hopeless, however LH was able to solve these games using a linear number of pivots. In conclusion, our results show that exact methods are inadequate for the 2nd and 3rd classes of games, so it is for these games that we are interested in the performance of approximation methods.

Approximation algorithms. Table 1 part B shows the running time and quality of approximation, respectively, when the algorithms were tested on games of size 2000×2000. Again, there is a clear split in the data between the "easy" games from GAMUT and the more challenging game classes. Note that, with only a handful of exceptions, the approximation algorithms were easily able to deal with games of this size. Only TS001 was observed to time out, and it timed out on 17% of the instances that it was tested on. This indicates that approximation algorithms can indeed be applied to games that cannot be tackled with exact methods. We summarize the performance of the algorithms as follows.

– **DMP.** This algorithm runs quickly and usually gives a poor approximation. In terms of quality, it is clearly outmatched by all of the other algorithms.
– **BBM1 and BBM2.** These algorithms were typically in the middle in terms of both approximation quality and running time. Only a handful of Blotto instances triggered the extra steps in BBM2, so these two algorithm are mostly identical.
– **Pure.** Whenever the game has a pure NE, this algorithm performs well, because it terminates once a pure NE has been found. Otherwise, it is among the slowest of the algorithms, because it is never faster than n^3, where n is the number of strategies. The quality of approximation results confirm that our new games succeed in avoiding pure strategy profiles that are close to being Nash equilibria.
– **TS2 and TS001.** The TS algorithm was the clear winner in terms of quality of approximation. The results show that the choice of δ can have a significant effect on the algorithm's characteristics. TS2 often terminates in a reasonable running time when compared to BBM, and it usually beats BBM significantly on quality of approximation. However, TS001 always beats TS2 in quality of approximation, and always provides the best approximations among all of the algorithms that we studied. This accuracy comes at the cost of speed, as there are many games upon which TS001 is slower than TS2.

Table 1. A shows the percentage of instances which did not time out on LH and SE for instances of various sizes (we used sizes 105, 300, 1035 for Blotto-3 games, 120, 364 and 969 for Blotto-4 games, 27, 57, 126 rows for Tournament and 100, 300, 1000 for all other games.) B and C show experiment results on games of size 2000 × 2000 (games of sizes 2016 for Blotto-3, 2024 for Blotto-4, 1999 for SGC, 200 rows for Tournament.) We display the average running time (in seconds) and average ε-NE (ε-WSNE for KS) across instances which did not time out (of the approximation algorithms only TS001 ever timed out - 17% of the time).

| | A: % completed within 15mins | | | | | | B: ε-Nash | | | | | | | | | | | | C: ε-WSNE | |
| | LH | | | SE | | | DMP | | BBM1 | | BBM2 | | Pure | | TS001 | | TS2 | | KS | |
	100	300	1000	100	300	1000	Time	ε	Time	ε	Time	ε	Time	ε	Time	ε	Time	ε	Time	ε
BertrandOligopoly	80	35	12	96	100	96	0.00	0.42	0.43	0.00	0.47	0.00	0.04	0.00	318.34	0.00	20.96	0.06	0.04	0.00
CournotDuopoly	100	100	68	100	100	68	0.00	0.25	0.42	0.00	0.43	0.00	28.14	0.00	1.95	0.00	2.94	0.00	28.14	0.00
GrabTheDollar	100	100	100	100	100	100	0.00	0.25	0.77	0.11	0.78	0.10	0.00	0.00	3.31	0.00	5.08	0.00	0.00	0.00
GuessTwoThirdAve	36	20	0	100	100	100	0.00	0.50	0.70	0.00	0.71	0.00	0.00	0.00	2.55	0.00	2.94	0.00	0.00	0.00
LocationGame	100	100	100	100	100	96	0.00	0.20	0.76	0.00	0.78	0.00	147.13	0.00	3.22	0.00	4.87	0.00	147.13	0.00
MinimumEffortGame	100	100	100	100	100	100	0.00	0.00	0.78	0.00	0.78	0.00	0.00	0.00	30.76	0.00	49.03	0.00	0.00	0.00
TravelersDilemma	100	75	56	100	100	100	0.00	0.16	0.72	0.00	0.78	0.00	0.00	0.00	233.38	0.00	134.10	0.00	0.00	0.00
WarOfAttrition	100	100	100	100	100	100	0.00	0.00	0.99	0.00	1.04	0.00	0.00	0.00	4.13	0.00	4.30	0.00	0.00	0.00
CovariantGame-1	100	50	4	82	62	24	0.00	0.21	59.04	0.01	57.43	0.01	146.50	0.02	125.35	0.00	572.97	0.00	198.30	0.01
CovariantGame-3	100	13	0	10	25	0	0.00	0.22	56.75	0.01	58.08	0.00	150.07	0.04	130.26	0.00	671.48	0.00	208.96	0.01
CovariantGame-5	100	0	0	0	0	0	0.00	0.27	58.76	0.01	61.78	0.00	149.42	0.11	235.68	0.00	727.18	0.00	208.35	0.01
CovariantGame-7	100	0	0	0	0	0	0.00	0.29	56.62	0.01	53.85	0.01	148.41	0.14	330.66	0.00	768.42	0.00	203.86	0.01
CovariantGame-9	100	19	0	0	0	0	0.00	0.32	54.88	0.00	57.90	0.00	153.06	0.21	529.19	0.00	743.63	0.00	215.01	0.01
RandomGame	100	73	17	88	80	56	0.00	0.35	56.38	0.03	59.06	0.03	89.94	0.03	228.91	0.00	550.99	0.00	109.11	0.00
Blotto-3-5	80	48	48	56	60	72	0.00	0.30	6.76	0.10	6.86	0.06	67.63	0.06	18.89	0.00	10.29	0.03	67.50	0.03
Blotto-3-7	68	48	60	48	56	56	0.00	0.37	5.75	0.11	5.95	0.11	74.69	0.10	22.92	0.00	10.30	0.04	75.22	0.04
Blotto-3-9	52	40	52	56	48	56	0.00	0.38	1.24	0.05	1.23	0.05	100.08	0.12	17.19	0.00	11.49	0.02	100.56	0.03
Blotto-4-5	76	44	32	40	24	40	0.00	0.38	1.41	0.13	1.39	0.13	117.29	0.10	24.33	0.00	9.28	0.04	120.78	0.10
Blotto-4-7	68	44	44	32	12	32	0.00	0.38	2.18	0.12	2.10	0.13	118.91	0.13	28.19	0.00	12.75	0.05	123.28	0.10
Blotto-4-9	76	52	36	8	12	16	0.00	0.40	2.12	0.10	2.07	0.18	164.55	0.18	19.89	0.00	13.25	0.03	167.79	0.11
SGC	100	100	100	0	0	0	0.00	0.25	3.46	0.00	2.87	0.00	121.55	0.25	57.02	0.00	63.53	0.00	119.16	0.00
Ranking	100	100	100	0	0	0	0.00	0.32	6.82	0.15	7.00	0.18	148.89	0.18	207.00	0.00	45.55	0.06	158.68	0.18
Tournament	16	12	12	16	0	0	0.23	0.50	1.95	0.06	1.96	0.06	374.25	1.00	39.37	0.00	61.92	0.01	388.14	0.07
Unit	100	96	44	0	0	0	0.00	0.39	57.70	0.01	57.77	0.01	153.08	0.00	46.74	0.00	32.96	0.00	211.65	0.00

Table 1 shows the results for the WSNE algorithms on the same set of games. Due to its $O(n^4)$ running time, we found that that KS+ timed out on all instances, so results for this algorithm are omitted. Recall that KS uses a preprocessing step to search for pure WSNE, and then solves an LP. We found that the preprocessing almost always provides the better approximation, but the search over pure strategies is the dominant component of KS's running time. So, there is a significant cost for targeting ϵ-WSNE over ϵ-Nash equilibria, since Pure is the slowest algorithm for ϵ-Nash equilibria, and KS is never faster than Pure.

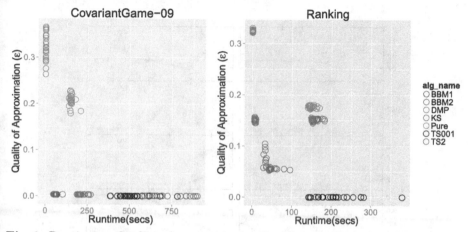

Fig. 1. Runtime vs Quality of approximation for CovariantGame-9 and Ranking for instances of size 2000×2000

Finally, we comment on the trade off between running time and quality of approximation. In Figure 1 we plot these two metrics against each other for CovariantGame-9 and Ranking, which provide a fairly representative sample of the results that we observed across the dataset. The points towards the lower left of the diagrams are those that minimize the running time for a given approximation guarantee. In order of accuracy, we typically see points from DMP, BBM1, TS2, and TS001 along this frontier.

The TS algorithm. Since our results indicate that the TS algorithm gives the best approximations, it is worth spending more time analysing this algorithm. Both the quality of approximation and the running time of the algorithm are affected by the choice of δ. We now give more detailed results on how this parameter affects the of the algorithm. To test the dependence on δ, we ran the TS algorithm on one hundred 400×400 instances of RandomGame for various values of δ in the range $(0, 0.14]$. The results of these experiments are displayed in Figure 2. The left side of the figure shows the results for a single game, while the right side of the figure shows the average results over all instances.

The first two rows show the runtime and quality of approximation, respectively. It can be seen that the algorithm does not scale smoothly with respect to δ, and instead there are discontinuities in both running time and quality of approximation. The explanation for these discontinuities can be found in the third and fourth rows, which show the number of rows and the size of the LP that is solved in each iteration, respectively. The third row shows that, as we would expect, the number of iterations increases as δ decreases. However, the data in the fourth row shows that the story is more complicated. The size of the LP that is formulated in each iteration increases as δ increases. Thus, although the number of iterations falls, the time per iteration gets larger.

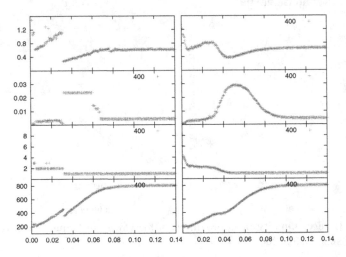

Fig. 2. TS performance plots of runtime (row 1), quality of approximation (row 2), number of iterations (row 3,) and average LP-Size (row 4) against δ. Left diagrams shows results for a single 400×400 instance of RandomGame, while right diagrams show averages over one hundred 400×400 instances of RandomGame.

4   Worst Case Examples

The best theoretical upper bound on the performance of the TS algorithm is that it produces a 0.3393-Nash equilibrium, but previous work has not been able to show a matching lower bound. As we have seen, the algorithm usually finds a very good approximation in practice. The Blotto games were the only class that challenged the algorithm, and even then the approximations were usually good. Figure 3 shows box and whisker plots for the quality of approximation of TS001 on the Blotto games that we considered. Almost all points were close to 0.01, and only a handful of instances were larger. The worst approximation that we found was a 0.14-Nash equilibrium, which is still far from the worst-case guarantee.

To test the limits of the TS algorithm, we used a genetic algorithm to try and find worst-case examples. More precisely, we used a genome that encoded a

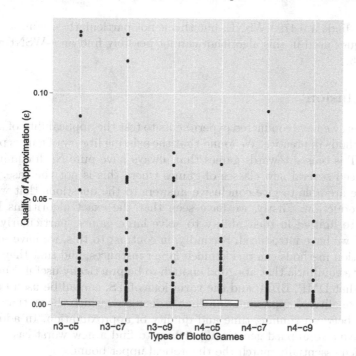

Fig. 3. Box and whisker plots for quality of ϵ-NE found by TS001 on Blotto games

5×5 game, and a fitness function that measured the quality of approximation found by TS001 on the game. The result of this was a 5×5 game[2] for which TS001 produces a 0.3385-Nash equilibrium. Note that this essentially matches the theoretical upper bound.

Although this example provides a good lower bound against TS001, we found that BBM2 produced a 0.024-NE when applied to this instance. For this reason, we decided to test the limits of our entire portfolio of algorithms. We used the same genetic algorithm, but this time the fitness function was the best of the approximations found by TS001, Pure, and BBM2. We produced a 5×5 game for which all of the above algorithms produced at best a 0.3189-Nash equilibrium (TS produced a 0.3189-NE, Pure produced a 0.324-NE, and BBM2 produced a 0.321-NE.)

Both of the games mentioned so far contain dominated strategies. While this does not invalidate the lower bounds, it is obviously undesirable. For this reason, we reran the experiments with a fitness function that penalizes dominated strategies. For TS, we found a game with no dominated strategies for which the algorithm produces a 0.3254-NE, and for the portfolio we found a game with no dominated strategies for which the portfolio finds a 0.3253-NE.

Finally, we applied the genetic algorithm to try and find a worst case example for KS+, but we were unsuccessful. We were able to produce a 5×5 game for

---

[2] All games found are given in the full version: http://arxiv.org/abs/1502.04980

which KS+ finds a 0.4799-WSNE, but this is not particularly useful, as some of the techniques used in this algorithm cannot possibly find an ϵ-WSNE where ϵ is better 0.5.

5   Conclusion

In this paper, we have conducted experiments to test the applicability of approximation methods in practice. We found that the existing library of games provided by GAMUT is biased towards games that always have pure Nash equilibria, so we introduced several new classes of games where this is not the case. Having done so, we are able to give conclusive answers to the questions that we posed in the introduction. Firstly, we have seen that the exact algorithms LH and SE are quite limited in their ability to solve large games, particularly on the games that we have introduced. Secondly, in contrast to this, we have seen that approximation methods can tackle much larger instances, and that they provide approximate equilibria that are good enough to be practically useful. Finally, we have seen that DMP, BBM, and the variations of TS, can all be useful depending on the quality of approximation that is required, which shows that there is a trade-off between running time and quality of approximation. In addition to this, we have also applied genetic algorithms to find a new worst case-example for TS which essentially match the theoretical upper bound.

This work has highlighted the need for a comprehensive library of games upon which game theoretic algorithms can be tested. In addition to the games that we have introduced, there are also a number of other areas that could be represented here. For example, there are many auction problems from which games could be derived. Our study has focussed on algorithms with provable guarantees on the quality of approximate equilibria found in polynomial time. One direction for further study would to be to consider exact algorithms as heuristics for finding approximate equilibria. For example, one could randomly sample supports and find the best approximate equilibria on these supports using linear programming, or one could run an algorithm like LH for a fixed time or number of steps and check how good the strategy profiles it traces are as approximate equilibria. It would be interesting to see the extent that genetic algorithms can be applied to this. For example, what happens when we try to make LH bad, for quality of approximation or for running time? For approximation results, we would be interested in the approximation that is found after the algorithm has taken a fixed number of steps, like linear, quadratic, or some other polynomial.

References

1. Anbalagan, Y., Norin, S., Savani, R., Vetta, A.: Polylogarithmic Supports Are Required for Approximate Well-Supported Nash Equilibria below 2/3. In: Chen, Y., Immorlica, N. (eds.) WINE 2013. LNCS, vol. 8289, pp. 15–23. Springer, Heidelberg (2013)
2. Bosse, H., Byrka, J., Markakis, E.: New algorithms for approximate Nash equilibria in bimatrix games. Theoretical Computer Science **411**(1), 164–173 (2010)

3. Chen, X., Deng, X., Teng, S.-H.: Settling the complexity of computing two-player Nash equilibria. Journal of the ACM 56(3), 14:1–14:57 (2009)
4. Codenotti, B., Rossi, S.D., Pagan, M.: An experimental analysis of Lemke-Howson algorithm (2008). CoRR, abs/0811.3247
5. Daskalakis, C., Goldberg, P.W., Papadimitriou, C.H.: The complexity of computing a Nash equilibrium. SIAM Journal on Computing 39(1), 195–259 (2009)
6. Daskalakis, C., Mehta, A., Papadimitriou, C.H.: A note on approximate Nash equilibria. Theoretical Computer Science 410(17), 1581–1588 (2009)
7. Fearnley, J., Goldberg, P. W., Savani, R., Sørensen, T. B.: Approximate well-supported nash equilibria below two-thirds (2012). CoRR, abs/1204.0707
8. Gatti, N., Patrini, G., Rocco, M., Sandholm, T.: Combining local search techniques and path following for bimatrix games (2012). CoRR, abs/1210.4858
9. Goldberg, L.A., Goldberg, P.W., Krysta, P., Ventre, C.: Ranking games that have competitiveness-based strategies. Theor. Comput. Sci. **476**, 24–37 (2013)
10. Goldberg, P.W., Papadimitriou, C.H., Savani, R.: The complexity of the homotopy method, equilibrium selection, and lemke-howson solutions. ACM Trans. Economics and Comput. **1**(2), 9 (2013)
11. Kontogiannis, S.C., Spirakis, P.G.: Well supported approximate equilibria in bimatrix games. Algorithmica **57**(4), 653–667 (2010)
12. Kontogiannis, S., Spirakis, P.: Approximability of symmetric bimatrix games and related experiments. In: Pardalos, P.M., Rebennack, S. (eds.) SEA 2011. LNCS, vol. 6630, pp. 1–20. Springer, Heidelberg (2011)
13. Lemke, C.E., Howson Jr, J.: Equilibrium points of bimatrix games. Journal of the Society for Industrial and Applied Mathematics **12**(2), 413–423 (1964)
14. Nudelman, E., Wortman, J., Shoham, Y., Leyton-Brown, K.: Run the gamut: a comprehensive approach to evaluating game-theoretic algorithms. In: AAMAS, pp. 880–887 (2004)
15. Porter, R., Nudelman, E., Shoham, Y.: Simple search methods for finding a nash equilibrium. Games and Economic Behavior, 642–662 (2008)
16. Sandholm, T., Gilpin, A., Conitzer, V.: Mixed-integer programming methods for finding nash equilibria. In: AAAI, pp. 495–501 (2005)
17. Savani, R., von Stengel, B.: Hard-to-solve bimatrix games. Econometrica **74**(2), 397–429 (2006)
18. Savani, R., von Stengel, B.: Unit vector games (2015). CoRR, abs/1501.02243
19. Tsaknakis, H., Spirakis, P.G.: An optimization approach for approximate Nash equilibria. Internet Mathematics **5**(4), 365–382 (2008)
20. Tsaknakis, H., Spirakis, P.G., Kanoulas, D.: Performance Evaluation of a Descent Algorithm for Bi-matrix Games. In: Papadimitriou, C., Zhang, S. (eds.) WINE 2008. LNCS, vol. 5385, pp. 222–230. Springer, Heidelberg (2008)

The Effect of Almost-Empty Faces on Planar Kandinsky Drawings

Michael A. Bekos[1], Michael Kaufmann[1],
Robert Krug[1(✉)], and Martin Siebenhaller[2]

[1] Wilhelm-Schickard-Institut Für Informatik, Universität Tübingen,
Tübingen, Germany
{bekos,mk,krug}@informatik.uni-tuebingen.de
[2] yWorks GmbH, Tübingen, Germany
martin.siebenhaller@yworks.com

Abstract. Inspired by the recently-introduced slanted orthogonal graph drawing model, we introduce and study planar Kandinsky drawings with *almost-empty faces* (i.e., faces that were forbidden in the classical Kandinsky model).

Based on a recent NP-completeness result for Kandinsky drawings by Bläsius et al., we present and experimentally evaluate (i) an ILP that computes bend-optimal Kandinsky drawings with almost-empty faces, and, (ii) a more efficient heuristic that results in drawings with relatively few bends. Our evaluation shows that the new model, in the presence of many triangular faces, not only improves the number of bends, but also the compactness of the resulting drawings.

1 Introduction

The Kandinsky model [13] is a well-established graph drawing model that is a special type of grid embeddings [15,17], which, however, can be employed to draw any graph (that is, of arbitrary vertex-degree) in an orthogonal style. In this model, two grids are present; a coarse one to accommodate the vertices and a fine one to route the edges. More precisely, a Kandinsky drawing $\Gamma(G)$ of a graph G is one in which (a) every vertex is drawn as a box centered at a point of the underlying coarse grid, (b) all vertex boxes are of uniform size, (c) every edge is drawn as a rectilinear polyline on the underlying fine edge-grid, and, (d) arbitrarily many edge-segments can be connected to each side of every vertex; see Figure 2a.

Due to its high importance in practical applications, several different variants of this model have been proposed and studied over the years (see, e.g., [3,6–8,10]). The classical orthogonal model studied by Tamassia [16] can also be seen as a restricted variant of the Kandinsky drawing model, where the graphs have maximum degree four and no two edges can be attached to the same side of a vertex.

This work has been supported by DFG grant Ka812/17-1.

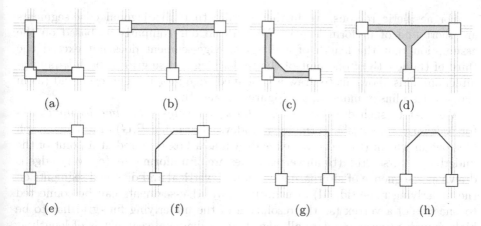

Fig. 1. (a)-(b) Illustration of empty faces: (a) empty L, and, (b) empty T; (c)-(d) Illustration of almost-empty faces: (c) almost-empty L, and, (d) almost-empty T; (e)-(f) Sample Kandinsky drawings: (e) without almost empty faces, and, (f) with almost empty faces, which results in height reduction. (e)-(f) Replacing a 90° bend by a pair of half-bends of 135°; (g)-(h) Illustration of a U-shaped edge drawn in the classical orthogonal model and in the new model.

Beside the area minimization, a typical objective that has received considerable attention is the bend minimization. For plane graphs, this problem was initially modeled as a min-cost flow problem [8,13]. Subsequently, it has been observed that the given algorithm needed additional constraints that could not be handled efficiently, and therefore efficient 2-approximations have been proposed [1,9]. The complexity of the problem was unknown for more than two decades until the NP-completeness result was recently found [4]. As in previous work, we only consider plane graphs.

Most of the known algorithms [1,9,13] for the Kandinsky drawing model heavily depend on the absence of *empty faces*; see Figures 1a and 1b. A common approach to avoid such faces is to adopt the so-called *bend-or-end property*. In the work of Fößmeier and Kaufmann [13], this property is the result of the min-cost flow formulation which assigns each angle of zero degrees between two edges to a specific bend on one of these edges. From a practical point of view, such faces are forbidden because it is too difficult to distinguish them in a drawing (as they are of almost zero area). In addition, the empty L (see Figure 1a) is not possible to be drawn without introducing vertex-edge crossings.

Inspired by recent work of Bekos et al. [2] on the so-called slanted orthogonal graph drawing model, we use intermediate diagonal edge-segments, which allow us to draw empty faces; see Figures 1c and 1d. In this way, we can create "empty faces" of non-zero area, making them acceptable in a drawing. We refer to such faces as *almost-empty faces*. To maintain a uniform approach in the way we draw the bends of the edges, we replace all 90° bends by pairs of *half-bends* of 135°; see Figures 1e and 1f.

For aesthetic reasons, we further require that all diagonal edge-segments are short and of uniform length. We perform our comparisons based on the assumption that the length of a diagonal edge-segment does not exceed one third of the length of one unit of the underlying coarse grid in both x- and y-direction. This allows us to draw a U-shaped edge in a 1×1 integer grid (as in classical Kandinsky models); see Figures 1g and 1h.

We refer to such drawings as *Kandinsky drawings with almost-empty faces* (or *podevsaef-drawings*[1]). Formally, a podevsaef drawing $\Gamma(G)$ of a plane graph G is one in which: (a) every vertex is drawn as a box centered at a point of the underlying coarse grid, (b) all vertex boxes are of uniform size, (c) every edge is drawn as a sequence of alternating horizontal, vertical and diagonal segments on the underlying fine-grid, (d) arbitrarily many edge-segments can be connected to one side of a vertex (so the resolution of the underlying fine-grid has to be high enough to accommodate all edges), (e) a diagonal segment is of length at most one third of the length of one unit of the underlying coarse grid in both x- and y-direction and is never incident to a vertex, (f) the minimum of the angles formed by two consecutive segments of an edge always is $135°$, which suggests that a bend in $\Gamma(G)$ is always incident to a diagonal segment and to either a horizontal or a vertical one. For sample podevsaef drawings produced by implementations of our algorithms refer to Figures 2b and 2c; the corresponding bend-optimal Kandinsky drawing of the same graph is given in Figure 2a.

Our goal is to find out how much we can save with respect to bends and area when allowing almost-empty faces in Kandinsky drawings. Note that almost-empty faces are always triangular. So, we expect that the improvements will be greater in graphs with many triangular faces. Since the recent NP-completeness result of Bläsius et al. [4] implies that also our problem is NP-complete, we take an experimental approach. To quantify the results of our experiments, we present an ILP-formulation, which extends the standard one of Eiglsperger et al. [11] and which results in bend-optimal podevsaef-drawings (see Section 2). By relaxing the bend-optimality constraint on the resulting drawings, we are able to present an efficient heuristic which results in podevsaef-drawings with relatively few bends (see Section 3). In Section 4 we experimentally evaluate the podevsaef drawing model and we also compare it with the classical Kandinsky drawing model. We conclude in Section 5 with open problems and future work.

2   Bend-Optimal Podevsaef Drawings of Planar Graphs

In this section, we present an approach that results in podevsaef drawings of minimum number of bends. Following standard practice, our approach consists of two phases; the *orthogonalization phase* (where the angles and the bends of the drawing are computed) and the *compaction phase* (where the actual coordinates for the vertices and the edges are computed); refer, e.g., to [16]. For the first

---

[1] The term is inspired by a term that also refers to Kandinsky drawings: *podevsnef drawings* [13], which stands for Planar Orthogonal Drawings with Equal Vertex Sizes and No Empty Faces.

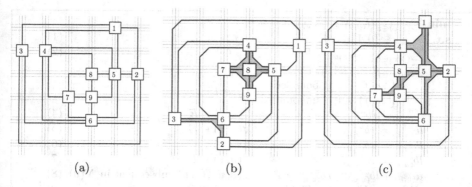

(a)   (b)   (c)

Fig. 2. A sample planar graph drawn (a) in the classical Kandinsky drawing model, (b) in the podevsaef drawing model using the ILP of Section 2, and, (c) in the podevsaef drawing model using the heuristic of Section 3. Almost-empty faces are drawn gray.

phase, we present a modification of a standard ILP-formulation of Eiglsperger et al. [11] that results in representations of minimum number of bends. In the compaction phase, we employ a transformation that allows us to use any known compaction algorithm for the original Kandinsky model.

Before we proceed with the description of our modification for the orthogonalization phase, we first quickly recall the ILP-formulation of Eiglsperger et al. [11]. For each edge $e = (u, v)$, variable $a_{(u,v)} \cdot 90°$ corresponds to the angle formed by edge e and its cyclic predecessor at vertex u. Clearly, $a_{(u,v)} \in \{0, 1, 2, 3, 4\}$. Since the sum of the angles around a vertex equals to $360°$, it follows that for each vertex $u \in V$, $\sum_{(u,v) \in N(u)} a_{(u,v)} = 4$ must hold, where $N(u)$ denotes the neighbors of u.

In order to count the number of left turns (or simply *left-bends*) along an edge $e = (u, v)$, three variables, $lb^u_{(u,v)}$, $lb^v_{(u,v)}$ and $lb_{(u,v)}$, are employed, which correspond to the left Kandinsky-bend (that is, the special bend resulting from the bend-or-end property) at vertex u, the left Kandinsky-bend at vertex v and the remaining left-bends of edge (u, v). For the right-bends, variables $rb^u_{(u,v)}$, $rb^v_{(u,v)}$ and $rb_{(u,v)}$ are defined similarly. Clearly, for reasons of symmetry $lb^u_{(u,v)} = rb^u_{(v,u)}$, $lb_{(u,v)} = rb_{(v,u)}$ and $lb^v_{(u,v)} = rb^v_{(v,u)}$ must hold. Note that variables $lb^u_{(u,v)}$, $lb^v_{(u,v)}$, $rb^u_{(u,v)}$ and $rb^v_{(u,v)}$ are binary, while variables $lb_{(u,v)}$ and $rb_{(u,v)}$ are non-negative integers.

Since only one Kandinsky-bend is allowed at each end of each edge, $lb^u_{(u,v)} + rb^u_{(u,v)} \leq 1$ must hold for each edge $(u, v) \in E$. For ease of notation, we denote by $l_{(u,v)}$ and $r_{(u,v)}$ the total number of left and right bends per edge, respectively, that is, $l_{(u,v)} = lb^u_{(u,v)} + lb_{(u,v)} + lb^v_{(u,v)}$ and $r_{(u,v)} = rb^u_{(u,v)} + rb_{(u,v)} + rb^v_{(u,v)}$. Since the sum of the angles formed at the vertices and at the bends of a bounded face f equals to $180 \cdot (p(f) - 2)$, where $p(f)$ denotes the number of such angles, it follows that $\sum_{(u,v) \in f} (a_{(u,v)} + l_{(u,v)} - r_{(u,v)}) = 2a(f) - 4$, where $a(f)$ denotes the number of vertex angles in f. If f is not bounded, the sum is increased by 8. The bend-or-end property (which guarantees the absence of empty faces) is

$$\min \sum_{(u,v)\in E}(l_{(u,v)} + r_{(u,v)})$$

$$\text{s.t.} \quad a_{(u,v)} \in \{0,1,\ldots,4\} \qquad\qquad \forall (u,v)\in E \qquad\qquad (1)$$

$$\sum_{(u,v)\in N(u)} a_{(u,v)} = 4 \qquad\qquad \forall u \in V \qquad\qquad (2)$$

$$\sum_{(u,v)\in f}(a_{(u,v)} + l_{(u,v)} - r_{(u,v)})$$
$$= \begin{cases} 2a(f) - 4; \ f \text{ bounded} \\ 2a(f) + 4; \ f \text{ unbounded} \end{cases} \forall f \in F \qquad\qquad (3)$$

$$lb^u_{(u,v)} + rb^u_{(u,v)} \le 1 \qquad\qquad \forall (u,v)\in E \qquad\qquad (4)$$

$$lb^u_{(u,v)} = rb^u_{(v,u)} \qquad\qquad \forall (u,v)\in E \qquad\qquad (5)$$

$$lb_{(u,v)} = rb_{(v,u)} \qquad\qquad \forall (u,v)\in E \qquad\qquad (6)$$

$$lb^v_{(u,v)} = rb^v_{(v,u)} \qquad\qquad \forall (u,v)\in E \qquad\qquad (7)$$

$$a_{(v,u)} + lb^v_{(v,w)} + rb^v_{(v,u)} \ge 1 \qquad \forall (v,w),(v,u) \text{ subsequent in } N(v) \ (8)$$

$$lb^u(u,v), rb^u(u,v) \in \{0,1\} \qquad\qquad \forall (u,v)\in E \qquad\qquad (9)$$

$$lb(u,v), rb(u,v) \in \mathbb{N} \qquad\qquad \forall (u,v)\in E \qquad\qquad (10)$$

Linear Program 1. The ILP of Eiglsperger et al. [11] for computing bend-optimal Kandinsky representations

implied by requiring $a_{(v,u)} + lb^v_{(v,w)} + rb^v_{(v,u)} \ge 1$, for all pairs of consecutive edges (v,w) and (v,u) around v. Of course, the objective function of the corresponding ILP-formulation must minimize the sum of all (i.e., either left or right) bends, that is $\min \sum_{(u,v)\in E}(l_{(u,v)} + r_{(u,v)})$. The complete program is given in Linear Program 1.

To enable the aforementioned ILP-formulation to use the almost-empty T and almost-empty L-shapes (see Figures 1a and 1b, respectively), we observe that a bend of a classical Kandinsky drawing always corresponds to a pair of half-bends in our drawing model. So, in our formulation we are working with bends (not half-bends), which we eventually replace with pairs of half-bends only in the compaction phase. In addition, we replace Constraint 8 of Linear Program 1 with new constraints that we describe in the following (refer to Constraint Sets 1 and 2). For each triangular face f, we introduce two binary variables, say T_f and L_f, that are set to 1 if and only if f is drawn using the almost-empty T or the almost-empty L-shape, respectively. We also employ a large constant M which we use to "activate" or "deactivate" constraints; a common trick used in ILPs. It is known, however, that the choice of the value for the constant M might significantly influence the time required to compute an optimal solution of an ILP [5]. Our experimental evaluation showed, however, that in our case the choice of the value of M did not have any particular effect on the computation time.

If variable T_f of face f is set to 1 (that is, f is drawn using the almost-empty T-shape), then Constraint 11 ensures that all angles of f are zero. Constraints 12 and 13 force f to have in total two right-bends on all edges (hence, the third edge of f must be bend-less). Constraint 14 ensures that no edge has a left-bend and Constraint 15 guaranteed that all edges have at most one bend in total. On the other hand, if T_f is set to 0, then all constraints are deactivated, so they impose no restriction on the edges.

The constraints for an L-shaped face f are similar. If variable L_f is set to 1, Constraints 16 and 17 ensure that there is one $90°$ angle in f. Constraints 18 and 19 force f to have exactly one right-bend and Constraint 20 makes sure that there are no left-bends. Again, setting L_f to zero trivially fulfills all these constraints and they pose no restriction on f.

Observe that the constraints for T-shaped and L-shaped faces exclude each other. So, there is no reason to add an extra constraint for this purpose. Since we intend to allow either the almost-empty T or the almost-empty L-shape, it follows that it suffices to replace Constraint 8 of Linear Program 1 with the following constraint for each triangular face f:

$$a_{(u,v)} + lb^v_{(v,w)} + rb^v_{(v,u)} + T_f + L_f \geq 1, \ \forall (v,w), (v,u) \text{ subsequent in } N(v)$$

$\sum_{e \in E_f} a_e \leq 0 + (1 - T_f) \cdot M$ (11)	$\sum_{e \in E_f} a_e \leq 1 + (1 - L_f) \cdot M$ (16)
$\sum_{e \in E_f} r_e \geq 2 - (1 - T_f) \cdot M$ (12)	$\sum_{e \in E_f} a_e \geq 1 - (1 - L_f) \cdot M$ (17)
$\sum_{e \in E_f} r_e \leq 2 + (1 - T_f) \cdot M$ (13)	$\sum_{e \in E_f} r_e \leq 1 + (1 - L_f) \cdot M$ (18)
$\sum_{e \in E_f} l_e \leq 0 + (1 - T_f) \cdot M$ (14)	$\sum_{e \in E_f} r_e \geq 1 - (1 - L_f) \cdot M$ (19)
$\forall e \in E_f : r_e \leq 1 + (1 - T_f) \cdot M$ (15)	$\forall e \in E_f : l_e \leq 0 + (1 - L_f) \cdot M$ (20)

Constraint Set 1. f is T-shaped   **Constraint Set 2.** f is L-shaped

In order to prove that the modified ILP-formulation results in podevsaef representations with minimum number of bends, we observe that if we set all T_f and L_f variables to 0, then all new constraints no longer affect the underlying equation system and all old constraints stay exactly as before. So, the original proof of correctness of Eiglsperger et al. [11] holds. On the other hand, it is not difficult to see that if a face is to be drawn either as an almost-empty T or as an almost-empty L-shape, then Constraint Sets 1 and 2, respectively, ensure that all angles and bends are correctly computed. So, the corresponding podevsaef representations are correctly computed.

As already stated, in the compaction phase, where the computed representation has to be transformed into an actual drawing (that is, the actual coordinates of the vertices and edges have to be computed), we employ a simple transformation that allows us to use any known algorithm for the compaction phase of the original Kandinsky model, e.g., [12]. The transformation is illustrated in Figure 3. More precisely, for an almost-empty T-shaped face a new auxiliary vertex is required (refer to the gray colored vertex in Figure 3a) and the angles around it follow directly from the T-shape. Similarly, for an almost-empty L-shaped face we simply ignore the bent edge involved (see Figure 3b).

Once all almost-empty T-shaped and L-shaped faces are transformed according to the rules of Figure 3, we proceed to draw the new graph using any known compaction algorithm for the original Kandinsky model. In the resulting drawing, the applied transformations can be easily reversed by introducing the missing edges of the L-shaped faces and replacing the auxiliary vertices of the T-shaped faces with the original edges.

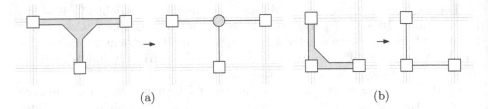

(a)   (b)

Fig. 3. (a) Transformation for T-shapes. (b) Transformation for L-shapes.

3   A Heuristic to Compute Podevsaef Drawings

As we will shortly see in Section 4, the running time needed to compute bend-optimal podevsaef drawings may be high, because of the ILP that computes the corresponding bend-optimal podevsaef representations. In this section, we present a significantly more efficient heuristic which given an orthogonal representation of minimum number of bends, computes a corresponding podevsaef drawing by trying to minimize the number of half-bends as possible. Our evaluation shows that the produced drawings are comparable to the optimal ones in terms of the total number of bends and the area requirement, which suggests that the proposed heuristic is useful for practical purposes.

The idea of our approach is to start from a classical bend-optimal orthogonal representation and try to modify the shape of as many triangular faces as possible to become T or L-shaped. To achieve this, we have identified several shapes that allow an easy transformation into the new almost-empty shapes; see Figure 4. Note that, these transformations do not require a drawing, but they are directly applicable to any orthogonal representation (not necessarily to bend-optimal ones). For each transformation the number of half-bends of the podevsaef representation either equals to twice (see Figures 4a, 4c, 4d) or is less than twice (see Figures 4b, 4e, 4f) the number of bends of the orthogonal representation. Note also the potential of saving gridlines; see Figure4f.

Once all triangular faces have been transformed according to the rules of Figure 4, we proceed with the compaction phase to obtain the final drawing (as described in the previous section). Alternatively, we could try to further reduce the number of bends by adding another orthogonalization step (as the transformed graph may allow a drawing with less bends). Particular attention must be payed on keeping the shape of the edges that have been transformed unchanged in subsequent steps. In particular, the angles around the grey colored vertex of Figure 3a must not change and its incident edges must not be bent. Similarly, one copes with the almost-empty L shape of Figure 3b.

In order to further improve the quality of the heuristic (in terms of the number of bends), we employ a preprocessing of the input, which according to our evaluation has proved to be very effective. When the input representation is computed in such a way that it is bend-optimal and simultaneously contains the maximum number of S-shaped edges, then the number of half-bends in the resulting podevsaef drawings tends to be reduced. This is because all transformations

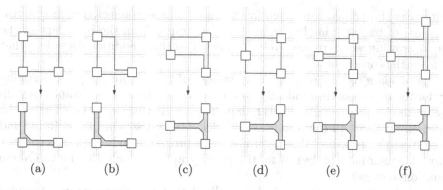

Fig. 4. Transformations into: (a),(b) L-shaped face; (c)-(f) T-shaped face

presented in Figure 4 that involve S-shaped edges require less half-bends than twice the corresponding number of bends (see Figures 4b, 4e, 4f). To achieve this, we introduce appropriate binary variables, which determine whether an edge is S-shaped, and employ them to modify (or, more precisely, weight) the objective function of the ILP of Eiglsperger et al. [11] so to "prefer" an S-shaped edge rather than two edges with a single bend each, whenever this is possible.

4 An Experimental Evaluation of the Podevsaef Drawing Model

In this section, we compare bend-optimal Kandinsky drawings computed with the ILP of Eiglsperger et al. [11] with bend-optimal and close-to-optimal podevsaef drawings computed with the algorithm of Section 2 and the heuristic of Section 3.

Experimental Setup: We implemented all aforementioned algorithms using Java and the yFiles library (http://www.yworks.com). The gurobi solver [14] was employed to solve the different linear programs. The experiment was performed on a desktop Linux machine with four cores at 2.5 GHz and 3 GB RAM.

As a test set for our experiment, we used three different graph-suites, each containing planar graphs of different densities (recall that the *density* of a graph is defined as the ratio of the number of its edges to the number of its vertices): (i) the *planar Rome graphs*, which form a collection of 3279 graphs with average density of 1.16 obtained from the graphdrawing.org website, (ii) the *planar North graphs*, which form a collection of 854 graphs with average density of 1.14 also obtained from the graphdrawing.org website, and, (iii) 940 randomly created triangulations with average density 2.82 which were uploaded to http://www.graph-archive.org and were created (based on the yFiles approach) as follows. Initially, an evenly distributed point set was created within a triangular region T. Then, the points were sorted from left to right. The first three points formed a triangle and each following point was connected with the visible points to its left. Finally, the points that were still on the boundary of the created drawing were appropriately connected to three additional vertices that reside on the corners of $\mathcal{T}$.

Due to space constraints, in the remainder of this section we present results only for the test set of the randomly created triangulations. These graphs contain only triangular faces. So, we expect that they will better show the effect of the almost-empty faces on the Kandinsky model. The test sets of planar Rome and planar North graphs show that no or not much progress can be expected without many triangular faces (only 12% for the Rome graphs and only 13% for the North graphs on average).

From our experiment, we quickly realized that the time required to solve the ILPs which are used to compute bend-optimal podevsaef representations increases rapidly with the number of triangular faces. So, we set a time-limit of 300 seconds in our experiment. If the solver was able to find a feasible solution within this time-limit, then the solution closest to the optimal one was used for our evaluation. Otherwise, the instance counted as failed and was excluded from the experiment. In total, we found just two faulty instances, both stemmed from the test set of the randomly created triangulations.

To obtain an input for our algorithms, we applied the combinatorial embedder of the yFiles graph library, which guarantees that if the input graph is planar, then the computed combinatorial embedding will be planar as well.

We are now ready to present the results of our evaluation. In all plots, the curve denoted by "Kand" stands for results for orthogonal drawings, while the curves denoted by "Pod" and "Heur" correspond to the results for bend-optimal and heuristically computed podevsaef drawings, respectively. Also, the values for a specific number of vertices were obtained by averaging over all instances with the same number of vertices.

Number of Bends: In Figure 5a the required number of bends is plotted against the number of vertices for the test set of the randomly generated triangulations. Since a bend of a classical Kandinsky drawing always corresponds to a pair of half-bends of a podevsaef drawing, in Figure 5a we plotted twice the number of orthogonal bends against the number of half-bends produced by our algorithms. As expected, the number of half-bends of bend-optimal (or, more precisely, close-to-bend-optimal) podevsaef drawings is significantly less than twice the number of bends of bend-optimal classical Kandinsky drawings, especially for graphs with relatively many vertices. As illustrated in Figure 5b, the reduction of the number of bends for both algorithms tends to be around 13% with respect to the classical Kandinsky drawings. The reason is that the graphs of our test set contain only triangular faces, which facilitates the bend-reduction under the podevsaef drawing model.

It is worth mentioning, though, that the drawings produced by the ILP of Section 2 and the ones produced by the heuristic of Section 3 are of comparable number of half-bends. More importantly, both seem to have the same tendency,

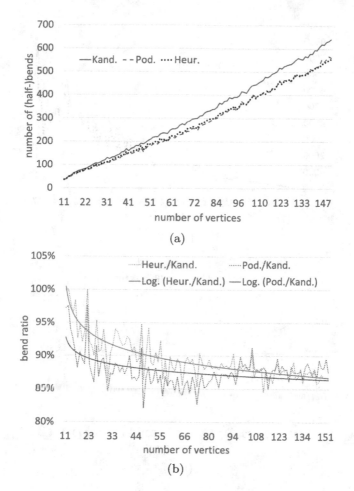

(a)

(b)

Fig. 5. Experimental results for the test set of the randomly created triangulations: (a) The total number of bends is plotted against the number of vertices, (b) the ratio and the logarithm of the ratio of the total number of bends of our algorithms to the total number of bends of bend-optimal Kandinsky drawings is plotted against the number of vertices.

as can be seen in Figure 5b. This justifies our claim that the heuristic is of practical importance.

For the test sets of planar Rome and planar North graphs the profit is significantly smaller (all algorithms seem to produce drawings with very similar number of bends on average) graphs are very sparse and with very few triangular faces.

Area Requirements: In Figure 6a the required area is plotted against the number of vertices for the test set of the randomly generated triangulations.

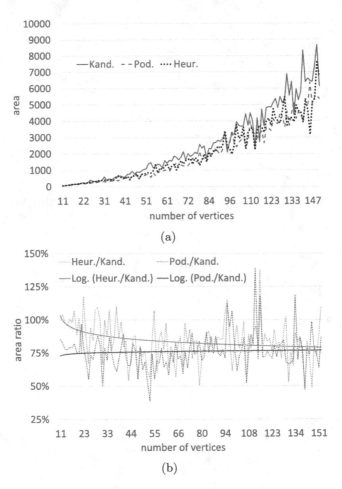

Fig. 6. Experimental results for the test set of the randomly created triangulations: (a) The area requirement is plotted against the number of vertices, (b) the ratio and the logarithm of the ratio of the area of our algorithms to the area of bend-optimal Kandinsky drawings is plotted against the number of vertices.

Again, for graphs with relatively many vertices the area required for podevsaef-drawings is less than the corresponding one for classical Kandinsky drawings. We also observed that the ILP of Section 2 and the heuristic of Section 3 seem to have comparable performance in terms of area requirements, which according to Figure 6b corresponds to an area reduction of around 20% with respect to the classical Kandinsky drawings. It is also worth mentioning that in several cases the drawings computed by the heuristic of Section 3 were more compact than those produced by the ILP of Section 2. However, both seem to have the same tendency in terms of the area requirements, as can be seen in Figure 6b. For the other two test sets of our experiment, the podevsaef drawings and the classical Kandinsky ones require comparable area.

Running Time: On the negative side, the time required by the ILP to compute bend-optimal podevsaef representations increases rapidly with the number of triangular faces of the graph. More precisely, in the test set of the randomly created triangulations, we observed that for graphs with more than 20 vertices the time required to compute an optimal podevsaef drawing exceeded the time-limit of 300 seconds. On the other hand, this negative behavior was difficult to be observed for the test sets of planar Rome and planar North graphs, where all instances could be solved within a few seconds. The reason is that these graphs are much more sparse and have very few triangular faces. This shows that the test set of the randomly created triangulations was the most demanding one in terms of running time.

On the positive side, however, it was almost always possible to compute at least a close-to-optimal solution within the time-limit we set (except for just two faulty cases, as we already mentioned). On the other hand, both the ILP of Eiglsperger et al. [11] and the heuristic of Section 3 seem to require comparable running times; i.e. in all test cases less than two seconds. Since the heuristic of Section 3 eventually produced drawings of comparable number of bends and area (with respect to the ones produced by the ILP of Section 2), it confirms our claim that it is more suitable for practical applications.

5 Conclusion and Open Problems

In this paper, we introduced and experimentally evaluated the podevsaef graph drawing model that is appropriate for drawing planar graphs of arbitrary vertex-degree. Since the problem of minimizing the total number of bends in this model turned out to be NP-complete, we modeled it as an ILP and we also presented a more efficient heuristic. Our evaluation showed that the suggested model is able to improve the quality of a classical Kandinsky drawing (in terms of total number of bends and area requirements) in the presence of many triangular faces. We strongly believe that our new model is of importance, as the Kandinsky drawing model is well-established and widely used in practical applications. Of course, our work is ongoing and raises several open problems:

1. A more sophisticated heuristic or a constant-factor approximation algorithm for computing close-to-optimal podevsaef drawings would be of interest.
2. A different approach that will allow for faster computation of optimal (in terms of the total number of bends) podevsaef drawings, especially when the input graph is triangulated, is also of interest.
3. We only considered the bend-minimization problem. Is there an efficient algorithm that results in podevsaef drawings of provable small drawing area for given plane graphs, especially for triangulations? Are there any non-trivial bounds that one could derive?
4. We considered only planar graphs. It is of interest to extend the proposed model also to the case of non-planar graphs. A reasonable research direction is again to adopt the slanted orthogonal drawing model, which restricts all edge-crossings to diagonal edge-segments at 45°.

References

1. Barth, W., Mutzel, P., Yıldız, C.: A new approximation algorithm for bend minimization in the kandinsky model. In: Kaufmann, M., Wagner, D. (eds.) GD 2006. LNCS, vol. 4372, pp. 343–354. Springer, Heidelberg (2007)
2. Bekos, M.A., Kaufmann, M., Krug, R., Näher, S., Roselli, V.: Slanted orthogonal drawings. In: Wismath, S., Wolff, A. (eds.) GD 2013. LNCS, vol. 8242, pp. 424–435. Springer, Heidelberg (2013)
3. Bertolazzi, P., Di Battista, G., Didimo, W.: Computing orthogonal drawings with the minimum number of bends. IEEE Transactions on Computers 49(8), 826–840 (2000)
4. Bläsius, T., Brückner, G., Rutter, I.: Complexity of higher-degree orthogonal graph embedding in the kandinsky model. In: Schulz, A.S., Wagner, D. (eds.) ESA 2014. LNCS, vol. 8737, pp. 161–172. Springer, Heidelberg (2014)
5. Chen, D.S., Batson, R.G., Dang, Y.: Applied Integer Programming: Modeling and Solution. Wiley (2010)
6. Di Battista, G., Didimo, W., Patrignani, M., Pizzonia, M.: Orthogonal and quasi-upward drawings with vertices of prescribed size. In: Kratochvíl, J. (ed.) GD 1999. LNCS, vol. 1731, pp. 297–310. Springer, Heidelberg (1999)
7. Di Battista, G., Eades, P., Tamassia, R., Tollis, I.G.: Graph Drawing: Algorithms for the Visualization of Graphs. Prentice Hall PTR (1998)
8. Duncan, C.A., Goodrich, M.T.: Graph drawing and cartography. In: Tamassia, R. (ed.) Handbook of Graph Drawing and Visualization, chap. 7, pp. 223–246. CRC Press (2013)
9. Eiglsperger, M.: Automatic Layout of UML Class Diagrams: A Topology-Shape-Metrics Approach. Ph.D. thesis, Universität Tübingen (2003)
10. Eiglsperger, M., Fekete, S.P., Klau, G.W.: Orthogonal Graph Drawing. In: Kaufmann, M., Wagner, D. (eds.) Drawing Graphs. LNCS, vol. 2025, pp. 121–171. Springer, Heidelberg (2001)
11. Eiglsperger, M., Fößmeier, U., Kaufmann, M.: Orthogonal graph drawing with constraints. In: Shmoys, D.B. (ed.) Symposium on Discrete Algorithms, pp. 3–11. ACM/SIAM (2000)
12. Eiglsperger, M., Kaufmann, M.: Fast compaction for orthogonal drawings with vertices of prescribed size. In: Mutzel, P., Jünger, M., Leipert, S. (eds.) GD 2001. LNCS, vol. 2265, pp. 124–138. Springer, Heidelberg (2002)
13. Fößmeier, U., Kaufmann, M.: Drawing high degree graphs with low bend numbers. In: Brandenburg, Franz J. (ed.) GD 1995. LNCS, vol. 1027, pp. 254–266. Springer, Heidelberg (1996)
14. Gurobi Optimization, I.: Gurobi optimizer reference manual (2014). http://www.gurobi.com
15. Leiserson, C.E.: Area-efficient graph layouts. In: Foundations of Computer Science, pp. 270–281. IEEE Computer Society (1980)
16. Tamassia, R.: On embedding a graph in the grid with the minimum number of bends. SIAM Journal of Computing 16(3), 421–444 (1987)
17. Valiant, L.G.: Universality considerations in VLSI circuits. IEEE Transaction on Computers 100(2), 135–140 (1981)

Graph Problems II

An Experimental Analysis of a Polynomial Compression for the Steiner Cycle Problem

Stefan Fafianie$^{(\boxtimes)}$ and Stefan Kratsch

University of Bonn, Bonn, Germany
{fafianie,kratsch}@cs.uni-bonn.de

Abstract. We implement and evaluate a *polynomial compression* algorithm for the STEINER CYCLE problem that was recently developed by Wahlström (STACS 2013). STEINER CYCLE is a generalization of HAMILTONIAN CYCLE and asks, given a graph $G = (V, E)$ and a set of k *terminals* $T \subseteq V$, whether there is a simple cycle containing T as well as an arbitrary number of further vertices of G. Wahlström's compression algorithm takes any such instance and in polynomial time produces an equivalent instance of size polynomial in k. The algorithm has several distinguishing features that make it interesting as a test subject for evaluating theoretical results on preprocessing: It uses Gaussian elimination on the Tutte matrix of (essentially) the input graph instead of explicit reduction rules. The output is an instance of an artificial matrix problem, which might not even be in NP, rather than STEINER CYCLE.

We study to what extend this compression algorithm is useful for actually speeding up the computation for STEINER CYCLE. At high level, we find that there is a substantial improvement of using the compression in comparison to outright running a $\mathcal{O}(2^k \cdot |V|^c)$ algebraic algorithm also due to Wahlström. This is despite the fact that, at face value, the creation of somewhat artificial output instances by means of nonstandard tools seems not all that practical. It does benefit, however, from being strongly tied into a careful reorganization of the algebraic algorithm.

1 Introduction

The present work provides an implementation and experimental evaluation of the recent *polynomial compression* algorithm for the STEINER CYCLE problem presented by Wahlström [20]. In the STEINER CYCLE problem we are given an undirected graph $G = (V, E)$ and a set $T \subseteq V$ of *terminals* with the question of whether there exists a simple cycle containing all vertices of T plus an arbitrary number of additional vertices. We are interested in the parameterized version with parameter k equal to the number of terminals in T.

Supported by the Emmy Noether-program of the German Research Foundation (DFG), research project PREMOD (KR 4286/1). Work done in part while both authors were at Technische Universität Berlin, Germany.

E. Bampis (Ed.): SEA 2015, LNCS 9125, pp. 367–378, 2015.
DOI: 10.1007/978-3-319-20086-6_28

STEINER CYCLE
Input: A graph $G = (V, E)$ and a set of terminals $T \subseteq V$.
Parameter: $k := |T|$.
Output: Is there a simple cycle C in G such that each $t \in T$ is in C?

STEINER CYCLE is NP-hard since it generalizes HAMILTONIAN CYCLE using $T = V$, but it can be solved in time $\mathcal{O}^*(2^k)$[1] [4,20]. The main result of Wahlström [20] is an efficient algorithm that transforms any given instance of STEINER CYCLE to an equivalent instance of size polynomial in the number k of terminals. Let us outline why such polynomial compression algorithms are interesting, and why we are interested in this particular one.

A polynomial compression algorithm is a relaxed variant of the notion of a *polynomial kernelization* from parameterized complexity (see Section 2 for formal definitions). Kernelization is a theoretical formalization of the pervasive notion of preprocessing in the sense of data reduction and instance simplification. Unlike practical applications of preprocessing, e.g., in the CPLEX package or within SAT solvers, which are of a mostly heuristic flavor, the notion of kernelization allows us to prove rigorous upper and lower bounds on *worst-case performance* for preprocessing. In the case of STEINER CYCLE, a polynomial kernelization with respect to parameter k would be required to return an equivalent instance of STEINER CYCLE of *total size polynomial in k*; a polynomial compression may instead return a polynomial in k sized instance of an arbitrary decision problem.[2]

There is a clear appeal of being able to complement the practical success of preprocessing by rigorously proven bounds. Unsurprisingly, there is quite a large gap between what is provable in theory vs. the empirical success of preprocessing in practice. Similarly, it is not always clear that the theoretical results are implementable with reasonable effort and, crucially, that the reduction in instance size also entails a substantial reduction in total runtime. Thus, experimental evaluation and checking practical feasibility seem like important complements to theoretical breakthroughs.

In this light, Wahlström's [20] polynomial compression for STEINER CYCLE is interesting for a couple of reasons: First, STEINER CYCLE is quite similar to a couple of other problems that all do not admit polynomial kernels/compressions. Second, the arguments used in the compression are quite unique in that they do not rely on reduction rules and create output instances of a fairly contrived matrix problem that might not even be in NP. Thus, there is not even the beginning of an idea for how to preprocess the problem by concrete reduction rules (as is typical for kernelization and also in heuristic preprocessing). Finally, it could be argued that STEINER CYCLE is not that suitable for being solved outright by heuristics since, assuming that we want to include all k terminals, there is no natural notion of a suboptimal (or approximate) solution or of a feasible starting solution. Thus, we might be more inclined to implement an exact algorithm for it. Let us give a few more details about the first two aspects.

---

[1] We use $\mathcal{O}^*$ notation to suppress polynomial factors.
[2] This can range from allowing additional annotations to, for example, returning a small instance of CNF-SATISFIABILITY instead.

Status of Similar Problems. Problems similar to STEINER CYCLE are for example k-CYCLE, where we seek a cycle of length exactly k, and the classic STEINER TREE problem, where we seek a *minimum size tree* that connects a given terminal set. Both problems can be solved in time $\mathcal{O}^*(c^k)$ [2,17] but neither admits a polynomial kernelization or compression unless the polynomial hierarchy collapses [7,11].[3] Furthermore, there is the DISJOINT PATHS problem, where the task is to connect k pairs of terminals by finding disjoint paths (of arbitrary length) that connects them. This problem is sufficiently general to be used as a subroutine for solving STEINER CYCLE; however, the best known algorithm takes time $\mathcal{O}^*(2^{2^{k^{10}}})$ [14] and the problem is known not to admit a polynomial kernelization [8]. Thus, the polynomial compression for STEINER CYCLE was a surprising result.

Overview of the Compression. Like the problems of finding a cycle of length at least k or a Steiner tree for k terminals (and several others), STEINER CYCLE can be solved by algebraic techniques that roughly follow the same paradigm: One defines an implicit multivariate polynomial of exponential size such that solutions to the instance correspond to the presence of particular monomials in the polynomial. Using cancellation arguments over finite fields of characteristic two (or a suitable group algebra) all other monomials vanish such that presence of "good" monomials can be determined by testing whether the polynomial is nonzero; this can be done via the Schwartz-Zippel-DeMillo-Lipton Lemma by evaluating the polynomial at random spots. Computing the value at a certain position usually comes down to dynamic programming or inclusion-exclusion, using a succinct presentation of the polynomial, e.g., a recursive definition thereof.

An important insight of Wahlström [20] is that STEINER CYCLE can also be solved using the polynomial obtained from the determinant of the Tutte matrix of (essentially) the input graph, when interpreting it as a polynomial over a finite field of characteristic two. To clarify, when expanding out the polynomial by formally computing the determinant one would still get exponentially many monomials, i.e., products of formal variables. However, the algorithm only needs to evaluate the polynomial at random points, for which determinant computations over a finite field suffice since we first plug in the values. Crucially, the algorithm needs $\mathcal{O}(2^k)$ determinant computations, which is the dominating factor in the overall runtime, but the involved matrices change only very little. The polynomial-time preprocessing makes use of this and replaces the roughly $n \times n$ matrix (for a graph with n vertices) by a $3k \times 3k$ matrix from which we can obtain all required determinant values, and which can be encoded in space polynomial in k. To this end, a small number of special variables are introduced, which are not randomly assigned and which remain present in the $3k \times 3k$ matrix. This makes for a curious form of preprocessing: The operations on the matrix (i.e., partial Gaussian elimination) are not known to correspond to explicit reduction

---

[3] We do not discuss lower bounds for kernelization in this work and instead refer to recent surveys on kernelization [15,16]. All mentioned lower bounds are conditioned on non-collapse of the polynomial hierarchy and all apply also to compressions.

rules on the input graph. Implicitly, the output is an instance of a contrived problem that asks whether certain evaluations of a given matrix sum up to a nonzero value; *it is not even known whether this problem is in* NP. Thus, the result is formally a polynomial *compression*, i.e., we are not concerned with the particular problem of the output instance. Since that problem might not be in NP, standard arguments for deriving a polynomial kernelization fail.[4]

Naturally, with this being the first example of such a behavior, it might seem dubious at first whether this could be of any help in practically solving instances of STEINER CYCLE. Indeed, it comes with the arguable drawback that we do not obtain a small instance of STEINER CYCLE on which we could run some other algorithm. Given the nature of the output problem, we are more or less bound to continue the determinant computations (barring other clever insights into its structure, but note that it is still *hard* for NP).

Our Work. We implement Wahlström's [20] $\mathcal{O}^*(2^k)$ time algorithm as well as the polynomial compression. Particular consideration is given to handling the finite field arithmetic (see also related work below) for which we have implemented several variants. Similarly, a crucial detail is the partial Gaussian elimination in the presence of the remaining formal variables since the single matrix entries become (small) polynomials too. (Full Gaussian elimination would yield a diagonal matrix with some entries being polynomials of length exponential in k.)

We perform an experimental evaluation of both the algorithm by itself as well as in combination with the preprocessing. At high level, the speed-up achieved by applying the preprocessing is sizable: Without it, in at most ten minutes,[5] we can solve instances with roughly 400 vertices, 16000 edges, and 11 terminals. Using the preprocessing this goes up to 2400 vertices, 576000 edges, and 16 terminals. We also test different choices for implementing the finite field arithmetic and, barring very small choices for the field, they all achieve very high success probabilities. A detailed presentation of experimental results and choice of problem instances can be found in Section 5.

Related Work. Similar experimental work on algebraic algorithms was performed by Björklund et al. [5] who focused on quickly extracting a solution for the k-PATH problem based on an algebraic decision algorithm for the problem. We recall that all algorithms based on these algebraic techniques are necessarily decision algorithms only, but are of course amenable to self-reduction (or more clever approaches [5]). Björjklund et al. also give detailed consideration to the implementation of the finite field arithmetics. More recently, Björklund et al. [6] presented an intricate engineering study for finding motifs in very large graphs. (A motif here means a connected k-vertex subgraph whose vertices exhibit exactly the colors from a specified multiset of size k.) We point out that both

---

[4] The argument is simple: Because STEINER CYCLE is NP-hard there must exist a polynomial-time many-one reduction from whatever output problem in NP that we have back to STEINER CYCLE; this at most causes a polynomial blow-up in the size.

[5] With an Intel Xeon E5-1620 processor and 64 GB main memory.

problems are known not to admit polynomial kernels and compressions, so the goals of these works are slightly different (impressive feasible input sizes notwithstanding). We expect, that some of the used techniques (like bit packing) should also be applicable for STEINER CYCLE, while other savings are specific to the way that the polynomial used for GRAPH MOTIF is defined.

Experimental work for kernelization seems somewhat rare. There are older works of Abu-Khzam et al. [1], who considered kernels for the well-studied VERTEX COVER problem, and Weihe's well known paper [21] about preprocessing for a train maintenance problem on data from the German railway network. More recently, Betzler et al. [3] studied the notion of partial kernelization for KEMENY RANK from both theoretical and experimental perspective.

Organization. We start off with some preliminaries in Section 2 and proceed with a description of Wahlström's algebraic algorithm and polynomial compression in Section 3. Implementation details are given in Section 4. The experimental results are presented in Section 5 with concluding remarks in Section 6.

2   Preliminaries

Graphs. We use standard graph notation, mostly following Diestel [10]. Let us define only a few slightly less common notions: A *cycle cover* of a directed graph $D = (V, A)$ is a set $\mathcal{C}$ of disjoint cycles in D such that each $v \in V$ is part of some cycle $C \in \mathcal{C}$. A cycle C in a graph $D = (V, A)$ is *reversible* if it has length at least 3 and for each arc $v_i v_j$ in C there is an arc $v_j v_i \in A$, i.e., if we can traverse C in the other direction. We call a cycle cover $\mathcal{C}$ *reversible* if at least one cycle in $\mathcal{C}$ is reversible. An *oriented* cycle cover of an undirected graph G is a cycle cover of the bidirectional graph corresponding to G.

The *Tutte matrix* A_G corresponding to a graph $G = (V, E)$ with $V = \{v_1, \ldots v_n\}$ is an $n \times n$ matrix such that

$$A_G(i,j) = \begin{cases} x_{ij} & \text{if } v_i v_j \in E \text{ and } i < j, \\ -x_{ji} & \text{if } v_i v_j \in E \text{ and } i > j, \\ 0 & \text{otherwise,} \end{cases}$$

where x_{ij} are indeterminates. The relation between the Tutte matrix and cycle covers, in particular over fields of characteristic two, is discussed in the following section. It is well known (and related to the cycle covers) that the Tutte matrix has full rank if and only if the underlying graph has a perfect matching.

Parameterized Complexity. A *parameterized problem* is a language $\mathcal{Q} \subseteq \Sigma^* \times \mathbb{N}$; the second component of instances $(x, k) \in \Sigma^* \times \mathbb{N}$ is called the *parameter.* A parameterized problem $\mathcal{Q} \subseteq \Sigma^* \times \mathbb{N}$ is *fixed-parameter tractable* (and in the class FPT) if there is an algorithm that, on input of $(x, k) \in \Sigma^* \times \mathbb{N}$, takes time $\mathcal{O}(f(k)|x|^c) = \mathcal{O}^*(f(k))$ and correctly determines whether $(x, k) \in \mathcal{Q}$; here f is an arbitrary computable function.

A *kernelization* for a parameterized problem $Q \subseteq \Sigma^* \times \mathbb{N}$ is an efficient algorithm that, on input of $(x, k) \in \Sigma^* \times \mathbb{N}$, takes time polynomial in $|x| + k$ and returns an equivalent instance (x', k') with $|x'| + k' \le h(k)$, where $h \colon \mathbb{N} \to \mathbb{N}$ is a computable function, also called the *size of the kernelization*. If h is polynomially bounded then we have a *polynomial kernelization*. The notion of (polynomial) compression is defined similarly except that the output is an instance for an arbitrary decision problem L, i.e., on input (x, k) the compression returns $x' \in \Sigma^*$ of size bounded by $h(k)$ such that $(x, k) \in Q$ if and only if $x' \in L$. This generalizes kernelization but, in most cases, the two notions behave almost exactly the same, e.g., results obtained via the main lower bound framework for polynomial kernelization also apply to polynomial compressions.

The Schwartz-Zippel-DeMillo-Lipton Lemma. The following well-known lemma is a mainstay of (probabilistic) algebraic algorithms. Intuitively, it posits that there is a randomized procedure for testing whether a given polynomial is nonzero by evaluating it in randomly chosen positions over a sufficiently large finite field.

Lemma 1 (Schwartz [19], Zippel [22], DeMillo and Lipton [9]). *Let $P(x_1, \ldots x_n)$ be a multivariate polynomial of total degree at most d over a field $\mathbb{F}$, and assume that P is not identically zero. Pick $r_1, \ldots, r_n$ uniformly at random from $\mathbb{F}$. Then $Pr(P(r_1, \ldots, r_n) = 0) \le d/|\mathbb{F}|$.*

3   Overview of the Algebraic Algorithm and Compression

We give a brief overview of the randomized $\mathcal{O}^*(2^k)$ time algebraic FPT algorithm and the polynomial compression for STEINER CYCLE by Wahlström [20], followed by outlining more of the details. For a complete description and proof of correctness see [20].

Algebraic Algorithm. At high level, the algorithm exploits a bijection between cycle covers in a graph and the naïve summation of the determinant of the Tutte matrix.[6] One may view the Tutte matrix as a succinct representation of the formal exponentially long multivariate polynomial that is obtained by computing the determinant. When taking the polynomial to be over a field of characteristic two, all terms corresponding to permutations with at least one cycle of length at least three will cancel. By an easy trick, long cycles that pass through terminals are still possible. By a "sieving argument", i.e., a clever summation over roughly 2^k slightly different determinant computations, similar to inclusion-exclusion, also terms with multiple long cycles that together contain all terminals will

---

[6] The determinant summation is over all $n!$ permutations $\pi \in S_n$ and the cycles in π correspond to cycles in the cover. Permutations with non-adjacent consecutive vertices are equal to zero due to a zero in the corresponding entry in the Tutte/adjacency matrix. Transpositions in π give cycles of length two consisting of a single edge, and fixed points in π correspond to loops, assuming the relevant edge entries are nonzero.

cancel (in the summation), i.e., only terms with a single long cycle containing all terminals will survive. Thus, the implicitly defined multivariate polynomial is nonzero if and only if there is a Steiner cycle and the instance is YES. This can be tested by Lemma 1 by performing point-wise evaluations rather than a formal expansion; these can be performed on the matrix level and come down to roughly 2^k determinant computations for a matrix with finite field entries.

Polynomial Compression. The crux for the compression is that the 2^k evaluations are very similar: They correspond to choosing "directions" for passing through the terminals. This can be captured by using k auxiliary variables that, by being zero or nonzero, control the presence of either direction of the edge in question. (Recall that each edge has two entries in the adjacency matrix, one for either direction.) This in turn means, that we can effectively *postpone* the 2^k evaluations and attempt to first extract the implicit polynomial on just the auxiliary variables obtained after replacing all other variables by random field elements. Of course, this polynomial could still have exponential size (in k), so we have to proceed differently.

The solution is to perform a partial Gaussian elimination on the matrix after replacing all but these k variables. Crucially, these variables occur only in the "top left $3k \times 3k$ corner" of the matrix. Thus, we can use column and row operations to bring the matrix from

$$\mathcal{M} = \begin{pmatrix} A & B \\ C & D \end{pmatrix} \text{ into block form } \begin{pmatrix} M & 0 \\ 0 & R \end{pmatrix}.$$

Here A and M are $3k \times 3k$, and D and R about $n \times n$. The special variables are only in A, hence row and column operations can bring D into diagonal form D' while changing only B and C. Then D' can be used as pivot elements to replace B and C by zeros, achieving block form; this does affect A, but only on the level of creating additive values that are simple field elements (no blow-ups of formal polynomials).

Finally, using that $\det(\mathcal{M}) = \det(M) \cdot \det(R)$ one may discard R and restrict the remaining computation to the $3k \times 3k$ matrix M. (Multiplication of one row of M by $\det(R)$ ensures that the exact same value is obtained, but for zero vs. nonzero this is immaterial.) The matrix M can be represented in $\mathcal{O}(k^3)$ bits, using a sufficiently large field (see [20]). This constitutes an instance relative to the artificial question of whether the summation of 2^k determinants over matrices obtained from M (by setting the k auxiliary variables) evaluates to nonzero.

More Details About Algorithm and Compression. For increased completeness, we clarify some of the previously omitted details in the full version of this paper. The reader is equally well encouraged to skip these or perhaps take the "full tour" by reading the original paper [20].

4   Implementation

In this section we discuss some of the implementation details. Efficient implementation of the finite field arithmetic is vital for the performance of the algorithms in practice. We focus mainly on this component since the other steps in the algorithms described in Section 3 are rather straightforward.

Finite Field Arithmetic. We consider a few variants for the implementation of arithmetic in finite fields of characteristic two. For ease of discussion we assume that we always work over a finite field $\mathbb{F}$ of size 2^w and have an appropriate polynomial g that is irreducible over $\mathbb{F}$. (We use a hard-coded list of irreducibles obtained from [13]). A common choice is to represent field elements as binary numbers with w digits. We use unsigned 64-bit integers, which allows us to support fields of size up to 2^{64}. Addition over $\mathbb{F}$ is done simply by XOR-ing two field elements. Up to this point, all of the variants are the same. We will now describe how the variants deal with multiplication and division of elements in $\mathbb{F}$.

In the `naïve` variant, we multiply two field elements with the well-known Peasant's algorithm which takes $\mathcal{O}(w)$ bit-shifts and additions. For division of a by b where $a, b \in \mathbb{F}$, we first calculate the multiplicative inverse of b and then multiply by a. Note that $b^{2^w} = b$, i.e., $b^{2^w-1} = 1$ and $b^{2^w-2} = \frac{1}{b}$. Then

$$\frac{1}{b} = b^{2^w-2} = \prod_{i=1}^{w-1} b^{2^i}$$

can be computed with $\mathcal{O}(w)$ multiplications by iterative squaring to obtain powers b^{2^i} and multiplying with a running result. This yields division in time $\mathcal{O}(w^2)$.

In the `table` variant, we precompute two $2^w \times 2^w$ tables in which we store the multiplication (resp. division) result of each pair of field elements. This yields constant time division and multiplication once these tables have been computed, which is feasible for small w. This approach was used by Björklund et al. [5].

In the `logtb` variant, we use a similar approach by [18] for slightly bigger values of w. Here, we precompute two length 2^w arrays in which we store the logarithm base g (resp. reverse logarithm base g) of each field element. Then, multiplication can be performed in constant time since $a \cdot b = g^{\log_g a + \log_g b}$. In a similar way, we also have division in constant time.

In the final variant, `clmul`, which was also used by [5], we use the `pclmulqdq` operation which is supported by some modern processors. This operation can be used to obtain the result of carry-less multiplication of two field elements. This result can be reduced modulo g by a constant number of bit-shifts and XOR's. A full description for multiplication of elements in $GF(2^{128})$ can be found in [12]. We modify this and obtain constant time multiplication for elements in $GF(2^{64})$. Division is performed similarly to the `naïve` variant, i.e., in time $\mathcal{O}(w)$.

$\mathcal{O}^*(2^k)$ Algorithm. We use the folklore $\mathcal{O}(n^3)$ algorithm to calculate the determinant of an $n \times n$ matrix A by using Gaussian elimination to transform A to a triangular matrix.

Polynomial Compression. We introduce a new $3k \times 3k$ matrix M' before the compression starts in which we store the multiplicative factors that result from introducing indeterminates $a_1, \ldots, a_k$. I.e., when multiplying $I_G[x, y] = c$ by a_i

we set $I_G[x, y]$ to 0 and $M'[x, y]$ to c since $c \cdot a_i = 0 \cdot c + a_i \cdot c$. Similarly, when multiplying $I_G[x, y] = c$ by $1 - a_i$ we set $I_G[x, y]$ to c and $M'[x, y]$ to c since $c \cdot (1 - a_i) = c - c \cdot a_i = c + c \cdot a_i$. We can then proceed with reducing I_G to block form and obtain orientations afterwards by, e.g., adding $M'[k + 2i - 1, i]$ to $M[k + 2i - 1, i]$, etc., if we set $a_i = 1$, or do nothing if we set $a_i = 0$.

During the compression phase, if the algorithm detects a non-zero column while partial Gaussian elimination is used to reduce I_G to blocks form, it halts and reports failure. Once the determinant of the lower-right block C is calculated, we multiply the first row of M (and M') by the result since this row has only a single non-zero value.

5   Experimental Results

In this section we present results of an experimental evaluation of the $\mathcal{O}^*(2^k)$ algorithm for STEINER CYCLE and the compression described in Section 3. In addition, we report on the impact of the finite field arithmetic implementation variants described in Section 4 and observe how the size of a finite field influences the probability of finding a correct answer. The algorithms have been implemented in C++ and the computations have been carried out with an Intel Xeon E5-1620 processor and 64 GB main memory. Each combination of algorithm and finite field arithmetic variant is given a maximum of 10 minutes of computation time for a given instance. This does not include time required to precompute multiplication (division) tables for the table variants, and logarithm tables (inverse logarithm) tables for the logtb variants since it is conceivable that these tables are precomputed once and stored in practice.

The experiments were performed on graphs of varying size and number of terminals. Multiple instances were used for each specific setting. In Table 1 and in Table 2 we exhibit the performance of the algorithm without the compression step, and with the compression step respectively. Results are shown for 3 different finite field arithmetic variants, where the size of the finite field is 2^w. For each graph size, we have generated 100 completely random instances (not including instances with 400 nodes, for which we have 50). In Table 1 we report the average time in milliseconds.[7] The number of times (percentage) where it is correctly determined whether there is a Steiner cycle is shown in the column marked with %.[8] In Table 2 we report under *comp* the average time used for compression and the percent of times that the compression succeeds. The average time for the 2^k evaluations of determinants after compression and the number of correct answers is shown under *eval*.

We report similar results for a different type of instance in the full version of this paper. These instances where generated by placing all terminals on a simple

---

[7] Dashes (—) indicate that the maximum time was reached before finding an answer.

[8] For random instances we could assume that the correct answer is given with $w = 64$ as the chance of failure is extremely small. However, for all of the random instances that we used we determined that the answer is YES (false positives cannot occur).

Table 1. Results for normal evaluation on random instances

			table $w = 8$		logtb $w = 16$		clmul $w = 64$	
$|V|$	$|E|$	$|T|$	ms	%	ms	%	ms	%
50	250	5	8	95	11	100	43	100
100	1000	7	290	93	422	100	1026	100
200	4000	9	9428	93	14829	100	24701	100
300	9000	10	62559	94	102113	100	143650	100
400	16000	11	292845	92	498462	100	—	

Table 2. Results for compression + evaluation on random instances

			table $w = 8$				logtb $w = 16$				clmul $w = 64$			
			comp		eval		comp		eval		comp		eval	
$|V|$	$|E|$	$|T|$	ms	%	ms	%	ms	%	ms	%	ms	%	ms	%
50	250	5	2	98	< 1	93	2	100	< 1	100	7	100	1	100
100	1000	7	16	98	1	91	18	100	1	100	45	100	6	100
200	4000	9	124	97	6	91	145	100	8	100	288	100	43	100
300	9000	10	414	94	16	91	488	100	22	100	880	100	109	100
400	16000	11	982	94	41	86	1170	100	57	100	2008	100	268	100

Table 3. Results for compression + evaluation on large YES instances

			table $w = 8$				logtb $w = 16$				clmul $w = 64$			
			comp		eval		comp		eval		comp		eval	
$|V|$	$|E|$	$|T|$	ms	%	ms	%	ms	%	ms	%	ms	%	ms	%
800	64000	13	8190	95	256	80	9416	100	348	100	14944	100	1555	100
1600	256000	15	76784	85	1478	60	90710	100	2056	100	131630	100	8543	100
2400	576000	16	279550	90	3461	90	331984	100	4874	100	464706	100	19726	100
3200	1024000	17	—				—				—			

cycle before adding random edges, thus guaranteeing that a solution is present. Results for the naïve variant with different field sizes are also available.

From the obtained results it is immediately clear that applying the compression yields superior running times. Furthermore, a correct answer is found in large fraction of cases, even when the field is of relatively small size. For fields of size 2^{16} a correct answer is already found in all of the observed computations. The naïve field arithmetic variant is outperformed by other variants, even for small finite fields. The other variants show a seemingly constant factor divergence in running times.

In Table 3 we explore the limit of what the compression algorithm can handle in 10 minutes. Here, we have generated 20 instances for each setting. The results seem quite impressive.

6 Concluding Remarks

We have implemented and tested the algorithm and polynomial compression for the STEINER CYCLE problem that was suggested in a theoretical work of Wahlström [20]. At high level, we found that the impact of using the preprocessing rather than just running the algorithm itself is sizable in that we can solve much larger instances within the same time bounds. First of all, this is a clear success for preprocessing and for the practical impact of kernelization research. It also shows that somewhat atypical kernelization or compression results are "nothing to be afraid off." Second, including the implementation work towards fast finite field arithmetic, this gives a fairly practical way of solving large instances of STEINER CYCLE.

Arguably, one reason for the perceived impact of Wahlström's compression for STEINER CYCLE is that, in hindsight, it benefits mainly from the reorganization of the $\mathcal{O}^*(2^k)$ determinant computations and harvesting the savings from not computing the same quantities again and again. Crucially, this hinges on the fact that the many evaluations can be expressed as different assignments of a small number of variables (which rest in the upper $3k \times 3k$ corner of the matrix). This of course poses the question of whether similar ideas can be used for other parameterized problems. Since the known lower bounds vs. polynomial kernelization also rule out polynomial compression, this would only be applicable to problems with known polynomial kernelization or, at least, with unknown kernelization status.

References

1. Abu-Khzam, F.N., Collins, R.L., Fellows, M.R., Langston, M.A., Suters, W.H., Symons, C.T.: Kernelization algorithms for the vertex cover problem: theory and experiments. In: Arge, L., Italiano, G.F., Sedgewick, R. (eds.) Proceedings of the Sixth Workshop on Algorithm Engineering and Experiments and the First Workshop on Analytic Algorithmics and Combinatorics, New Orleans, LA, USA, January 10, 2004, pp. 62–69. SIAM (2004)
2. Alon, N., Yuster, R., Zwick, U.: Color-coding. J. ACM **42**(4), 844–856 (1995)
3. Betzler, N., Bredereck, R., Niedermeier, R.: Partial kernelization for rank aggregation: theory and experiments. In: Raman, V., Saurabh, S. (eds.) IPEC 2010. LNCS, vol. 6478, pp. 26–37. Springer, Heidelberg (2010)
4. Björklund, A., Husfeldt, T., Taslaman, N.: Shortest cycle through specified elements. In: Rabani, Y. (ed.) Proceedings of the Twenty-Third Annual ACM-SIAM Symposium on Discrete Algorithms, SODA 2012, Kyoto, Japan, January 17–19, 2012, pp. 1747–1753. SIAM (2012)

5. Björklund, A., Kaski, P., Kowalik, L.: Fast witness extraction using a decision oracle. In: Schulz, A.S., Wagner, D. (eds.) ESA 2014. LNCS, vol. 8737, pp. 149–160. Springer, Heidelberg (2014)
6. Björklund, A., Kaski, P., Kowalik, L., Lauri, J.: Engineering motif search for large graphs. In: Brandes, U., Eppstein, D. (eds.) Proceedings of the Seventeenth Workshop on Algorithm Engineering and Experiments, ALENEX 2015, San Diego, CA, USA, January 5, 2015, pp. 104–118. SIAM (2015)
7. Bodlaender, H.L., Downey, R.G., Fellows, M.R., Hermelin, D.: On problems without polynomial kernels. J. Comput. Syst. Sci. 75(8), 423–434 (2009)
8. Bodlaender, H.L., Thomassé, S., Yeo, A.: Kernel bounds for disjoint cycles and disjoint paths. Theor. Comput. Sci. 412(35), 4570–4578 (2011)
9. DeMillo, R.A., Lipton, R.J.: A probabilistic remark on algebraic program testing. Inf. Process. Lett. 7(4), 193–195 (1978)
10. Diestel, R.: Graph theory (graduate texts in mathematics) (2005)
11. Dom, M., Lokshtanov, D., Saurabh, S.: Kernelization lower bounds through colors and ids. ACM Transactions on Algorithms 11(2), 13 (2014)
12. Gueron, S., Kounavis, M.: Efficient implementation of the galois counter mode using a carry-less multiplier and a fast reduction algorithm. Information Processing Letters 110(14), 549–553 (2010)
13. Hansen, T., Mullen, G.L.: Primitive polynomials over finite fields. Mathematics of Computation 59(200), 639–643 (1992)
14. Kawarabayashi, K.: An improved algorithm for finding cycles through elements. In: Lodi, A., Panconesi, A., Rinaldi, G. (eds.) IPCO 2008. LNCS, vol. 5035, pp. 374–384. Springer, Heidelberg (2008)
15. Kratsch, S.: Recent developments in kernelization: A survey. Bulletin of the EATCS 113 (2014)
16. Lokshtanov, D., Misra, N., Saurabh, S.: Kernelization – preprocessing with a guarantee. In: Bodlaender, H.L., Downey, R., Fomin, F.V., Marx, D. (eds.) Fellows Festschrift 2012. LNCS, vol. 7370, pp. 129–161. Springer, Heidelberg (2012)
17. Nederlof, J.: Fast polynomial-space algorithms using möbius inversion: improving on steiner tree and related problems. In: Albers, S., Marchetti-Spaccamela, A., Matias, Y., Nikoletseas, S., Thomas, W. (eds.) ICALP 2009, Part I. LNCS, vol. 5555, pp. 713–725. Springer, Heidelberg (2009)
18. Plank, J.S., et al.: A tutorial on reed-solomon coding for fault-tolerance in raid-like systems. Softw., Pract. Exper. 27(9), 995–1012 (1997)
19. Schwartz, J.T.: Fast probabilistic algorithms for verification of polynomial identities. Journal of the ACM (JACM) 27(4), 701–717 (1980)
20. Wahlström, M.: Abusing the Tutte Matrix: an algebraic instance compression for the K-set-cycle problem. In: 30th International Symposium on Theoretical Aspects of Computer Science (STACS 2013). Leibniz International Proceedings in Informatics (LIPIcs), vol. 20, pp. 341–352. Schloss Dagstuhl-Leibniz-Zentrum fuer Informatik (2013)
21. Weihe, K.: Covering trains by stations or the power of data reduction. In: Proceedings of Algorithms and Experiments (ALEX), pp. 1–8 (1998)
22. Zippel, R.: Probabilistic algorithms for sparse polynomials. In: Ng, E.W. (ed.) Symbolic and Algebraic Computation. LNCS, vol. 72, pp. 216–226. Springer, Heidelberg (1979)

On the Quadratic Shortest Path Problem

Borzou Rostami[1]([✉]), Federico Malucelli[2],
Davide Frey[3], and Christoph Buchheim[1]

[1] Fakultät für Mathematik, TU Dortmund, Dortmund, Germany
brostami@mathematik.tu-dortmund.de
[2] Department of Electronics, Information, and Bioengineering,
Politecnico di Milano, Milan, Italy
[3] INRIA-Rennes Bretagne Atlantique, Rennes, France

Abstract. Finding the shortest path in a directed graph is one of the most important combinatorial optimization problems, having applications in a wide range of fields. In its basic version, however, the problem fails to represent situations in which the value of the objective function is determined not only by the choice of each single arc, but also by the combined presence of pairs of arcs in the solution. In this paper we model these situations as a Quadratic Shortest Path Problem, which calls for the minimization of a quadratic objective function subject to shortest-path constraints. We prove strong NP-hardness of the problem and analyze polynomially solvable special cases, obtained by restricting the distance of arc pairs in the graph that appear jointly in a quadratic monomial of the objective function. Based on this special case and problem structure, we devise fast lower bounding procedures for the general problem and show computationally that they clearly outperform other approaches proposed in the literature in terms of their strength.

Keywords: Shortest Path Problem · Quadratic 0–1 optimization · Lower bounds

1 Introduction

The Shortest Path Problem (SPP) is among the best studied combinatorial optimization problems on graphs. It arises frequently in practice in a variety of settings and often appears as a subproblem in algorithms for other combinatorial optimization problems. In a directed network with arbitrary given lengths, the SPP is the problem of finding a directed path from an origin node s to a target node t with shortest total length. Many classical algorithms such as Dijkstra's labeling algorithm [7] and Bellman-Ford's successive approximation algorithm [2] have been developed to solve the problem.

The basic SPP fails to model situations in which the value of a linear objective function is not the only interesting parameter in the choice of the optimal solution. Such problems include situations in which the choice of the shortest path is constrained by parameters such as the variance of the cost of the path,

© Springer International Publishing Switzerland 2015
E. Bampis (Ed.): SEA 2015, LNCS 9125, pp. 379–390, 2015.
DOI: 10.1007/978-3-319-20086-6_29

or cases in which the objective function takes into account not only the cost of each selected arc but also the cost of the interactions among the arcs in the solution. We call such a problem Quadratic Shortest Path Problem (QSPP).

The first variant of the SPP studied in the literature that is directly related to QSPP is probably that of Variance Constrained Shortest Path [14]. The problem seeks to locate the path with the minimum expected cost subject to the constraint that the variance of the cost is less than a specified threshold. The problem arises for example in the transportation of hazardous materials. In such cases a path must be short but it must also be subject to a constraint that the variance of the risk associated with the route is less than a specified threshold. Possible approaches to solving the Variance Constrained Shortest Path problem involve a relaxation in which the quadratic variance constraint is incorporated into the objective function, thus yielding a QSPP problem. In this case, the quadratic part of the objective function is determined by the covariance matrix of the coefficient's probability distributions. In [13] the authors develop a multi-objective model to minimize both the expected travel time of a path and its variance. Then they solve the multi-objective optimization problem by combining the linear and quadratic objective functions into a single quadratic shortest path problem. More generally, such problem may arise in all situations in which the costs associated with each arc consist of stochastic variables. Different types of cost functions on the stochastic shortest path problem have been studied in the literature for which we refer the reader to [11].

A different type of applications arises from research on network protocols. In [10], the authors study different restoration schemes for self-healing ATM networks. In particular, the authors examine line and end-to-end restoration schemes. In the former, link failures are addressed by routing traffic around the failed link, in the latter, instead, traffic is rerouted by computing an alternative path between source and target. Within their analysis, the authors point out the need to solve a QSPP to address rerouting in the latter scheme. Nevertheless, they do not provide details about the algorithm used to obtain a QSPP solution.

Recently, Amaldi et al. [1] introduced new combinatorial optimization problems called *reload cost paths, tours, and flows* which have several applications in transportation networks, energy distribution networks, and telecommunication networks. In the reload cost problems, one is given a graph whose every edge is assigned a color and there is a reload cost when passing through a node on two edges that have different colors. Therefore, the reload cost path problem is a special case of the QSPP in which the objective function takes into account only the reload cost of consecutive arcs with different colors. The authors proved that the reload cost path problem is polynomially solvable.

All problems described above involve variants of the shortest-path problem in which the cost associated with each arc is integrated by a contribution associated with the presence of pairs of arcs in the solution. Such a contribution can be expressed by a quadratic objective function on binary variables associated with each arc, and leads to the definition of a QSPP. To best of our knowledge, there is no previous research dealing directly with solution methods nor complexity studies of the QSPP. Buchheim and Traversi [4] proposed a generic framework for solving binary quadratic programming problems by computing quadratic global underestimators of the objective function that are separable but not necessarily convex. In their computational experiments, they solve some special classes of quadratic $0 - 1$ problems including the QSPP.

In this paper we analyze the complexity of the QSPP and study different special cases of the problem which can be solved in polynomial time. We then develop efficient lower bounding schemes which build a classical SPP or a new special QSPP from the original problem in order to obtain lower bounds. It turns out that the new bounds outperform all lower bounding schemes proposed in the literature so far [4].

2   Problem Formulation and Complexity

Given a directed graph $G(V, A)$, a source node $s \in V$, a target node $t \in V$, a cost function $c : A \to \mathbb{R}^+$, which maps every arc to a non-negative cost, and a cost function $q : A \times A \to \mathbb{R}^+$ that maps every pair of arcs to a non-negative real cost, we denote by $\delta^-(i) = \{j \in V \mid (j, i) \in A\}$ and $\delta^+(i) = \{j \in V \mid (i, j) \in A\}$ the set of predecessor and successor nodes for any given $i \in V$. Defining a binary variable x_{ij} indicating the presence of arc (i, j) on the optimal path, the QSPP is represented as:

$$
\text{QSPP:} \quad z^* = \min \sum_{(i,j),(k,l) \in A} q_{ijkl} x_{ij} x_{kl} + \sum_{(i,j) \in A} c_{ij} x_{ij} \tag{1}
$$
$$
\text{s.t.} \quad x \in X_{st}, \; x \text{ binary.}
$$

Here the feasible region, X_{st}, is exactly the same as that associated with the standard shortest-path problem, i.e.,

$$
X_{st} = \left\{ 0 \leq x \leq 1 : \sum_{j \in \delta^+(i)} x_{ij} - \sum_{j \in \delta^-(i)} x_{ji} = b(i) \quad \forall i \in V \right\}.
$$

Note that $b(i) = 1$ for $i = s$, $b(i) = -1$ for $i = t$, and $b(i) = 0$ for $i \in V \setminus \{s, t\}$.

Theorem 1. *QSPP is strongly NP-hard.*

Proof. Let us consider the general form of the Quadratic Assignment Problem (QAP) on a complete bipartite graph $G = (U, V, E)$ with nodes $U \cup V$, undirected arcs E, a linear cost c, and a quadratic cost q. We may assume that nodes in U and V are both numbered $1, \ldots, m$. We show that this generic instance of the QAP can be reduced to a corresponding instance of QSPP in polynomial time. To this end, we define an QSPP instance on a graph $\tilde{G} = (\tilde{V}, \tilde{A})$ and map each feasible QAP assignment onto a feasible path in $\tilde{G}$, where $\tilde{V}$ and $\tilde{A}$ are defined as follows:

$$
\tilde{V} = (U \times V) \cup \{s, t\}, \quad \text{and} \quad \tilde{A} = A_s \cup A_+ \cup A_t,
$$

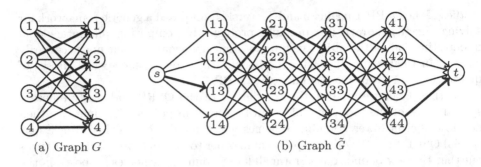

(a) Graph G (b) Graph $\tilde{G}$

Fig. 1. Graph G and $\tilde{G}$. Bold lines in Graph G and $\tilde{G}$ illustrate a feasible assignment for the QAP and its corresponding unique feasible path for QSPP, respectively.

where

$$A_s = \{(s, (1, i)) : i \in V\}, \quad A_t = \{((m, i), t) : i \in V\}, \text{ and}$$
$$A_+ = \{((i, j), (i+1, k)) : i \in U \setminus \{m\}, \; j, k \in V, \; j \neq k\}.$$

Each node $(i, j) \in U \times V$ corresponds to an edge in the original QAP instance, we will use the notation $u((i, j)) := i$ and $v((i, j)) := j$ in the following.

Figure 1 shows the graphs G and $\tilde{G}$ with $m = 4$. With reference to this figure, $u((i, j))$ represents the column of node (i, j) when the graph $\tilde{G}$ is arranged on a grid as shown. Moreover, it represents the index of the first of the two QAP nodes corresponding to (i, j) in the bipartite graph on the left. Analogously, $v((i, j))$ represents the row in the grid and the index of the second QAP node in the bipartite graph.

The graph structure resulting from the above transformation has a number of nodes equal to $m^2 + 2$ and a number of arcs equal to $m^3 - 2m^2 + 3m$, which makes the reduction polynomial.

Moreover, this construction maps each feasible assignment $\pi : U \to V$ in G to a unique feasible path in $\tilde{G}$ as follows: the first arc of the path is $(s, (1, \pi(1)))$, the next arcs are $((i, \pi(i)), (i+1, \pi(i+1)))$ for $i = 1, \ldots, m-1$, and the final arc is $((m, \pi(m)), t)$. By construction and since $\pi(i) \neq \pi(i+1)$ for all $i = 1, \ldots, m-1$, all arcs in this path exist in $\tilde{G}$. Vice versa, every path in $\tilde{G}$ uniquely determines a function $\pi : U \to V$ by setting $\pi(u(w)) = v(w)$ for all $w \in U \times V$ belonging to the path. However, this function is not necessarily a feasible QAP assignment, as different nodes of U may be mapped to the same node of V. This problem is easily addressed by appropriately generating the cost matrix as we show next.

The linear cost vector is defined in Equation (2). The cost for any arc pointing to node e is given by the cost of the arc from $u(e)$ to $v(e)$ in the QAP.

$$\tilde{c}_{fe} = \begin{cases} c_{u(e)v(e)} & e \neq t \\ 0 & e = t. \end{cases} \tag{2}$$

The assignment of quadratic costs to pairs of arcs in $\tilde{G}$ is defined according to Equation (3). In general, the cost $\tilde{q}_{fehw}$ corresponding to the pair

$(f, e), (h, w) \in \tilde{A}$ is equal to the cost $q_{u(e)v(e)u(w)v(w)}$ in the original problem. However, Equation (3) includes an additional constraint to prevent the creation of paths corresponding to infeasible QAP solutions, where two distinct nodes in U are assigned to the same node in V.

$$\tilde{q}_{fehw} = \begin{cases} q_{u(e)v(e)u(w)v(w)} & e \neq t \wedge w \neq t \wedge v(e) \neq v(w) \\ 0 & e = t \vee w = t \\ \infty & \text{otherwise.} \end{cases} \tag{3}$$

The last case in Equation (3) thus makes sure that any optimal solution of QSPP in graph $\tilde{G}$ defines a feasible assignment π in graph G, so that there is a one-to-one correspondence between the feasible assignments in G and the directed paths in $\tilde{G}$ with finite weight, as explained above. It is easy to verify that by construction the cost remains the same under this transformation.

As the QAP problem is strongly NP-hard [12] and the numbers defined in the transformation all have polynomial values (infinite costs can be replaced by an appropriate polynomial value M), the result follows. □

3   The Adjacent Quadratic Shortest Path Problem

In this section, we consider special cases of the QSPP where the quadratic part of the cost function has a *local* structure, meaning that each pair of variables appearing jointly in a quadratic term in the objective function corresponds to a pair of arcs lying close to each other. We start with the Adjacent QSPP (AQSPP), where interaction costs of all non-adjacent pair of arcs are assumed to be zero. Therefore, only the quadratic terms of the form $x_{ij}x_{kl}$ with $j = k$ and $i \neq l$ or with $j \neq k$ and $i = l$ have nonzero objective function coefficients. The AQSPP can be viewed as a generalization of the Reload Cost path introduced by Amaldi et al. [1].

In order to solve the AQSPP, we propose a polynomial-time algorithm based on a transformation that reduces the original problem on graph $G = (V, A)$ to the classical shortest path problem in an auxiliary directed graph $G' = (V', A')$. For this, we may assume w.l.o.g. that there is no direct arc from s to t in G. Now define

$$V' = \{\langle s, s \rangle\} \cup \{\langle i, j \rangle : (i, j) \in A\} \cup \{\langle t, t \rangle\},$$
$$A' = \{(\langle i, j \rangle, \langle j, k \rangle) : \langle i, j \rangle, \langle j, k \rangle \in V'\},$$

where $\langle s, s \rangle$ and $\langle t, t \rangle$ represent nodes s and t, respectively, while all the other nodes in G' correspond to the arcs in the original graph G. Next, we associate each arc $(\langle i, j \rangle, \langle j, k \rangle) \in A'$ with a weight w defined as:

$$w(i, j, k) = \begin{cases} c_{jk} + q_{ijjk} & \langle i, j \rangle \neq \langle s, s \rangle \wedge \langle j, k \rangle \neq \langle t, t \rangle \\ c_{jk} & \langle i, j \rangle = \langle s, s \rangle \\ 0 & \langle j, k \rangle = \langle t, t \rangle \end{cases}$$

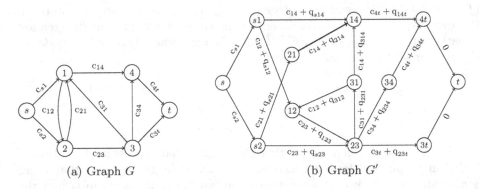

(a) Graph G (b) Graph G'

Fig. 2. Graph $G = (V, A)$ and its auxiliary graph $G' = (V', A')$

Since G' contains $|A| + 2$ nodes and $\delta^+(s) + \delta^-(t) + \sum_{i \neq s,t}(\delta^-(i)\delta^+(i))$ arcs, it can be constructed in polynomial time. In Figure 2 we present an example of a graph G and the corresponding auxiliary graph G'.

Let $c(P) = \sum_{(i,j) \in P} c_{ij} + \sum_{(i,j),(j,k) \in P} q_{ijjk}$ be the cost of any $s - t$ path P in G, and $w(P') = \sum_{e \in P'} w_e$ be the cost of any $\langle s, s \rangle - \langle t, t \rangle$ path P' in G'. The following lemma is a straightforward result implied by the construction of G'.

Lemma 1. *For any $s - t$ path P in G there exists an $\langle s, s \rangle - \langle t, t \rangle$ path P' in G' with $c(P) = w(P')$, and vice versa.*

Proof. For a given $s - t$ path $P \subseteq A$ in G, the path $P' \subseteq A'$ is defined as follows: an arc $(\langle i, j \rangle, \langle j, k \rangle)$ belongs to P' if and only if $(i,j), (j,k) \in P \cup \{(s,s),(t,t)\}$. The path P can be computed from P' accordingly. □

This immediately implies the following

Theorem 2. *An optimal solution for AQSPP in graph G can be obtained by solving a classical shortest path over G'.*

Corollary 1. *For any given source node s and target node t, the AQSPP on graph G can be solved in $O(\min\{|A|^2, |V|^3\} + |A| \log |A|)$ time.*

Proof. Using Dijkstra's algorithm, the running time is $O(|A'| + |V'| \log |V'|)$, where $|A'|$ can be both restricted by $|A|^2$, as each edge in G' corresponds to a pair of edges in G, and by $|V|^3$, as it is defined by three nodes in G. □

If the vertex degrees in G are bounded by Δ, a bound of $O(\Delta^2|V| + |A| \log |A|)$ on the running time can be obtained.

These results hold for the case of a fixed source s and target t. Let us now consider the single-source AQSPP which finds the minimum AQSPP from a given source s to each vertex $v \in V$. To solve the problem we again consider the graph G', but since t is not specified, we do not add node $\langle t, t \rangle$, nodes $\langle k, t \rangle \forall k$, and the arcs incident to these nodes. Then we use Dijkstra's algorithm to find the shortest path $P^*_{\langle s,s \rangle \langle i,j \rangle}$ from the source node $\langle s, s \rangle$ to all the other nodes

$\langle i, j \rangle$ of G'. For any target node $t \in V$, the solution of AQSPP can then be obtained by computing

$$\min\{w(P^*_{\langle s,s \rangle \langle i,t \rangle}) : \langle i, t \rangle \in A'\}. \tag{4}$$

The total running time for solving the single-source AQSPP is thus again given by $O((\min\{|A|^2, |V|^3\} + |A| \log |A|))$, since the additional total running time needed to solve (4) for all $t \in V$ is $O(|A'|)$ and thus dominated but the running time of the first phase.

Motivated by the results of Theorem 2, we can generalize the Adjacent QSPP to an r-Adjacent QSPP by defining the concept of r-adjacency.

Definition 1. *Given a fixed positive integer r, the graph $G = (V, A)$ and two arcs (i, j) and (k, l) in A, we say that (i, j) and (k, l) are r-adjacent in G if there exists a directed path of length at most r containing both arcs.*

We can now define the r-Adjacent QSPP (r-AQSPP) as a more general case of the AQSPP where objective function coefficients of the quadratic terms $x_{ij}x_{kl}$ of non-r-adjacent arcs $(i, j), (k, l) \in A$ are assumed to be zero. With this definition, the AQSPP agrees with the 2-Adjacent QSPP.

Therefore, for any fixed positive integer number $r \geq 2$, we can apply the aforementioned graph construction to transform an r-AQSPP to an $(r - 1)$-AQSPP, where the 1-AQSPP is equivalent to the classical shortest path problem. For fixed r, this leads to a polynomial time algorithm for the r-AQSPP. However, the running time increases exponentially with r. Clearly, for large enough r, the r-AQSPP agrees with the general QSPP and is thus NP-hard by Theorem 1.

4   Lower Bounding Schemes

In this section, we propose lower bounding schemes for the general case of QSPP based on a simple observation on the structure of the problem combined with the polynomial solvability of the AQSPP. The methods are based on the Gilmore-Lawler (GL) procedure. The GL procedure is one of the most popular approaches to find a lower bound for the QAP proposed by Gilmore [8] and Lawler [9] and has been adapted to many other quadratic 0–1 problems in the meantime [5].

For each arc $e = (i, j) \in A$, potentially in the solution, we consider the minimum interaction cost of e in a path from s to t. In other words, we compute the shortest among the paths from s to t which contain arc e, using the ij-th column of the quadratic cost matrix as the cost vector. Let P_e be such a subproblem for a given arc $e \in A$:

$$P_e : \quad z_e = \min\left\{ \sum_{f \in A} q_{ef} x_f : x \in X_{st}, \ x_e = 1 \right\} \quad \forall e \in A. \tag{5}$$

The value z_e is the best quadratic contribution to the QSPP objective function where arc e is in the solution. One possible way to solve P_e is to consider it as a minimum cost flow problem with two origins s and j and two destinations i

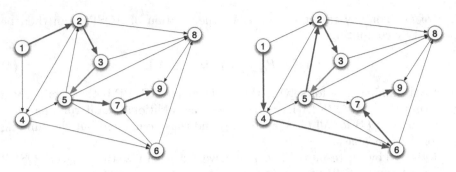

Fig. 3. Possible solutions to P_e when $e = (3, 5)$, $s = 1$ and $t = 9$

and t in a network without arc e. Thus a solution to P_e can be found by solving a minimum-cost-flow problem with two units of cost to be transferred between two sources s and j and two destinations i and t in a graph $\bar{G} = (V, A \setminus \{e\})$. However, this represents a relaxation of P_e: in particular, it admits solutions that consist of the union of a path from s to t that does not contain arc e and a cycle containing e. The resulting solution will then have either of the two forms depicted in Figure 3.

To avoid the situations presented in Figure 3, one can modify the shortest path algorithms to include any given fixed arc $e = (i, j)$. The main idea is to compute the shortest path from s to i, add arc e to the path, and compute the shortest path from j to t. In addition we set to infinity the weights of all the arcs incident to t when computing the path from s to i. This prevents node t from being included in this path.

Once z_e has been computed for each $e \in A$, the GL bound is given by the solution to the following shortest path problem:

$$LB_{GLT} = \min \left\{ \sum_{e \in A} (z_e + c_e) x_e : x \in X_{st} \right\}.$$

The popularity of the GL approach for computing lower bounds stems from its low computational cost. However, for some quadratic 0–1 problems the obtained bounds deteriorate quickly as the size of the problem increases [6]. In the following subsections we propose two novel approaches to improve the GL lower bound for the QSPP.

4.1   A Generalized Gilmore-Lawler Type Bound

We consider a generalization of the GL (GGL) procedure which considers the minimum interaction cost not only of one arc but of two consecutive arcs. More precisely, for each two consecutive arcs $e = (i, j), f = (j, k) \in A$, potentially in the solution, we consider a subproblem P_{ef} to compute the shortest among the paths from s to t which contains these two arcs, i.e.,

$$P_{ef}: \quad z_{ef} = \min \left\{ \sum_{h \in A} \hat{q}_{ef}^h x_h : x \in X_{st}, \ x_e = x_f = 1 \right\} \quad \forall e, f \in S_{2A},$$

where S_{2A} is the set of all 2-adjacent arcs in G, and $\hat{q}$ is defined as follows:

$$\hat{q}_{ef}^h = \begin{cases} \frac{1}{2}(q_{eh} + q_{fh}) & i \neq s, k \neq t \\ q_{eh} + \frac{1}{2} q_{fh} & i = s, k \neq t \\ \frac{1}{2} q_{eh} + q_{fh} & i \neq s, k = t. \end{cases}$$

Similar to problem P_e, the solution to P_{ef} can be easily found by either solving a minimum-cost-flow problem or applying a modified version of the shortest path algorithms. Then the GGL bound is defined to be the solution of the following AQSPP:

$$LB_{GGL} = \min \left\{ \sum_{e \in A} c_e x_e + \sum_{e,f \in S_{2A}} z_{ef} x_e x_f : x \in X_{st} \right\}.$$

By the results of Section 3, the value of LB_{GGL} can be computed in polynomial time. It is now easy to show

Theorem 3. *LB_{GGL} is a lower bound for QSPP; that is $LB_{GGL} \leq z^*$.*

Proof. Let P be any $s - t$ path in G, consisting of edges $e_1, \ldots, e_k$. Then the cost of P is

$$c(P) = \sum_{i=1}^{k} c_{e_i} + \sum_{i,j=1}^{k} q_{e_i e_j} = \sum_{i=1}^{k} c_{e_i} + \sum_{i=1}^{k-1} \sum_{j=1}^{k} \hat{q}_{e_i e_{i+1}}^{e_j} \geq \sum_{i=1}^{k} c_{e_i} + \sum_{i=1}^{k-1} z_{e_i e_{i+1}}.$$

By definition, the latter expression is bounded from below by LB_{GGL}.   □

Note that this approach can be easily generalized by using the r-Adjacent QSPP in order to obtain lower bounds. Clearly, as r is increased, the resulting bound will converge towards the optimal solution. However, the running time for computing the bound grows exponentially in r. Parameter r can thus be used to balance running time and quality of the bound.

4.2 An Iterated Gilmore-Lawler Type Bound

Next, we present Iterated GL (IGL), an iterative bounding procedure inspired by the one proposed in [6] for the QAP. We start by defining a new cost matrix using the reduced costs associated with the dual problem of P_e.

$$\bar{q}_{ef} = q_{ef} + (\lambda_e)_k - (\lambda_e)_l - (\mu_e)_f \quad \forall f = (k, l) \in A \tag{6}$$

where λ_e is the optimal dual-solution vector associated with X_{st}, and μ_e is the one associated with constraint $x \leq 1$. Using this matrix, and (5), we reformulate the QSPP by shifting some of the quadratic costs to the linear part.

$$\text{RQSPP:} \quad \overline{z}^* = \min \quad \sum_{e,f \in A} \overline{q}_{ef} x_e x_f + \sum_{e \in A} (c_e + z_e) x_e$$

$$\text{s.t.} \quad x \in X_{st}, \; x \text{ binary.}$$

(7)

The use of the reduced costs as the quadratic-cost matrix balances the increased linear costs making $RQSPP$ equivalent to $QSPP$ as shown by the following theorem. The proof is omitted due to space restrictions.

Theorem 4. *Problems QSPP and RQSPP are equivalent.* ☐

The theorem allows us to iterate the procedure by applying (6) to the reformulated problem. This results in a sequence of equivalent QSPP instances $(Q_0, Q_1, \ldots, Q_k$ with $Q_0 = \text{QSPP})$, each characterized by a stronger impact of linear costs than the previous ones, and thus providing a better bound. Note that the GLT bound is obtained by considering only the linear portion of the objective function in the first iteration.

5   Computational Results

In this section, we present our computational experiments to evaluate the strength of the lower bounds for the QSPP presented in Section 4. We compare the results of the GLT, GGL, and IGL procedures with three other methods considered in [4]: the first is the root bound calculated by Cplex 12.4 when applied to the problem formulation (1). The other approaches, called QCR and OSU, are general approaches for solving quadratic 0-1 programming problems. The QCR (quadratic convex relaxation) method reformulates quadratic 0-1 programming with linear constraints into an equivalent 0-1 program with a convex quadratic objective function, where the reformulation is chosen such that the resulting lower bound is maximized. For this, an appropriate semidefinite program is solved [3]. The OSU (optimal separable underestimators) approach computes quadratic global underestimators of the objective function that are separable but not necessarily convex [4]. To evaluate and compare all methods, we use the random instances with $|V| = 100, 121, 144, 169, 196, 225$ on grid graphs generated in [4]. The linear and quadratic costs are generated uniformly at random in $\{1, \ldots, 10\}$. Given a pair of arcs (i, j) and (k, l), their associated quadratic costs is equal to $q = q_{ijkl} + q_{klij}$. Since in each subproblem of our lower bounding schemes, each of these two values are processed separately, we consider a redistribution of the quadratic cost $q_{ijkl} = q_{klij} = q/2$. Table 1 presents the results. The first two columns give the problem sizes and the optimal objective values. Columns three to eight give the lower bound values obtained by Cplex, QCR, OSU, GLT, GGL, and IGL respectively. The last five columns of the table present the percentage gap closed by QCR, OSU, GLT, GGL, and IGL over Cplex with respect to the optimum. The formula we used to compute the relative gap closed by a lower bound LB over the lower bound of Cplex (LBc) is $100 \times (LB - LBc)/(OPT - LBc)$.

Table 1. Lower bound comparison for QSPP. The best results are in boldface.

Instance		Lower bound						Impv. vs. Cplex (%)				
n	Opt.	Cplex	QCR	OSU	GLT	GGL	IGL	QCR	OSU	GLT	GGL	IGL
100	621	200	489	357	434	528	511	68.8	37.2	55.5	**77.9**	73.8
100	635	211	501	323	419	511	512	68.3	26.4	49.1	70.7	**70.9**
100	636	217	498	367	449	532	530	56.4	35.7	55.3	**75.1**	74.7
100	661	209	491	359	447	537	534	62.3	33.1	52.6	**72.5**	71.9
100	665	233	504	367	453	549	545	62.7	31.1	50.9	73.1	**73.2**
Ave.								63.7	32.7	52.7	**73.9**	72.7
121	813	253	609	420	531	658	663	63.5	29.8	49.6	72.3	**73.2**
121	788	251	593	417	518	630	631	63.6	30.9	49.7	70.5	**70.7**
121	795	225	592	384	530	643	645	64.3	27.8	53.5	73.4	**73.6**
121	782	236	619	402	518	629	648	70.1	30.4	51.6	71.9	**75.4**
121	767	228	582	404	536	650	644	65.6	32.6	57.1	**78.2**	77.1
Ave.								65.4	30.3	52.3	73.2	**74.0**
144	959	271	714	479	623	767	775	64.3	30.2	51.1	72.1	**73.2**
144	963	282	707	524	627	768	764	62.4	35.3	50.6	**71.3**	70.7
144	900	259	687	491	592	730	735	66.7	36.1	51.9	73.4	**74.2**
144	960	236	698	481	625	758	766	63.8	33.8	53.7	72.1	**73.2**
144	976	289	701	479	632	773	772	59.9	27.6	49.9	**70.4**	70.3
Ave.								63.4	32.6	51.4	71.9	**72.3**
169	1159	335	805	586	730	899	891	57.0	30.4	47.9	**68.4**	67.4
169	1178	333	821	590	759	940	920	57.7	30.4	50.4	**71.8**	69.4
169	1164	325	822	558	733	883	876	59.2	27.7	48.6	**66.5**	65.6
169	1110	301	805	568	729	887	875	62.2	33.0	52.9	**72.4**	70.9
169	1115	322	842	567	737	918	897	65.5	30.8	52.3	**75.1**	72.5
Ave.								60.3	30.5	50.4	**70.1**	69.2
196	1363	364	959	680	841	1055	1064	59.5	31.6	47.7	69.1	**70.1**
196	1367	357	963	660	859	1058	1056	60.0	30.8	49.7	**69.4**	69.2
196	1320	334	934	651	820	1040	1009	60.8	32.1	50.0	**72.6**	69.4
196	1347	348	982	661	862	1058	1062	63.4	31.3	51.4	71.1	**71.4**
196	1344	354	949	704	868	1070	1043	60.1	35.3	51.9	**72.3**	69.5
Ave.								60.8	32.2	50.1	**70.9**	69.9
225	1551	367	1094	729	965	1199	1200	61.4	30.5	50.5	70.2	**70.3**
225	1588	412	1099	806	987	1223	1211	58.4	33.5	48.8	**68.9**	67.9
225	1561	419	1067	762	937	1169	1168	56.7	30.0	45.3	**65.6**	65.5
225	1569	386	1061	744	938	1173	1146	57.1	30.2	46.6	**66.5**	64.2
225	1582	389	1084	791	978	1223	1203	58.2	33.6	49.3	**69.9**	68.2
Ave.								58.4	31.6	48.1	**68.2**	67.2

The results show that Cplex provides by far the worst lower bounds. The GLT lower bound is better than the OUS bound, but both are outperformed by QCR, GGL, and IGL. GGL and IGL provide very similar bounds and clearly outperform QCR. Moreover, our purely combinatorial approach allows us to compute the GLT, GGL, and IGL bounds quickly, while the QCR bound requires solving a semidefinite program, which is often time-consuming in practice even if theoretically possible in polynomial time. Moreover, allowing a longer running time for our GGL approach, we could also improve our bounds by using the 3-Adjacent QSPP.

6   Conclusion

In this paper, we have investigated the quadratic variant of the shortest path problem. We have analyzed its complexity and studied polynomially solvable cases of the problem obtained by allowing only products of adjacent arcs in the objective function. We have proposed efficient procedures to compute strong lower bounds that are based on the well-known Gilmore-Lawler approach combined with the polynomial solvability of the SPP and AQSPP. Our future research will concentrate on combining the GGL procedure with some different reformulation techniques to improve the lower bounds, and an integration of these lower bounds into a branch-and-bound scheme.

Acknowledgments. The first author has been supported by the German Research Foundation (DFG) under grant BU 2313/2.

References

1. Amaldi, E., Galbiati, G., Maffioli, F.: On minimum reload cost paths, tours, and flows. Networks **57**(3), 254–260 (2011)
2. Bellman, R.: On a Routing Problem. Quarterly of Applied Mathematics **16**, 87–90 (1958)
3. Billionnet, A., Elloumi, S., Plateau, M.C.: Improving the performance of standard solvers for quadratic 0–1 programs by a tight convex reformulation: The QCR method. Discrete Applied Mathematics **157**(6), 1185–1197 (2009)
4. Buchheim, C., Traversi, E.: Quadratic 0–1 optimization using separable underestimators. Tech. rep., Optimization Online (2015)
5. Caprara, A.: Constrained 0–1 quadratic programming: Basic approaches and extensions. European Journal of Operational Research **187**(3), 1494–1503 (2008)
6. Carraresi, P., Malucelli, F.: A new lower bound for the quadratic assignment problem. Operations Research **40**(1-Supplement-1), S22–S27 (1992)
7. Dijkstra, E.W.: A note on two problems in connexion with graphs. Numerische Mathematik **1**(1), 269–271 (1959)
8. Gilmore, P.C.: Optimal and suboptimal algorithms for the quadratic assignment problem. Journal of the Society for Industrial & Applied Mathematics **10**(2), 305–313 (1962)
9. Lawler, E.L.: The quadratic assignment problem. Management science **9**(4), 586–599 (1963)
10. Murakami, K., Kim, H.S.: Comparative study on restoration schemes of survivable atm networks. In: INFOCOM 1997. Sixteenth Annual Joint Conference of the IEEE Computer and Communications Societies. Proceedings IEEE, vol. 1, pp. 345–352. IEEE (1997)
11. Nikolova, E., Kelner, J.A., Brand, M., Mitzenmacher, M.: Stochastic shortest paths via quasi-convex maximization. In: Azar, Y., Erlebach, T. (eds.) ESA 2006. LNCS, vol. 4168, pp. 552–563. Springer, Heidelberg (2006)
12. Sahni, S., Gonzalez, T.: P-complete approximation problems. J. ACM **23**(3), 555–565 (1976)
13. Sen, S., Pillai, R., Joshi, S., Rathi, A.K.: A mean-variance model for route guidance in advanced traveler information systems. Transportation Science **35**(1), 37–49 (2001)
14. Sivakumar, R.A., Batta, R.: The variance-constrained shortest path problem. Transportation Science **28**(4), 309–316 (1994)

A Solution Merging Heuristic for the Steiner Problem in Graphs Using Tree Decompositions

Thomas Bosman[✉]

VU University Amsterdam, Amsterdam, The Netherlands
t.n.bosman@student.vu.nl

Abstract. Fixed parameter tractable algorithms for bounded treewidth are known to exist for a wide class of graph optimization problems. While most research in this area has been focused on exact algorithms, it is hard to find decompositions of treewidth sufficiently small to make these algorithms fast enough for practical use. Consequently, tree decomposition based algorithms have limited applicability to large scale optimization. However, by first reducing the input graph so that a small width tree decomposition can be found, we can harness the power of tree decomposition based techniques in a heuristic algorithm, usable on graphs of much larger treewidth than would be tractable to solve exactly. We propose a solution merging heuristic to the Steiner Tree Problem that applies this idea. Standard local search heuristics provide a natural way to generate subgraphs with lower treewidth than the original instance, and subsequently we extract an improved solution by solving the instance induced by this subgraph. As such the fixed parameter tractable algorithm becomes an efficient tool for our solution merging heuristic. For a large class of sparse benchmark instances the algorithm is able to find small width tree decompositions on the union of generated solutions. Subsequently it can often improve on the generated solutions fast.

Keywords: Combinatorial optimization · Steiner Tree Problem · Tree decomposition

1 Introduction

Treewidth, tree decomposition and related graph decomposition concepts have been studied extensively as a means for finding theoretically efficient algorithms for optimization problems in graphs. For graphs of bounded treewidth, polynomial time algorithms can be found for a large number of graph optimization problems. However, due to large constants hidden in the time complexity as well as (super)exponential dependency on the treewidth, in practice these algorithms are often too slow to solve optimization problems. Though heuristic methods for finding tree decompositions of small width have been developed, most applications of tree decompositions are in speeding up exact algorithms. Little work has been done in using tree decompositions as a tool for high performance heuristic optimization algorithms.

© Springer International Publishing Switzerland 2015
E. Bampis (Ed.): SEA 2015, LNCS 9125, pp. 391–402, 2015.
DOI: 10.1007/978-3-319-20086-6_30

To the best of our knowledge the only work in combinatorial optimization exploring this avenue is the tour merging algorithm for the Traveling Salesman Problem (TSP) by Cook and Seymour [3], using the related concept of branch decomposition. In their paper they describe an algorithm that first generates a pool of high quality solutions to the TSP using a local search heuristic with different starting points. In the merging phase, the graph union of these solutions is then taken to produce a sparse subgraph of the original graph. This makes the computation of a low width branch decomposition feasible, which they then use to quickly find the optimal solution to the TSP instance induced by this sparse subgraph. Experimental results showed a fair improvement over the best solution found, in a small amount of additional time.

In this paper we report experimental results applying the same paradigm described in [3] on the Steiner Tree Problem in Graphs (STP). A set of locally optimal solutions is generated to create a sparse subgraph, and subsequently tree decomposition is used to quickly solve the restricted instance to optimality. The main difference with the technique by Cook and Seymour is that we allow the algorithm to discard some of the generated solutions, if it helps finding a tree decomposition of sufficiently small width on the graph union of the remaining solutions. Though this hurts solution quality in some cases, the improvement in running time warrants this trade off.

For generating solutions we use a multistart heuristic by Ribeiro et al. [6] available under the name *Bossa*. The instances induced by the generated solutions are solved using dynamic programming (DP), for which we use a fairly recent tree decomposition based implementation by Fafianie et al. [4]. We compare the performance of our algorithm to the *path relinking* solution merging strategy proposed in [6] which is part of the Bossa implementation.

Experimental results show that our method is very promising. Test runs on sparse benchmark sets showed up to an average 6-fold improvement of the optimality gap provided by the best generated solution, within only one or two percent of the running time of the solution generating phase. On the other hand for dense graphs it often wasn't possible to find a combination of local solutions within our predefined treewidth limit. By using a fast greedy heuristic for finding tree decompositions however, it takes little time to identify this, and therefore the overhead of running the merging algorithm is negligible in such situations.

It should be noted that Bossa is no longer competitive in terms of performance. As pointed out by an anonymous reviewer, a heuristic by Polzin and Daneshmand [8, Chapter 4] was shown to give similar or better solutions compared to Bossa in a fraction of the time on established benchmarks.

However, a very recent advancement by Pajor et al. [9] indicates that our results are still highly relevant. They present an improved multistart heuristic and experimental results indicate that this implementation outperforms the heuristic by Polzin and Daneshmand again. The proposed heuristic has a strong similarity to Bossa. Though some structural changes yield better quality solution pools in the same number of iterations, most of the performance gain is actually achieved by faster implementations of the same local search techniques. Since the improved implementation could be directly plugged in as a solution generator for our method, we expect the positive results to carry over when replacing the multistart heuristic with the improved version of Pajor et al., though further experiments are need to confirm this.

The rest of this article is organized as follows. In Section 2 basic notation is introduced, we give a formal definition of treewidth and we discuss a greedy algorithm for treewidth that plays an important part in the performance of our algorithm. In Section 3 we describe the heuristic for selecting solutions for merging and briefly discuss the algorithms used for generating solutions and solving the instance induced by those solutions. In Section 4 we report experimental results on a variety of benchmark instances.

2   Preliminaries

We denote an undirected graph G with vertex set $V(G)$ and edge set $E(G)$, or V and E when no confusion is possible. Together with a weight function $w : E \to \mathbb{R}$ we have a weighted graph. In this paper we assume graphs to be simple: no loops or parallel edges are allowed. A graph union $G \cup G'$ is equal to the graph found by taking the union of both the vertex and the edge sets of the operands. The neighbours of v in G are denoted $N_G(v) = \{u \in V(G) | (u,v) \in E(G)\}$.

The subject of this paper is the classical Steiner Tree Problem. This famous NP-Complete graph optimization problem should need no introduction to the reader but we include a formal definition for completeness:

*Problem 1 (*STEINER TREE PROBLEM*).*
Given a connected weighted graph $G = (V, E, w)$ with non-negative weights and a set of terminal vertices $Q \subseteq V$, find a minimum weight subgraph T of G such that all terminal vertices are pairwise connected.

Treewidth. The concept of treewidth is a graph invariant that indicates how *tree-like* a graph is. It is derived from the tree decomposition, a transformation that projects a general graph onto a tree. The formal definition is as follows:

Definition 1 (Tree decomposition). *A tree decomposition of a graph $G = (V, E)$ is a tree $\bar{T} = (I, F)$, where each node $i \in I$ is labelled with a vertex set $X_i \subset V$, called a bag, satisfying the following conditions:*

1. $\bigcup_{i \in I} X_i = V$
2. *for all $(v, w) \in E$ there is an X_i such that $\{v, w\} \subset X_i$*
3. *if $v \in X_i$ and $v \in X_j$ then $v \in X_k$ for all k on the path between i and j in the tree $\bar{T}$*

The width $w(\bar{T})$ of a tree decomposition is equal to the size of the largest bag minus one. The treewidth of a graph $\mathrm{tw}(G)$ is the smallest width over all possible tree decompositions of G. As finding the treewidth of a graph is NP-complete no polynomial time exact algorithms exists unless $P = NP$ [2].

Algorithm 1. GreedyDegree(Graph G)

---
while $|V(G)| \geq 0$ **do**
    $v \leftarrow$ a minimum degree vertex in G
    $\pi_i \leftarrow v$
    add edges to G such that $N_G(v)$ is a clique
    remove v from G
return $\pi = (\pi_1, ..., \pi_n)$

Heuristic approaches come in many shapes, including local search techniques and heuristics derived from exact algorithms, see Bodlaender and Koster [1]. We will use a simple but very effective greedy heuristic described in [1].

Algorithm 1, GREEDYDEGREE, constructs an *elimination order*, which is a permutation of vertices of the graph. It does so by iteratively choosing the minimum degree vertex, adding edges between all its neighbours, and then removing it from the graph. These last two steps are called *vertex elimination*.

Any elimination order can be used to construct a (unique) tree decomposition in linear time (see [1, pg.5]). For convenience we will directly treat the output of Algorithm 1 as a tree decomposition. The width of the tree decomposition produced by Algorithm 1 is equal to the highest degree of any vertex at the moment it is eliminated from the graph [1].

In this paper we often abuse language by referring to the width found by GREEDYDEGREE(G) as the treewidth of G, especially when G is a graph induced by a set of solutions. Of course this is just an upper bound, but since we never solve for exact treewidth in our algorithm, it is not necessary to make the distinction when from context it is clear that we mean the width of the tree decomposition found.

3   The Algorithm

The basic outline of our approach consists of three steps. Let an instance of the STP be denoted by STP(G, Q).

1. Generate as set $\mathcal{S}$ of locally optimal solutions for STP(G, Q).
2. Pick a subset $\mathcal{U} \subseteq \mathcal{S}$ such that $\bar{T} = $ GREEDYDEGREE($G_\mathcal{U}$) has width($\bar{T}$) $\leq m$, where $G_\mathcal{U} = \bigcup_{T \in \mathcal{U}} T$.
3. Solve STP($G_\mathcal{U}, Q$) using DP guided by the decomposition $\bar{T}$ found in 2.

The DP implementation we used for the last step, more on that in Section 3.2, has running time linear in $|V|$, but exponential in the treewidth. The multistart heuristic used to generate locally optimal solution is an implementation of a hybrid greedy randomized adaptive search procedure (GRASP) for the STP (see section 3.3). As the first and last steps are basically black box routines with respect to the solution merging heuristic, we will first explain how we construct a suitable subset of solutions.

3.1   Selecting Solutions

In the implementation of tour merging for the TSP in [3] a fixed number of solutions is generated, and quite some time is spent on finding a good branch decomposition. If the algorithm can not find a decomposition of sufficiently small width, the merging heuristic is deemed intractable and returns no solution.

Our method is a little different. We also generate a fixed number of heuristic solutions, and limit the width of the tree decomposition deemed acceptable to proceed with the DP step. However we allow more flexibility by accepting a subset of solutions such that GREEDYDEGREE finds a decomposition of width at most m on their graph union.

An initial approach to finding a good subset of solutions is motivated by the idea that if we cannot use all solutions, we give priority to those with the highest quality. Let S be the set of solutions generated in step 1 and $f(T) : S \to \mathbb{R}$ their weights. Initially we sort the solutions in ascending order of $f(T)$ and apply Algorithm 2. This keeps iteratively adding solutions to the graph union as long as the limit m is not violated by the decomposition found by GREEDYDEGREE. In a sense the algorithm finds a *maximal* subset of solutions, that is, no solution can be added without breaching the width limit.

Algorithm 2. GREEDYSTEINERUNION(S, m)

Input:
   S. List of Solutions, ordered
   m: Maximum treewidth
Output:
   U: List of solutions, such that $tw \left(\bigcup_{T \in U} T \right) \leq m$
   procedure GREEDYSTEINERUNION(S, m)
       $U \leftarrow \{S(1)\}$
       for $i = 2$ **to** $|S|$ **do**
           $U' \leftarrow U \cup \{S(i)\}$
           $\ell \leftarrow$ GREEDYDEGREEWIDTHMAXM$(\bigcup_{T \in U'} T, m)$
           if $\ell \leq m$ **then**
               $U \leftarrow U'$
       return U
   procedure GREEDYDEGREEWIDTHMAXM(G, m)
       $\ell \leftarrow 0$
       while $\ell \leq m \wedge |V(G)| \geq 0$ **do**
           $v \leftarrow$ a minimum degree vertex in G
           $\ell \leftarrow \max\{degree(v), \ell\}$
           add edges to G such that $N_G(v)$ is a clique
           remove v from G
       return ℓ

This procedure usually gives reasonably good improvements in the DP step if the number of solutions rejected by Algorithm 2 is low.

However, if only small sets of solutions stay within width limit m, and there are consequently many possible maximal solution sets, the chance of the greedy procedure finding a good set from the possible alternatives is small. Specifically, experiments showed that increasing the width limit m may often result in a decrease in the eventual solution quality, a highly undesirable result.

To improve the robustness of the solution picking step we introduce the randomized *ranking* procedure described in Algorithm 3. This procedure is akin to a simulation of step 2 and step 3 of the solution merging algorithm with a lower width limit k, where we shuffle the solutions instead of sorting them by $f(T)$. We use the value of the solution found in each iteration to adjust the rank of all solutions that were picked by Algorithm 2 in that iteration.

Algorithm 3. RANKINGPROCEDURE$(\mathcal{S}, f(T), k, r)$

Input:
    $\mathcal{S}$: List of Solutions
    $f(T)$: Map giving the weigth of every Steiner Tree T in $\mathcal{S}$
    k: Maximum treewidth
    r: Number of random ranking iterations
Output:
    $f_A(T)$: Map assigning an adjusted value to every solution in $\mathcal{S}$
    procedure RANKINGPROCEDURE$(\mathcal{S}, f(T), k, r)$
        $Z_T \leftarrow \{f(T)\}, \forall T \in \mathcal{S}$
        for r iterations **do**
            Shuffle the order of $\mathcal{S}$ at random
            $\mathcal{U} \leftarrow$ GREEDYSTEINERUNION$(\mathcal{S}, k)$
            $\hat{z} \leftarrow$ weight of Steiner Tree T found by DP on the graph $G = \bigcup_{T \in \mathcal{U}} T$
            add $\hat{z}$ to all sets Z_T for which $T \in \mathcal{U}$
        return $f_A(T) \leftarrow \sum_{z' \in Z_T} \frac{z'}{|Z_T|}, \forall T \in \mathcal{S}$

The adjusted values $f_A(T)$ can be interpreted as a metric for how promising the inclusion of a solution T is in terms of the improvement found in step 3. These values are then used to sort the solutions before a final run of Algorithm 2 with maximum width m. This yields a much more robust algorithm as in experiments we never observed an increase in m resulting in a decrease in solution quality.

Experimental results indicate that the execution time of the DP grows roughly with 10^m where m is the width of the tree decomposition. Therefore if we run Algorithm 3, for example, with $k = m - 2$ for 10 iterations, its execution time is still expected to be an order of magnitude smaller than directly running the DP once on a graph with decomposition of width m. A byproduct is that we can check more combinations of solutions for improvement. In fact, sometimes the best solution found during the execution of Algorithm 3 is better than the final solution found on the graph union with maximum width m, even after ranking according to the adjusted values. However, this does not happen too often and in general it pays off to execute a last iteration with the higher limit m.

Taking it all together the steps for picking the set of solutions are:

- find $f_A(T) = $ RANKINGPROCEDURE$(S, f(T), k, r)$
- sort the solutions S ascending according to $f_A(T)$
- find $\mathcal{U} = $ GREEDYSTEINERUNION(S, m)

The graph union of $\mathcal{U}$ and its tree decomposition are then used as input for the final DP run in step 3.

3.2 Dynamic Programming

A recent implementation of dynamic programming for the STP was introduced in [4]. It uses the GREEDYDEGREE algorithm to find a decent tree decomposition of the input graph, and then proceeds with a novel dynamic programming algorithm that reduces the search space in every stage by removing entries that cannot affect optimality. We will not reproduce the formal dynamic program here, for which we refer to the paper.

However, the idea is that the DP is guided by a tree decomposition, such that the size of the state space is governed by the number of partitions of the vertex sets in each bag. In the paper multiple methods for reducing the size of the search space are proposed and implemented in the corresponding software. We use the default *classic* DP however, as the relative speed ups are not large enough to make a significant contribution in our implementation.

3.3 Greedy Randomized Adaptive Search Procedure

A Hybrid for the STP was introduced in [6] for which the code is publicly available under the name Bossa [7].

Using a simple multistart approach, in which a construction heuristic is started from different nodes to produce a solutions that is then improved to a local optimum, does not work particularly well for the STP. For reasons that seem to be inherent to the problem most construction heuristics usually produce the same or a few different solutions even for widely different starting points.

To still be able to improve on deterministic heuristics, the Hybrid GRASP algorithm in Bossa employs a variety of techniques to force the algorithm to explore different areas of the search space. These include multiple different construction heuristics, randomization in the local search procedure and weight perturbations. This makes the Hybrid GRASP particularly useful for our algorithm, as it can generate a set of good but disjoint solutions. For a full explanation of these techniques please see the paper.

The Bossa code also includes a solution merging heuristic called Path Relinking, which can be used in combination with GRASP. We use it to compare the performance of our algorithm.

4   Results

The algorithm was implemented in JAVA integrating the existing JAVA code from [4] for the merging part and using system calls and text files to interface with the binary executable of Bossa, to generate the solutions pool. Though working with text files gave some overhead, this effect was insignificant as the time spent on read/write operations was usually small compared to the computation time.

All experiments were run in a single thread on 16 core Intel Xeon E5-2650 v2 @ 2.6 GHz and 64GB of ram. At any time no more than 15 processes were running to make sure one core was free for background processes. The maximum heap space for the JAVA Virtual Machine was set to 1GB for all instances.

For all experiments, in the solution generation phase 16 solutions were generated with GRASP, where each run of the GRASP was set to 8 iterations and with a different random seed. We set the maximum treewidth for the final DP to 10 and the maximum treewidth for the ranking procedure to 8, with 20 iterations of random shuffling. In all experiments where GRASP alone solved an instance to optimality, the instance was dropped from the test set.

4.1   Benchmarks

I640. An initial test was run on the last 50 instances of the classic I640 benchmark set available through the SteinLib [5] repository. All instances are randomly generated. This benchmark is a little outdated in that nowadays most instances can quickly be solved to optimality, but the clear distinction in parameters with which the instances were generated facilitates an easy analysis of the results.

All instances in the benchmark set have 640 vertices, but differ in edge densities and number of terminals. For most instances the optimal value is known, in the other cases we used the best known

Table 1. Results for I640

Instance	$|Q|$	$|E|$	Gap %			Time (s)			#Trees
			GRASP	SMH	Impr. %	GRASP	SMH	Rel.	
201	50	960	0.28	0.28	0.0	9.4	2.1	0.2	13
204	50	960	0.44	0.00	100.0	8.8	0.4	0.1	5
205	50	960	0.02	0.02	0.0	8.0	4.0	0.5	11
211	50	4135	3.00	2.99	0.6	11.2	1.3	0.1	2
212	50	4135	3.23	2.95	8.7	11.2	3.0	0.3	2
213	50	4135	1.69	1.49	11.9	11.6	2.3	0.2	2
214	50	4135	2.22	2.22	0.0	9.2	5.1	0.6	2
215	50	4135	1.72	1.60	7.2	8.6	2.2	0.3	2
231	50	1280	0.09	0.01	92.3	5.1	6.7	1.3	6
232	50	1280	1.04	0.04	96.1	4.6	5.6	1.2	7
233	50	1280	0.62	0.00	100.0	5.0	6.2	1.2	6
234	50	1280	1.22	0.32	73.7	5.0	2.8	0.6	9
235	50	1280	0.74	0.39	46.8	4.5	4.7	1.1	3
241	50	40896	2.25	2.25	0.0	31.9	5.3	0.2	2
242	50	40896	1.79	1.79	0.0	31.6	4.5	0.1	2
243	50	40896	2.02	2.02	0.0	37.0	5.1	0.1	2
244	50	40896	1.62	1.62	0.0	30.8	5.1	0.2	2
245	50	40896	1.54	1.54	0.0	32.3	5.1	0.2	2
301	160	960	0.14	0.00	100.0	6.3	5.5	0.9	7
302	160	960	0.27	0.20	26.2	6.4	7.3	1.1	10
303	160	960	0.25	0.15	39.6	6.9	4.9	0.7	9
304	160	960	0.62	0.39	36.1	7.3	2.9	0.4	9
305	160	960	0.66	0.29	55.5	7.5	2.4	0.3	3
311	160	4135	1.49	1.49	0.0	12.6	1.7	0.1	1
312	160	4135	1.91	1.91	0.0	11.9	1.9	0.2	1
313	160	4135	1.49	1.49	0.0	12.0	1.9	0.2	1
314	160	4135	1.52	1.52	0.0	11.6	1.8	0.2	1
315	160	4135	1.63	1.63	0.0	12.0	1.3	0.1	1
331	160	1280	0.53	0.49	8.3	7.2	3.1	0.4	2
332	160	1280	0.84	0.84	0.0	8.1	2.0	0.2	2
333	160	1280	0.87	0.86	1.1	7.9	2.2	0.3	2
334	160	1280	1.00	1.00	0.0	6.7	1.5	0.2	1
335	160	1280	0.82	0.50	38.6	8.2	11.1	1.4	2
341	160	40896	0.70	0.70	0.0	68.3	3.0	0.0	1
342	160	40896	0.62	0.62	0.0	65.3	2.9	0.0	1
343	160	40896	0.53	0.53	0.0	60.2	2.7	0.0	1
344	160	40896	0.53	0.53	0.0	56.4	2.9	0.1	1
345	160	40896	0.60	0.60	0.0	58.8	1.9	0.0	1

upper bound as an approximation to find the optimality gap. This is only the case for instances I640-311−I640-315.

The results are in Table 1. Next to the instance name the number of terminals and edges is shown. The optimality gaps of GRASP and our solution merging heuristic (SMH) are given as a percentage of the optimal value. The column Impr.% gives the percentage improvement of the optimality gap by SMH compared to GRASP. The running time for SMH does not include GRASP. The column Rel. gives the time spent on SMH relative to the time spent on GRASP. The last column, #Trees is the number of local solutions that were eventually accepted after the sorting procedure(see Algorithm 3) in the solution union for the SMH.

The table clearly reveals the difference in performance between sparse and denser graphs. For instances with less than 1280 edges SMH usually gives a good improvement, even solving the instances to optimality in three instances, yet for none of the most dense instances an improvement was found. This is also reflected in the number of solutions that were used by the algorithm in the final run of the merging phase. There is a clear inverse relation between the density and the number of solutions the algorithm can merge while keeping treewidth within limits. As results e.g. instance 201 and 205 show, a high number of solutions merged does not guarantee improvement, although apparently it is a good indicator. Also the running time of the SMH relative to GRASP is usually lower when no improvement can be found.

As stated before most of the instances in the I640 are not particularly hard to solve with todays hardware. To get a better view of the power of the SMH algorithm we wanted to apply it to some bigger instances. The most notouriously hard test set in Steinlib is the PUC testset, of which most instances have no known optimal solution after more than 13 years in the field. No results are plotted but for completeness that SMH gave poor results on these instances: for all but the smallest instances we were not able to find any combination of solutions within width limit. We don't know if this is because our greedy tree decomposition works particularly bad for these graphs, or because high treewidth is an inherent property of the graph. In any case most instances from PUC are denser than the second highest density instances from I640, for which SMH was already hardly able to show improvement.

Fortunately there are some other test sets in the Steinlib repository that are big enough to justify the use of our merging heuristic but not so dense as to make it run into trouble because of the treewidth limit. Results on these test sets are discussed in the rest of this section.

To compare performance we also ran the path relinking algorithm (PR) from Bossa. The path relinking algorithm is itself a solution merging heuristic which comes in two flavours. On standard settings it first tries these different flavours and then picks the one that seems to perform best. For our experiments we forced it to use the random relink heuristic, as this turned out to perform best on all tested instances, and the initial run that determines the best settings takes a considerable amount of time. This makes for a more fair comparison. For more information see [6].

ES1000(0)FST. The ES1000FST test set contains 15 instances of randomly generated points on a grid, with L1 distances as edge weights. Each instance has 1000 terminal vertices and between 2500 and 2900 vertices in total. Due to preprocessing techniques applied to the graphs these instances only have between 3600 and 4500 edges, making them very sparse.

The results for GRASP, GRASP+SMH and GRASP+PR are shown in Table 2. Again, for SMH and PR the time does not include the initial GRASP iterations. Results are averaged over all instances. The number of instances for which the algorithms produced the best solution among all produced solutions for that instance is given by the row #best.

Table 2. Results on ES1000FST

	GRASP	SMH	PR
Opt Gap %	0.392	0.061	0.109
Time (s)	402	6	493
#Best	0	14	1

Table 3. Results on ES10000FST

	GRASP	SMH
Opt Gap %	0.441	0.189
Total CPU Time(s)	194485	310
Wall Clock Time(s)	12155	310

Overall the SMH seems to perform better on these instances. Its also nice to note that on average 15.4 solutions were used in the final DP run of the merging phase. This is probably caused by the inherently low treewidth of the instances. However, the treewidth of these instances is not so low that direct use of DP would be feasible. As the treewidth found by GREEDYDEGREE for the ES1000FST instances had a minimum of 14 and an average of 22, running the tree decomposition based DP on the original instances would take ages.

We also tested on the single instance in the ES10000FST set. This graph is created in a similar way but with a factor 10 more terminals and vertices. Because this instance is so large that only running the 128 GRASP iterations needed for the SMH takes more than two cpu days, we did not compare it with path relinking in a sequential run. Instead grasp was run on 16 cores in parallel and the solution merging heuristic was run on a single core thereafter. Results are in Table 3. Both the wallclock time (time untill all threads where finished) and the total computation time summed over all threads are shown. Though not the most spectacular improvement in optimality gap it shows the good scaling properties of SMH. The relative time spent on SMH compared to GRASP has about the same ratio as seen in the ES1000FST instances when we compare wallclock time, yet it is an order of magnitude smaller when we compare the total CPU time. We need to notice however that for this instance SMH was only able to use 3 solutions, as the treewidth of the entire solution pool combined was rather high at 22.

LIN. The LIN test set from Steinlib has very similar properties as the ES1000FST set. These instances are generated from placing rectangles of different sizes in a plane, such that their corners become vertices and there edges graph edges. Though no preprocessing is done on these instances, it still makes for a very sparse graph, with no vertex having a degree more than 4. After dropping instances that were solved to optimality by GRASP, only the last 13

Table 4. Results on LIN

	GRASP	SMH	PR
Opt Gap %	0.33	0.09	0.13
Time(s)	336	2	157
#Best	1	10	6

Table 5. Results on ALUT/ALUE

	GRASP	SMH	PR
Opt Gap %	0.27	0.07	0.11
Time(s)	1386	7	1057
#Best	1	8	4

instances remained. The number of vertices of these instances are in the range 3700-39000.

Results are in Table 4. Again the SMH performs very good compared to PR, in smaller amount of time. In all cases the merging phase could use all 16 solutions, often producing a union well below the treewidth limit.

ALUT/ALUE. The last test sets we ran experiments on are the ALUT and ALUE sets from Steinlib, which we combined because of their strong similarities. The structural properties of these instances are very much like the LIN test set. However, these come from very-large-scale-circuit (VLSI) applications. The results is a grid graph with rectangular holes in it. This graph again has a maximum degree of 4. After dropping instances which GRASP solved to optimality, 10 instances remained ranging in number of vertices between 3500 and 37000 and a number of terminals between 68 and 2344. Because of the fairly large size of some of these instances, we put a maximum on the running time for the combination of GRASP and merging heuristic of 3.5 hours. This gave a timeout for PR on the largest instance, so we took the best found solution up to that point. To compare, SMH only took 40 seconds to run for this instance, while GRASP took about 2 hours.

One of the nice properties of using the tree decomposition based approach is that for graphs with a regular structure such as with the last two sets we tested on, the size of the graph does not seem to matter much for the treewidth of the union of solutions, while the DP runs linear in the number of vertices. In the experiment run on the ALUT/ALUE sets, for all but two of the remaining instances the merging phase was able to use all 16 generated solutions. The two exceptions, where only 15 solutions were used, were the largest instance, and surprisingly, the smallest instance. This illustrates that observation quite well.

5   Conclusions and Suggestions for Further Research

Experimental results showed that a tree decomposition based exact algorithm can be employed as an efficient means to merge local heuristic solutions to the STP on sufficiently sparse graphs. As we have seen in results on the ALUT/ALUE test set, the sparse structure natural to VLSI derived graphs is exactly that at which our heuristic performs well. As VLSI is one of the major applications of the STP, this makes the heuristic practically relevant.

As mentioned in the introduction the algorithm we used to generate solutions is no longer state of the art. In theory any algorithm capable of generating distinct locally optimal solutions could be employed with our algorithm. We plan to investigate the competitiveness of our solution merging heuristic when combined with a faster implementation such as [9] for generating solutions in preparation of a journal version of this paper.

That fixed parameter tractable algorithms can be used as a heuristic solution merging technique for the TSP had been shown in [3], while we established results for the STP. It seems likely that there are more optimization problems where this technique can be used. A minimal requirement seems to be that any feasible solution has low value for the chosen parameter. However, whether a low width decomposition can be found on a combination of local solutions depends on the instance, and in the case of the STP the density of the input graph seems a good indicator for that. It would be interesting to see if such a characterization is possible for other optimization problems that have low width solutions.

As a final remark, in our algorithm we managed the treewidth of the solution union by discarding solutions. A simple extension would be to use an iterative scheme to reduce treewidth of the solution pool: first run the solution merging heuristic on (small) subsets of the generated solutions to generate a new solution pool with less solutions, and repeat until all solutions are within the treewidth limit. It seems likely this could further improve the performance.

Acknowledgments. I would like to thank N. Olver and L. Stougie for their feedback and an anonymous reviewer for helpful comments.

References

1. Bodlaender, H.L., Koster, A.M.C.A.: Treewidth computations I. Upper bounds, Information and Computation **208**, 259–275 (2010)
2. Bodlaender, H.L., Fomin, F.V., Kratsch, D., Koster, A.M.C.A., Thilikos, D.M.: On exact algorithms for treewidth. ACM Trans. Algor. **9**(1), 12 (2012)
3. Cook, W., Seymour, P.: Tour merging via branch-decomposition. INFORMS Journal on Computing **15**(3), 233–248 (2003)
4. Fafianie, S., Bodlaender, H.L., Nederlof, J.: Speeding Up Dynamic Programming with Representative Sets. In: Gutin, G., Szeider, S. (eds.) IPEC 2013. LNCS, vol. 8246, pp. 321–334. Springer, Heidelberg (2013)
5. Koch, T., Martin, A., Voß, S.: Steinlib, an updated library on Steiner tree problems in graphs. Technical Report ZIB-Report 00–37, Konrad-Zuse Zentrum fur Informationstechnik Berlin (2000). http://steinlib.zib.de/
6. Ribeiro, C.C., Uchoa, E., Werneck, R.F.: A hybrid GRASP with perturbations for the Steiner problem in graphs. INFORMS Journal on Computing **14**(3), 228–246 (2002)
7. Uchoa, E., de Aragao, M.P., Werneck, R., Ribeiro, C.C.: Bossa (2002). http://www.cs.princeton.edu/rwerneck/bossa/
8. Polzin, T.: Algorithms for the Steiner problem in Networks. Ph.d. Thesis, Universität des Saarlandes (2003)
9. Pajor, T., Uchoa, E., Werneck, R.F.: A Robust and Scalable Algorithm for the Steiner Problem in Graphs arXiv preprint (2014). arXiv:1412.2787

Author Index

Printed in the United States
By Bookmasters